PLANT PIGMENTS, FLAVORS AND TEXTURES: THE CHEMISTRY AND BIOCHEMISTRY OF SELECTED COMPOUNDS

Plant Pigments, Flavors and Textures: The chemistry and Biochemistry of Selected Compounds

Editor:

Dr. Sajid Mehmood, Ph.D

Delve Publishing

www.delvepublishing.com

Plant Pigments, Flavors and Textures: The chemistry and Biochemistry of Selected Compounds

Editor: **Dr. Sajid Mehmood, Ph.D**

708 3rd Avenue, 6th Floor
New York, NY 10017
United States of America

www.delvepublishing.com

ISBN: 978-1-68095-896-6

Library of Congress Control Number: 2016951497

About the Editor

Dr. Sajid Mehmood, Ph.D

Dr. Sajid Mehmood is Associate Professor of Biochemistry and Molecular Biology. He has done his PhD under Indigenous Scholarship program funded by HEC, Government of Pakistan. He is fellow/member of different national and international scientific societies. He has served as research fellow in Denmark Technical University, Copenhagen, Denmark, Assistant and Associate Professor in University of Gujrat, Pakistan. He has published 51 scholarly papers in journal of international repute with h index 10 and more than 400 citations. He has vast experience in teaching and research.

Contents

List of Contributors

Arturo Solís Herrera
Human Photosynthesis® Study Center, Aguascalientes, Mexico

Paulina Kuczynska
Faculty of Biochemistry, Biophysics and Biotechnology, Department of Plant Physiology and Biochemistry, Jagiellonian University, Gronostajowa 7, Krakow 30-387, Poland

Malgorzata Jemiola-Rzeminska
Faculty of Biochemistry, Biophysics and Biotechnology, Department of Plant Physiology and Biochemistry, Jagiellonian University, Gronostajowa 7, Krakow 30-387, Poland
Małopolska Centre of Biotechnology, Gronostajowa 7A, Krakow 30-387, Poland

Kazimierz Strzalka
Faculty of Biochemistry, Biophysics and Biotechnology, Department of Plant Physiology and Biochemistry, Jagiellonian University, Gronostajowa 7, Krakow 30-387, Poland
Małopolska Centre of Biotechnology, Gronostajowa 7A, Krakow 30-387, Poland

Nilangani N Harris
New Zealand Institute for Plant & Food Research Limited, Private Bag 11- 600, Palmerston North, New Zealand
Commonwealth Scientific and Industrial Research Organization, Ecosystem Sciences, Urrbrea, South Australia 5064, Australia.

John Javellana
New Zealand Institute for Plant & Food Research Limited, Private Bag 11- 600, Palmerston North, New Zealand

Kevin M Davies
New Zealand Institute for Plant & Food Research Limited, Private Bag 11- 600, Palmerston North, New Zealand

David H Lewis
New Zealand Institute for Plant & Food Research Limited, Private Bag 11- 600, Palmerston North, New Zealand

Paula E Jameson
School of Biological Sciences, University of Canterbury, Private Bag 4-800, Christchurch, New Zealand

Simon C Deroles
New Zealand Institute for Plant & Food Research Limited, Private Bag 11- 600, Palmerston North, New Zealand

Kate E Calcott
New Zealand Institute for Plant & Food Research Limited, Private Bag 11- 600, Palmerston North, New Zealand
Victoria University of Wellington, Wellington 6140, New Zealand

Kevin S Gould
Victoria University of Wellington, Wellington 6140, New Zealand

Kathy E Schwinn
New Zealand Institute for Plant & Food Research Limited, Private Bag 11- 600, Palmerston North, New Zealand

Bernhard Kräutler
Institute of Organic Chemistry and Centre of Molecular Biosciences, University of Innsbruck, Innrain 80/82, A-6020 Innsbruck, Austria.

H Munawaroh
Chemistry of Department, Faculty of Mathematics and Natural Sciences, Sebelas Maret University, Jl. Ir. Sutami, Kentingan, Surakarta 57126, Central Java, Indonesia

G Fadillah
Chemistry of Department, Faculty of Mathematics and Natural Sciences, Sebelas Maret University, Jl. Ir. Sutami, Kentingan, Surakarta 57126, Central Java, Indonesia

L N M Z Saputri
Chemistry of Department, Faculty of Mathematics and Natural Sciences, Sebelas Maret University, Jl. Ir. Sutami, Kentingan, Surakarta 57126, Central Java, Indonesia

Q A Hanif
Chemistry of Department, Faculty of Mathematics and Natural Sciences, Sebelas Maret University, Jl. Ir. Sutami, Kentingan, Surakarta 57126, Central Java, Indonesia

R Hidayat
Chemistry of Department, Faculty of Mathematics and Natural Sciences, Sebelas Maret University, Jl. Ir. Sutami, Kentingan, Surakarta 57126, Central Java, Indonesia

S Wahyuningsih
Chemistry of Department, Faculty of Mathematics and Natural Sciences, Sebelas Maret University, Jl. Ir. Sutami, Kentingan, Surakarta 57126, Central Java, Indonesia

Rasika Sunnadeniya
Department of Molecular Biosciences, The University of Texas at Austin, Austin, Texas, United States of America

Alexander Bean
Department of Molecular Biosciences, The University of Texas at Austin, Austin, Texas, United States of America

Matthew Brown
Department of Molecular Biosciences, The University of Texas at Austin, Austin, Texas, United States of America

Neda Akhavan
Department of Molecular Biosciences, The University of Texas at Austin, Austin, Texas, United States of America

Gregory Hatlestad
Department of Molecular Biosciences, The University of Texas at Austin, Austin, Texas, United States of America

Antonio Gonzalez
Department of Molecular Biosciences, The University of Texas at Austin, Austin, Texas, United States of America

V. Vaughan Symonds
Institute of Fundamental Sciences, Massey University, Palmerston North, New Zealand

Alan Lloyd
Department of Molecular Biosciences, The University of Texas at Austin, Austin, Texas, United States of America

Williamson D.D
Department of Chemistry, University of the West Indies, Mona Campus, Kingston 7, Jamaica.

Benkeblia N
Department of Life Sciences, University of the West Indies, Mona Campus, Kingston 7, Jamaica.

Romain Gastineau
FR CNRS 3473 IUML, Mer-Molécules-Santé (MMS), Université du Maine, Ave O. Messiaen, 72085 Le Mans cedex 9, France

François Turcotte
Institut des Sciences de la mer de Rimouski, Université du Québec à Rimouski, 310 des Ursulines, Rimouski, QC G5L 3A1, Canada

Jean-Bernard Pouvreau
EA 1157, Laboratoire de Biologie et Pathologie Végétales (LBPV), Université de Nantes, SFR 4207 QUASAV, 44322 Nantes, France

Michèle Morançais
FR CNRS 3473 IUML, Mer-Molécules-Santé (MMS), Université de Nantes, 44322 Nantes, France

Joël Fleurence
FR CNRS 3473 IUML, Mer-Molécules-Santé (MMS), Université de Nantes, 44322 Nantes, France

Eko Windarto
FR CNRS 3473 IUML, Mer-Molécules-Santé (MMS), Université du Maine, Ave O. Messiaen, 72085 Le Mans cedex 9, France

Fiddy Semba Prasetiya
FR CNRS 3473 IUML, Mer-Molécules-Santé (MMS), Université du Maine, Ave O. Messiaen, 72085 Le Mans cedex 9, France

Sulastri Arsad
FR CNRS 3473 IUML, Mer-Molécules-Santé (MMS), Université du Maine, Ave O. Messiaen, 72085 Le Mans cedex 9, France

Pascal Jaouen
FR CNRS 3473 IUML, UMR-CNRS 6144 (GEPEA), Université de Nantes, CRTT 37 Bd de l'Université, F-44602 Saint-Nazaire, France

Mathieu Babin
Institut des Sciences de la mer de Rimouski, Université du Québec à Rimouski, 310 des Ursulines, Rimouski, QC G5L 3A1, Canada

Laurence Coiffard
FR CNRS 3473 IUML, Mer-Molécules-Santé (MMS), Université de Nantes, 44322 Nantes, France

Céline Couteau
FR CNRS 3473 IUML, Mer-Molécules-Santé (MMS), Université de Nantes, 44322 Nantes, France

Jean-François Bardeau
UMR CNRS 6283, Institut des Molécules et Matériaux du Mans (IMMM), Université du Maine, Ave O. Messiaen, 72085 Le Mans cedex 9, France

Boris Jacquette
FR CNRS 3473 IUML, Mer-Molécules-Santé (MMS), Université du Maine, Ave O. Messiaen, 72085 Le Mans cedex 9, France

Vincent Leignel
FR CNRS 3473 IUML, Mer-Molécules-Santé (MMS), Université du Maine, Ave O. Messiaen, 72085 Le Mans cedex 9, France

Yann Hardivillier
FR CNRS 3473 IUML, Mer-Molécules-Santé (MMS), Université du Maine, Ave O. Messiaen, 72085 Le Mans cedex 9, France

Isabelle Marcotte
Department of Chemistry, Université du Québec à Montréal, P.O. Box 8888, Downtown Station, Montréal, QC H3C 3P8, Canada

Nathalie Bourgougnon
Université de Bretagne-Sud, Laboratoire de Biotechnologie et Chimie Marines, Campus de Tohannic, F-56017 Vannes, France

Réjean Tremblay
Institut des Sciences de la mer de Rimouski, Université du Québec à Rimouski, 310 des Ursulines, Rimouski, QC G5L 3A1, Canada

Jean-Sébastien Deschênes
Institut des Sciences de la mer de Rimouski, Université du Québec à Rimouski, 310 des Ursulines, Rimouski, QC G5L 3A1, Canada

Hope Badawy
UMR CNRS 6283, Institut des Molécules et Matériaux du Mans (IMMM), Université du Maine, Ave O. Messiaen, 72085 Le Mans cedex 9, France

Pamela Pasetto
UMR CNRS 6283, Institut des Molécules et Matériaux du Mans (IMMM), Université du Maine, Ave O. Messiaen, 72085 Le Mans cedex 9, France

Nikolai Davidovich
Karadag Nature Reserve of the National Academy of Sciences, p/o Kurortnoe, Feodosiya 98188, Crimea

Gert Hansen
Department of Biology, Marine Biological Section, University of Copenhagen, Øster Farimagsgade 2D, 1353 Copenhagen, Denmark

Jens Dittmer
UMR CNRS 6283, Institut des Molécules et Matériaux du Mans (IMMM), Université du Maine, Ave O. Messiaen, 72085 Le Mans cedex 9, France

Jean-Luc Mouget
FR CNRS 3473 IUML, Mer-Molécules-Santé (MMS), Université du Maine, Ave O. Messiaen, 72085 Le Mans cedex 9, France

Dar NA
Department of Environmental Science, University of Kashmir, Hazratbal, Srinagar, India

Pandit AK
Department of Environmental Science, University of Kashmir, Hazratbal, Srinagar, India

Ganai BA
Department of Environmental Science, University of Kashmir, Hazratbal, Srinagar, India

Michael L. Schwieterman
Plant Molecular and Cellular Biology Program, University of Florida, Gainesville, Florida, United States of America
Plant Innovation Program, University of Florida, Gainesville, Florida, United States of America

Thomas A. Colquhoun
Plant Molecular and Cellular Biology Program, University of Florida, Gainesville, Florida, United States of America
Department of Environmental Horticulture, University of Florida, Gainesville, Florida, United States of America
Plant Innovation Program, University of Florida, Gainesville, Florida, United States of America

Elizabeth A. Jaworski
Department of Environmental Horticulture, University of Florida, Gainesville, Florida, United States of America
Plant Innovation Program, University of Florida, Gainesville, Florida, United States of America

Linda M. Bartoshuk
College of Dentistry, University of Florida, Gainesville, Florida, United States of America
Plant Innovation Program, University of Florida, Gainesville, Florida, United States of America

Jessica L. Gilbert
Horticultural Sciences Department, University of Florida, Gainesville, Florida, United States of America
Plant Innovation Program, University of Florida, Gainesville, Florida, United States of America

Denise M. Tieman
Horticultural Sciences Department, University of Florida, Gainesville, Florida, United States of America
Plant Innovation Program, University of Florida, Gainesville, Florida, United States of America

Asli Z. Odabasi
Food Science and Human Nutrition Department, University of Florida, Gainesville, Florida, United States of America
Plant Innovation Program, University of Florida, Gainesville, Florida, United States of America

Howard R. Moskowitz
Moskowitz Jacobs Inc., White Plains, New York, United States of America

Kevin M. Folta
Plant Molecular and Cellular Biology Program, University of Florida, Gainesville, Florida, United States of America
Horticultural Sciences Department, University of Florida, Gainesville, Florida, United States of America
Plant Innovation Program, University of Florida, Gainesville, Florida, United States of America

Harry J. Klee
Plant Molecular and Cellular Biology Program, University of Florida, Gainesville, Florida, United States of America
Horticultural Sciences Department, University of Florida, Gainesville, Florida, United States of America
Plant Innovation Program, University of Florida, Gainesville, Florida, United States of America

Charles A. Sims
Food Science and Human Nutrition Department, University of Florida, Gainesville, Florida, United States of America
Plant Innovation Program, University of Florida, Gainesville, Florida, United States of America

Vance M. Whitaker
Food Science and Human Nutrition Department, University of Florida, Gainesville, Florida, United States of America
Gulf Coast Research and Education Center, University of Florida, Wimauma, Florida, United States of America
Plant Innovation Program, University of Florida, Gainesville, Florida, United States of America

David G. Clark
Plant Molecular and Cellular Biology Program, University of Florida, Gainesville, Florida, United States of America
Department of Environmental Horticulture, University of Florida, Gainesville, Florida, United States of America
Plant Innovation Program, University of Florida, Gainesville, Florida, United States of America

Machewad GM
Department of Food and Industrial Microbiology, College of Food Technology, Marathwada Krishi Vidyapeeth, Parbhani, India

Ghatge P
Department of Food Chemistry and Nutrition, College of Food Technology, Marathwada Krishi Vidyapeeth, Parbhani, Maharashtra

Chappalwar V
Department of Food Chemistry and Nutrition, College of Food Technology, Marathwada Krishi Vidyapeeth, Parbhani, Maharashtra

Jadhav B
Department of Food Chemistry and Nutrition, College of Food Technology, Marathwada Krishi Vidyapeeth, Parbhani, Maharashtra

Chappalwar A
Department of Livestock Product Technology, College of Veterinary and Animal Sciences, Maharashtra Animal and Fishery Sciences University, Parbhani, Maharashtra

Dixa Singh
Department of Chemistry, H.N.B. Garhwal University Campus-Pauri (Uttarakhand), India

V. S. Singh
Department of Chemistry, H.N.B. Garhwal University Campus-Pauri (Uttarakhand), India
Mst. Jesmin Sultana
Department of Materials Science and Engineering, Rajshahi University, Rajshahi, Bangladesh

Fazle Rabbi Shakil Ahmed
Department of Pharmacy, Rajshahi Science and Technology University, Natore, Bangladesh

Chander Parkash
Indian Agricultural Research Institute, Regional Station, Katrain (Kullu Valley), H.P.-175129 India

S.S. Dey
Indian Agricultural Research Institute, Regional Station, Katrain (Kullu Valley), H.P.-175129 India

Raj Kumar
Indian Agricultural Research Institute, Regional Station, Katrain (Kullu Valley), H.P.-175129 India

M.R. Dhiman
Indian Agricultural Research Institute, Regional Station, Katrain (Kullu Valley), H.P.-175129 India

Munish Dhiman
Indian Agricultural Research Institute, Regional Station, Katrain (Kullu Valley), H.P.-175129 India

Satish Kumar
Indian Agricultural Research Institute, Regional Station, Katrain (Kullu Valley), H.P.-175129 India

Vinod Kumar
Indian Agricultural Research Institute, Regional Station, Katrain (Kullu Valley), H.P.-175129 India

Jingfeng Huang
Institute of Agricultural Remote Sensing & Information Application, Zijingang Campus, Zhejiang University, Hangzhou, China

Chen Wei
Institute of Agricultural Remote Sensing & Information Application, Zijingang Campus, Zhejiang University, Hangzhou, China
Zhejiang Meteorological Service Center, Hangzhou, China

Yao Zhang
Institute of Agricultural Remote Sensing & Information Application, Zijingang Campus, Zhejiang University, Hangzhou, China

George Alan Blackburn
Lancaster Environment Centre, Lancaster University, Lancaster, United Kingdom

Xiuzhen Wang
Institute of Remote Sensing and Earth Sciences, Hangzhou Normal University, Hangzhou, China

Chuanwen Wei
Institute of Agricultural Remote Sensing & Information Application, Zijingang Campus, Zhejiang University, Hangzhou, China

Jing Wang
Institute of Agricultural Remote Sensing & Information Application, Zijingang Campus, Zhejiang University, Hangzhou, China

Neeranuch Chairungsi
Department of Chemistry, Faculty of Science, Chiangmai University, Chiang Mai, 50200, Thailand

Kanlaya Jumpatong
Department of Chemistry, Faculty of Science, Chiangmai University, Chiang Mai, 50200, Thailand

Weerachai Phutdhawong
Department of Chemistry, Faculty of Science, Maejo University, Chiang Mai, 50292, Thailand
Duang Buddhasukh
Department of Chemistry, Faculty of Science, Chiangmai University, Chiang Mai, 50200, Thailand

Nahid Tamanna
Graduate Program in Biological Sciences, University of Manitoba, Winnipeg, MB, Canada R3T 2N2

Niaz Mahmood
Graduate Program in Biochemistry and Medical Genetics, University of Manitoba, Winnipeg, MB, Canada R3E 0J9

Ammara Ahad
Center of Excellence in Molecular Biology, University of the Punjab, Lahore, Pakistan

Aftab Ahmad
Center of Excellence in Molecular Biology, University of the Punjab, Lahore, Pakistan

Salah ud Din
Center of Excellence in Molecular Biology, University of the Punjab, Lahore, Pakistan

Abdul Q. Rao
Center of Excellence in Molecular Biology, University of the Punjab, Lahore, Pakistan

Ahmad A. Shahid
Center of Excellence in Molecular Biology, University of the Punjab, Lahore, Pakistan

Tayyab Husnain
Center of Excellence in Molecular Biology, University of the Punjab, Lahore, Pakistan

Preface

Pigments or biochrome are the substances present in all living matter and play basic roles in the development of living organisms including plants. Plants use these pigments for the variety of functions including photosynthesis, signaling, to ward off herbivores and attracting pollinating and dispersal agents. The most common plant pigments are chlorophylls, carotenoids, betalains and anthocyanins. In fact, chlorophylls are the most important organic compounds, essential for photosynthesis. Carotenoids presents in plants are usually play their roles in both nutrition and photosynthesis and are necessary for the survival of both mammals and plants. They also served as precursors of plant hormone abscisic acid in higher plants. Another role of carotenoids is to protect photosynthetic organisms against potentially harmful photooxidative processes. Betalains are nitrogen containing water soluble compound, belonging to the families of Caryophyllales, except for Caryophyllaceae and Molluginaceae. Betalains are classified into yellow betaxanthins and red betacyanins and are immonium conjugates of betalamic acid. Anthocyanins are water soluble vacuolar pigments belonging to flavonoids. They appear in plants as blue, red or purple depending on the pH and synthesized through phenylpropanoid pathway. As these plant pigments are of many roles including structural to physiological functions in cells therefore special consideration has been given in this book to their salient characteristics, some biosynthesis and their chemical and biochemical role in different fields. Special attention is also given to specific role of these plant pigments in bioassays regarding the human health. Methodological applications for these pigments are also deeply demonstrated in few chapters including the In silico study for diversifying the molecular pathway of pigment formation.

Editor

Dr. Sajid Mehmood, Ph.D

CHAPTER 1

THE BIOLOGICAL PIGMENTS IN PLANTS PHYSIOLOGY

Arturo Solís Herrera

Human Photosynthesis® Study Center, Aguascalientes, Mexico

ABSTRACT

The physiology of the plants has come to be a formidable challenge. Despite best efforts, the mysteries remain, most of the processes identified cannot be played in vitro, and to date it cannot be said that knowledge about the complex biochemical processes of the plant are understood properly. The need to produce food is an emergency in the world, so any knowledge that will allow us to advance in this sense is important. Our discovery about the intrinsic property of melanin transform light energy into chemical energy by means of the dissociation of the molecule of water, such as chlorophyll in plants, represents a turning point in relation to the chemical reaction so far considered the most important in the world: photosynthesis.

INTRODUCTION

A pigment is defined as any substance capable of absorbing light [1], so it could be considered pigments almost all substances. Usually, substances absorb very specific wavelengths, and emit a characteristic electromagnetic signal that allows us to identify them. Color is the property of electromagnetic radiation with a wavelength between 300 - 400 and 700 - 800 nm. The color of individual chemical substances derived from their interaction with white light. If the total radiation falling on a chemical compound passes through or is reflected from it, the substance has a white color. If white light on contact with a substance is completely absorbed, appears black. Most often chemical compounds selectively absorb within certain radiation ranges and the color of a given compound is complementary to the absorbed radiation. This allows us to infer that the energy inherent in the light produced a change in substance or molecule that it absorbed, so far in accordance with the law of conservation of energy, whose first postulate says that energy, defined as everything that produces a change is not created or destroyed, only is thus transformed. Therefore, the light energy absorbed by the compound or molecule, brought about a change in it, and the new molecule issued part of the absorbed energy, in the form of electromagnetic signal, for example, light with different wavelength to which was originally absorbed.

Nature is full of examples in this regard, and the plants are no exception. And also in them it happens that the wavelengths of light can be, wholly or partly; absorbed, transmitted or reflected. And many compounds containing a plant that

Citation: Herrera, A. S. (2015).The Biological Pigments in Plants Physiology. Agricultural Sciences 6:1262- 1271.http://dx.doi.org/10.4236/as.2015.610121.

they can classified as pigments, for example: porphyrins, carotenoids, anthocyanins, etc. However, there are two that have powerfully drawn attention: the chlorophyll and recently the melanin. And the reason is that both molecules have very peculiar behaviors, we could call them unique in nature. This is: molecules both are able to dissociate the water molecule. And it is the attention, because we are against the way in which nature transforms light energy into chemical energy. It is possible that there are other molecules that can do so, but the two examples most known or at least shown are chlorophyll, which until a decade ago was considered the only molecule with such capacity, and recently, melanin.

Chlorophyll, absorbs the ends of the visible spectrum, and with that power separates the liquid water in its gaseous components, hydrogen and molecular oxygen; and it carried out at room temperature and in an irreversible manner, then ejects the oxygen into the atmosphere. The chemical reaction can be written in the following way.

$$2H_2O_{(liquid)} \rightarrow 2H_{2\,(gas)} + O_{2\,(gas)}$$

It is the attention that the reaction happens so quickly and at room temperature inside the chlorophyll's leaves, as in the laboratory we can dissociate water, but requires warm water at 2000°C and the products that we get, they are not so ordered as in chlorophyll and melanin. This reaction is impelled by the extremes of visible light (violet and red) Another interesting detail is that part of the energy that is released when water dissociates, is transported by molecular hydrogen, which was to be expected since it is the main carrier of energy in the entire universe. On the other hand, molecular oxygen, which is very stable and toxic, is expelled into the atmosphere by the plant, and since it has always been present in the equation, the plant optimizes it, given its high toxicity. In melanin (Greek: melanos; dark), the reaction is similar, but quicker and thousands of times more efficient, because the melanin absorbs electromagnetic full spectrum and in addition is capable not only of separate water into its components but also re-shape the molecule, this is: gaseous components takes her to the liquid state again [2]. This reaction happens inside melanin, and can be located both intra and extracellularly. The chemical reaction is written thus:

$$2H_2O_{(liquid)} \rightarrow 2H_{2\,(gas)} + O_{2\,(gas)} \rightarrow 2H_2O_{(liquid)} + 4e^-$$

Or so:

$$2H_2O_{(liquid)} \leftrightarrow 2H_{2\,(gas)} + O_{2\,(gas)} + 4e^-$$

For every two molecules of water re-formed, are generated 4 high energy electrons, which are electrons whose probability cloud is located farthest from the atomic nucleus in question as a low-energy electron or stable state. High energy electrons or excited are exchanged easily, making it almost impossible to identify the atom or molecule to which they belong.

MELANIN, THE META-CHLOROPHYLL

Melanin is what lies behind the chlorophyll, which explains the misunderstood chlorophyll, which remedied the deficiencies; in few words, is that even explains the origin of life. Chemicals Processes are known in plants, which the mere dissociation of water by chlorophyll cannot explain, could tell that the physiology of the plant is full of mysteries until now misunderstood. The energy that emanates from the chlorophyll does not provide a satisfactory explanation for so many and so many observed phenomena, both in the root, stem and leaves. Only the circulation of the SAP not has been explained to date. The so-called cohesion-tension theory that try to explain SAP transport by the xylem, from the roots of the plant to the leaves through the wood tubules, is to date; a theory.

The mysterious forces that drive the viscous solutions through the trunk, from the root to the leaves, in order, in a sequence that is not understood, no doubt require energy, cannot be explained by the chlorophyll, glucose or ATP, but the energy that emanates from the melanin comes to fill a hole that will allow us to advance in the understanding of the intricate biochemical processes that affect the life of the plant. Melanin in humans had passed as unnoticed as the lignin in plants, the main interest at present was to modify the composition of lignin, which is the melanin in the plants; as a basic strategy to increase the digestibility of cell walls in order to produce biodiesel [3]. It is expected that as you go knowing the unsuspected intrinsic property of melanin transform light energy into chemical energy by means of the water molecule dissociation, the interest becomes more reasonable, because the omnipresence of melanin explains the germination of the seeds, as it was a mystery as the seed detected level appropriate water to flourish (Figures 1-3).

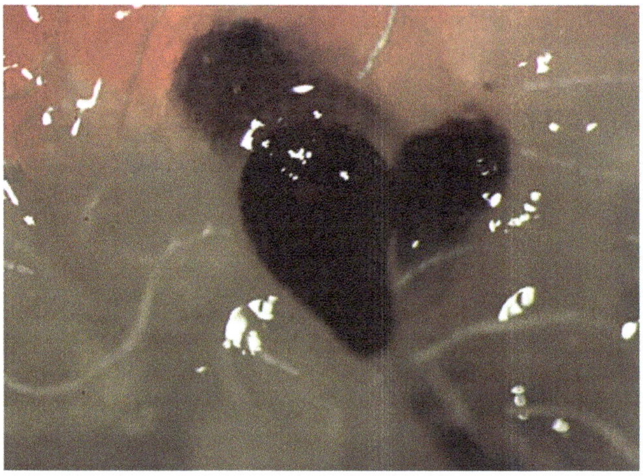

Figure 1. Melanin is present in practically all of the seeds, and the pitaya is no exception. When the amount of water is scarce, power that can produce melanin also is, by what the mechanisms leading to the emergence of seed not can be driven, when the amount of water is adequate, then the energy that emanates from the melanin will be adequate to promote each and every one of the mechanisms that lead to the germination of the seeds (16×).

Figure 2. Seed color depends on the amount of melanin, granule size, the orientation of the molecule, the combination with other compounds, as well as the nature of the structures that are in, but, either way, its main function is the same: production of chemical energy (16×).

Figure 3. The same principle applies in all living organisms, apparently is the only or at least the best system to convert light energy into chemical energy. In Sunflower seed we can see melanin. The magnification is 25×.

The similarity of the surface of the iris of the human eye (Figure 4, Figure 5) with the surface of some seeds, for example, the peach, are given by the notable content of melanin in both structures. The concept that photosynthetic membrane is the microscopic powerhouse of our biosphere, now is broken into small pieces. Molecular hydrogen and high energy electrons coming from the dissociation of water, due to the melanin molecule, constitute the true universal biological engine. The oxidation of water as the basis of molecular biology is more true than ever, but as the melanindone, chlorophyll only is an additional mechanism that optimizes some biochemical processes, as does the hemoglobin in human body, as it is a molecule that also has the amazing ability to dissociate the water molecule irreversibly, and was to be expected, because the biochemical differences between chlorophyll and hemoglobin are really minimal. Because of the huge differences between melanin and chlorophyll, it is that, based on this last, it has not been possible to explain the metabolism of plants. To date, the questions are many, despite the best efforts of researchers and Governments that support research.

BRIEF HISTORY OF THE DISCOVERY

Since melanin absorbs all kinds of energy and dissipates it separating the water molecule, their behavior is unique in nature, because it does not emit any electromagnetic, visible or invisible signal that will allow us to characterize it. As a result, the study of melanin in the laboratory has been the date; a formidable challenge [4]. So it is understandable that the finding of its intrinsic property of transform light energy into chemical energy, by means of the water molecule dissociation, came from another type of study, in this case, of a clinical trial; because we observed incidentally melanin acting in one of its multiple location that is in nature. Why is that we consider important to a brief description of the how it came to knowledge. In 1990, we started an observational, descriptive, study about the three main causes of blindness in Mexico, and that is no coincidence that is the same around the world: Glaucoma, diabetic retinopathy, and macular degeneration related to age.

Our working hypothesis was trying to characterize the morphology of the blood vessels of the optic nerve, very small structure in humans, whose average diameter is equivalent to twelve human hair together. Our aim was to try to find vascular anatomical changes that could eventually serve as indicators of early disease and therefore allowed us early treatments.

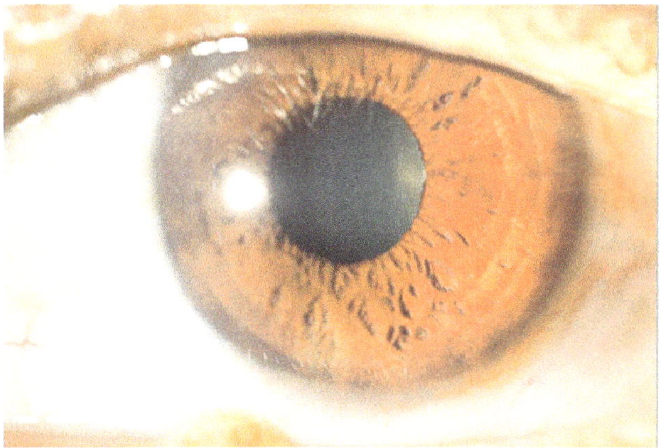

Figure 4. Melanin, wherever you are or whatever that he is called, has the same function, namely the transformation of light, visible and invisible power; in energy chemistry through dissociation and further re-formating of the water molecule. The effect of melanin on the nearest atoms and molecules is surprisingly consistent, as evidenced the huge resemblance between the surface of the iris of the eye and the surface of some seeds, bark of trees, etc.

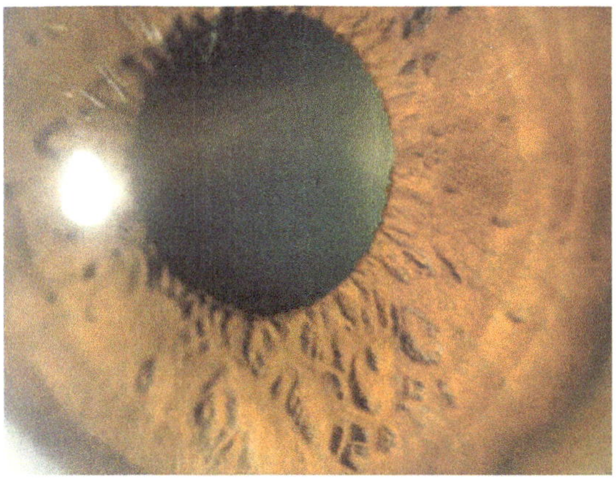

Figure 5. To greater amplification, the resemblance is even greater between the surface of the iris and some seeds, including with the bark of some trees. Is conceivable that at higher magnifications, come a time that would be difficult to discern which is which.

However, the magnifications required to carry out properly the characterization of vascular anatomical pattern, are in the order of 20× to 40×. And to these amplifications, we started to notice the omnipresence of the melanin in the vicinity of the optic disc (Figure 6). And the insistence of nature in place melanin in the entirety of the almost 6000 patients who checked in the span of 12 years which lasted the study (1990-2002), struck us mightily, because nature just insists on important things. So a few months of starting the study, instead of one, there were two main variables in study: blood vessels and melanin.

We were sure that melanin should play an important biological role, but we didn't understand that. Eventually we began to detect an antagonism between the blood vessels and the melanin that was becoming more and more evident. Some time later, we had a conclusion: more melanin, fewer blood vessels and vice versa (Figure 7).

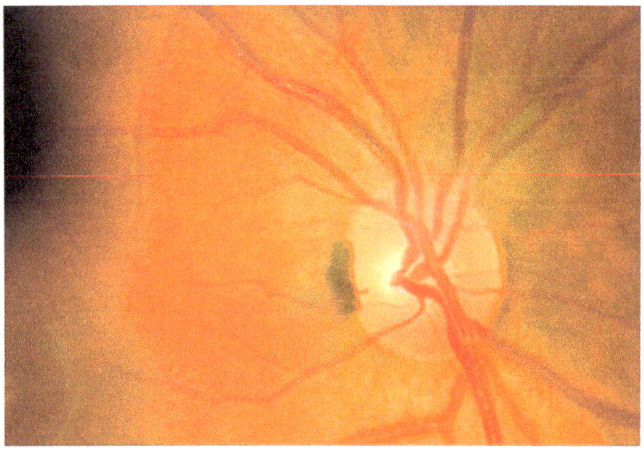

Figure 6. The picture shows the optic nerve of the right eye, with its characteristic orange colour, its oval shape, vessels entering and leaving; and in the meridian of the 9 there is a patch of dark color which corresponds to melanin.

Figure 7. PICTURED are the retinal vessels, and a dark spot that surrounds them partially and which corresponds to the melanin. The effect on the vessels is somewhat subtle, because in this case only observed a light pale around the melanin. In people with fair skin angiogenesis inhibitor effect of melanin is still more subtle, and in dark-skinned people is almost impossible to detect by the amount of melanin.

What remained was to try to explain it. In principle, we did not know which predominated on which. On the one hand, melanin does not possess something like organelles that could produce some kind of transmitters, e.g. peptides, neither nor has something that could be interpreted as receivers. In relation to the blood vessels, these have rough endoplasmic reticulum, so it can synthesize factors that function as transmitters of information, but melanin has no receptors. And something that was surprising was that the melanin antiangiogenic effect is very powerful, because it is present all the time, which was at odds with some kind of factor, for example: Pigment epithelium derived factor or Vascular endothelial growth factor; because the biological variability in the same factor and the corresponding receiver significantly reduces its effect, perhaps in 60%.

And the effect of melanin is surprisingly consistent, then it had to be something different from the known factors. By then, we realized that molecular oxygen levels were different between tissues, according to the amount of pigment. A greater amount of melanin, higher levels of molecular oxygen and vice versa. So the oxygen was the answer, because it is the best anti-angiogenic agent referred to. We have been solving the unknowns, but now appeared a new: where does oxygen came from? The difference in oxygen levels between the pigmented tissues and the less pigmented, in our laboratory conditions, in the human eye, it goes from 34% to 54%, a huge difference.

We review several molecules, various tissues, looking for changes consistent with a constant oxygen donation, and did not find anything, it seemed that not passing anything, but a molecule or tissue that had to submit so much oxygen, , sooner or later, will has any perceptible change, and nothing. Finally, we had only one option: the water containing the eye and that is something more than 95% of the body. And it was the perfect candidate, because in water is hardly notice the changes. But the water not releasing oxygen for free, have to pay it, i.e. energy is required. So that left us with an unknown: where does the energy so the water yield their oxygen atoms? And after several unsuccessful attempts, the answer appeared in front of a Blackboard full of cluttered biochemical equations, was February 2002; "He is dissociating water" I thought, and to write the equation on the Blackboard, suddenly appeared the order in the rest of the equations.

$$2H_2O \rightarrow 2H_2 + O_2$$

She was the equation written, but as we did laboratory tests, we found that apart from the elevated oxygen levels, also had flow of electrons; which was easy to explain because the reaction is reversible in appropriate conditions. So, eventually we perfect the reaction in the following manner:

$$2H_2O \rightarrow 2H_2 + O_2 \rightarrow 2H_2O + 4e^-$$

Melanin releases energy in the form of symmetrical in all directions. Something like increased energy spheres (Figure 8). The position of the granules of melanin into the cell eucarionte is strategic since it is located mainly in the perinuclear space, allowing him to be the source of energy of the nucleus, which does not possess neither mitochondria nor ATP; and at the same time it floods the cell cytoplasm with these growing spheres of energy, which are distributed throughout the entire cell, following the laws of simple diffusion (Figure 9). The process of transformation of light energy into chemical energy seems to be the same in all living organisms, including insects (Figure 10).

Figure 8. Melanin drawn to the Center, and the areas surrounding it, mean the different concentrations of H_2 and O_2 resulting of the dissociation of water, also the different concentrations of water reformed, and at the same time, also with their corresponding 4 highenergy electrons for every two molecules of re-formed water.

Figure 9. The scheme represents the cell membrane, mitochondria, Golgi apparatus, rough endoplasmic reticulum, which, like the granules of melanin (melanosomes) surrounds completely to the cell nucleus.

Figure 10. Melanin is also present in the common moth, and its main function is also transform light energy into chemical energy through dissociation and re-forming of the water molecule.

MELANIN AND PLANTS

The downward movement of sugary plant fluids in phloem requieres energy, indeed, and melanin is the explanation. Mistery about how plants move fluids long distances in their phloem is easiest of understand taking in account the energy that comes from melanin. That sucrose is either used as an energy source for its own metabolism is an idea very difficult to explain. So far, there are only hypothesis, in example, hypothesis of how water rises up in the xylem to the tops of tall trees, how sucrose and a proton (H*) are pumped into the cell; how water comes and build a pressure, how the sap can move a very significant distance; how sucrose is pumped out, how water is pumped out, how sucrose can be splitted to its monomers and thence to other hexoses. Let us remember that any chemical reaction involves exchange of energy. The metabolism of plants as animals is poorly understood, since the power source was not known. Glucose is a special molecule that organisms use as a source of biomass as a source of carbon chains, but even there, and both animals and plants take the necessary energy from water. Energy can be defined as everything that produces a change, and the word metabolism means continuous change. The metabolic needs of the plant are continuous, incessant, both day and night; and it is the way in which melanin delivered energy, in the form of incessant, continuous, both day and night. Therefore, the mysteries about life in the plants will have to look in the trunk, not so much in the leaves (Figure 11).

Figure 11. The chlorophyll in plants, as well as the hemoglobin in humans, possess the ability to dissociate the water molecule in an irreversible manner, using the light energy it contains. But both molecules are only an extension of the master molecule that is melanin [5].

CONCLUSIONS

Both plants and humans are enveloped in a dense cloud of mysteries, at least in regards to your metabolism. It is no coincidence that of the 7000 cellular reactions described in the literature, only about 200 are described in the same way in the different sources of information, there are significant disputes in the remaining 6800. And it is that cell biology has puzzled trying to concatenate the glucose as a source of energy and at the same time of carbon chains; which is to say that a wonderful molecule not only brings carbon atoms to organisms but also carries the necessary energy for its own metabolism. This wrong concept is deeply rooted in both the scientist and public in general. But glucose just provides the building-blocks of the 99% of the biomolecules that mold our organism. Thereby the body knows glucose very well, so is able to split it, to combine it with other elements or molecules, and in a very precise way; even more, the cell is able to metabolize glucose to CO2 and water; but the energy that supposedly glucose contains is a myth. Therefore, that biology at the molecular level has been stagnant. It is not possible to extract more energy from glucose containing that. However, if from now forward we take into account the unexpected role of melanin as an important source bioenergetics, the role of the glucose will refine and allows us to re-order our concepts in a manner more consistent with nature, where everything depends on the generation and distribution of energy. The discovery that the main source of energy comes from lignin [6] and chlorophyll, represents a substantial change in the study of photosynthesis in plants. We will have to reconsider many metabolic pathways so far accepted dogmatically but remain without plausible explanation, whereas to date theories. A significant advance in the knowledge will be able to be here, and it is a sharp change of direction to redirect the attention toward the trunk of the plants, rather than the leaves; but this will allow us to better understand the mysteries of the physiology of plants.

REFERENCES

1. Hoffman, J. and Puszynski, A. (2009) Chemical Engineeering and Chemical Process Technology. Vol. V. Pigments and Dyestuffs. Encyclopedia of Life Support Systems (EOLSS).
2. Solis Herrera, A., Arias Esparza, M.C., Solís Arias, R.I., et al. (2010) The Unexpected Capacity of Melanin to Dissociate the Water Molecule Fill the Gap between the Life before and after ATP. Biomedical Research, 21, 224-227.
3. Anderson, A.N., Tobimatsu, Y., Ciesielki, P.N., Ximenes, E., Ralph, J., Donohoe, B.S., Ladisch, M. and Chapple, C. (2015) Manipulation of Guaiacil and Syringyl Monomer Byosynthesis in an Arabidopsis Cinnamyl Alcohol Dehydrogenase Mutant Results in Atypical Lignin Byosynthesis and Modified Cell Wall Structure. The Plant Cell, 27, 2195- 2209. http://dx.doi.org/10.1105/tpc.15.00373
4. Hill, H.Z. (1992) The Function of Melanin or Six Blind People Examine an Elephant. BioEssays, 14, 49-56. http://dx.doi.org/10.1002/bies.950140111
5. Solís-Herrera, A. And Solís-Arias, M.P. (2013) The Odyssey of Atmospheric Oxygen in Their Futile Attempt to Reach the Interior of the Cell. International Journal of Health Research and Innovation, 1, 37-52.
6. Solís-Herrera, A. And Arias-Esparza, M.C. and Solís Arias, M.P. (2015) Photosynthesis in Humans, Chapter 12 of the Book: Photosynthesis, Functional Genomics, Physiological Processes and Environmental Issues. Edited by Nafees Khan. Nova Publishers, New York, 257-279.

CHAPTER 2

PHOTOSYNTHETIC PIGMENTS IN DIATOMS

Paulina Kuczynska [1], Malgorzata Jemiola-Rzeminska [1,2] and Kazimierz Strzalka [1,2]

[1] Faculty of Biochemistry, Biophysics and Biotechnology, Department of Plant Physiology and Biochemistry, Jagiellonian University, Gronostajowa 7, Krakow 30-387, Poland

[2] Małopolska Centre of Biotechnology, Gronostajowa 7A, Krakow 30-387, Poland

ABSTRACT

Photosynthetic pigments are bioactive compounds of great importance for the food, cosmetic, and pharmaceutical industries. They are not only responsible for capturing solar energy to carry out photosynthesis, but also play a role in photoprotective processes and display antioxidant activity, all of which contribute to effective biomass and oxygen production. Diatoms are organisms of a distinct pigment composition, substantially different from that present in plants. Apart from light-harvesting pigments such as chlorophyll a, chlorophyll c, and fucoxanthin, there is a group of photoprotective carotenoids which includes β-carotene and the xanthophylls, diatoxanthin, diadinoxanthin, violaxanthin, antheraxanthin, and zeaxanthin, which are engaged in the xanthophyll cycle. Additionally, some intermediate products of biosynthetic pathways have been identified in diatoms as well as unusual pigments, e.g., marennine. Marine algae have become widely recognized as a source of unique bioactive compounds for potential industrial, pharmaceutical, and medical applications. In this review, we summarize current knowledge on diatom photosynthetic pigments complemented by some new insights regarding their physico-chemical properties, biological role, and biosynthetic pathways, as well as the regulation of pigment level in the cell, methods of purification, and significance in industries.

INTRODUCTION

Diatoms are becoming more and more prominent microalgae. More advanced knowledge about them has also enhanced their importance and usefulness in commercial and industrial applications such as biofuels, pharmaceuticals, health foods, biomolecules, materials relevant to nanotechnology, and as bioremediators of contaminated water [1]. The singularity of these organisms is physiological in nature, involving, for example, novel metabolic pathways and compounds, but is also due to their importance in the evolutionary history of eukaryotes as well as their ecological success. They are mainly associated with the silicon metabolism engaged in the biogeochemical cycling of Si in the sea. The siliceous structures in their cell wall create the unique morphotypes that are used as taxonomic keys [2]. Another interesting process is the ornithine-urea cycle which is absent in green algae and plants and is essential for diatom growth and their contribution to marine productivity [3]. Furthermore, these organisms are rich in bioactive compounds capable of antiviral activity, including naviculan [4] and neuroexcitatory amino

Citation: Kuczynska, P., Jemiola-Rzeminska, M., & Strzalka, K. (2015). Photosynthetic Pigments in Diatoms. Marine Drugs, 13(9), 5847–5881. http://doi.org/10.3390/md13095847.

acid derivative domoic acid [5], not to mention the cytotoxic and blood platelet inhibitory activity caused by adenosine [6]. There is a high diversity of beneficial diatom cell components in lipids and pigments, whose amount in the cell may be partially regulated by certain abiotic stresses or genetic modifications of metabolic pathways. For example, the overexpression of endogenous Δ5 desaturase leads to an accumulation of eicosapentaenoic acid, a fatty acid known to have a variety of health benefits including anti-inflammatory effects and better neuronal functioning [7]. On the other hand, enhanced biosynthesis of strong antioxidant fucoxanthin was obtained through low light and nitrogen treatment [8]. Although not all the compounds in diatoms are known, due to the importance of these organisms, there is continual research to find, identify, and examine their properties. In the marine diatom *Haslea ostrearia*, a water soluble blue pigment named marennine was identified and further experiments have shown its allelopathic, antioxidant, antibacterial, antiviral, and growth-inhibiting properties [9]. Another diatom, *Haslea karadagensis*, also has a blue pigment. Although it has quite different absorption maxima from those of marennine, its similar bioactivity has come to be called marennine-like [10]. While these unusual pigments occur only in selected species, there are some which are included in the more general group of photosynthetic pigments and are common among diatoms. These also have great benefits in medicine, pharmacy, cosmetics, food, and supplements. However, the pigment profile of diatoms is quite different than that found in plants and some algae. Chlorophyll a (Chl *a*) and chlorophyll c (Chl *c*) together with fucoxanthin (Fx) are components of fucoxanthin-chlorophyll protein (FCP) complexes which replace plants light harvesting complexes (LHC) in performing the light-harvesting function [11]. Furthermore, the three carotenoids—β-carotene (β-car), diadinoxanthin (Ddx), and diatoxanthin (Dtx)—are known to play an important role in photoprotection and, additionally, violaxanthin (Vx), antheraxanthin (Ax), and zeaxanthin (Zx) may be engaged in this process. This article is focused on the above-described photosynthetic pigments which are essential for diatom life and which are commonly used in various industries. Despite the fact that studies in this field have been carried out for many years, a lot of aspects that require further analysis still remain. What follows below represents the current knowledge of their physical and chemical properties, their biosynthetic pathways, their regulation of pigment level, as well as their localization in the cell, their role in photosynthesis and photoprotection, methods for their identification and purification, and their significance in industries.

PHYSICAL AND CHEMICAL PROPERTIES OF PHOTOSYNTHETIC PIGMENTS OF DIATOMS

Diatoms contain two types of pigments involved in light harvesting and photoprotection: chlorophylls and carotenoids. Chlorophylls trap light energy—blue and red portions of the electromagnetic spectrum, in particular, which are used in photosynthesis. Generally, chlorophylls can be defined as a magnesium coordination complex of cyclic tetrapyrroles containing a fifth isocyclic ring, referred to as porphyrin, with a long-chain isoprenoid alcohol ester group. Being the highly conservative structural motif of Chls, however, the phytyl chain is not present in the majority of Chl *c* pigments found in diatoms. The second group of pigments, carotenoids, are engaged mainly in photoprotection; however, Fx also participates in light harvesting. They are comprised of carotenes (hydrocarbons) and their oxygenated derivatives, xanthophylls. Extensive data compiled for 47 of the most important chlorophylls and carotenoids found in marine algae were published by Jeffrey and Vesk [12].

Chlorophylls

Several kinds of chlorophylls are found in photosynthetic organisms; however, only two forms occur in diatoms: Chl *a* and, identified in various algae, Chl *c*. The predominant Chl *a* plays a central role in the photochemical energy conversion of the majority of photosynthesizing organisms, while Chl *c* participates effectively in photosynthesis as an accessory pigment, similar in its functional activity to the Chl *b* of higher plants. From among different forms of Chls *c* which were described in diatoms, the most abundant are Chl $c_1$and c_2 (Figure 1). The distinct structure of a Chl *c* brings changes in the absorption spectrum to produce a strong Soret (blue) absorption band in comparison with a weak band in the red region. The ratios of band I (at ~630 nm) to band II (at ~580 nm) are >1 for Chl c_1-like chromophores, ~1 for Chl c_2-like chromophores, and <1 for Chl c_3-like chromophores [13].

Among 51 species (71 isolates) of tropical and sub-tropical diatoms from 13 out of 22 families examined by Stauber and co-workers [14], Chl c_2 was present in all the diatoms tested and occurred together with Chl $c_1$in 88% of them. Where Chl c_1 was absent or occurred in trace amounts only, it was usually replaced by a Chlc_3, identified by Fookes and Jeffrey as (7-methoxycarbonyl)-Chl c_2 [15]. The presence of a methoxycarbonyl group (–$COOCH_3$) at C-7 (ring B) explains the difference in molecular weight (653 *m*/*z*) compared to Chl c_2(609 *m*/*z*), as well as the significant decrease in absorption band I (Qy) intensity compared to the Chl c_1 andc_2 values [16]. Exceptions are *Nitzschia closterium* (CS-114), with merely Chl c_2, and *Nitzschia bilobata*(CS-47), which contains all three Chls (c_1, c_2, and c_3) in approximately equal amounts. Five species that have Chls c_1 and c_2 also contain Chl c_3 in trace quantities. Additionally, diatoms may possess the Chl c_2-P.gyrans-type found, for example, in *Pseudo-nitzschia multiseries* [16]. Finally, traces of (DV)-PChlide with a propionic acid side chain at C-17 instead of acrylic acid are commonly present in diatoms. An overview of the Chl *c* distribution in microalgae classes is given by Jeffrey and Vesk [12].

In the aquatic environment, pigments may undergo degradation in response to chemical, photochemical, and biological processes. Studies of marine organic matter from the central equatorial Pacific were able to weigh the overall reactivity of various biochemical classes and found that pigments, especially chlorophylls, were the most labile compounds [17]. However, possessing nitrogen, chlorophylls are more prone to being salvaged during senescence and biological breakdown than carotenoids. Although the photooxidation of chlorophyll was studied almost exclusively in terms of the porphyrin moiety of the molecule, its unsaturated chain is also susceptible to a reaction with singlet oxygen or the hydroxy and peroxy radicals which are generated during Chl photodegradation. It has been demonstrated by Rontani and co-workers [18] that several free and esterified oxidized isoprenoid compounds are produced during the photodegradation of

Chl*a* in seawater. Moreover, they suggested that in the dead cells of *Phaeodactylum tricornutum*, the Chl phytyl chain will be reduced to up to one half of its initial concentration after prolonged exposure to high light.

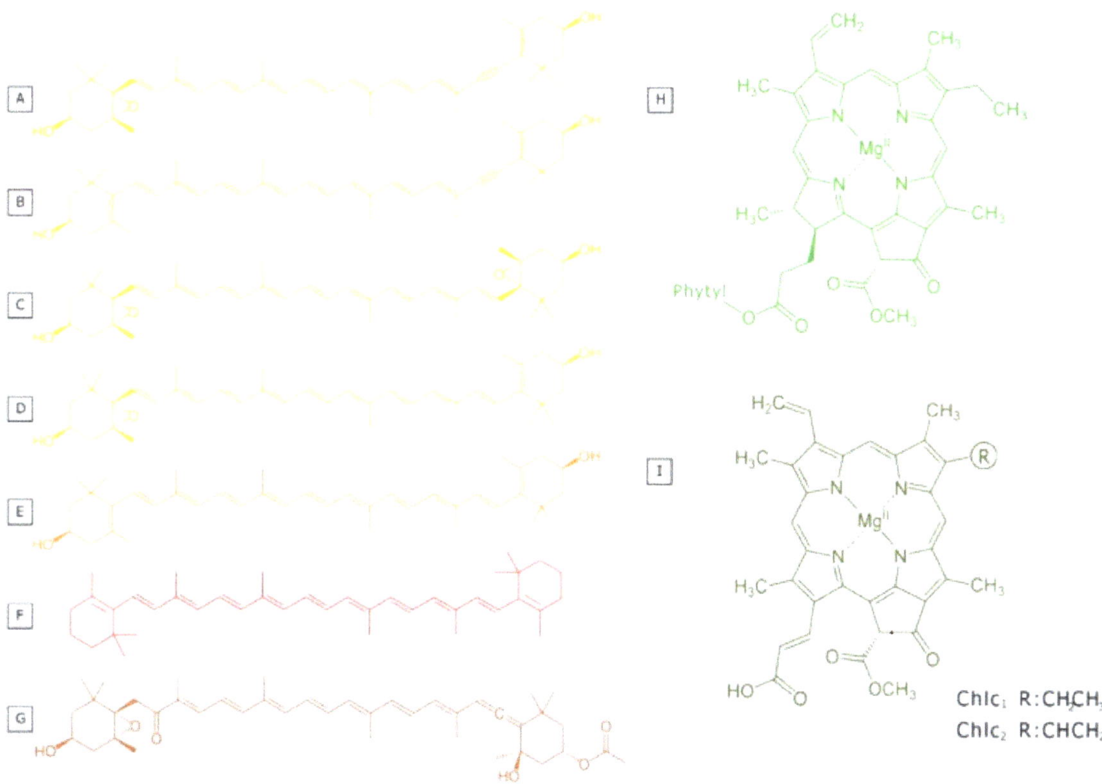

Figure 1: Structural formula of photosynthetic pigments in diatoms including *all-trans* carotenoids: (**A**) diadinoxanthin; (**B**) diatoxanthin; (**C**) violaxanthin; (**D**) antheraxanthin; (**E**) zeaxanthin; (**F**) β-carotene; (**G**) fucoxanthin; and chlorophylls: (**H**) chlorophyll a (**I**) chlorophyll c.

Carotenoids

More than 700 types of carotenoids were identified in nature [19]. They are commonly synthesized by plants, algae, and some micro-organisms. Seven kinds of carotenoids were found in diatoms with β-car as an example of carotenes, as well as Fx, Dtx, Ddx, Zx, Ax, and Vx, which represent xanthophylls (Figure 1). Although derivatives of all mentioned pigments, including isomers and products of degradation, may occur in the cell, the all-*trans*-isomers are the most abundant and functionally active forms. The presence of a conjugated polyene chain in carotenoids may be the cause of carotenoid instability, which is related to their susceptibility to oxidation, E/Z isomerization by heat, light, and chemicals. As a result of the *cis*-isomerization of a chromophore's double bond there is a slight loss of color, as well as a small hypsochromic shift and a hypochromic effect, accompanied by the appearance of a *cis* peak about 142 nm below the longest wavelength absorption maximum of the *trans*-carotenoid when measured in hexane [16]. Moreover, the geometrical structures of *cis* and *trans* carotenoids, which are differentially oriented in the thylakoid membrane, have an impact on membrane physical properties. Intercalation of carotenoids changes permeability for the oxygen and other small molecules, which is associated with their protective activity [20,21]. The activation energy for allenic (R/S) isomerization in carotenoids is higher than that associated with geometrical (E/Z) isomerization. Consequently, E/Z stereomutation of Fx can occur in sunlight in the absence of a catalyst, whereas allenic isomerization was shown to occur only to a very low extent. Recently, data has been published on the iodine or diphenyl diselenide-mediated photoisomerization of Fx, which, in view of the increased reaction rate with UVA radiation, support the radical mechanism of R/S isomerization [22]. Additionally, some carotenoids are highly unstable in the presence of acid. Under weak acidic conditions, 5,6-epoxide is readily rearranged into furanoid 5,8-epoxide. On the other hand, the lability of Fx towards alkali has been established, which precludes the use of saponification in the isolation procedure of carotenoid mixture containing Fx. First, chromophoric changes upon treatment of this allenic carotenoid with a weak base (K_2CO_3) were reported and, then, two products were identified: (i) isofucoxathinol as the kinetically controlled product and (ii) fucoxanthinol hemiketal, with a shorter chromophore, as the thermodynamically controlled product [23]. Unlike chlorophylls, carotenoids are often broken down into a colorless compound by the destruction of the long chain of alternating double bonds and cannot be detected by regular pigment analysis [24].

Carotenoids exhibit intense absorption between 400 and 500 nm. The conjugation length and type of the functional groups that are attached to the ionone rings terminating the polyene chain largely determine the absorption properties of the carotenoids [17,25]. The absorption spectra of carotenoids are markedly solvent-dependent, which should be considered when analyzing pigment extracts by high performance liquid chromatography with a photodiode array detector (HPLC-DAD), because in most cases spectra are taken in mixed solvents. The λ_{max} values of carotenoids in hexane or petroleum ether are practically the same as in diethyl ether, methanol, ethanol, and acetonitrile, and are higher by 2–6 nm in acetone, 10–20 nm in chloroform, 10–20 nm in dichloromethane, and 18–24 nm in toluene; see [16] for ultraviolet and visible absorption data for common carotenoids. To give an idea of the spectral fine structure, the values of %III/II are also given along with the λ_{max} values.

The main light-harvesting carotenoid in diatoms is Fx, which is also abundant in brown algae. Unusually, small amounts of a 19′-butanoyloxyfucoxanthin-like pigment, in addition to Fx, were found in one diatom species, *Thalassiothrix heteromorpha*, as reported in [26]. Fx has an allenic bond, a conjugated carbonyl, a 5,6-monoepoxide, and acetyl groups that contribute to the unique structure and spectral properties of the molecule (Figure 1). In contrast to other carotenoids, its broad absorption band (between 460 and 570 nm) covers much of the gap left by chlorophyll in the green region. Diatoms also possess the β-car of carotenes as well as two xanthophylls, Ddx and Dtx, which are also asymmetric molecules, containing an acetylenic group at one of the ionone rings. Moreover, three other xanthophylls, characteristic of higher plants, Vx, Ax, and Zx, may also occur (Figure 1) [24,27]. However, these carotenoids accumulate only under specific conditions, e.g., during long-time illumination with strong light (see Section 8).

BIOSYNTHESIS PATHWAYS

The biosynthetic pathways of both chlorophylls and carotenoids in plants have been extensively studied and complete information in this field is available. This opens up many opportunities for studies on genetically modified organisms and for *in vitro* approaches using recombinant proteins. In diatoms, some pathway points remain unclear and the production of cell lines with an enhanced amount of a selected photosynthetic pigment is currently difficult. Most of the genes which encode enzymes in these steps of the pathway were sought out by genome alignment, but they have not been identified. The whole genomes of two diatom species, *Phaeodactylum tricornutum* [28] and *Thalassiosira pseudonana* [29], were sequenced, but few analogues of the genes which occur in plants or algae were found in these diatoms.

Two pathways, methylerythritol phosphate (MEP) and mevalonate (MEV), of the early steps of carotenoid biosynthesis have been described. Their occurrence is not clear but a few studies show that it depends on the taxon or the growth rate. However, the products of both are dimethylallyl diphosphate (DMAPP) and its isomer, isopentenyl pyrophosphate (IPP) [14,16,30]. The next steps on the pathway to lycopene synthesis are the conversion of DMAPP to geranylgeranyl pyrophosphate (GGPP), which is catalyzed by GGPP synthase, then to phytoene by phytoene synthase (PSY), afterwards to ζ-carotene, which is catalyzed by phytoene desaturase (PDS), and, finally, the product of ζ-carotene desaturase (ZDS) is lycopene [30]. The biosynthesis pathway from lycopene to xanthophylls is presented in Figure 2. Firstly, lycopene as a long and straight molecule is cyclized by lycopene β-cyclase (LCYB) to β-car, having two β-ionone rings at both ends of the yield. In the next step, xanthophyll is first formed, and this reaction requires hydroxylation. However, a gene encoding β-carotene hydroxylase (BCH) was not found in the diatom genome and another one that is similar to LUT1 has been proposed as a putative enzyme to make the formation of Zx from β-car possible [31]. Two further light-dependent and reversible reactions lead to Vx formation via the intermediate product Ax. Both are catalyzed by Vx de-epoxidases (VDEs) in high light conditions, but reverse reactions are catalyzed by Zx epoxidases (ZEPs) in low light or in the dark [32]. In *T. pseudonana*, two VDEs and two ZEPs were identified [33], while in *P. tricornutum*, three isoforms of ZEPs (ZEP1, ZEP2, ZEP3) and VDEs (VDE, VDL1, VDL2) were found [31]. Alternatively, Vx might be formed from Zx through β-cryptoxanthin (Cx) and β-cryptoxanthin-epoxide (CxE) [27]. Further steps which lead to the formation of Fx, Ddx, and Dtx are still unclear because of missing data about the enzymes engaged in the process. However, to date, two models of possible conversions from Vx to Fx have been presented. The first model was described by Lohr and Wilhelm [27], who proposed Vx as a precursor of Ddx, Dtx, and Fx, with the reaction from Vx to Dtx as well as to Fx proceeding via Ddx. This hypothesis was confirmed experimentally using norflurazon, which inhibits the *de novo* synthesis of carotenoids and which was used after the accumulation of Vx. In low light, an increase in the Fx level was detected. The other model is based on speculation about the chemical properties of these xanthophylls and neoxanthin (Nx) was regarded as a precursor of both Ddx and Fx [34]. Nx is the most likely candidate for enabling the formation of the acetylenic bond in Ddx or the allenic double bond in Fx. However, the formation of Fx requires two modification steps: the ketolation of Nx and the acetylation of an intermediate, probably fucoxanthinol. To support one of these hypotheses, the identification of the enzymes is necessary. The most powerful approach in this field is to look for genes encoding the proteins of interest on databases. Unfortunately, the gene encoding Nx synthase (NXS), which catalyzes the conversion of Vx to Nx in *Arabidopsis thaliana*, was not found in brown seaweeds. Moreover, Nx has not been detected either. However, LCYB shares a 64% amino acid identity with NXS and is proposed to be engaged in Nx production, although no LCYB-like NXS in brown seaweeds was identified [35]. It is important to reveal the whole xanthophyll biosynthetic pathway because of the many opportunities to further studies and also to prepare transgenic organisms with an increased xanthophyll level.

The chlorophyll biosynthetic pathway has been extensively studied in higher plants and also in some groups of algae, but in the case of diatoms, it has been poorly investigated. Nevertheless, the main aspects are similar in all photosynthetic organisms. Three general steps are needed for Chl synthesis: aminolevulinic acid formation, its transformation into Mg-porphyrins, and protochlorophyllide conversion to Chl [36]. The initial steps rely on tetrapyrrole cyclization, the insertion of Mg leading to diviny-PChlide *a* formation, and its reduction to PChlide *a*. Afterwards, light-independent PChlide oxidoreductase (DPOR) and a light-dependent enzyme (LPOR) catalyze PChlide hydrogenation to Chlide and further steps lead to Chl *a*formation. In four diatom species, more than one isoform of LPOR was found [37]. The final step is the introduction of phytol residue, which is associated with the MEP pathway also used in carotenoid formation [38]. In higher plants, the chlorophyll cycle relies on conversion between Chl *a* and Chl *b*, allowing the adjustment of their ratio to light conditions [39]; however, in diatoms, instead of Chl *b*, Chl *c* (Chl c_1, Chl c_2, and rarely Chl c_3) was identified [13,14] (see also Section 2). The molecular structure of Chl *c* may suggest that PChlide is its precursor in the biosynthetic pathway where oxidation and dehydration are required, but no enzyme carrying out these steps has been described [40]. In general, data about Chl biosynthesis and the enzymes which catalyze each step are still poorly understood.

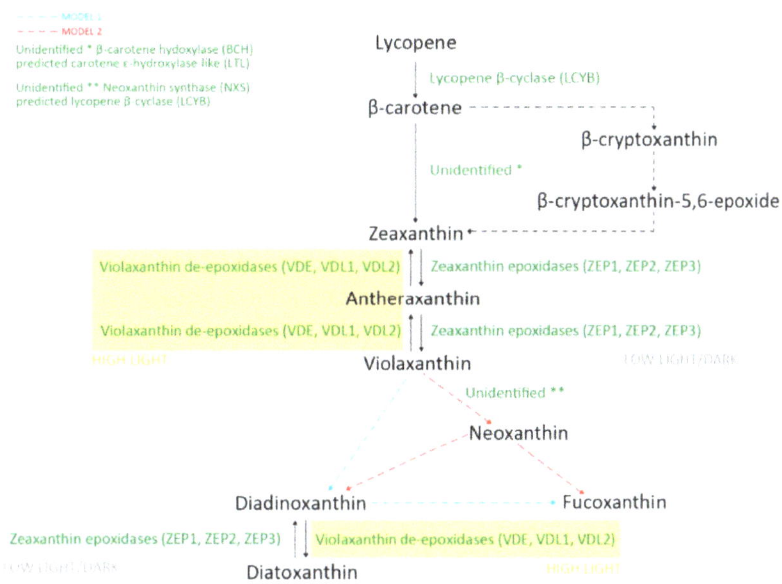

Figure 2: Biosynthetic pathway of photosynthetic carotenoids in the diatom*Phaeodactylum tricornutum* from lycopene to fucoxanthin and diatoxanthin.

REGULATION OF PIGMENT LEVEL IN THE CELL

Changes in photosynthetic pigment levels are usually a fast response to environmental conditions because they are engaged in basic processes such as photosynthesis and photoprotection, which are essential for cell life (Table 1). Diatom cell physiology displays some significant differences that are observed during growth, including an exponential, stationary, and declining phase. In the exponential phase of growth, the metabolism is the highest, especially that of amino acids, whereas in the declining phase, catabolism is predominant and is related to an increase in metabolites like terpens and putrescine [41]. Experiments which concern cell responses to stress factors and pigment measurements should be performed during the exponential phase of growth.

Table 1: Changes in pigment content (Chl *a*: chlorophyll *a*; Chl *c*: chlorophyll *c*; β-car: β-carotene; Fx: fucoxanthin; Ddx: diadinoxanthin; Dtx: diatoxanthin; Vx: violaxanthin; Ax: antheraxanthin; Zx: zeaxanthin) in diatoms in response to selected stress conditions. The down arrow represents a decrease of pigment content, the up arrow is the opposite, and const means no changes.

Conditions	Species	Changes in Pigment Content								
		Chl *a*	**Chl *c***	**β-Car**	**Fx**	**Ddx**	**Dtx**	**Vx**	**Ax**	**Zx**
HL (140 μmol photons $m^{-2} \cdot s^{-1}$) in comparison to LL (40 μmol photons $m^{-2} \cdot s^{-1}$), 16 h light/8 h dark photoperiod [51]	*Cyclotella meneghiniana*	N/A	*const*	*const*	*const*	↑	↑	N/A	N/A	N/A
Iron-replete medium (12 μM) compared to iron-reduced medium (1 μM) in HL (140 μmol photons $m^{-2} \cdot s^{-1}$) [52]	*Cyclotella meneghiniana*	↑	N/A	N/A	N/A	↓		N/A	N/A	N/A
Iron-replete medium (12 μM) compared to iron-reduced medium (1 μM) in LL (40 μmol photons $m^{-2} \cdot s^{-1}$) [52]	*Cyclotella meneghiniana*	↑	N/A	N/A	N/A	*const*		N/A	N/A	N/A
HL (300 μmol photons $m^{-2} \cdot s^{-1}$) compared to LL (50 μmol photons $m^{-2} \cdot s^{-1}$), 14 h light/10 h dark photoperiod [53]	*Phaeodactylum tricornutum*	N/A	↓	↑	↓	↑	↑	N/A	N/A	N/A
B-HL (450 PFD) compared to BR-HL (450 PFD in R:B ratio 0.25) [43]	*Pseudonitzschia multistriata*	↓	↑	*const*	↓	↓	↓	N/A	N/A	N/A
B-LL (250 PFD) compared to BR-LL (250 PFD in R:B ratio 0.25) [43]	*Pseudonitzschia multistriata*	↓	↑	*const*	↓	↓	↓	N/A	N/A	N/A
B-LL (24 (10 absorbed) μmol photons $m^{-2} \cdot s^{-1}$) compared to W-LL (40 (10 absorbed) μmol photons $m^{-2} \cdot s^{-1}$) [54]	*Phaeodactylum tricornutum*	↑	*const*	*const*	*const*	*const*	N/A	↓	N/A	N/A
R-LL (41 (10 absorbed) μmol photons $m^{-2} \cdot s^{-1}$) compared to W-LL (40 (10 absorbed) μmol photons $m^{-2} \cdot s^{-1}$) [54]	*Phaeodactylum tricornutum*	↑	↓	↑	↓	↓	N/A	↓	N/A	N/A
HL (1250 μmol photons $m^{-2} \cdot s^{-1}$) in comparison to LL (40 μmol photons $m^{-2} \cdot s^{-1}$), 12 h light/12 h dark photoperiod [55]	*Phaeodactylum tricornutum*	↓	*const*	*const*	*const*	↓	↑	N/A	N/A	N/A
6 days acclimated to shift from BL (24 (10 absorbed) μmol photons $m^{-2} \cdot s^{-1}$) to RL (40 (10 absorbed) μmol photons $m^{-2} \cdot s^{-1}$) [45]	*Phaeodactylum tricornutum*	↑	N/A	N/A	N/A	↓	N/A	N/A	N/A	N/A
6 days acclimated to shift from RL (40 (10 absorbed) μmol photons $m^{-2} \cdot s^{-1}$) to BL (24 (10 absorbed) μmol photons $m^{-2} \cdot s^{-1}$) [45]	*Phaeodactylum tricornutum*	*const*	N/A	N/A	N/A	↑	N/A	N/A	N/A	N/A
14 days dark storage culture [56]	*Thalassiosira weissflogii*	↓	↓	↓	↓	↓		N/A	N/A	N/A
HL (700 μmol photons $m^{-2} \cdot s^{-1}$) in comparison to LL (40 μmol photons $m^{-2} \cdot s^{-1}$), 16 h light/8 h dark photoperiod [24]	*Cyclotella meneghiniana*	N/A	N/A	N/A	N/A	↓	↑	*const*	↑	↑
high nitrogen culture (18 mM) compared to low nitrogen culture (6 mM) in LL (100 μmol photons $m^{-2} \cdot s^{-1}$) [8]	*Odontella aurita*	N/A	N/A	N/A	↑	N/A	N/A	N/A	N/A	N/A

B-HL: high blue light; B-LL: low blue light; BR-HL: high blue/red light; BR-LL: low blue/red light; HL: high light; LL: low light; PFD: photon flux density; R-LL: low red light; W-LL: low white light; N/A: not available

Pigment level is mostly regulated by light conditions, resulting in fast conversions of them, usually without any changes in gene expression, although long-term acclimation causes changes in the transcript level of the genes which encode proteins in biosynthetic pathways. The effect of white and blue-green light on pigment content during the exponential and stationary growth phases of three diatom species was studied [42] and it was reported that Chl *a* content decreased by nearly 50% in the stationary phase and carotenoid content increased in blue-green light. In high light, the amount of Chl *a*, Fx, and β-car decreased, but with unrelated variations in Chl *c*, and it was also reported that the Chl *c* content in FCP complexes might be regulated by light, but not that of Chl *a* and Fx [43]. Under low light, FCP trimers contain mainly Lhcf5 proteins with a high Fx:Chl *c* ratio, whereas trimers in high light contain mainly Lhcf4 proteins with a low Fx:Chl *c* ratio. A different variation of Chl *a*, β-car, and Fx is correlated with a decreasing number of PSII units in high light. This photoacclimation strategy described in species growing in variable light conditions enables the efficient regulation of photosystem structures to the amount of absorbed energy [43,44]. The spectral composition of light plays a crucial role in growth rate, photoprotective mechanisms, and photosynthetic efficiency, and thereby in pigment content. In blue light-acclimated *P. tricornutum* cells, the pool of xanthophyll cycle pigments was higher than those growing in red light, but in neither case was Dtx detected [45]. Studies on the long-term dark incubation of diatoms in the sediment from a tidal mud flat showed that after one year of microscope observations, fluorescence was detected. Pigment analysis including Fx, Chl *c*, Chl *a*, Dtx, and Ddx showed a rapid decrease in the first weeks and, then, slower changes, and although Ddx was not detected after two months, Dtx remained steady over time [46]. It is not only light conditions that have an impact on pigment levels, but also nutrient limitation along with the heavy metals which are more and more abundant in the environment [47]. Iron limitation results in a wide range of changes in diatoms, including in gene expression in factors which regulate pigment content. As many as 20 genes of iron-responsive regulation were described in *P. tricornutum* [48]. It has been reported that under iron limitation, there was an increase in the transcript level of 16 genes involved in Chl biosynthetic pathways, although the Chl content decreased [49]. With an increased cadmium concentration, a rapid decrease in Chl *a* as well as carotenoids was observed in epiphytic diatoms of *Myriophyllum triphyllum* [50]. Although the pigment composition is similar in each diatom species, the ratio between them is different and highly variable. Unfortunately, a comparison of pigment content in different diatom species growing under optimal or stress conditions is very complicated because of the variety of ways in which it is calculated. This is very often done in relation to Chl*a* content, but the absolute content in dry or wet weight is also measured, or the percentage of each pigment is measured.

LOCALIZATION IN THE CELL

Diatom cells contain either a few small chloroplasts or one large chloroplast [57]. In diatoms, the thylakoid membranes, where pigments responsible for the absorption of light for photosynthesis are located, are not differentiated into granal and stromal lamellae, *i.e.*, granal stacking is absent [58]. Instead, the diatom thylakoid membranes are arranged into groups of three loosely stacked lamellae which span through the whole length of the chloroplast (Figure 3) [59]. Consequently, no lateral heterogeneity in the distribution of photosystems (PS) I and PS II has been detected so far. The organization of the LHC proteins, recently reviewed in detail by Gundermann and Büchel [60], also exhibit differences when compared to the LHCs of higher plants; in diatoms, FCP complexes are the main light-harvesting antennae. Moreover, the attribution of the different FCPs to the two photosystems and/or their supramolecular structure remains unknown. However, when comparing plant LHCs with FCPs the most obvious difference is in pigmentation.

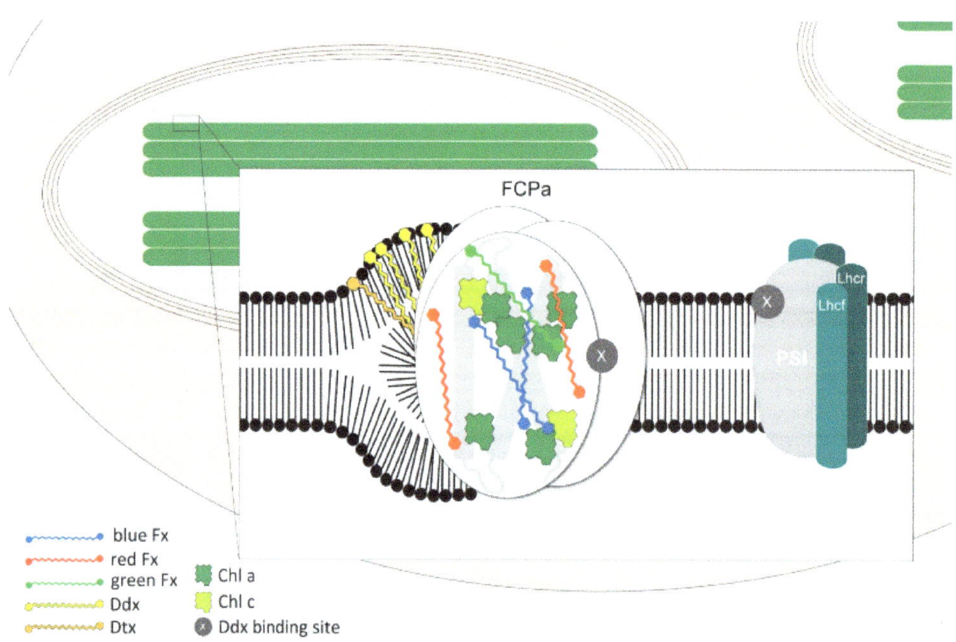

Figure 3: Simplified model of diatom thylakoid membrane showing the localization of photosynthetic pigments within FCP, PS I, and those localized within an monogalactosyldiacylglycerol (MGDG) shield surrounding the FCP. See the text for more information. Based on Gundermann and Büchel [60].

Fx is found in much higher amounts in FCPs than carotenoids are in LHCII: the molar Chl/carotenoid ratio is almost 1:1 and 14:4, respectively [52,61]. Upon binding to the protein, Fx undergoes extreme bathochromic shifts, and since it depends strongly on the polarity of the protein environment, several populations can be distinguished, *i.e.*, Fx red, Fx green, and Fx blue (Figure 3) [62,63,64]. In diatoms, the Ddx pool is heterogeneous. Recently, three different pools of diadinoxanthin cycle pigments were proposed. Two of them are bound to special antenna proteins within PS I and FCP, respectively, and since their turnover is very low, they play no direct role in the Ddx cycle [27]. The largest pool which would be localized within an MGDG shield surrounding the FCP (Figure 3) [65,66] is convertible to Dtx during the Ddx cycle. This spatial separation has been explained in terms of the probable functional heterogeneity of these pigments. The protein-bound diadinoxanthin cycle pigments would participate in the non-photochemical quenching (NPQ) mechanism, while the lipid-associated ones would essentially play an antioxidant function, scavenging $^1O_2$and peroxylipids. Interestingly, Alexandre and co-workers [51] reported that the additional Ddx molecules observed when cells are grown in high light conditions adopt a more twisted conformation than the lower levels of Ddx present when the cells are grown in low light conditions. They conclude that this pool of Ddx is more tightly bound to a protein-binding site, which must differ from the one occupied by the Ddx present in low light conditions. In the thylakoid membranes of diatoms, other xanthophylls like Vx, Ax, and Zx could also occur [24,27]. However, these carotenoids accumulate only under specific conditions, e.g., during long-term illumination with strong light. Moreover, it has been shown that Vx can be either a direct or an indirect (through the formation of Ddx) precursor of Dtx (see Section 3).

A structural homology model based on LHCII structure and spectroscopic analyses was recently published by Gundermann and Büchel [60]. It is based on the previously postulated one by Eppard and Rhiel [67] with five conserved Chl *a* binding sites (a602, a603, a610, a612, and a613; nomenclature according to Liu *et al.*[68]) and modified after Premvardhan *et al.* [62], who identified two further conserved binding sites: a614 and b609. The model accounts for some major requirements listed in [60] as follows: (i) the pigment content of FCPa and FCPb is based on 2 Chl *c* per monomer, *i.e.*, the FCPs contain 6–8 Chl *a*: 2 Chl *c*: 5–6 Fx in total; (ii) the Chl *a* molecules cannot be arranged in a way as to favor excitonic interactions; (iii) Fx does not transfer energy to Chl *c* and Fx and Chl *c* should thus be at an appropriate distance and/or orientation; (iv) Chl *c* to Chl *a* transfer is extremely fast, *i.e.*, Chl *c* has to be in close proximity to a Chl *a* molecule. For further details the reader is referred to the text [60].

PIGMENTS INVOLVED IN PHOTOSYNTHESIS

Diatoms are responsible for about 40% of marine productivity and, because the oceans cover about 70% of the Earth's surface, they make a great contribution to global productivity [69]. However, the marine environment, especially for planktonic species, is changing continuously because of water turbulence. The cells are exposed to varied light intensities as well as light spectrums, depending on the depth (Figure 4). The cell morphology and physiology are strongly connected to the types of habitats [70,71,72], which have a great importance, especially in ecological studies based on organisms in natural environments. The most common diatom species with their typical habitats and lifestyles are described in Table 2. Consequently, diatoms are well adapted to these conditions through efficient photon accumulation and CO_2 uptake and also through fast response to strong light to prevent photodamage. Moreover, according to Falkowski and Knoll [73], the ecological success of diatoms as one of the most important groups of planktonic species is related to their pigment profile, including Chl *a*, Chl *c*, β-car, Ddx, Dtx, and Fx, which enables them to harvest light more efficiently than do green [74,75] and red algae [74]. Chlorophylls play a light-harvesting role and are also known for their electron transfer function. This applies to Chl c, which serves exclusively as an antenna pigment. However, free monomeric Chls can easily transmute into an excited state, inducing free radicals. Therefore, the closely localized carotenoids Ddx and Dtx in diatoms are able to quench it efficiently [36].

Studies on acclimation to different light conditions, not only in terms of the quantity but also the quality of light, have been conducted. It was observed that during daily insolation the Chl content increased and it was accompanied by a decrease in the Fx content [78]. Both pigments are present in FCP complexes and have different absorption maxima, which results in efficient photon accumulation. Although the ratio between light harvesting and photoprotective pigments is balanced and, therefore, photosynthetic efficiency is lowered under high irradiance, because the amount of alternative electrons is kept to a minimum under fluctuating light, diatoms can convert energy into biomass efficiently [78]. Additionally, there is valuable data about the adaptation of diatom cells to their environment, achieved through experiments with selected light spectra. Blue light has a significant impact on diatoms, resulting in the production of more cells with greater photosynthetic activity in comparison to white light [79]. Moreover, planktonic diatoms which float in the water column are exposed to changes in the ratio between blue and red light. Recently, Jungandreas and co-workers [45] found that a shift from blue to red light results in a slightly diminished growth rate within 48 h, whereas a shift from red to blue light completely abolished growth within 24 h. Furthermore, metabolic reorganization based on the carbon allocation in diatom cells is the first response in acclimation to new light conditions [45]. Beyond light stress, another group of factors such as nutrient availability have a substantial impact on photosynthetic efficiency. Silicon is a special inorganic compound in diatoms, occurring abundantly in their cell wall. The energy for silicification is more linked to respiration than to photosynthesis and, therefore, it has no effect on oxygen production but leads to decreased utilization of carbohydrates during biomineralization [2]. It has been reported that silicon limitation results in increased Chl *a* content [80,81], but its addition had no significant effect on Chl accumulation [82]. Nevertheless, silicate importance for photosynthetic efficiency is similar to other nutrients [83], including iron, which plays an important role in marine primary productivity. A decreased Chl content, lowered growth rate, up-regulated expression of genes engaged in early steps of carotenoid biosynthesis, and enhanced NPQ were observed under iron deficiency [49], which altogether resulted in a down-regulation of photosynthesis, nitrate assimilation, and mitochondrial electron transport.

Table 2: Ecological specification of the most common diatom taxa, cell morphology, colony lifestyle, habitats. The presence of species in ecological region is variable and dependent on, e.g., the season, nutrient availability, salinity, and conductivity; however, the most frequent occurrence is specified. One group is benthic species including epiphytic (attached to plants), epilithic (attached to rock surfaces), epipelic (on mud), and epipsammic (on sand) species and the second is pelagic diatoms (free living in the water column) [76,77].

Species	Morphology	Colony-Forming	Lifestyle	Habitat
Actinella punctata	eunotioid	yes	benthic	acidic, humic lakes, and ponds
Actinocyclus normanii	centric	no	planktonic	coasts, brackish waters, sediment core
Amphora minutissima	asymmetrical biraphid	no	benthic	marine habitats, often epiphytic
Bacillaria paradoxa	nitzschioid	yes	benthic	marine, brackish, and freshwaters
Campylodiscus hibernicus	surirelloid	no	benthic	epipelon in fresh, brackish, marine waters
Cocconeis pediculus	monoraphid	no	benthic	planktonic, epiphytic, epilithic habitats
Cyclotella distinguenda	centric	no	planktonic	preferentially alkaline waters
Cymbella amplificata	asymmetrical biraphid	yes	benthic	oligotrophic waters
Diatoma vulgaris	araphid	yes	benthic	fresh and brackish water
Discostella stelligera	centric	no	planktonic	primarily in lakes and large rivers
Distrionella incognita	araphid	yes	benthic	alkaline lakes and streams
Epithemia turgida	epithemioid	no	benthic	epiphyte on coarse filamentous algae
Eucocconeis alpestris	monoraphid	no	benthic	the littoral zone of oligotrophic lakes
Eunotia exigua	eunotioid	N/A	benthic	moist soils, wet walls, streams, waterfalls
Fragilaria crotonensis	araphid	yes	planktonic	mesotrophic lakes, water column
Gyrosigma acuminatum	symmetrical biraphid	no	benthic	primarily an epipelic species
Navicula reinhardtii	symmetrical biraphid	no	benthic	fresh water, slightly brackish
Nitzschia regula	nitzschioid	no	benthic	cold-water, ponds, and streams
Phaeodactylum tricornutum	fusiform, triradiate, oval	no	planktonic	marine coastal waters
Pinnularia rabenhorstii	symmetrical biraphid	no	benthic	cold oligotrophic waters in the mountains
Pleurosira laevis	centric	yes	benthic	naturally saline or polluted waters
Thalassiosira weissflogii	centric	N/A	planktonic	primarily in marine waters

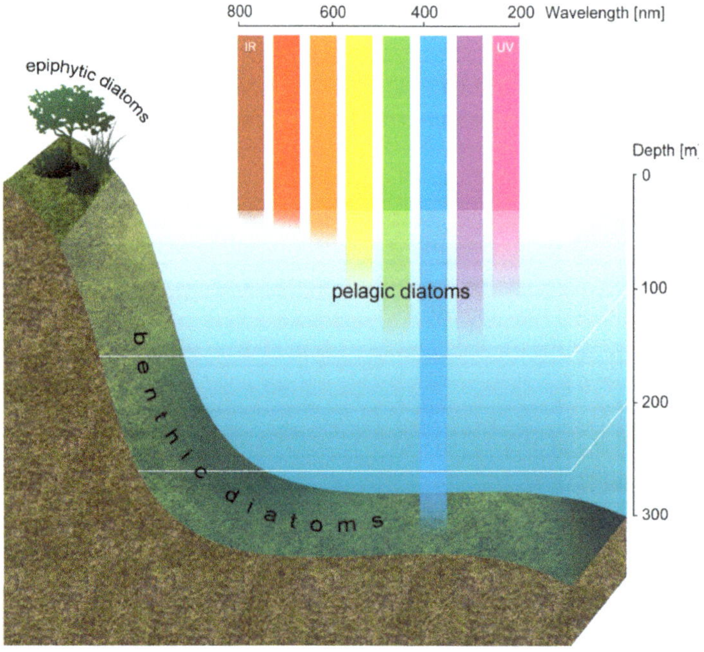

Figure 4: Spectra composition of light depending on the depth.

Although photosynthetic pigments are related to processes depending on light, dark phase reactions are equally important. Diatoms might perform both C_3 as well as C_4 photosynthesis and, indeed, both types of initial products were found in one species, and it was reported that the CO_2 concentration has no effect on the transcripts of genes engaged in this process [84]. As in other photosynthetic organisms, the enzyme responsible for carbon fixation is RubisCO, which also has the ability to fix oxygen. Additionally, diatoms' cytoplasmic membrane is highly permeable and a lot of CO_2 is lost by diffusion. However, diatoms are well adapted to low CO_2 concentrations through an efficient CO_2 concentration mechanism based on the active transport of carbon from the cytoplasm into the chloroplast [85]. As described above, despite environmental fluctuations, diatoms are well-adapted marine organisms with great importance to global productivity.

PIGMENTS INVOLVED IN PHOTOPROTEC-TION

Light conditions change significantly depending on the daily cycle, season, habitat, and environment. These conditions determine the life of photosynthetic organisms that have to react to differing situations, accumulate photons efficiently in low light, and dissipate excesses of energy in high light. Excess light might lead to photoinhibition, inactivation of photosystems, and induce the formation of reactive oxygen species (ROS), resulting in photodamage and wide-ranging changes in the cells caused by oxidative stress. To avoid this, phototrophs have developed photoprotective mechanisms such as NPQ and the xanthophyll cycle.

The NPQ relies on the dissipation of excessive excitation energy as heat and might be detected as a quenching of Chl *a* fluorescence. Various mechanisms have been postulated to describe this process; however, all of them require structural changes in photosynthetic antenna that allow a transition from light harvesting to an energy dissipation state [86]. The role of the proton gradient between thylakoid lumen and chloroplast stroma, as well as the presence of xanthophylls and the LHCII structure, have all been thoroughly investigated in higher plants (for a review see [86]), but there has also been significant progress in this field when it comes to diatoms. Studies carried out so far indicate the importance of Dtx in NPQ that is observed only in the presence of this pigment. Contrary to plants and green algae, the proton gradient across the thylakoid membrane in diatoms is not sufficient to induce NPQ [87]. Furthermore, the decrease in the proton gradient does not result in NPQ relaxation, which is correlated with Dtx epoxidation [88].

Six types of xanthophyll cycles have been described in photosynthetic organisms and two of them occur in diatoms [89]. Typical of this group is the diadinoxanthin cycle (DD cycle, see Figure 5), which exists in algae such as diatoms, phaeophytes, dinophytes, and haptophytes. The two pigments engaged in this process, Ddx and Dtx, are normally present in their cells, although their amount depends on the light intensity (seeSection 8). In high light, Ddx, which is a monoepoxide xanthophyll, is converted to epoxy-free Dtx by Ddx de-epoxidase (DDE, also called VDEs, three isoforms were even identified); in low light or dark conditions there is the reverse reaction, leading to Ddx formation which is catalyzed by Dtx epoxidase (DEP, also called ZEPs, including three isoforms). Overlong light stress leads to the formation of Zx, one of three xanthophylls involved in the violaxanthin cycle (VAZ cycle), as shown in Figure 6, which occurs mostly in higher plants, mosses, and lichens, but in several conditions also observed in diatoms [24]. It relies on a cyclic conversion of di-epoxy Vx and epoxy-free Zx via an intermediate product, mono-epoxy Ax. De-epoxidation of Vx is catalyzed by VDEs and epoxidation by ZEPs. Nevertheless, it is still not known which of the three isoforms of both enzymes identified in *P. tricornutum* are engaged in the DD or VAZ cycle.

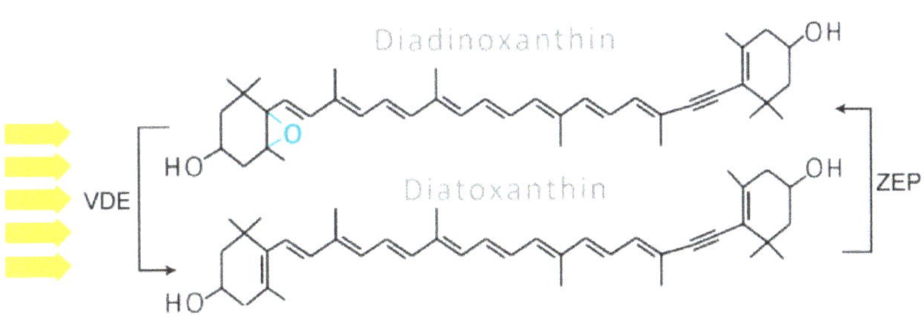

Figure 5: The diadinoxanthin cycle: in high light, diadinoxanthin with one epoxy group is converted to epoxy-free diatoxanthin by violaxanthin de-epoxidase (VDE); the reverse reaction is observed in low light and dark and is catalyzed by zeaxanthin epoxidase (ZEP).

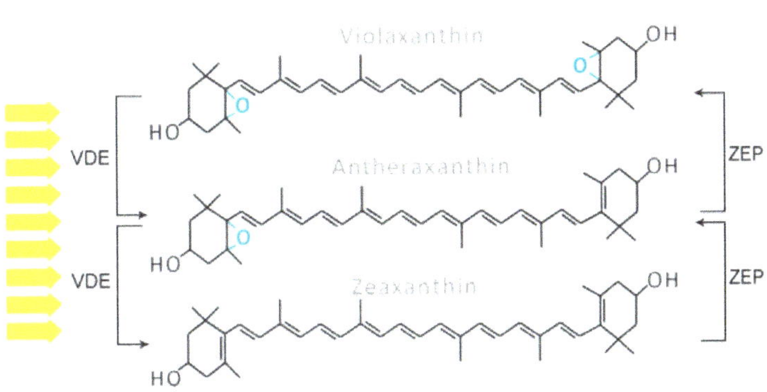

Figure 6: The violaxanthin cycle: under high light, violaxanthin (which is normally a precursor of fucoxanthin) is converted to zeaxanthin via the intermediate antheraxanthin and this reaction is catalyzed by violaxanthin de-epoxidase (VDE), whereas in low light and dark, two single

steps of oxygenation catalyzed by zeaxanthin epoxidase (ZEP) lead to violaxanthin formation.

Studies on ZEPs and VDEs in this species have shown that their transcript levels vary (the lowest for ZEP3 and VDL1) and are stimulated by white and blue light [31], but there is no data on their correlation with the type of xanthophyll cycle. Although there are similarities of these enzymes in diatoms and higher plants based on structure, substrate, and cosubstrate requirements, the properties and the mechanism of light-dependent activation/deactivation are not yet understood. Epoxidation in diatoms under low light is faster than those in higher plants and green algae [88]. Although it is known that this reaction in higher plants occurs in both light and dark conditions, in diatoms, the presence of a proton gradient between thylakoid lumen and chloroplast stroma in high light almost completely inhibits epoxidation [88,90]. In *P. tricornutum*, the rate of Ddx de-epoxidation is higher than that of Vx, whereas Dtx and Zx epoxidation rates are similar [24]. Moreover, VDEs in diatoms are active at higher pH values than those in plants and need a lower ascorbate concentration as a cosubstrate [91,92]. Although the photoprotective role of both xanthophyll cycles as well as the ability of Dtx and Zx to quench excited Chl and free radicals are known, the reason why they exist together in diatom cells and the activation mechanism of each remain unclear. Indeed, Zx might be formed from both Vx and β-car and de-epoxidation, and *de novo* carotenoid synthesis could play a substantial role in its accumulation under high light [24]. It is interesting that in *Chlamydomonas reinhardtii*with no functional xanthophyll cycle, the accumulation of Zx was observed, which indicates the crucial role of β-car as a substrate [93] and diminishes the importance of the VAZ cycle. Moreover, the possibility that Zx may not be an obligate precursor of Vx was put forward when Cx and CxE were detected in *P. tricornutum*, although conversions between β-car, Cx, CxE, Vx, Ax, and Zx require detailed studies [27]. A comparative analysis of the photoprotective properties of Zx and Dtx is needed. The quenching efficiency of Dtx has been tested and it was shown that this pigment is a better quencher in low light because additional molecules of Dtx and Ddx bind to FCP, and while not participating in NPQ, they are precursors of Fx [94]. This may suggest that Zx is necessary to avoid photodamage in high light and could explain its accumulation.

METHODS OF PIGMENT ANALYSIS AND IDENTIFICATION

Photosynthetic pigments are used as indicators of the relative abundance of major microalgal taxonomic groups and offer a means for assessing changes or differences in the relative abundance of phototropic functional groups in mixed assemblages [95]. This approach provides measurements of the quantitative changes in the relative abundance of major microalgal groups under similar environmental conditions and habitats. As reported by Schagerl and Kunzl [96], within the year 2005, in nine international journals, 15 different concentrations and six solvents were used, whereas pigment quantification was done by measuring pigment extracts by means of fluorometry (35%) and spectrophotometry (35%), followed by HPLC (25%). Given the fact that Chl *a* is extensively used as an indirect measure of overall algal biomass, and other pigments like Chl *c* or carotenoids have been successfully applied as markers for specific taxonomic groups within algal communities [96] and the references therein, the varieties of methods make comparisons very difficult. Although the determination of chlorophylls is routinely conducted by spectrophotometric [97] or spectrofluorometric [98] methods, the results may be erroneous due to the fact that the absorption and emission bands of Chl *a* overlap with those of other Chls and the degradation products of chlorophylls are neither detected nor determined along with their parent chlorophylls. Moreover, the determination of individual carotenoids is difficult to achieve by these methods. Thus, pigment analysis by HPLC has become the favorite tool for marine researchers.

Extraction of Pigments

The best extraction technique is still a matter for debate. The efficiency of pigment extraction may vary depending on the properties of the solvent, the duration of the extraction, cell concentration, species of algae, and whether mechanical disruption is used. Physical methods of sample disruption such as grinding, bath sonication, high power sonication, or soaking are in use and the most suitable solvents currently applied include acetone, methanol, and non-volatile *N,N*-dimethylformamide (DMF) (see [96] and the references therein). The need for the complete extraction of all pigments present in the sample, compatibility with chromatographic technique, and stability of the pigments are to be considered before the choice is made.

The application of various extraction solvents has been thoroughly discussed by Wright and co-workers [99], of which both acetone and methanol are widely applied in the extraction of algal pigments. Although the highest extraction efficiency is generally achieved by using methanol as an extraction solvent in combination with mechanical disruption, the stability of pigments in methanol is low. It has often been claimed to promote allomerization of Chl *a* [100,101] and, accordingly, Mantoura and Llewellyn [102] found that methanol led to the formation of Chl *a*-derivative products. On the other hand, there is evidence that certain chlorophylls and carotenoids are more thoroughly extracted with methanol [103] or dimethyl sulfoxide [104]. Altogether, acetone seems to be the best choice when there is no information on the specific composition of the sample. While acetone and methanol have the same polarity index, acetone has greater eluotrophic strength than methanol for carbon-rich substrates [105]. Moreover, it has been proved to cause fewer artifacts than methanol and DMF as the derivatives of Chl *a* (mainly Chlide *a*, Chl *a* allomer and epimer) were found at only up to 5% of measured Chl *a* [96]. The extraction efficiency of acetone increases considerably when, prior to extraction, freeze-drying is applied, which breaks up the protein matrix of membranes, creating accessibility for the extraction solvent even with more recalcitrant algae [100,106]. Additionally, freeze-drying inhibits the activity of chlorophyllase, the enzyme found in many marine taxa [101,107], which catalyzes the hydrolysis of the bond linking the phytol chain to the propionic acid residue of Chls *a* and *b*, thereby converting Chl to its corresponding chlorophyllide [108,109]. Since chlorophyllase is mainly active in an aqueous environment [108], it appears that dehydration by freeze-drying creates conditions that inhibit enzyme activity. To conclude, it is worth remembering that algal pigments are extracted differentially by various solvents, and there is no single combination of solvent and extraction time that is best for all species [103].

HPLC Analysis of Pigments

HPLC was applied to phytoplankton studies in the early 1980s [110]. All current methods use reverse phase separation where compounds are resolved primarily on the basis of their polarity. Up until now, different columns have been tested, including octadecylsilica (ODS) C18 and octylsilica (OS) C8, as well as C30 stationary phase, with different diameters, lengths, and particle sizes, which provide different degrees of effectiveness in the resolution of key pigments. Pigment separation and resolution can be also tuned by the column temperature, mobile phase composition, and the gradient program. For the resolution of acidic chlorophylls, buffering, ion-pairing, or an ion suppression reagent is required, with ammonium acetate as the one of choices.

A comprehensive monograph where both modern analytical methods and their applications to biological oceanography were reviewed is that of Jeffery and co-workers from 1997 [111] and 1999 [112]. Nineteen HPLC methods published between 1983 and 1998 have been reported. In short, the work has been started by Mantoura and Llewellyn [102], who developed a reverse-phase HPLC technique for the rapid separation and quantification of 17 carotenoids and 14 chlorophylls together with their breakdown products from acetone extracts of algal culture and natural waters. They also suggested that the presence of an ion-pairing reagent during the chromatographic separation was essential to obtain good resolution for chloropigments (chlorophylls, chlorophyllides, and phaeophorbides). A modified technique was described by Wright and Shearer [113] in which photosynthetic pigments from phytoplankton were separated by HPLC using a lineal elution gradient from 90% acetonitrile to ethyl acetate. Although a better resolution was obtained by the latter method for carotenoid pigments, it does not resolve the polar chlorophylls c_1, c_2, c_3, and Mg-2,4 divinyl phaeoporphyrin a_5 monomethyl ester (Mg2,4D), which coelute as a single peak, or adequately resolve the chlorophyllides *a* and *b*. A more comprehensive method was published in 1991 [114], and has been used thereafter for routine work. Further improvement has been achieved by replacing ammonium acetate with either a pyridine or tetrabutylammonium (TBAA) modifier. In particular, Zapata and co-workers [115], using a gradient from aqueous methanol/acetonitrile to methanol/acetonitrile/acetone with a pyridine modifier (as the acetate, pH 5.0), succeeded in resolving, on a C8 column, seven polar Chl *c* derivatives, the Chl *a*/DV Chl *a* pair, and, partially, Chl *b*/DV Chl *b*. The method of Van Heukelem and Thomas [116] is based on an aqueous methanol to methanol gradient at 60 °C with a TBAA modifier (pH 6.5). The effectiveness of two HPLC methods (a monomeric C18 column with a high ion strength mobile phase composed of methanol, acetonitryl, and ethyl acetate by Kraay and co-workers [117] (Figure 7 and Figure 8) and a monomeric C8 column with a pyridine-containing mobile phase [116]) in the separation, identification, and quantification of pigments were compared by Mendes and co-workers [118], using phytoplankton, microphytobenthos, and algal cultures. Although the C8 method allows the separation of chlorophylls c_1, c_2, and Mg-2, 4-divinyl phaeoporphyrin a_5 monomethyl ester, as well as the pair Chl *a*/DV Chl *a*, the C18 method had a shorter elution program and a lower solvent flow rate, making it cheaper and faster than the C8 method. The excessive separations feasible with C30 columns have proven useful for unique applications such as pigment isolation [116], although the long analysis times are a deterrent if routine separations of complex mixtures are required.

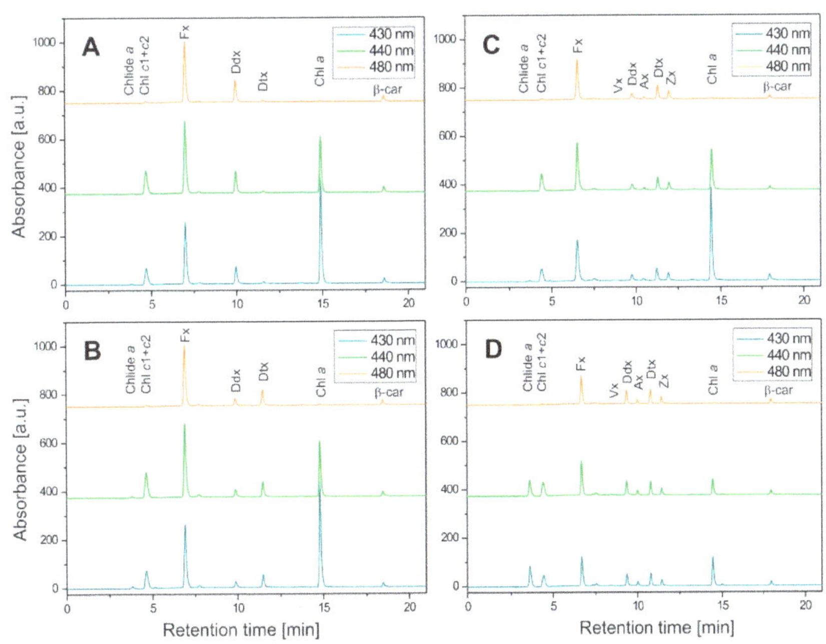

Figure 7: Examples of HPLC chromatograms of absorbance recorded at 430, 440, and 480 nm obtained with the method by Kraay and co-workers [117] for pigment extracts from the diatom *P. tricornutum*: (**A**) 1 h dark incubated cells growing in LL; (**B**) 1 h HL illuminated cells growing in LL; (C) 1 h HL illuminated cells growing in ML; (D) 48 h HL illuminated cells growing in LL. LL: white light with the intensity of 100 µmol $m^{-2} \cdot s^{-1}$ in a 16 h light/8 h dark photoperiod; ML: white light with the intensity of 700 µmol $m^{-2} \cdot s^{-1}$ in a 6 h light/18 h dark photoperiod; HL: white light with the intensity of 1250 µmol $m^{-2} \cdot s^{-1}$.

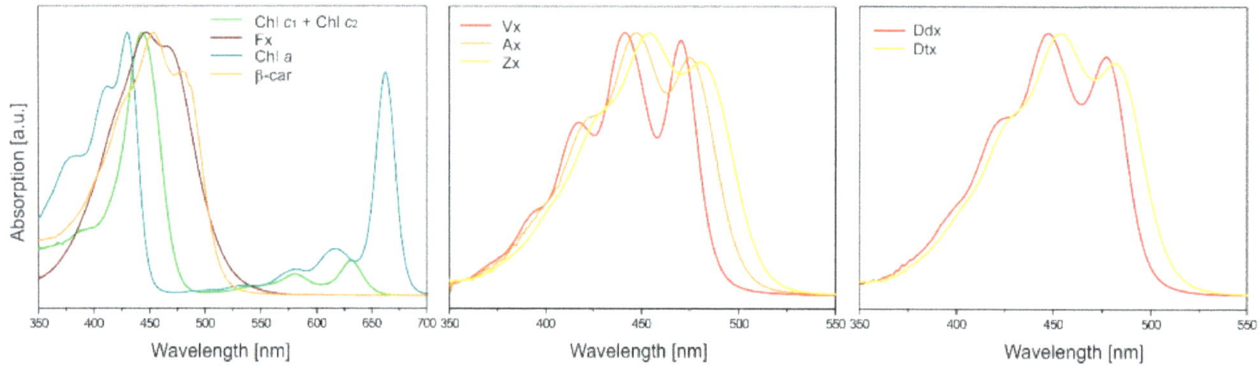

Figure 8: Absorption spectra of photosynthetic pigments recorded during HPLC-DAD analysis performed on extracts from the diatom *P. tricornutum*with the method by Kraay and co-workers [118]. The spectra were normalized at λ_{max}.

For the detection and quantitative analysis of pheophytin *a*, pheophorbide *a*, and their derivatives, the wavelengths of 420–430 nm are useful, while the region of 450–480 nm detects all carotenoids and Chl *c*without interference from Chl *a* derivatives. Peaks can be quantified either using internal or external standards. The first approach ensures higher accuracy as it accounts for any changes resulting from evaporation or dilution as well as providing a check on the injection status. Commercially available internal standards in common use are ethyl 8′-apo-β-carotenoate, 8′-apo-β-carotenal, vitamin E, and canthaxanthin, the first of which attracts special interest due to its stability and non-occurrence in natural systems. Table 3shows the elution order of photosynthetic pigments from the diatom *P. tricornutum*, analyzed with the method by Kraay and co-workers [117], and their visible absorption characteristics. The absorbance spectra of more algal pigments and their elution order are available in the literature [53,102,114,118,119].

Table 3: Elution order of photosynthetic pigments from the diatom *P. tricornutum* and their visible absorption characteristics.

Pigment	Literature Data				HPLC Data *	
	Solvent	λ_{max} (nm)	E (L $g^{-1}\cdot cm^{-1}$)	Reference	λ_{max} (nm)	Retention Time (min)
Chlorophyll c_1	acetone (90%)	443	318	[120]		
Chlorophyll c_2	acetone (90%)	444	374	[120]		
Chlorophyll $c_1 + c_2$	N/A	N/A	N/A	N/A	443	4.7
Fucoxanthin	petrol ether	449	165	[121]	447	7.0
	ethanol	450	114	[122]		
	acetone	443	165	[123]		
Violaxanthin	N/A	443	N/A	[114]	441	8.4
Diadinoxanthin	acetone	448	224	[124]	447	9.9
	methanol	445	225	[124]		
	hexane	446	211	[124]		
Anteraxanthin	ethanol	446	235	[125]	447	10.8
Diatoxanthin	acetone	454	N/A	[114]	454	11.3
Zeaxanthin	ethanol	452	254	[126]	453	11.9
Chlorophyll *a*	acetone	662	88.15	[127]	429	14.9
β-carotene	etanol	453	262	[128]	454	18.7
	hexane	453	259.2	[128]		
	acetone	454	250	[129]		

*: data obtained with the method by Kraay andco-workers [117] for pigment extracts from the diatom *P. tricornutum*.

Pigment identification can be confirmed by performing LC/MS and comparing the resulting mass spectra with published fragment ion abundance and/or mass spectral libraries of pigment standards.

SIGNIFICANCE OF DIATOM PIGMENTS

Marine algae are rich sources of structurally diverse bioactive compounds with various biological activities. Among them, natural pigments have received particular attention as they exhibit beneficial activities such as antioxidant, anticancer, anti-inflammatory, anti-obesity, anti-angiogenic, and neuroprotective effects important in the fields of food, cosmetics, and pharmacology [130]. Although chemically synthesized pigments are widely available, natural products are more attractive to consumers as they exhibit higher bioavailability and bioactivity. Since bioreactors allow successful cultivation of different species of marine microorganisms, including diatoms, the latter are considered a good source of natural pigments. A relatively fast growth rate along with with low costs is by far the best advantage; however, it has to be taken into account that pigments for food or pharmaceutical applications need to be of a consistently high quality that is safe, traceable, measurable, and readily extractable, which makes the technology more demanding. Nevertheless, the great health benefits and nutraceutical properties represented by diatom pigments make them interesting for pharmaceutical companies.

Carotenoids, especially β-car as the most efficient natural quencher, are involved in the scavenging of ROS, singlet molecular oxygen, and peroxyl radicals, so they may protect against lipid peroxidation [131]. β-car is also an effective stimulator of gap junctional communication, which is associated with cell growth, differentiation, and apoptosis and, together with α-tocopherol, inhibits the lipid peroxidation better than on its own [132]. The anticancer activity of β-car was confirmed, e.g., for lung cancer, although the risk of cancer increased when high doses of a dietary supplement rich in this carotenoid were consumed by smokers [133]. Lutein and Zx are responsible for the coloration of the macula lutea, which is a part of the retina, the area of maximal visual acuity; thus, both pigments are extremely important in ophthalmology and are also used in protection against age-related macular degeneration [134]. This photoprotection arises from their function as filters for damaging blue light and as quenchers of excited oxygen. Strong antiproliferative activity of Vx on breast cancer cells was demonstrated, which makes this pigment a potent compound of drugs for the induction of apoptosis [135]. Although diatoms contain a few photosynthetic pigments including chlorophylls and carotenoids, Fx has the greatest significance in these organisms. This xanthophyll is present in large amounts in diatoms and brown algae, whereas Zx, Ax, Vx, and β-car are sourced mainly from plants where they are present abundantly. Nevertheless, two other xanthophylls, Dtx and Ddx, are also specific to diatoms and might be obtained from them. Unfortunately, the biological activity of their extracts, health benefits, and importance in the diet are still unknown and should be a major motivation for future research.

Fx has great commercial importance because of its varied and numerous bioactivities, such as antioxidant, anticancer, anti-inflammatory, anti-obesity, anti-angiogenic, neuroprotective, photoprotective, and prevention of osteoporosis [130]. The extracts of a few seaweed species such as *Hyaleucerea fusiformis*, *Cladosiphon okamuranus*, *Undaria pinnatifida*, and *Sargassum fulvellum* showed strong DPPH radical scavenging activity [136]. Fx may inhibit intracellular ROS formation, DNA damage, and apoptosis induced by $H_2O_2$and its antioxidant activity is comparable to that of α-tocopherol [136]. The anti-inflammatory activity of Fx is associated with its ability to reduce the nitric oxide level, prostaglandin E2, tumor necrosis factor-α, interleukin-1β, interleukin-6, and histamine [137,138]. Strong antiproliferative activity is exhibited by Fx which may induce the apoptosis of human leukemia cell line HL-60, and reduce the viability of PC-3 human prostate cancer cells and human colon cancer cell lines Caco-2, HT-29, and DLD-1 [139,140] (for a review see [136]). Moreover, Fx is a promising compound that might be used in obesity prevention through the upregulation of UCP1 which promotes energy expenditure by thermogenesis [141]. It was reported that Fx mixed with pomegranate seed oil reduced body weight, body fat, and liver fat content [142].

METHODS OF PURIFICATION OF DIATOM PIGMENTS

From among photosynthetic pigments occurring in diatoms, neither chlorophylls nor β-car are of interest to be extracted and purified on a preparative scale since they are more abundant in plants. On the contrary, the broad application potential of Fx has prompted researchers to explore the commercial production of this photosynthetic pigment from algae. To find an efficient source of Fx, brown macroalgae, such as *Laminaria japonica*, *Eisenia bicyclis*, and *Undaria pinnatifida* [143,144], were studied; they are traditional foods in Southeast Asia and some European countries. However, the Fx concentration from these macroalgae ranged from 0.02 to 0.58 $mg \cdot g^{-1}$ in fresh samples and 0.01 to 1.01 $mg \cdot g^{-1}$ in dried samples [26], which makes the production of Fx from brown macroalgae not commercially viable. Since the reported Fx concentration in microalgae is one to three orders of magnitude greater than that found in macroalgae, ranging from 2.24 to 18.23 $mg \cdot g^{-1}$ [26], diatoms are considered a promising source of Fx for various commercial applications. In particular, *P. tricornutum* was found to contain Fx as a main carotenoid. Studies by the Rebolloso-Fuentes research group [145] showed that 60% of the total carotenoids in the acetone extract was Fx, which gives Fx a concentration of 1.81 mg/g dw. Xia and co-workers [8] investigated the marine diatom *Odontella aurita*and determined that it can accumulate high concentrations of Fx (>20 $mg \cdot g^{-1}$ of dry weight). The Fx concentration of 3.33–21.67 $mg \cdot g^{-1}$ and the volumetric concentration of 18.15–79.56 $mg \cdot L^{-1}$ obtained in this study have set the records for algae-based Fx production. A dataset containing the Fx concentrations in samples of different macroalgae and microalgae is available in [8,26].

There are numerous methods used to recover and purify Fx from brown algae, including: (i) centrifugal partition chromatography; (ii) pressurized liquid extraction; (iii) enzymatic pre-treatment followed by co-solvent extraction; (iv) supercritical carbon dioxide with ethanol as co-solvent; and (v) traditional solvent extraction followed by chromatographic methods, as reviewed by Gomez-Loredo in [146]. However, the costs associated with new technologies and equipment are major drawbacks to their implementation. Thus, there is a need to develop low cost, simple, and scalable strategies for the recovery of Fx as one of the high value compounds occurring naturally in abundance in diatoms. The results show that Fx extraction efficiency, as with other phytoplankton pigments, is highly dependent on the extraction conditions—the solvent type, in particular (see Section 8). Kim and co-workers [26] have assessed a number of extraction procedures to investigate the effect of solvent type, extraction time, temperature, and extraction method (maceration, ultrasound-assisted extraction, Soxhlet extraction, and pressurized liquid extraction) on Fx extraction efficiency. Among the investigated solvents, ethanol provided the best Fx extraction yield (15.71 mg g^{-1}freeze-dried sample weight), whereas its content in the extracts produced by the different methods was quite constant (15.42–16.51 $mg \cdot g^{-1}$ freeze-dried sample weight). The pigment extracts obtained from freeze-dried powder of *P. tricornutum* extracted with 100% ethanol at room temperature for 3 h were then subjected to silica-gel adsorption chromatography with an eluent of n-hexane:acetone (7:3, v/v). These results are in accordance with findings reported by Xia and co-workers [8], who additionally pointed to the fact that selecting a proper ratio of solvent to dry algal biomass (v/w) is important, as it may affect the quantity and quality of Fx. In particular, they showed that an ethanol/dry biomass ratio of 20:1 is sufficient for the effective extraction of Fx from this microalga. In this work, the crude pigment extracts were subjected to open silica gel column chromatography with n-hexane:acetone (6:4, v/v) as the eluting solvent system. An orange-red colored Fx-rich fraction containing Fx, *cis*-Fx, Ddx, and Dtx was separated by the column. Further purification of Fx, whose purity was estimated as 86.7% in the mixture, was carried out by a prep-HPLC yielding a final Fx purity >97%.

Alternatively, the two-phase solvent system of n-hexane:ethanol:water with a volume ratio of 10:9:1 was determined to be the best system for the separation of Fx and lipids from extracts of *Isochrysis galbana*. Under these conditions, Fx was fractionated in the hydroalcohol phase apart from the hexane phase containing lipids [147]. Recently, Gómez-Loredo and co-workers [146] proposed a different potential strategy for the recovery and partial purification of Fx. In their study, the partition behavior of Fx in aqueous two-phase systems (ATPS) composed of ethanol and potassium phosphate was evaluated. The optimal extraction conditions were found to be un-acidified methanol (2.5% w/v for *P. tricornutum*) or ethanol (5% w/v for *P. tricornutum*) as a solvent with orbital agitation at 250 rpm for 1 h at 20 °C. The best results were observed in systems with a tie-line length, TLL 50% w/w, and a volume ratio of 1 for *P. tricornutum*methanolic extracts (95.36% recovery and 66.01% purity).

To the best of our knowledge, there is no method published so far to obtain Ddx and Dtx on a preparative scale, except for a semi-preparative HPLC approach. Thus, the method based on column chromatography recently elaborated in our laboratory (see Acknowledgments) seems to be promising for the potential application of these xanthophylls naturally found in diatoms.

CONCLUSIONS

Apart from chlorophylls *a* and *c*, and fucoxanthin, which play a light-harvesting function, a group of photosynthetic pigments in diatoms comprises those engaged in photoprotection, including β-carotene, diatoxanthin, diadinoxanthin, violaxanthin, antheraxanthin and zeaxanthin. Physical and chemical properties of diatom pigments have been extensively studied, which contributed to a more complete understanding of the function they play in cells as well as enabled explanation of their health promoting activity and beneficial effects. There are, however, still further studies required to gain the knowledge on Chls *c* to a level comparable with the well-studied Chls *a* and *b*. Moreover, it is of great importance to explore biosynthetic pathways of Chls *c* and xanthophylls by identifying the enzymes involved and the specificity of their reactions. There is a growing body of studies on diatom pigments, which refers to their photosynthetic and photoprotective functions and explains the mechanisms of these processes. Nevertheless, there are still many questions to be answered in the field of xanthophyll cycle and non-photochemical quenching research. Photoacclimation and pigment level variation are also commonly studied, which is related to the fact that marine environment is highly changeable. Furthermore, analysis of diatom pigments is widely used in global oceanography to assess phytoplankton biomass, productivity, community structure and ecological processes. Natural pigments, including those derived from algae have received particular attention as they exhibit beneficial activities important in terms of their commercial and industrial applications. In this respects, diatoms seem to be a promising source of unique bioactive compounds, with fucoxanthin, diadinoxanthin and diatoxanthis as representatives of photosynthetic pigments.

Acknowledgments

This work was supported by project No. 2011/01/M/NZ1/01170. Faculty of Biochemistry, Biophysics and Biotechnology of Jagiellonian University is a partner of the Leading National Research Center (KNOW) supported by the Ministry of Science and Higher Education.

A method of diadinoxanthin and diatoxanthin purification has been reported as

(1) Patent application No. P. 412178, priority date 29.04.2015, authors: Paulina Kuczynska and Malgorzata Jemiola-Rzeminska;

(2) Patent application No. P. 412177, priority date 29.04.2015, authors: Paulina Kuczynska and Malgorzata Jemiola-Rzeminska.

Malgorzata Jemiola-Rzeminska would like to express her sincere gratitude to Susann Schaller-Laudel and Reimundt Goss for the cooperation and all the scientific support she has received from the Institute of Biology I, Plant Physiology, University of Leipzig, Germany.

Abbreviations

Ax: antheraxanthin; ATPS: aqueous two-phase systems; BCH: β-carotene hydroxylase; BHT: butylated hydroxytoluene; Chl *a*: chlorophyll a; Chl *b*: chlorophyll b; Chl *c*: chlorophyll c; Cx: β-cryptoxanthin; CxE: β-cryptoxanthin epoxide; DD cycle: diadinoxanthin cycle; Ddx: diadinoxanthin; DMAPP: dimethylallyl diphosphate; DMF: *N,N*-dimethylformamide; DPOR: light-independent PChlide oxidoreductase; Dtx: diatoxanthin; FCP: fucoxanthin-chlorophyll protein; Fx: fucoxanthin; GGPP: geranylgeranyl pyrophosphate; HPLC: high performance liquid chromatography; IPP: isopentenyl pyrophosphate; ICT: intramolecular charge transfer; LC/MS: liquid chromatography/mass spectrometry; LCYB: lycopene β-cyclase; LHC: light harvesting complex; LPOR: light-dependent PChlide oxidoreductase; MEP: methylerythritol phosphate; MEV: mevalonate; MGDG: monogalactosyldiacylglycerol; NPQ: non-photochemical quenching; Nx: neoxanthin; NXS: neoxanthin synthase; ODS: octadecylsilica; OS: octylsilica; PChlide: protochlorophillide; PDS: phytoene desaturase; PS I: photosystem I; PSY: phytoene synthase; ROS: reactive oxygen species; RubisCO: ribulose bisphosphate carboxylase-oxygenase; TBAA: tetrabutylammonium; TLL: tie-line length; VAZ cycle: violaxanthin cycle; VDEs (VDE, VDL1, VDL2): violaxanthin de-epoxidases; Vx: violaxanthin; ZDS: ζ-carotene desaturase; ZEPs (ZEP1, ZEP2, ZEP3): zeaxanthin epoxidases; Zx: zeaxanthin; β-car: β-carotene.

AUTHOR CONTRIBUTIONS

Paulina Kuczynska made a substantial contribution to conception and design of paper; analyzed data and wrote parts relating to biological aspects of this review including physiology, ecology, biosynthesis and genetics; prepared Figure 1, Figure 2, Figure 3, Figure 4, Figure 5 and Figure 6, Table 1 and Table 2; participated in revising the manuscript in response to the reviewers' suggestions. Malgorzata Jemiola-Rzeminska analyzed information and wrote parts concerning physical, chemical, structural and methodical aspects of this review; prepared Figure 7 and Figure 8 and Table 3; contributed to revising critically for important intellectual content and made necessary modifications in response to the reviewers' comments. Kazimierz Strzalka contributed to proofreading and final approval of the version to be published.

CONFLICTS OF INTEREST

The authors declare no conflict of interest.

REFERENCES

1. Bozarth A., Maier U.G., Zauner S. Diatoms in biotechnology: Modern tools and applications. Appl. Microbiol. Biotechnol. 2009;82:195–201. doi: 10.1007/s00253-008-1804-8.
2. Martin-Jézéquel V., Hildebrand M., Brzezinski M.A. Silicon metabolisim in diatoms: Implications for growth. J. Phycol. 2000;36:821–840. doi: 10.1046/j.1529-8817.2000.00019.x.
3. Allen A.E., Dupont C.L., Oborník M., Horák A., Nunes-Nesi A., McCrow J.P., Zheng H., Johnson D.A., Hu H., Fernie A.R., et al. Evolution and metabolic significance of the urea cycle in photosynthetic diatoms. Nature. 2011;473:203–207. doi: 10.1038/nature10074.
4. Lee J.-B., Hayashi K., Hirata M., Kuroda E., Suzuki E., Kubo Y., Hayashi T. Antiviral sulfated polysaccharide from Navicula directa, a diatom collected from deep-sea water in Toyama Bay. Biol. Pharm. Bull. 2006;29:2135–2139. doi: 10.1248/bpb.29.2135.
5. Perl T.M., Bédard L., Kosatsky T., Hockin J.C., Todd E.C., Remis R.S. An outbreak of toxic encephalopathy caused by eating mussels contaminated with domoic acid. N. Engl. J. Med. 1990;322:1775–1780. doi: 10.1056/NEJM199006213222504.
6. Prestegard S.K., Oftedal L., Nygaard G., Skjaerven K.H., Knutsen G., Døskeland S.O., Coyne R.T., Herfindal L. Marine benthic diatoms contain compounds able to induce leukemia cell death and modulate blood platelet activity. Mar. Drugs. 2009;7:605–623. doi: 10.3390/md7040605.
7. Peng K.-T. Delta 5 Fatty Acid Desaturase Upregulates the Synthesis of Polyunsaturated Fatty Acids in the Marine Diatom *Phaeodactylum tricornutum*. J. Agric. Food Chem. 2014;62:8773–8776. doi: 10.1021/jf5031086.
8. Xia S., Wang K., Wan L., Li A., Hu Q., Zhang C. Production, characterization, and antioxidant activity of fucoxanthin from the marine diatom odontella aurita. Mar. Drugs. 2013;11:2667–2681. doi: 10.3390/md11072667.
9. Gastineau R., Turcotte F., Pouvreau J.B., Morançais M., Fleurence J., Windarto E., Arsad S., Prasetiya F.S., Jaouen P., Babin M., et al. Marennine, promising blue pigments from a widespread Haslea diatom species complex. Mar. Drugs. 2014;12:3161–3189. doi: 10.3390/md12063161.
10. Gastineau R., Davidovich N.A., Bardeau J.F., Caruso A., Leignel V., Hardivillier Y., Rince Y., Jacquette B., Davidovich O.I., Gaudin P., et al. *Haslea karadagensis* (Bacillariophyta): A second blue diatom, recorded from the Black Sea and producing a novel blue pigment. Eur. J. Phycol. 2012;47:469–479. doi: 10.1080/09670262.2012.741713.
11. Gelzinis A., Butkus V., Songaila E., Augulis R., Gall A., Büchel C., Robert B., Abramavicius D., Zigmantas D., Valkunas L. Mapping energy transfer channels in fucoxanthin-chlorophyll protein complex.Biochim. Biophys. Acta Bioenerg. 2015;1847:241–247. doi: 10.1016/j.bbabio.2014.11.004.
12. Jeffrey S., Vesk M. Introduction to marine phytoplankton and their pigment signatures. In: Jeffrey S., Mantoura R., Wright S., editors. Phytoplankton Pigments in Oceanography: Guidelines to Modern Methods. UNESCO Publishing; Paris, France: 1997. pp. 37–84.
13. Zapata M., Garrido J., Jeffrey S. Chlorophyll c Pigments: Current Status. In: Grimm B., Porra R., Rüdiger W., Scheer H., editors. Chlorophylls and Bacteriochlorophylls. Springer Netherlands; Dordrecht, The Netherlands: 2006. pp. 39–53.
14. Stauber J.L., Jeffrey S.W. Photosynthetic pigments in fifty-one species of marine diatoms. J. Phycol.1988;24:158–172. doi: 10.1111/j.1529-8817.1988.tb04230.x.
15. Fookes C., Jeffrey S. The structure of chlorophyll c3, a novel marine photosynthetic pigment. J. Chem. Soc. Chem. Commun. 1989;23:1827–1828. doi: 10.1039/c39890001827.
16. Britton G. UV/visible spectroscopy. In: Liaaen-Jensen S., Britton G., Pfander H., editors. Carotenoids: Spectroscopy, Vol 1B. Birkhäuser Verlag; Basel, Switzerland: 1995. pp. 13–63.
17. Zigmantas D., Hiller R.G., Sharples F.P., Frank H.A., Sundstrom V., Polivka T. Effect of a conjugated carbonyl group on the photophysical properties of carotenoids. Phys. Chem. Chem. Phys. 2004;6:3009–3016. doi: 10.1039/b315786e.
18. Rontani J.-F., Beker B., Raphel D., Baillet G. Photodegradation of chlorophyll phytyl chain in dead phytoplanktonic cells. J. Photochem. Photobiol. A Chem. 1995;85:137–142. doi: 10.1016/1010-6030(94)03891-W.
19. Britton G., Liaaen-Jensen S., Pfander H. In: Carotenoids. 1st ed. Britton G., Liaaen-Jensen S., Pfander H., editors. Birkhäuser Verlag; Basel, Switzerland: 2004.
20. Subczynski W.K., Markowska E., Sielewiesiuk

J. Effect of polar carotenoids on the oxygen diffusion-concentration product in lipid bilayers. An EPR spin label study. Biochim. Biophys. Acta Biomembr.1991;1068:68–72. doi: 10.1016/0005-2736(91)90061-C.

21. Berglund A.H., Nilsson R., Liljenberg C. Permeability of large unilamellar digalactosyldiacylglycerol vesicles for protons and glucose—Influence of α-tocopherol, β-carotene, zeaxanthin and cholesterol. Plant Physiol. Biochem. 1999;37:179–186. doi: 10.1016/S0981-9428(99)80032-1.

22. Refvem T., Strand A., Kjeldstad B., Haugan J.A., Liaaen-Jensen S. Stereoisomerization of Allenic Carotenoids—Kinetic, Thermodynamic and Mechanistic Aspects. Acta Chem. Scand. 1999;53:114–123. doi: 10.3891/acta.chem.scand.53-0114.

23. Haugan J.A., Englert G., Liaaen-Jensen S., Frimpong-Manso S., Springborg J., Wang D.-N., Christensen S.B. Algal Carotenoids. 50. Alkali Lability of Fucoxanthin—Reactions and Products. Acta Chem. Scand.1992;46:614–624. doi: 10.3891/acta.chem.scand.46-0614.

24. Lohr M., Wilhelm C. Algae displaying the diadinoxanthin cycle also possess the violaxanthin cycle.Proc. Natl. Acad. Sci. USA. 1999;96:8784–8789. doi: 10.1073/pnas.96.15.8784.

25. Frank H.A., Cogdell R.J. Carotenoids in photosynthesis. Photochem. Photobiol. 1996;63:257–264. doi: 10.1111/j.1751-1097.1996.tb03022.x.

26. Kim S.M., Jung Y.J., Kwon O.N., Cha K.H., Um B.H., Chung D., Pan C.H. A potential commercial source of fucoxanthin extracted from the microalga *Phaeodactylum tricornutum*. Appl. Biochem. Biotechnol. 2012;166:1843–1855. doi: 10.1007/s12010-012-9602-2.

27. Lohr M., Wilhelm C. Xanthophyll synthesis in diatoms: Quantification of putative intermediates and comparison of pigment conversion kinetics with rate constants derived from a model. Planta. 2001;212:382–391. doi: 10.1007/s004250000403.

28. Bowler C., Allen A.E., Badger J.H., Grimwood J., Jabbari K., Kuo A., Maheswari U., Martens C., Maumus F., Otillar R.P., et al. The *Phaeodactylum* genome reveals the evolutionary history of diatom genomes. Nature. 2008;456:239–244. doi: 10.1038/nature07410.

29. Armbrust E.V., Berges J.A., Bowler C., Green B.R., Martinez D., Putnam N.H., Zhou S., Allen A.E., Apt K.E., Bechner M., et al. The genome of the diatom *Thalassiosira pseudonana*: Ecology, evolution, and metabolism. Science. 2004;306:79–86. doi: 10.1126/science.1101156.

30. Bertrand M. Carotenoid biosynthesis in diatoms. Photosynth. Res. 2010;106:89–102. doi: 10.1007/s11120-010-9589-x.

31. Coesel S., Oborník M., Varela J., Falciatore A., Bowler C. Evolutionary origins and functions of the carotenoid biosynthetic pathway in marine diatoms. PLoS ONE. 2008;3:1–16. doi: 10.1371/journal.pone.0002896.

32. Stransky H., Hager A. Das Carotinoidmuster und die Verbreitung des lichtinduzierten Xanthophyllcyclus in verschiedenen Algenklassen. Arch. Mikrobiol. 1970;73:315–323. doi: 10.1007/BF00412298.

33. Montsant A., Allen A.E., Coesel S., Mangogna M., Siaut M., Heijde M., Jabbari K., Maheswari U., Rayko E., Vardi A., et al. Identification and comparative genomic analysis of signaling and regulatory components in the diatom *Thalassiosira pseudonana*. J. Phycol. 2007;43:585–604. doi: 10.1111/j.1529-8817.2007.00342.x.

34. Dambek M., Eilers U., Bretenbach J., Steiger S., Büchel C., Sandmann G. Biosynthesis of fucoxanthin and diadinoxanthin and function of initial pathway genes in *Phaeodactylum tricornutum*. J. Exp. Bot.2012;63:5607–5612. doi: 10.1093/jxb/ers211.

35. Mikami K., Hosokawa M. Biosynthetic pathway and health benefits of fucoxanthin, an algae-specific xanthophyll in brown seaweeds. Int. J. Mol. Sci. 2013;14:13763–13781. doi: 10.3390/ijms140713763.

36. Grimm B., Porra R.J., Rüdiger W., Scheer H. In: Chlorophylls and Bacteriochlorophylls: Biochemistry, Biophysics, Functions and Applications. 1st ed. Grimm B., Porra R.J., Rüdiger W., Scheer H., editors. Volume 25 Springer Netherlands; Dordrecht, The Netherlands: 2006.

37. Hunsperger H.M., Randhawa T., Cattolico R.A. Extensive horizontal gene transfer, duplication, and loss of chlorophyll synthesis genes in the algae. BMC Evol. Biol. 2015;15:1–19.

38. Disch A., Schwender J., Muller C., Lichtenthaler H.K., Rohmer M. Distribution of the mevalonate and glyceraldehyde phosphate/pyruvate pathways for isoprenoid biosynthesis in unicellular algae and the cyanobacterium Synechocystis PCC 6714. Biochem. J. 1998;333:381–388. doi: 10.1042/bj3330381.

39. Ito H., Ohtsuka T., Tanaka A. Conversion of chlorophyll b to chlorophyll a via 7-hydroxymethyl chlorophyll. J. Biol. Chem. 1996;271:1475–1479. doi: 10.1074/jbc.271.3.1475.

40. Porra R.J. Recent Progress in Porphyrin and Chlorophyll Biosynthesis. Photochem. Photobiol.1997;65:492–516. doi: 10.1111/j.1751-1097.1997.tb08596.x.

41. Vidoudez C., Pohnert G. Comparative metabolomics of the diatom *Skeletonema marinoi* in different growth phases. Metabolomics. 2012;8:654–669. doi: 10.1007/s11306-011-0356-6.

42. Sánchez-Saavedra M.P., Voltolina D. Effect of photon fluence rates of white and blue-green light on growth efficiency and pigment content of three diatom species in batch cultures. Cienc. Mar. 2002;28:273–279.

43. Brunet C., Chandrasekaran R., Barra L., Giovagnetti V., Corato F., Ruban A.V. Spectral radiation dependent photoprotective mechanism in the diatom Pseudo-nitzschia multistriata. PLoS ONE. 2014;9:1–10. doi: 10.1371/journal.pone.0087015.

44. Lepetit B., Goss R., Jakob T., Wilhelm C. Molecular dynamics of the diatom thylakoid membrane under different light conditions. Photosynth.

Res. 2012;111:245–257. doi: 10.1007/s11120-011-9633-5.

45. Jungandreas A., Costa B.S., Jakob T., von Bergen M., Baumann S., Wilhelm C. The acclimation of*Phaeodactylum tricornutum* to blue and red light does not influence the photosynthetic light reaction but strongly disturbs the carbon allocation pattern. PLoS ONE. 2014;9:1–14.

46. Veuger B., van Oevelen D. Long-term pigment dynamics and diatom survival in dark sediment. Limnol. Oceanogr. 2011;56:1065–1074. doi: 10.4319/lo.2011.56.3.1065.

47. Järup L. Hazards of heavy metal contamination. Br. Med. Bull. 2003;68:167–182. doi: 10.1093/bmb/ldg032.

48. Yoshinaga R., Niwa-Kubota M., Matsui H., Matsuda Y. Characterization of iron-responsive promoters in the marine diatom *Phaeodactylum tricornutum*. Mar. Genom. 2014;16:55–62. doi: 10.1016/j.margen.2014.01.005.

49. Allen A.E., Laroche J., Maheswari U., Lommer M., Schauer N., Lopez P.J., Finazzi G., Fernie A.R., Bowler C. Whole-cell response of the pennate diatom *Phaeodactylum tricornutum* to iron starvation. Proc. Natl. Acad. Sci. USA. 2008;105:10438–10443. doi: 10.1073/pnas.0711370105.

50. Sivaci R.E., Sivaci A., Eroglu S. Changes in photosynthetic pigments and species diversity of epiphytic diatoms on *Myriophyllum triphyllum* exposed to cadmium. ScienceAsia. 2013;39:100–104. doi: 10.2306/scienceasia1513-1874.2013.39.100.

51. Alexandre M., Gundermann K., Pascal A., van Grondelle R., Büchel C., Robert B. Probing the carotenoid content of intact Cyclotella cells by resonance Raman spectroscopy. Photosynth. Res.2014;119:273–281. doi: 10.1007/s11120-013-9942-y.

52. Beer A., Juhas M., Büchel C. Influence of different light intensities and different iron nutrition on the photosynthetic apparatus in the diatom *Cyclotella meneghiniana* (bacillariophyceae) J. Phycol.2011;47:1266–1273. doi: 10.1111/j.1529-8817.2011.01060.x.

53. Brotas V., Plante-Cuny M.R. The use of HPLC pigment analysis to study microphytobenthos communities. Acta Oecol. 2003;24:109–115. doi: 10.1016/S1146-609X(03)00013-4.

54. Costa B.S., Jungandreas A., Jakob T., Weisheit W., Mittag M., Wilhelm C. Blue light is essential for high light acclimation and methylation and chromatin patterning photoprotection in the diatom *Phaeodactylum tricornutum*. J. Exp. Bot. 2013;64:483–493. doi: 10.1093/jxb/ers340.

55. Domingues N., Matos A.R., da Silva J.M., Cartaxana P. Response of the Diatom *Phaeodactylum tricornutum* to photooxidative stress resulting from high light exposure. PLoS ONE. 2012;7:1–6. doi: 10.1371/journal.pone.0038162.

56. Katayama T., Murata A., Taguchi S. Responses of pigment composition of the marine diatom*Thalassiosira weissflogii* to silicate availability during. Plankt. Benthos Res. 2011;6:1–11. doi: 10.3800/pbr.6.1.

57. Lavaud J. Fast regulation of photosynthesis in diatoms: Mechanisms, evolution and ecophysiology.Funct. Plant Sci. Biotechnol. 2007;1:267–287.

58. Gibbs S. The comparative ultrastructure of the algal chloroplast. Ann. N. Y. Acad. Sci. 1970;175:454–473. doi: 10.1111/j.1749-6632.1970.tb45167.x.

59. Pyszniak A., Gibbs S. Immunocytochemical localization of photosystem I and the fucoxanthinchlorophyll a/c light-harvesting complex in the diatom *Phaeodactylum tricornutum*. Protoplasma.1992;166:208–217. doi: 10.1007/BF01322783.

60. Gundermann K., Büchel C. In: The Structural Basis of Biological Energy Generation. 1st ed. Hohmann-Marriott M., editor. Springer Netherlands; Dordrecht, The Netherlands: 2014.

61. Papagiannakis E., van Stokkum I.H.M., Fey H., Büchel C., van Grondelle R. Spectroscopic characterization of the excitation energy transfer in the fucoxanthinchlorophyll protein of diatoms.Photosynth. Res. 2005;86:241–250. doi: 10.1007/s11120-005-1003-8.

62. Premvardhan L., Sandberg D.J., Fey H., Birge R.R., Büchel C., van Grondelle R. The charge-transfer properties of the S2 state of fucoxanthin in solution and in fucoxanthin chlorophyll-a/c2 protein (FCP) based on stark spectroscopy and molecular-orbital theory. J. Phys. Chem. B. 2008;112:11838–11853. doi: 10.1021/jp802689p.

63. Premvardhan L., Bordes L., Beer A., Büchel C., Robert B. Carotenoid structures and environments in trimeric and oligomeric fucoxanthin chlorophyll a/c2 proteins from resonance raman spectroscopy. J. Phys. Chem. B. 2009;113:12565–12574. doi: 10.1021/jp903029g.

64. Premvardhan L., Réfrégiers M., Büchel C. Pigment organization effects on energy transfer and Chl a emission imaged in the diatoms *C. meneghiniana* and *P. tricornutum in vivo*: A confocal laser scanning fluorescence (CLSF) microscopy and spectroscopy study. J. Phys. Chem. B. 2013;117:11272–11281. doi: 10.1021/jp402094c.

65. Gundermann K., Schmidt M., Weisheit W., Mittag M., Büchel C. Identification of several sub-populations in the pool of light harvesting proteins in the pennate diatom *Phaeodactylum tricornutum*.Biochim. Biophys. Acta Bioenerg. 2013;1827:303–310. doi: 10.1016/j.bbabio.2012.10.017.

66. Lepetit B., Volke D., Gilbert M., Wilhelm C., Goss R. Evidence for the existence of one antenna-associated, lipid-dissolved and two protein-bound pools of diadinoxanthin cycle pigments in diatoms. Plant Physiol. 2010;154:1905–1920. doi: 10.1104/pp.110.166454.

67. Eppard M., Rhiel E. The genes encoding light-harvesting subunits of *Cyclotella cryptica*(Bacillariophyceae) constitute a complex and heterogeneous family. Mol. Gen. Genet. 1998;260:335–345. doi: 10.1007/s004380050902.

68. Liu Z., Yan H., Wang K., Kuang T., Zhang J., Gui L., An X., Chang W. Crystal structure of

spinach major light-harvesting complex at 2.72 A resolution. Nature. 2004;428:287–292. doi: 10.1038/nature02373.

69. FalkowskiP.G.BiogeochemicalControlsandFeedbackson Ocean Primary Production. Science.1998;281:200–206. doi: 10.1126/science.281.5374.200.

70. Stevenson R.J., Peterson C.G. Variation in benthic diatom (Bacillariophyceae) immigration with habitat characteristics and cell morphology. J. Phycol. 1989;25:120–129. doi: 10.1111/j.0022-3646.1989.00120.x.

71. PadisákJ.,Soróczki-PintérÉ.,ReznerZ.Sinkingproperties of some phytoplankton shapes and the relation of form resistance to morphological diversity of plankton—An experimental study. Hydrobiologia.2003;500:243–257. doi: 10.1023/A:1024613001147.

72. Verleyen E., Hodgson D.A., Leavitt P.R., Sabbe K., Vyverman W. Quantifying habitat-specific diatom production: A critical assessment using morphological and biogeochemical markers in Antarctic marine and lake sediments. Limnol. Oceanogr. 2004;49:1528–1539. doi: 10.4319/lo.2004.49.5.1528.

73. Falkowski P.G., Knoll A.H. Evolution of Primary Producers in the Sea. In: Falkowski P., Knoll A.H., editors. Evolution of Primary Producers in the Sea. Academic Press; New York, NY, USA: 2007. p. 456.

74. Haxo F.T., Blinks L.R. Photosynthetic action spectra of marine algae. J. Gen. Physiol. 1950;33:389–422. doi: 10.1085/jgp.33.4.389.

75. Nicklisch A. Growth and light absorption of some planktonic cyanobacteria, diatoms and Chlorophyceae under simulated natural light fluctuations. J. Plankton Res. 1998;20:105–119. doi: 10.1093/plankt/20.1.105.

76. Spaulding S.A., Lubinski D.J., Potapova M. Diatoms of the United States. (accessed on 6 July 2015). Available online: http://westerndiatoms.colorado.edu.

77. Johansen J.R. Morphological variability and cell wall composition of *Phaeodactylum tricornutum*(Bacillariophyceae) Great Basin Nat. 1991;51:310–315.

78. Wagner A.H., Jakob T., Wilhelm C., Wagner H. Balancing the energy captured under fluctuating light conditions to. New Biotechnol. 2005;169:95–108.

79. Holdsworth E.S. Effect of growth factors and light quality on the growth, pigmentation and photosynthesis of two diatoms, Thalassiosira gravida and *Phaeodactylum tricornutum*. Mar. Biol.1985;86:253–262. doi: 10.1007/BF00397512.

80. Darley W., Volcani B. Role of silicon in diatom metabolism. A silicon requirement for deoxyribonucleic acid synthesis in the diatom *Cylindrotheca fusiformis* Reimann and Lewin. Exp. Cell Res. 1969;58:334–342. doi: 10.1016/0014-4827(69)90514-X.

81. Lombardi A., Wangersky P. Influence of phosphorus and silicon on lipia class production by the marine diatom *Chaetoceros gracilis* grown in turbidostat cage cultures. Mar. Ecol. Prog. Ser. 1991;77:39–47. doi: 10.3354/meps077039.

82. Brzezinski M.A., Baines S.B., Balch W.M., Beucher C.P., Chai F., Dugdale R.C., Krause J.W., Landry M.R., Marchi A., Measures C.I., et al. Co-limitation of diatoms by iron and silicic acid in the equatorial Pacific. Deep. Res. II Top. Stud. Oceanogr. 2011;58:493–511. doi: 10.1016/j.dsr2.2010.08.005.

83. Lippemeier S., Hartig P., Colijn F. Direct impact of silicate on the photosynthetic performance of the diatom *Thalassiosira weissflogii* assessed by on- and off-line PAM fluorescence measurements. J. Plankton Res. 1999;21:269–283. doi: 10.1093/plankt/21.2.269.

84. Roberts K., Granum E., Leegood R.C., Raven J.A. C3 and C4 pathways of photosynthetic carbon assimilation in marine diatoms are under genetic, not environmental, control. Plant Physiol. 2007;145:230–235. doi: 10.1104/pp.107.102616.

85. Hopkinson B.M., Dupont C.L., Allen A.E., Morel F.M.M. Efficiency of the CO_2-concentrating mechanism of diatoms. Proc. Natl. Acad. Sci. USA. 2011;108:3830–3837. doi: 10.1073/pnas.1018062108.

86. Goss R., Lepetit B. Biodiversity of NPQ. J. Plant Physiol. 2014;172:13–32. doi: 10.1016/j.jplph.2014.03.004.

87. Lavaud J., Rousseau B., Etienne A.L. In diatoms, a transthylakoid proton gradient alone is not sufficient to induce a non-photochemical fluorescence quenching. FEBS Lett. 2002;523:163–166. doi: 10.1016/S0014-5793(02)02979-4.

88. Goss R., Ann Pinto E., Wilhelm C., Richter M. The importance of a highly active and ΔpH-regulated diatoxanthin epoxidase for the regulation of the PS II antenna function in diadinoxanthin cycle containing algae. J. Plant Physiol. 2006;163:1008–1021. doi: 10.1016/j.jplph.2005.09.008.

89. García-Plazaola J.I., Matsubara S., Osmond C.B. The lutein epoxide cycle in higher plants: Its relationships to other xanthophyll cycles and possible functions. Funct. Plant Biol. 2007;34:759–773. doi: 10.1071/FP07095.

90. Mewes H., Richter M. Supplementary ultraviolet-B radiation induces a rapid reversal of the diadinoxanthin cycle in the strong light-exposed diatom *Phaeodactylum tricornutum*. Plant Physiol.2002;130:1527–1535. doi: 10.1104/pp.006775.

91. Jakob T., Goss R., Wilhelm C. Unusual pH-dependence of diadinoxanthin de-epoxidase activation causes chlororespiratory induced accumulation of diatoxanthin in the diatom *Phaeodactylum tricornutum*. J. Plant Physiol. 2001;158:383–390. doi: 10.1078/0176-1617-00288.

92. Grouneva I., Jakob T., Wilhelm C., Goss R. Influence of ascorbate and pH on the activity of the diatom xanthophyll cycle-enzyme diadinoxanthin de-epoxidase. Physiol. Plant. 2006;126:205–211. doi: 10.1111/j.1399-3054.2006.00613.x.

93. Baroli I., Do A.D., Yamane T., Niyogi K.K. Zeaxanthin accumulation in the absence of a functional xanthophyll cycle protects *Chlamydomonas reinhardtii* from photooxidative stress. Plant Cell. 2003;15:992–1008. doi: 10.1105/tpc.010405.

94. Schumann A., Goss R., Jakob T., Wilhelm C. Investigation of the quenching efficiency of diatoxanthin in cells of *Phaeodactylum tricornutum* (Bacillariophyceae) with different pool sizes of xanthophyll cycle pigments. Phycologia. 2007;46:113–117. doi: 10.2216/06-30.1.

95. Pinckney M., Micheli F. Microalgae on seagrass mimics: Does epiphyte community structure differ from live seagrasses. J. Exp. Mar. Biol. Ecol. 1998;221:59–70. doi: 10.1016/S0022-0981(97)00115-9.

96. Schagerl M., Künzl G. Chlorophyll a extraction from freshwater algae—A reevaluation. Biol. Bratisl.2007;62:270–275. doi: 10.2478/s11756-007-0048-x.

97. Strickland J., Parsons T. In: A Practical Handbook of Seawater Analysis. 2nd ed. Stevenson J.C., editor. Volume 167 Fisheries Research Board of Canada; Ottawa, ON, Canada: 1972.

98. Holm-Hansen O., Lorenzen C., Holmes R., Strickland J. Fluorometric determination of chlorophyll. J. Cons. Perm. Int. Explor. Mer. 1965;30:3–15. doi: 10.1093/icesjms/30.1.3.

99. Wright S.W., Jeffrey S.W., Mantoura R.F.C. Evaluation of methods and solvents for pigment extraction. In: Wright S.W., Jeffrey S.W., Mantoura R.F.C., editors. Phytoplankton Pigments in Oceanography: Guidelines to Modern Methods. UNESCO; Paris, France: 1997. pp. 261–282.

100. Chen N., Bianchi T., Bland J. Novel decomposition products of chlorophyll-a in continental shelf (Louisiana shelf) sediments: Formation and transformation of carotenol chlorine esters. Geochim. Cosmochim. Acta. 2003;67:2027–2042. doi: 10.1016/S0016-7037(02)01297-8.

101. Jeffrey S. Profiles of photosynthetic pigments in the ocean using thin-layer chromatography. Mar. Biol.1974;26:101–110. doi: 10.1007/BF00388879.

102. Mantoura R., Llewellyn C. The rapid determination of algal chlorophyll and carotenoid pigments and their breakdown products in natural waters by reverse-phase high-performance liquid chromatography. Anal. Chim. Acta. 1983;151:297–314. doi: 10.1016/S0003-2670(00)80092-6.

103. Bowles N., Paerl H., Tucker J. Effective solvents and extraction periods employed in phytoplankton carotenoi d and chlorophyll determination. Can. J. Fish. Aquat. Sci. 1985;42:1127–1131. doi: 10.1139/f85-139.

104. Shoaf W.T., Lium B.W. Improved extraction of chlorophyll a and b from algae using dimethyl sulfoxide. Limnol. Oceanogr. 1976;21:926–928. doi: 10.4319/lo.1976.21.6.0926.

105. Stock R., Rice C.B.F. Chromatographic Methods. 1st ed. Springer US; New York, NY, USA: 1974.

106. Buffan-Dubau E., Carman K.R. Extraction of benthic microalgal pigments for HPLC analyses. Mar. Ecol. Prog. Ser. 2000;204:293–297. doi: 10.3354/meps204293.

107. Jeffrey S., Hallegraeff G. Chlorophyllase distribution in ten classes of phytoplankton: A problem for chlorophyll analysis. Mar. Ecol. Prog. Ser. 1987;35:293–304. doi: 10.3354/meps035293.

108. Barrett J., Jeffrey S.W. Chlorophyllase and formation of an atypical chlorophyllide in marine algae.Plant Physiol. 1964;39:44–47. doi: 10.1104/pp.39.1.44.

109. Draziewicz M. Chlorophyllase: Occurrence, functions, mechanisms of action, effects of external and internal factor. Photosynthesis. 1994;30:321–331.

110. Rowan K. Photosynthetic Pigments of Algae. Cambridge University Press; Cambridge, UK: 1989.

111. Jeffrey S.W., Mantoura R.F.C., Wright S.W. In: Phytoplankton Pigments in Oceanography: Guidelines to Modern Methods. Jeffrey S., Mantoura R., Wright S., editors. UNESCO Publishing; Paris, France: 1997.

112. Jeffrey S., Wright S., Zapata M. Recent advances in HPLC pigment analysis of phytoplankton. Mar. Freshw. Res. 1999;50:879–896. doi: 10.1071/MF99109.

113. Wright S., Shearer J. Rapid extraction and HPLC of chlorophylls and carotenoids from marine phytoplankton. J. Chrom. 1984;294:281–295. doi: 10.1016/S0021-9673(01)96134-5.

114. Wright S.W. Improved HPLC method for the analysis of chlorophylls and carotenoids from marine phytoplankton. Mar. Ecol. Prog. Ser. 1991;77:183–196. doi: 10.3354/meps077183.

115. Zapata M., Rodríguez F., Garrido J.L. Separation of chlorophylls and carotenoids from marine phytoplankton: A new HPLC method using a reversed phase C8 column and pyridine-containing mobile phases. Mar. Ecol. Prog. Ser. 2000;195:29–45. doi: 10.3354/meps195029.

116. Van Heukelem L., Thomas C. Computer-assisted high-performance liquid chromatography method development with applications to the isolation and analysis of phytoplankton pigments. J. Chromatogr. A.2001;910:31–49. doi: 10.1016/S0378-4347(00)00603-4.

117. Kraay G.W., Zapata M., Veldhuis M.J.W. Separation of chlorophylls c1c2, and c3 of marine phytoplankton by reversed-phase-C18-high-performance liquid chromatography. J. Phycol. 1992;28:708–712. doi: 10.1111/j.0022-3646.1992.00708.x.

118. Mendes C.R., Cartaxana P., Brotas V. HPLC determination of phytoplankton and microphytobenthos pigments: Comparing resolution and sensitivity of a C18 and a C8 method. Limnol. Oceanogr. Methods.2007;5:363–370. doi: 10.4319/lom.2007.5.363.

119. Jeffrey S., Mantoura R., Bjørnland T. Data for the identification of 47 key phytoplankton pigments. In: Jeffrey S., Mantour R., Wright S., editors. Phytoplankton Pigments in Oceanography: Guidelines to Modern Methods. UNESCO Publishing; Paris, France: 1997. pp. 447–559.

120. Jeffrey S.W. Preparation and some properties of crystalline chlorophyll c1 and c2 from marine algae. Biochim. Biophys. Acta. 1972;279:15–33. doi: 10.1016/0304-4165(72)90238-3.

121. Jensen A. Algal carotenoids, I. Fucoxanthin

monoacetate. Acta Chem. Scand. 1961;15:1604–1628. doi: 10.3891/acta.chem.scand.15-1604.

122. Antia N.J. The optical activity of fucoxanthin. Can. J. Chem. 1965;43:302–303. doi: 10.1139/v65-034.

123. Haugan J.A., Liaaen-Jensen S. Improved isolation procedure for fucoxanthin. Phytochemistry.1989;28:2797–2798. doi: 10.1016/S0031-9422(00)98091-9.

124. Johansen J.E., Svec W.A., Liaaen-Jensen S. Carotenoids of the Dinophyceae. Phytochemistry.1974;13:2261–2271. doi: 10.1016/0031-9422(74)85038-7.

125. Hager A., Meyer-Bertenrath T. Die Isolierung und quantitative Bestimmung der Carotinoide und Chlorophylle von Blättern, Algen und isolierten Chloroplasten mit Hilfe dünnschichtchromatographischer Methoden. Planta. 1966;69:198–217. doi: 10.1007/BF00384873.

126. Davies B.H. Carotenoids. In: Goodwin T., editor. Chemistry and Biochemistry of Plant Pigments Vol. II. Academic Press; London, UK: 1976. pp. 38–165.

127. Jeffrey S.W., Humphrey G.F. New spectrophotometric equations for determining chlorophylls a, b, c1 and c2 in higher plants, algae and natural phytoplankton. Biochem. Physiol. Pflanz. 1975;167:191–194.

128. Isler O., Lindlar H., Montavon M., Rüegg R., Zeller P. Die technische Synthese von beta-Carotin. Helv. Chim. Acta. 1956;39:249–259. doi: 10.1002/hlca.19560390128.

129. Hiyama T., Nishimura M., Chance B. Determination of carotenes by thin-layer chromatography. Anal. Biochem. 1969;29:339–342. doi: 10.1016/0003-2697(69)90318-2.

130. Pangestuti R., Kim S.K. Biological activities and health benefit effects of natural pigments derived from marine algae. J. Funct. Foods. 2011;3:255–266. doi: 10.1016/j.jff.2011.07.001.

131. Stahl W., Sies H. Bioactivity and protective effects of natural carotenoids. Biochim. Biophys. Acta Mol. Basis Dis. 2005;1740:101–107. doi: 10.1016/j.bbadis.2004.12.006.

132. Palozza P., Krinsky N.I. β-Carotene and α-tocopherol are synergistic antioxidants. Arch. Biochem. Biophys. 1992;297:184–187. doi: 10.1016/0003-9861(92)90658-J.

133. Omenn G.S., Goodman G.E., Thornquist M.D., Balmes J., Cullen M.R., Glass A., Keogh J.P., Meyskens F.L., Valanis B., Williams J.H., et al. Risk factors for lung cancer and for intervention effects in CARET, the Beta-Carotene and Retinol Efficacy trial. J. Natl. Cancer Inst. 1997;89:325–326. doi: 10.1093/jnci/89.4.325-a.

134. Krinsky N.I., Landrum J.T., Bone R.A. Biologic mechanisms of the protective role of lutein and zeaxanthin in the eye. Annu. Rev. Nutr. 2003;23:171–201. doi: 10.1146/annurev.nutr.23.011702.073307.

135. Pasquet V., Morisset P., Ihammouine S., Chepied A., Aumailley L., Berard J.B., Serive B., Kaas R., Lanneluc I., Thiery V., et al. Antiproliferative activity of violaxanthin isolated from bioguided fractionation of Dunaliella tertiolecta extracts. Mar. Drugs. 2011;9:819–831. doi: 10.3390/md9050819.

136. Peng J., Yuan J.P., Wu C.F., Wang J.H. Fucoxanthin, a marine carotenoid present in brown seaweeds and diatoms: Metabolism and bioactivities relevant to human health. Mar. Drugs. 2011;9:1806–1828. doi: 10.3390/md9101806.

137. Heo S.-J., Ko S.-C., Kang S.-M., Kang H.-S., Kim J.-P., Kim S.-H., Lee K.-W., Cho M.-G., Jeon Y.-J. Cytoprotective effect of fucoxanthin isolated from brown algae *Sargassum siliquastrum* against H2O2-induced cell damage. Eur. Food Res. Technol. 2008;228:145–151. doi: 10.1007/s00217-008-0918-7.

138. Kim K.-N., Heo S.-J., Yoon W.-J., Kang S.-M., Ahn G., Yi T.-H., Jeon Y.-J. Fucoxanthin inhibits the inflammatory response by suppressing the activation of NF-κB and MAPKs in lipopolysaccharide-induced RAW 264.7 macrophages. Eur. J. Pharmacol. 2010;649:369–375. doi: 10.1016/j.ejphar.2010.09.032.

139. Hosokawa M., Wanezaki S., Miyauchi K., Kurihara H., Kohno H., Kawabata J., Odashima S., Takahashi K. Apoptosis-Inducing Effect of Fucoxanthin on Human Leukemia Cell Line HL-60. Food Sci. Technol. Res. 1999;5:243–246. doi: 10.3136/fstr.5.243.

140. Asai A., Sugawara T., Ono H., Nagao A. Biotransformation of fucoxanthinol into amarouciaxanthin a in mice and HepG2 cells: Formation and cytotoxicity of fucoxanthin metabolites. Drug Metab. Dispos.2004;32:205–211. doi: 10.1124/dmd.32.2.205.

141. Maeda H., Hosokawa M., Sashima T., Funayama K., Miyashita K. Fucoxanthin from edible seaweed,*Undaria pinnatifida*, shows antiobesity effect through UCP1 expression in white adipose tissues. Biochem. Biophys. Res. Commun. 2005;332:392–397. doi: 10.1016/j.bbrc.2005.05.002.

142. Abidov M., Ramazanov Z., Seifulla R., Grachev S. The effects of XanthigenTM in the weight management of obese premenopausal women with non-alcoholic fatty liver disease and normal liver fat.Diabetes Obes. Metab. 2010;12:72–81. doi: 10.1111/j.1463-1326.2009.01132.x.

143. Jaswir I., Dedi N., Salleh H.M., Taher M., Miyashita K., Ramli N. Analysis of fucoxanthin content and purification of all-*trans*-fucoxanthin from Turbinaria turbinata and *Sargassum plagyophyllum* by SiO2 open column chromatography and reversed phase HPLC. J. Liq. Chromatogr. 2013;36:1340–1354.

144. Kim S.M., Shang Y.F., Um B.H. A preparative method for isolation of fucoxanthin from *Eisenia bicyclis* by centrifugal partition chromatography. Phytochem. Anal. 2011;22:322–329. doi: 10.1002/pca.1283.

145. Rebolloso-Fuentes M.M., Navarro-Pérez A., García-Camacho F., Ramos-Miras J.J., Guil-Guerrero J.L. Biomass nutrient profiles of the microalga Nannochloropsis. J. Agric. Food Chem. 2001;49:2966–2972. doi: 10.1021/jf0010376.

146. Gómez-loredo A., Benavides J., Rito-palomares M. Partition behavior of fucoxanthin in ethanol-potassium phosphate two-phase systems. J. Chem. Technol. Biotechnol. 2014;89:1637–1645. doi: 10.1002/jctb.4514.

147. Kim S., Kang S.-W., Kwon O.-N., Chung D., Pan C.-H. Fucoxanthin as a major carotenoid in Isochrysis aff. galbana: Characterization of extraction for commercial application. J. Korean Soc. Appl. Biol. Chem. 2012;55:477–483. doi: 10.1007/s13765-012-2108-3.

CHAPTER 3

BETALAIN PRODUCTION IS POSSIBLE IN ANTHOCYANINPRODUCING PLANT SPECIES GIVEN THE PRESENCE OF DOPA-DIOXYGENASE AND L-DOPA

Nilangani N Harris[1,4], John Javellana[1], Kevin M Davies[1], David H Lewis[1], Paula E Jameson[2], Simon C Deroles[1], Kate E Calcott[1,3], Kevin S Gould[3] and Kathy E Schwinn[1]

[1] New Zealand Institute for Plant & Food Research Limited, Private Bag 11- 600, Palmerston North, New Zealand

[2] School of Biological Sciences, University of Canterbury, Private Bag 4-800, Christchurch, New Zealand

[3] Victoria University of Wellington, Wellington 6140, New Zealand

[4] Commonwealth Scientific and Industrial Research Organization, Ecosystem Sciences, Urrbrea, South Australia 5064, Australia.

ABSTRACT

Background

Carotenoids and anthocyanins are the predominant non-chlorophyll pigments in plants. However, certain families within the order Caryophyllales produce another class of pigments, the betalains, instead of anthocyanins. The occurrence of betalains and anthocyanins is mutually exclusive. Betalains are divided into two classes, the betaxanthins and betacyanins, which produce yellow to orange or violet colours, respectively. In this article we show betalain production in species that normally produce anthocyanins, through a combination of genetic modification and substrate feeding.

Results

The biolistic introduction of DNA constructs for transient overexpression of two different dihydroxyphenylalanine (DOPA) dioxygenases (DODs), and feeding of DOD substrate (L-DOPA), was sufficient to induce betalain production in cell cultures of Solanum tuberosum (potato) and petals ofAntirrhinum majus. HPLC analysis showed both betaxanthins and betacyanins were produced. Multi-cell foci with yellow, orange and/or red colours occurred, with either a fungal DOD (from Amanita muscaria) or a plant DOD (from Portulaca grandiflora), and the yellow/orange foci showed green autofluorescence characteristic of betaxanthins. Stably transformed Arabidopsis thaliana (arabidopsis) lines

Citation: Harris, N. N., Javellana, J., Davies, K. M., Lewis, D. H., Jameson, P. E., Deroles, S. C., Schwinn, K. E. (2012). Betalain production is possible in anthocyanin-producing plant species given the presence of DOPA-dioxygenase and L-DOPA. BMC Plant Biology, 12, 34. http://doi.org/10.1186/1471-2229-12-34.

containing 35S: AmDOD produced yellow colouration in flowers and orange-red colouration in seedlings when fed L-DOPA. These tissues also showed green autofluorescence. HPLC analysis of the transgenic seedlings fed L-DOPA confirmed betaxanthin production.

Conclusions

The fact that the introduction of DOD along with a supply of its substrate (L-DOPA) was sufficient to induce betacyanin production reveals the presence of a background enzyme, possibly a tyrosinase, that can convert L-DOPA to cyclo-DOPA (or dopaxanthin to betacyanin) in at least some anthocyanin-producing plants. The plants also demonstrate that betalains can accumulate in anthocyanin-producing species. Thus, introduction of a DOD and an enzyme capable of converting tyrosine to L-DOPA should be sufficient to confer both betaxanthin and betacyanin production to anthocyanin-producing species. The requirement for few novel biosynthetic steps may have assisted in the evolution of the betalain biosynthetic pathway in the Caryophyllales, and facilitated multiple origins of the pathway in this order and in fungi. The stably transformed 35S: AmDOD arabidopsis plants provide material to study, for the first time, the physiological effects of having both betalains and anthocyanins in the same plant tissues.

BACKGROUND

The variety of colours observed in flowers, fruits and vegetative tissues in plants are due to the presence of chromogenic plant secondary metabolites [1,2]. These pigments serve diverse functions including photosynthesis and the protection of the photosynthetic machinery, attraction of pollinators and seed dispersers, and protection against biotic and abiotic stresses [1,3]. In addition to their biological functions, plant pigments are also of much interest regarding their possible beneficial effects on human health, their use as natural colorants and their aesthetic value in ornamental and food crops [4]. Non-chlorophyll plant pigments predominantly belong to two groups: flavonoids and carotenoids. Within the flavonoids, anthocyanins are the most significant type, providing a range of colours including orange, red, pink, mauve, purple and blue. However, in certain families within the order Caryophyllales, another class of pigments, the betalains, replaces the anthocyanins [2,5,6]. Betalains are only present in the order Caryophyllales and some fungi. They occur in most families of the Caryophyllales, but species of at least two families accumulate anthocyanin pigments instead [7]. The basis of this differentiation is unknown, but may represent an initial evolution of betalain biosynthesis in an ancestor of the core Caryophyllales and then its subsequent loss on different occasions [7].

No plant has yet been found that produces both betalain and anthocyanin pigments [5-8]. This mutually exclusive nature of the betalain and anthocyanin production in the plant kingdom is a curious phenomenon and the evolutionary and biochemical mechanisms for this restriction are unknown [5-7].

There are two major types of betalains, the red-purple betacyanins and the yellow/orange betaxanthins, both of which accumulate in the vacuole. The betaxanthins also emit green autofluorescence, which is not seen with the betacyanins [9-11]. While the production of flavonoids and carotenoids has been extensively studied and metabolically engineered in a variety of species, betalain biosynthesis has yet to be fully characterised [1,2]. The betalain biosynthetic pathway is relatively simple with putatively only a few reactions that are enzyme catalysed (Figure 1). The initial biosynthetic step is the hydroxylation of tyrosine to L-3,4-dihydroxyphenylalanine (DOPA), attributed to the activity of a tyrosinase, although the exact role (if any) of tyrosinase in betalain synthesis has yet to be resolved [5,6,12,13] Cleavage of the cyclic ring of L-DOPA by DOPA-4,5-dioxygenase (DOD) forms an unstable *seco*-DOPA intermediate, which is thought to spontaneously convert to betalamic acid. The formation of betaxanthins occurs spontaneously from the condensation of betalamic acid with amines/amino acids [14], probably in the vacuole. The classic model of betacyanin biosynthesis involves the condensation of betalamic acid with (most commonly) *cyclo*-DOPA, which again is a likely spontaneous step. In this model, *cyclo*-DOPA is formed from L-DOPA through an oxidation reaction that also has been attributed to tyrosinase activity [5,6,13] The conversion proceeds via an unstable dopaquinone intermediate, which spontaneously cyclizes to form *cyclo*-DOPA. Betacyanins are generally *O*-glycosylated (at the C-5 or C-6) and frequently subsequently acylated. The aglycone product of betalamic acid and *cyclo*-DOPA condensation is termed betanidin and the 5-*O*-glucosylated form betanin, as with the anthocyanidin/anthocyanin convention. The timing of the glycosylation, regarding whether it occurs on *cyclo*-DOPA or betanidin, has been debated [15,16].

The only betalain biosynthetic enzymes characterised at the molecular level to date are for DOD and some of the glycosyltransferase activities. The first plant *DOD* gene characterised was from *Portulaca grandiflora*, and it defined a novel plant gene family of non-haem dioxygenases [2,11]. The plant dioxygenase is phylogenetically unrelated to DOD from the fungi *Amanita muscaria*, although the fungal sequence can complement betalain production in flowers of a *P. grandiflora dod* mutant [17]. The plant DOD carries out only the 4,5-cleavage of DOPA to yield betalamic acid, whereas the fungal enzyme can also conduct 2,3-extradiol cleavage of DOPA, yielding the yellow pigment muscaflavin [18]. Three cDNAs encoding proteins with relevant *O*-glycosylation activity have been identified, two that use both betanidin and flavonoids as substrates [19,20] and one that uses *cyclo*-DOPA [16].

Studies seeking to understand the genetic basis of the mutual exclusion of betalains and anthocyanins have focused on establishing the extent of the retention of the anthocyanin biosynthetic pathway in betalain producing species [21-23]. Flavonoids are present in betalain producing species, including flavonols and proanthocyanidins, and functional genes have been identified for the flavonoid biosynthetic enzymes chalcone synthase, dihydroflavonol 4-reductase and anthocyanidin synthase [21-23]. This suggests that the lack of anthocyanin production in betalain-producing species may be due to a lack of transcriptional activation of all the necessary biosynthetic genes [7,23], although a hypothesis based on repressive interaction between anthocyanin and betalain metabolites and the biosynthetic enzymes has also been suggested [24].

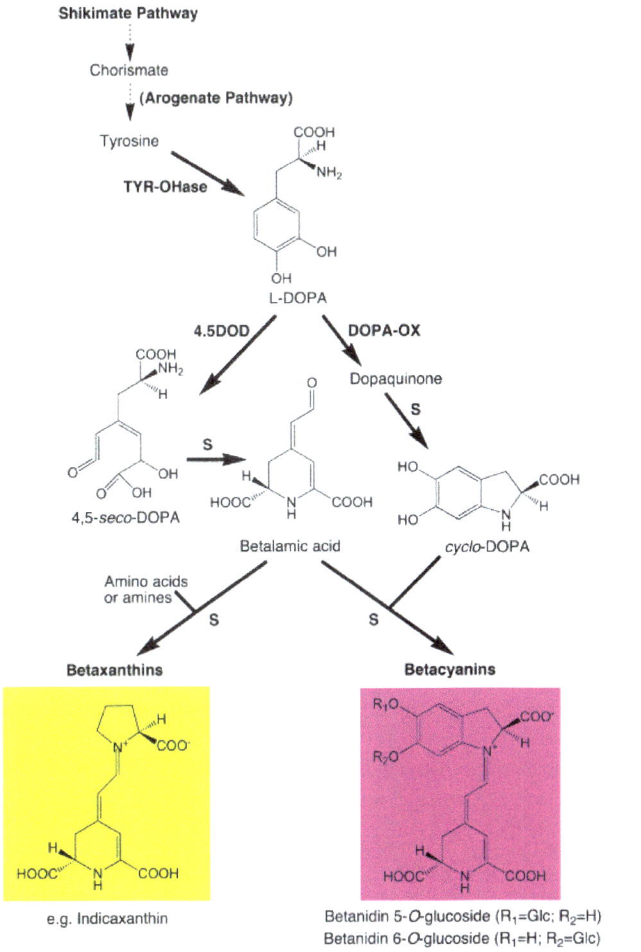

Figure 1: Schematic of the proposed biosynthesis pathway of betalains. Abbreviations are; 4,5DOD, DOPA-4,5-dioxygenase; DOPA-OX, DOPA oxidase; S, spontaneous conversion; TY-OHase, tyrosine hydroxylase. The steps for formation of the betacyanin glycosides are not shown.

Our aim in this study was to determine whether betalain production is possible in anthocyanin-producing species. Using genetic transformation and feeding of pathway intermediates, we have examined what the minimum number is of biosynthetic steps that must be introduced into an anthocyanin-producing species to allow betalain production, and whether betalains can accumulate to significant levels in such species. Stable or transient transgene expression was used with DOD cDNAs from the fungus *A. muscaria* and the plant *P. grandiflora*, introduced into *Arabidopsis thaliana* (arabidopsis) plants, *Solanum tuberosum* (potato) cell cultures and *Antirrhinum majus* (antirrhinum) petals. These are representatives of the asterids and rosids, the two major clades of eudicots.

RESULTS AND DISCUSSION

Betalain biosynthesis in potato cell cultures by transient expression of *DOD*

Potato cell suspension cultures were transformed using particle bombardment with *35S: green fluorescent protein* (*GFP*) or constructs having either the *P. grandiflora DOD* cDNA (*35S: PgDOD*) or the *A. muscaria DOD* (*35S: AmDOD*) driven by the CaMV35S promoter, and examined for betalain production following feeding with L-DOPA. In cells transformed with *35S: GFP*, GFP was detected in the cells 24 h after biolistic transformation but no pigmentation was apparent (data not shown). In the cells transformed with either *35S: PgDOD* or *35S: AmDOD* and fed L-DOPA, pigmented multi-celled clusters were apparent within 24 h post-bombardment (Figure 2). Both *35S: AmDOD* and *35S: PgDOD* resulted in yellow and orange cell clusters. In addition, *35S: AmDOD* also produced cells with red pigmentation. The yellow and red pigmentation suggests that betaxanthin and betacyanin production, respectively, had been conferred to the cells. Indeed, observation of the yellow pigmented areas under blue light showed the autofluorescence characteristic of betaxanthin pigments (data not shown).

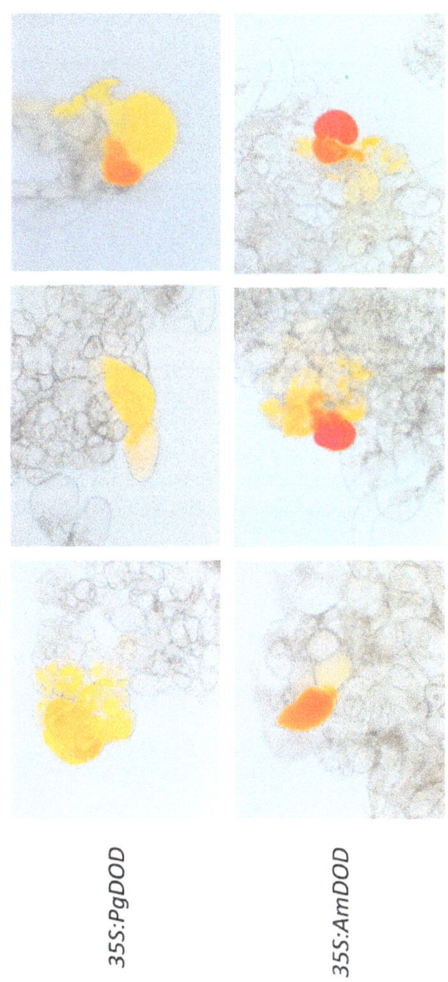

Figure 2: Pigment production in potato cell suspension cultures biolistically transformed with *35S: PgDOD* or *35S: AmDOD* and fed 10 mM L-DOPA. Examples are shown of cells with the resulting yellow to orange (*35S: PgDOD*) or yellow to red (*35S: AmDOD*) pigmentation.

Betalain Biosynthesis in Antirrhinum Petals by Transient Expression of *DOD*

The adaxial surface of dorsal petals of antirrhinum was transformed using particle bombardment with *35S: PgDOD* and examined for betalain production with or without infiltration of the petals with L-DOPA. Petals biolistically transformed with *35S: GFP* vector and fed with L-DOPA were used as an additional control. Antirrhinum lines having mutations in flavonoid production were used to provide anthocyanin-free petal backgrounds upon which to observe any pigment production. No pigments were visible in *35S: GFP* shot tissue or in petals transformed with *35S: PgDOD* but not infiltrated with L-DOPA, although positive GFP foci were apparent (Figure 3). The *35S: PgDOD* petals that were infiltrated with L-DOPA had numerous, multi-celled yellow foci 24 h after infiltration (48 h after bombardment) (Figure 3), indicating betaxanthin production. The yellow foci did indeed have the strong green autofluorescence typical of betaxanthins when observed under blue light (Figure 3). When using the *35S: AmDOD* construct, both yellow/orange and pink/red multi-celled foci occurred (Figure 4), although the pink/red foci were present only inconsistently (data not shown). Similarly, pink foci were sometimes present in some replicate experiments using *35S: PgDOD* (data not shown). The pink colouration indicates betacyanin production. The pink regions did not show strong autofluorescence while the central cells did, indicating possible betacyanin accumulation around a central region containing both betacyanins and betaxanthins.

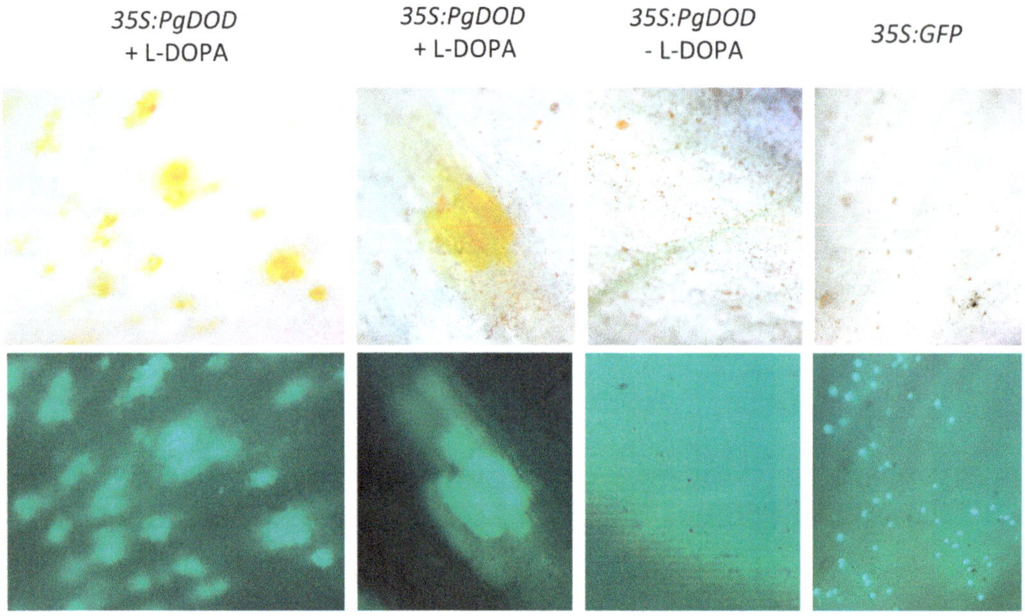

Figure 3: Betaxanthin production in antirrhinum petals transiently transformed with *35S: PgDOD* or *35S: GFP*. Plasmid constructs containing *35S: PgDOD* or *35S: GFP* were introduced into the adaxial epidermis of antirrhinum petals using particle bombardment. At 24 h after bombardment some 35S: PgDOD bombarded petals were infiltrated with 10 mM L-DOPA, while the other petals were infiltrated with water, and incubated for a further 24 h before observation. Representative petals are shown as viewed under white light (upper row) or blue light (lower row). The yellow, multi-cell foci for 35S: PgDOD petals fed with L-DOPA are shown at two magnifications.

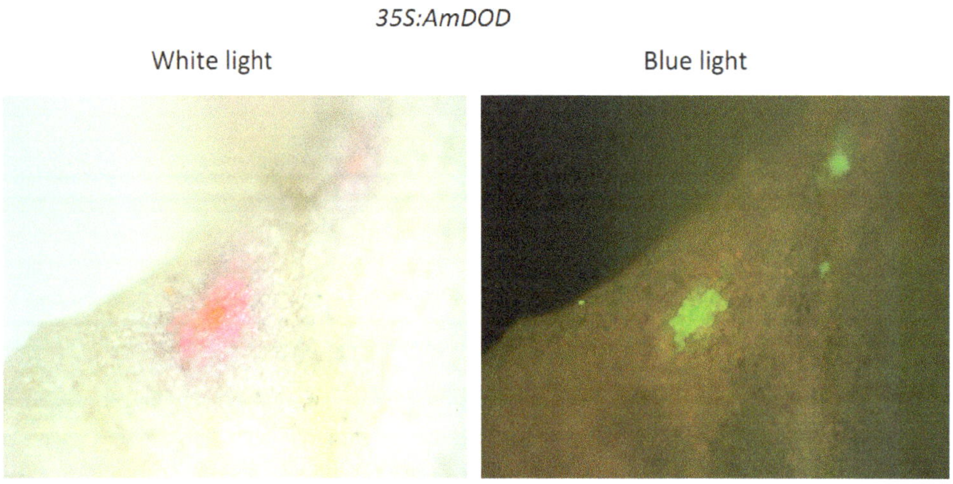

Figure 4: Pigment production in antirrhinum petals transiently transformed with *35S: AmDOD*. A plasmid construct containing *35S: AmDOD* was introduced into the adaxial epidermis of antirrhinum petals using particle bombardment. At 24 h after bombardment the bombarded petals were infiltrated with 10 mM L-DOPA and incubated for a further 24 h before observation. A representative petal is shown as viewed under white light (left) or blue light (right).

HPLC analysis (LC-DAD) was used to examine the nature of the yellow pigments produced following bombardment with *35S: PgDOD*. The ridge region of the petals was chosen for HPLC analysis, to ensure a similar region was sampled in each case. Betaxanthins and betacyanins have absorbance maxima at around 470 nm and 538 nm, respectively [25], and peak profiles were examined at these two wavelengths (Figure 5). No betalain-related peaks were detected with HPLC analysis of the *35S: GFP* petal tissue or with *35S: PgDOD* petals that were not infiltrated with L-DOPA. Petals transformed with *35S: PgDOD* and fed L-DOPA showed distinct peaks at both 470 nm and 538 nm. The compounds represented by the peaks were putatively identified by comparison of HPLC retention times and spectral data against those of a standard extract (from beetroot, Figure 5A and 5E) and reported spectral data [14,25,26]. The beetroot extract showed the expected peaks for the betaxanthin vulgaxanthin I (Peak 1) and the betacyanins betanin (betanidin 5-*O*-glucoside, Peak 2) and isobetanin (Peak 3) (Table 1). The *35S: PgDOD* L-DOPA fed samples showed a small amount of betanin (Peak 2; Figure 5F) and an unknown peak (Peak 5, 21.02 min; Figure 5B) that was possibly dopaxanthin (Table 1). Generally, the chromatograms at 538 nm showed the same patterns as at 470 nm except that vulgaxanthin I (Peak 1) and the putative dopaxanthin peak (Peak 5) were no longer detected, providing further evidence that Peak 5 is indeed a betaxanthin.

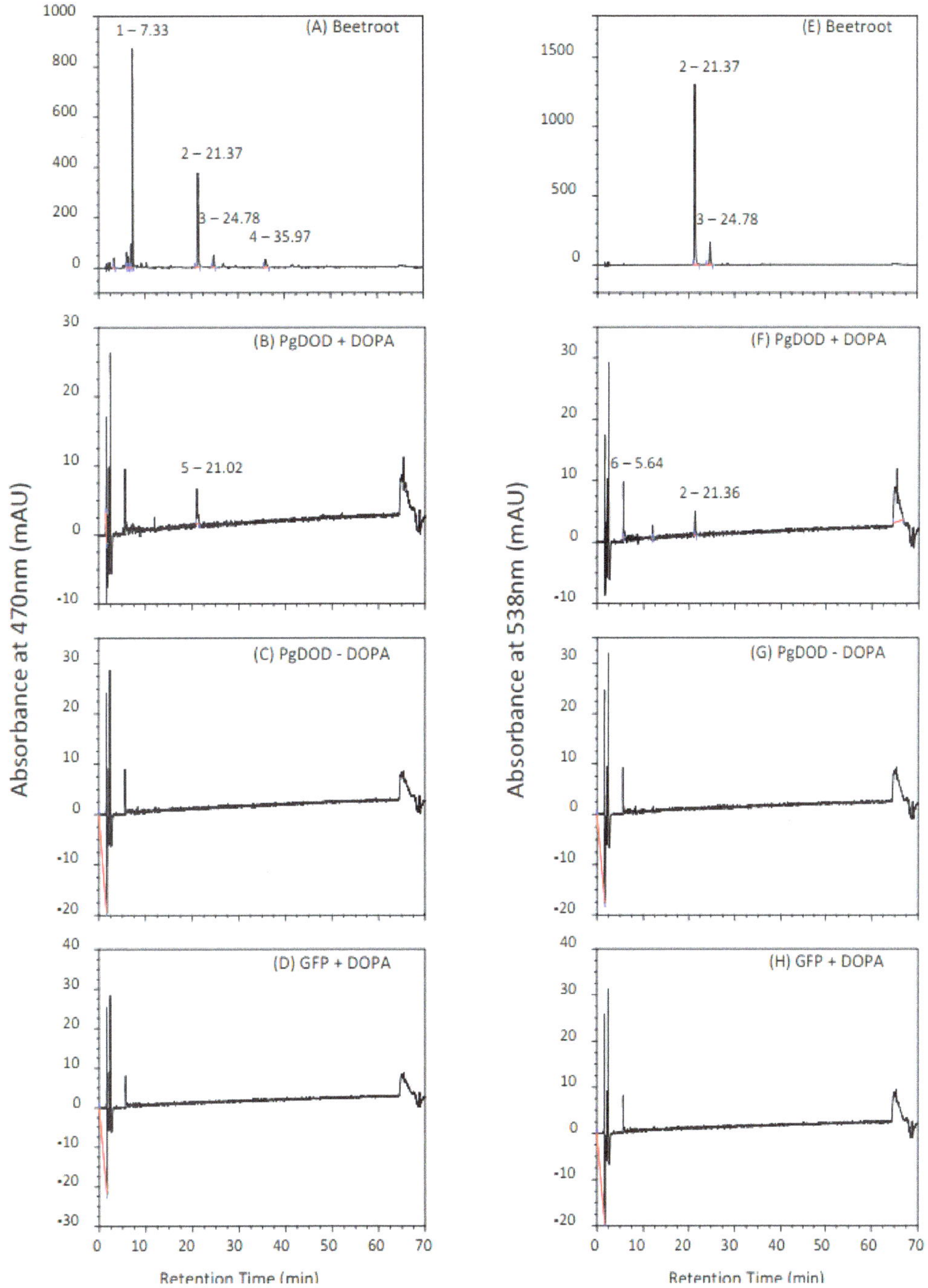

Figure 5: HPLC analysis of pigments produced in antirrhinum petals transiently transformed with *35S: PgDOD* or *35S: GFP*. HPLC analysis was conducted on the *35S: PgDOD* and *35S: GFP*antirrhinum petal material shown in Figure 3. Absorbance was monitored at 470 nm for betaxanthins or 538 nm for betacyanins. Chromatograms are for extracts from beetroot root (A and E), 35S: PgDOD antirrhinum infiltrated with L-DOPA (B and F), 35S: PgDOD antirrhinum without L-DOPA infiltration (C and G), 35S: GFP antirrhinum infiltrated with L-DOPA (D and H). The retention times of the major peaks are shown, and these were tentatively identified as per Table 1.

Table 1: Retention times and spectral maxima of the major betalain pigments detected in beetroot tissue (Peaks 1 to 3) or antirrhinum petal tissue bombarded with *35S: PgDOD* and infiltrated with L-DOPA (Peaks 4 and 5).

Peak[a]	Betalain	Retention Time (min)	λ_{max} (nm)
1	Vulgaxanthin I	7.33	259, 469
2	Betanin	21.37	269, 534
3	Isobetanin	24.78	269, 534
4	Unknown betaxanthin	35.97	265, 464
5	Unknown betaxanthin[b]	21.02	241, 470

HPLC absorbance traces are shown in Figure 4

[a]The peak present at 5.64 min in antirrhinum tissue was present in all samples, both control and transgenic; its spectral data is not consistent with that of betalain pigments therefore it was excluded from the analysis.

[b]Possibly dopaxanthin

Antirrhinum can produce yellow pigments in the petal face and throat naturally. These pigments are aurones and a product of the flavonoid pathway. In addition to the spectral data (aurones have spectra maxima in the range 390-430 nm), there is other evidence that the yellow pigments observed following *35S: PgDOD*bombardment and L-DOPA feeding are unlikely to be aurones; aurones are not present normally in the dorsal petals of the antirrhinum line used (JI19), and no yellow pigments were observed without infiltration with L-DOPA.

Despite a lack of the red colouration seen following bomardment, the *35S: PgDOD* antirrhinum samples analysed by HPLC did indeed contain small amounts of betacyanin (betanin; Figure 5F). Also, the presence of betacyanin is inferred in the *35S: AmDOD* expressing potato cells, given the red colouration of some foci. How betacyanin production can occur in non-betalain species through the introduced DOD acting on the supplied L-DOPA is not clear. The product of DOD action on L-DOPA is betalamic acid (Figure 1). Betaxanthins can be produced by spontaneous reactions of the DOD reaction end-product, betalamic acid, with amino acids or amines. However, betacyanin formation requires *cyclo*-DOPA, the formation of which from L-DOPA by oxidation has been attributed to the activity of a tyrosinase [6]. As tyrosinases are frequently present in plant cells it is possible that an endogenous enzyme with activity on L-DOPA occurs in most non-betalain producing species. Alternatively, betacyanin could be formed from endogenous tyrosinase activity on dopaxanthin [12]. The betacyanin detected in antirrhinum petals was likely betanin, an *O*-glycosylated betacyanin. This indicates that endogenous glycosyltransferases can act on the novel betacyanin substrates. Two *O*-glucosyltransferases have been characterized with activity on betanidin [19,20]. Both show sequence similarity to the *O*-glycosyltransferases involved in anthocyanin biosynthesis, and indeed, the betanidin 5-*O*-glucosyltransferase has activity with both betacyanins and flavonoids [20]. Thus, it may be the case that the endogenous flavonoid *O*-glucosyltransferases of antirrhinum are also able to act on betanidin (and/or *cyclo*-DOPA).

Biolistic transformation of both potato and antirrhinum with either *35S: PgDOD* or *35S: AmDOD* resulted in muli-celled foci producing betalains. It is unlikely that this would be due to movement of the DOD enzyme between cells as, similar to GFP (Figure 3), it is too large for passive intercellular movement. As the final betalain pigments accumulate within the vacuole, this suggests that some of the precursors migrate between cells. This is similar to the results of Mueller *et al.* [17] when they biolistically introduced *AmDOD* into different *P. grandiflora* mutant backgrounds. Single coloured cells seen within 18 h after bombardment developed into multi-cell foci by 48 h after bombardment [17]. It was suggested that betalains could have diffused through plasmodesmata to neighbouring cells. However, in contrast to these results, when complementation of the *P. grandiflora dod* mutant was conducted with PgDOD, only single cell foci occurred [11]. Furthermore, cell-specific betalain production is commonly observed in plants, such as in the epidermal cells of petals. In the case of our results, it is possible that high L-DOPA levels produced from tissue feeding allowed the movement of betalamic acid not just to the vacuole but also to neighbouring cells.

Betalain Pigment Production by Stable Transformation of Arabidopsis

Stably transformed arabidopsis plants were produced through *Agrobacterium*-mediated transformation with*35S: AmDOD*. T2 generation seedlings were checked for expression of the *35S: AmDOD* transgene (Additional file 1) and selected lines examined for their potential to produce betalains when fed with L-DOPA. When whole seedlings were fed L-DOPA, novel pigment production was visible within 12 h after feeding (Figure 6A-C), including in the etiolated root tissues and the hypocotyls. The colour ranged from pale yellow through to orange and dark orange-red. Under blue light, the pigments showed the green autofluorescence characteristic of betaxanthins (Figure 6D, E). When detached inflorescences from mature plants were fed L-DOPA pale yellow pigmentation was seen in all tissues 24 h after feeding, including the stem, petals and siliques, and this was accompanied by strong autofluorescence (Figure 7). *35S: AmDOD*seedlings or inflorescences not fed L-DOPA did not produce visible pigments after treatment and did not show significant autofluorescence under blue light (Figure 6F, G and Figure 7).

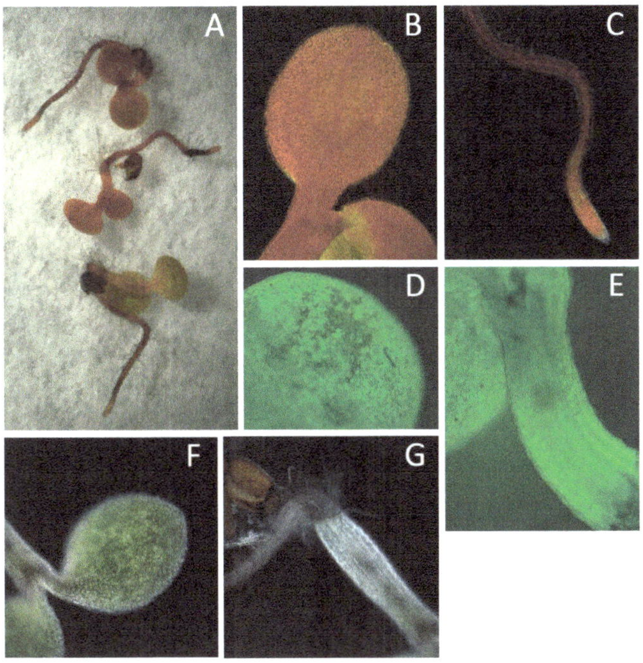

Figure 6: Betaxanthin production in arabidopsis seedlings stably transformed with *35S: AmDOD*. *35S: AmDOD* seedlings were grown on moistened filter disks either with (**A** to **E**) or without (**F** and**G**) 10 mM L-DOPA. Those fed L-DOPA accumulated yellow to orange/red pigments (A to C) and under blue light showed the autofluorescence typical of betaxanthins, as illustrated for the cotyledon (D) and hypocotyl (E). The seedlings not fed L-DOPA did not produce coloured pigments in either their cotyledons (F) or hypocotyls (G).

Figure 7: Pigment production in inflorescences of arabidopsis plants stably transformed with *35S: AmDOD*. Inflorescences from *35S: AmDOD*arabidopsis line 6 were infiltrated with either water or 10 mM L-DOPA for 24 h and examined under white light (upper panel) or blue light (lower panel). Those fed L-DOPA accumulated yellow to orange pigments and under blue light showed the autofluorescence typical of betaxanthins. The inflorescences not fed L-DOPA did not produce visible coloured pigments and did not show autofluorescence.

HPLC analysis was used to examine the nature of the pigments produced following L-DOPA feeding of the*35S: AmDOD* arabidopsis. Seedlings were fed L-DOPA and the entire seedling sampled for chemical analysis. HPLC analysis revealed several peaks present in the L-DOPA fed *35S: AmDOD* tissue but not in the control tissue (Figure 8). Four of these peaks were present in sufficient quantity to confirm that their spectral data and retention times were those characteristic of betaxanthins (Table 2).

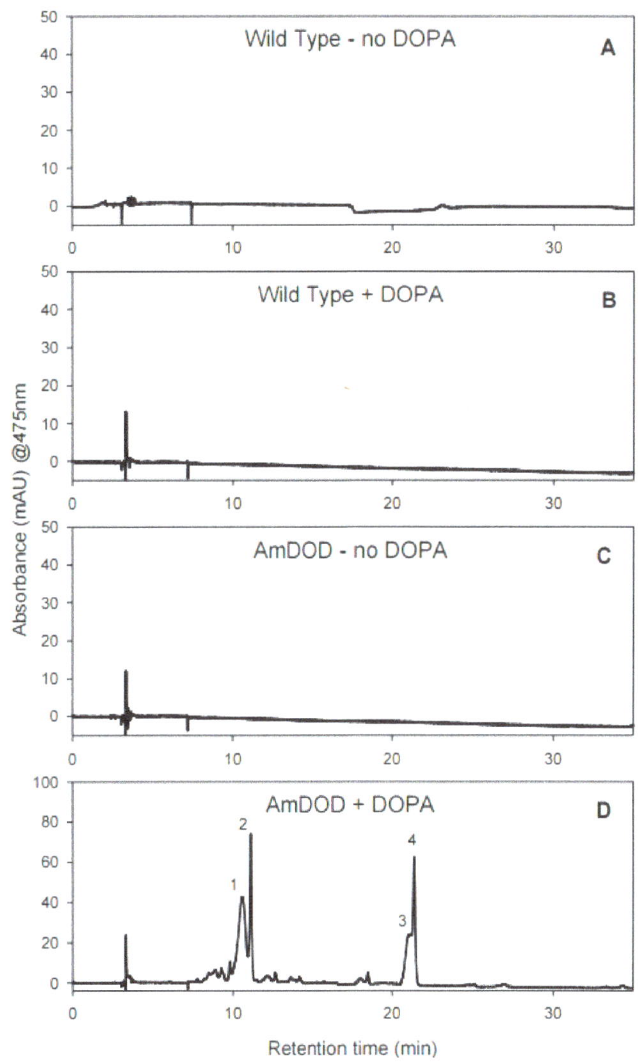

Figure 8: HPLC analysis of pigments produced in arabidopsis seedlings stably transformed with *35S: AmDOD*. Seedlings, representatives of which are shown in Figure 6, were grown on moistened filter paper with or without the addition of 10 mM L-DOPA and the pigments present examined using HPLC. Samples from wild-type (non-transgenic) arabidopsis and from beetroot root were included for comparison. Absorbance was monitored at 475 nm for betaxanthins. The retention times of the major peaks are shown, and these were identified as betaxanthins per Table 2.

Table 2: Retention times and spectral maxima of the major peaks detected by HPLC using samples from *35S: AmDOD* plants fed 10 mM L-DOPA

Peak	Betalain[a]	Retention Time (min)	λ_{max} (nm)
1	Vulgaxanthin I[b]	10.4	268, 468
2	Unknown betaxanthin	11.6	262, 470
3	Unknown betaxanthin	20.4	264, 472
4	Unknown betaxanthin	21.7	262, 468

[a]Spectral data for all four peaks was generally consistent with data reported previously for betaxanthins, but no specific betaxanthin assignments were made.

[b]Peak one has a retention time and spectrum consistent with that of Vulgaxanthin I, as reported for extracts from beetroot [14] and as compared against a beetroot extract run in the same experimental set.

The presence of betaxanthins matched the strong autofluorescence observed in the seedlings under blue light. The failure to detect the presence of betacyanin was a surprise given the orange/red colouration of the seedlings and the ability of *DOD* to confer both betaxanthin and betacyanin production when transiently expressed in antirrhinum petals. In accordance with the HPLC results, while the seedlings did have an orange-red colour that would suggest betacyanin production, the autofluorescence indicated strong betaxanthin presence. Furthermore, the fed inflorescences off the mature plants did not have a red colouration. Thus it appears that a high level of betaxanthin production in the particular background colouration of the seedlings resulted in the orange-red colour. It is unlikely that anthocyanins are contributing to the orange-red colouration of seedlings, as L-DOPA treatment of non-transgenic seedlings did not induce visible pigment production.

CONCLUSIONS

We report here methods to enable production of high-levels of betalains in the non-betalain, anthocyanin-producing, species antirrhinum, arabidopsis and potato, which represent both of the major clades of the eudicots. Betalains are not normally observed in anthocyanin-producing species, for unknown reasons.

The introduction of a single enzyme (DOD) and feeding of that enzyme's substrate (L-DOPA) were sufficient to induce both betaxanthin and betacyanin production. Although many steps in betalain production are suggested to be spontaneous, it is somewhat surprising that betacyanin production occurred. It indicates the background presence of an enzyme able to convert L-DOPA to *cyclo*-DOPA, or dopaxanthin to betacyanin/betanidin, and *O*-glycosyltransferases with activity against betanidin. Thus, it is likely that introduction of just two enzymes, DOD and the (currently uncharacterised) enzyme for conversion of tyrosine to L-DOPA, should be sufficient to confer betaxanthin and betacyanin production to anthocyanin-producing species. The requirement for few novel biosynthetic steps may have assisted in the initial evolution of this pathway in the Caryophyllales. A relatively simple mechanism for betalain biosynthesis would also support the multiple origins of the pathway in the Caryophyllales and in some fungi [7].

Given the extensive knowledge available on anthocyanin production and function in arabidopsis, the stably transformed *35S: AmDOD* arabidopsis plants should provide excellent material to study, for the first time, the physiological effects of having both betalains and anthocyanins in the same plant tissues. The large collection of mutant lines in arabidopsis should also allow aspects of betalain biosynthesis, such as vacuolar transport, to be examined.

METHODS

Plant and cell Culture Material

Antirrhinum majus plants were grown under standard glasshouse conditions in Palmerston North, New Zealand. The glasshouse was heated at 15°C and vented at 25°C, without supplementary lighting. Two antirrhinum lines were used. Transformation with *35S: PgDOD* used line JI19, which produces aurones and flavones but not anthocyanins in the petals, as it is lacking in flavanone 3-hydroxylase activity (homozygous*incolorataII*). 35S: AmDOD transformation used either JI19 or *rosea*^{dorsea}, which carries a mutation in the R2R3MYB gene *Rosea1* and lacks anthocyanin production in the adaxial epidermis of the petals [27].*Arabidopsis thaliana* transformation used the Colombia ecotype. Potato cell cultures were obtained by placing 2 cm^2 pieces of potato callus in 50 mL of modified liquid MS media [28] per cotton-plugged 250 mL flask. These cultures were incubated at room temperature (23-27°C) with shaking on an orbital shaker at 100 rpm in the dark. Subculturing was carried out every two weeks with 3 mL of the cell culture used to inoculate a fresh 50 mL of liquid media.

PCR Cloning of *P. Grandiflora* DOD cDNA and Vector Construction

A cDNA for the ORF of *DOD* was RT-PCR amplified from mRNA prepared from *P. grandiflora* betalain pigmented petals using primers designed to the published sequence [11]. The forward primer was 5'-AGTCAGAATCCATGGGTGTTGGGAAGGAA-3', with the first ATG matching the initiation ATG for DOD, and the reverse primer was 5'-AGTCATCTAGAATCATATGGAAGTGAACT-3', which incorporated an XbaI site. Standard PCR conditions were used with Taq Polymerase (New England Biolabs, Massachusetts, USA) and cycling parameters of 94°C for 4 min, then 30 cycles of 94°C for 30 s, 60°C for 30 s and 72°C for 1 min, with a final extension phase of 72°C for 7 min. The products were cloned into pGEM-T-Easy (Promega, Wisconsin, USA) and confirmed as *PgDOD* by DNA sequencing. The cDNA was then excised as an EcoRI/XbaI fragment and ligated into EcoRI/XbaI digested *pART7* [29] to from the vector pPN314. The *35S: AmDOD* construct for particle bombardment, containing the *A. muscaria* 4,5-DOPA dioxygenase cDNA driven by the CaMV35S promoter, was *pNcoDod* [17] (courtesy of Dr. Willibald Schliemann, Leibniz-Institute of Plant Biochemistry, Germany). The binary vector *pPN166* for transformation into arabidopsis was constructed by taking the CaMV35S-cDNA region from *pNcoDod* as a PvuII fragment and cloning it into the NotI site (after flushing by end-filling) in the binary vector *pMLBART*(a gift from Dr Bart Janssen, Plant & Food Research). The *35S: GFP* construct was pPN93. Vectors were verified by restriction fragment analysis and/or DNA sequencing.

Biolistic Transformation

Antirrhinum particle bombardment experiments were performed as described in Shang *et al.* [30], with the following variations; the pressure setting was 300 or 400 kPa, the shooting distance 11-12 cm, and petals were bombarded twice. The final DNA concentration for DOD constructs was 1 or 2 µg DNA per mg of 1.0 µm gold particles. Controls included gold particles alone (no DNA) and *35S: GFP* (added at 0.4 µg DNA per mg of 1.0 µM gold particles). Prior to bombardment the petals were surface sterilised by immersion for 15 min in 10% (v/v) bleach containing a few drops of Tween-20, followed by three rinses in sterile water. The adaxial surface of the dorsal petals was bombarded. After bombardment, the plant materials were then cultured on 1/2 MS medium under 20 - 50 µmol $m^{-2} s^{-1}$ light from Osram 36 W grolux fluorescent tubes (16 h photoperiod) at 25°C. At least two flowers were used for each construct per experiment, and each experiment was repeated at least twice.

Potato cell suspension cultures for particle bombardment-mediated transformation were prepared by filtering 3 mL of culture onto sterile filter paper and sub-culturing them on a solid media in tubs for 48 h prior to transformation. The biolistic parameters were the same as those used for the antirrhinum petals with exception that they were only bombarded once. A sterilised metal grid was placed over the cells, on top of the culture tub, to prevent displacement of the cells from the helium in-flow.

Arabidopsis Transformation

The floral spray method developed by Clough and Bent [31] was used for the transformation of arabidopsis. Seeds were harvested from the *Agrobacterium*-inoculated plants and transgenics selected by spraying germinating seedlings with glufosinate herbicide (Basta, Bayer Crop Sciences). Positive transformants were grown through to production of T2 seeds following self-pollination.

L-DOPA FEEDING

Infiltration of antirrhinum petals with L-DOPA was carried out 24 h after biolistic transformation. The petals were placed in 10 mM L-DOPA solution and a brief vacuum (30-60 s) applied, until the solution boiled vigorously and the petals became translucent. Following infiltration excess solution was blotted off and the petals incubated on 1/2 MS medium under the same conditions as post-bombardment. Control petals were handled in the same manner but with sterile water substituted for L-DOPA. Petals were observed for betalain production after a further 24 h. Transformed potato cell cultures were fed with 10 mM L-DOPA solution by dispensing 1 mL of the solution on to the filter disks supporting the cells. The L-DOPA solution was applied to half of the samples immediately after transformation.

Germinated arabidopsis seedlings were fed with 10 mM L-DOPA by transfer of the filter paper containing the seedlings on to a plate containing L-DOPA solution. Seedlings were observed for betalain pigment production after a further 12 h. For feeding of inflorescences, six inflorescences were collected from *35S: AmDOD* line 6 and cut to 1 cm long. Three inflorescences were immersed in 10 mM L-DOPA solution and the other three immersed in water, and both sets were left for 2 min. The inflorescences were then placed into 1.5 ml Eppendorf tubes with either 0.5 ml of 10 mM L-DOPA solution or water, and the tubes left open on the bench for 24 h.

PCR Analysis

RT- PCR analysis for *DOD* transgene expression in *35S: AmDOD* arabidopsis plants used total RNA extracted from leaf tissue following the RNeasy protocol of the Qiagen RNeasy mini kit. Five rosette leaves were sampled from each of six individual T2 lines, as well as a non-transgenic wild type line. Following DNAse I treatment 500 ng of RNA from each sample was used to generate cDNA using the Roche Transcriptor First Strand cDNA Synthesis Kit. PCR used Taq polymerase (Roche, New Zealand) and the following primers: DODF1 5'-CATACTACCATGTCCACCAAG-3', DODF2 5'-AGCACTGCTTCTATATCGTC-3', Act2S 5'-TCCCTCAGCACATTCCAGCAGAT-3', Act2AS 5'-AACGATTCCTGGACCTGCCTCATC-3'. The Actin primers correspond to the arabidopsis *Actin2* gene (AT3G18780). The thermocycling conditions were 94°C for 2 min and 25 cycles of 94°C for 20 s, 55°C for 30 s and 72°C for 30 s. The PCR products were separated on a 1% (w/v) TBE agarose gel containing ethidium bromide and visualised using UV-illumination. PCR products were also cloned and sequenced to confirm that the target PCR product was being amplified.

HPLC Analysis

For arabidopsis analysis whole seedlings were extracted. For antirrhinum analysis, to ensure that similar petal regions were being compared, the ridge region of the biolistically transformed petals was excised away from the remainder of the petals for chemical analysis. The ridge is a distinct raised area that divides the lobes and throat of the dorsal petals. Ridge tissue from between three and six petals was pooled to give total sample fresh weights of 60 to 300 mg. Absorbance at 470 nm and 538 nm was used for detection of betaxanthins or betacyanins, respectively. Beetroot (20 mg freeze dried sample of red root) was used as a standard source of betalains, and strong peaks with the expected retention times and spectra data were observed.

Each sample was extracted three times in 1 ml of 80% (v/v) methanol containing 50 mM sodium ascorbate, as described in

Schliemann *et al.* [14], with one overnight extraction at 4°C. The samples were centrifuged for 4 min at 10,000 rpm, the supernatant removed and the pellet re-extracted in the next 1 ml of 80% methanol with ascorbate. The supernatants once removed, were combined to give the crude extract. The extract was dried in vacuo on a Savant SC210 Speedvac to near dryness, then freeze dried overnight to complete dryness. Extracts were resuspended in MilliQ water and made up to a final volume of 500 μl. The extracts were syringe filtered through a 0.45 um nylon filter as per Svenson *et al.* [26] and the pigments analysed by high performance liquid chromatography (HPLC). The analysis of the antirrhinum samples used a Dionex 3000 Ultimate solvent delivery system with a Phenomenex Luna (5 μm, 150 × 4.6 mm) RP-18 endcapped column (column temperature 30°C) and a Dionex 3000 Diode Array Detector (DAD). Elution (1 ml min^{-1}) was performed using a solvent system comprising solvent A [1.0% formic acid in water] and solvent B [80% acetonitrile in water] and a linear gradient starting with 100% A, decreasing to 80% A at 62 mins, and then a linear gradient to 100% B at 67 mins, remaining at 100% B for a further 3 min, then returning to initial conditions. Betaxanthins were detected at 470 nm and betacyanins at 538 nm [25]. The analysis of the arabidopsis samples used the solvent system of Schliemann *et al.* [14]. It was conducted on a Hewlett Packard HP 1100 with two Merck Chromolith analytical columns and a C-18 guard column.

Authors' Contributions

NNH conducted the AmDOD biolistic experiments in antirrhinum, produced and contributed to the analysis of the arabidopsis plants, and contributed to writing of the manuscript; JJ made pPN314 and conducted the biolistic experiments in potato; KMD conducted the PgDOD biolistic experiments in antirrhinum and wrote the main manuscript draft; DHL conducted the HPLC analysis of biolistically transformed material and contributed to the manuscript draft; PEJ contributed to supervision of JJ and NNH and project design; SCD assisted with cell culture experiments and supervision of JJ; KEC conducted PCR analysis and inflorescence feeding for the arabidopsis plants; KSG contributed to supervision of KEC. KES conceived and coordinated the study, supervised NNH and JJ, and contributed to analysis of the results and writing of the manuscript. All authors read and approved the final manuscript.

Supplementary Material

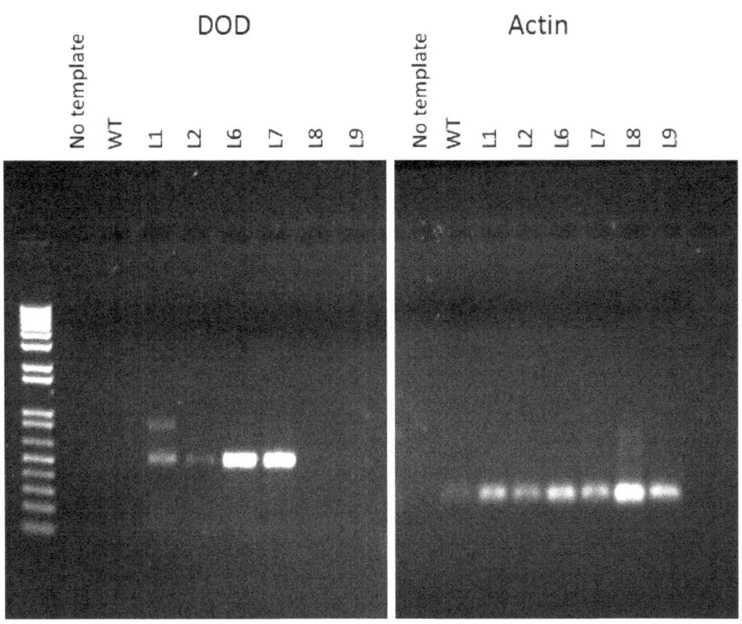

Additional file 1: PCR analysis for *DOD* transgene expression in *35S: AmDOD* arabidopsis plants. Total RNA was extracted from six lines of *35S: AmDOD* arabidopsis, as well as a non-transgenic wild type line, and analysed for *DOD* transcript levels using RT-PCR. PCR primers for an endogenous actin gene were used as a positive control for RNA/cDNA integrity. PCR products were separated on a 1% (w/v) agarose gel containing ethidium bromide and visualised using UV-illumination. (PPT 356 kb).

Acknowledgements

We thank Ian King and Julie Ryan for care of plants, Steve Arathoon and John Harris for technical assistance, Willibald Schliemann for the *A. muscaria* 4,5-DOPA dioxygenase cDNA vector, Andrew Gleave for *pART7*, Bart Janssen for *pMLBART*, Simon Coupe for *pPN93*, and the New Zealand Foundation for Research, Science, and Technology contract C02X0805 for supporting KES, SCD and KMD. NNH thanks The Todd Foundation, The Freemasons, NZ Federation of Graduate Women and the J. Skipworth Scholarship for their financial support. KEC thanks Victoria University for her PhD scholarship grant.

REFERENCES

1. Grotewold E. The genetics and biochemistry of floral pigments. Ann Rev Plant Biol. 2006;57:761–780. doi: 10.1146/annurev.arplant.57.032905.105248.

2. Tanaka Y, Sasaki N, Ohmiya A. Biosynthesis of plant pigments: anthocyanins, betalains and carotenoids. Plant J. 2008;54:733–749. doi: 10.1111/j.1365-313X.2008.03447.x.
3. Gould KS, Lister C. In: Flavonoids: Chemistry, Biochemistry and Applications. Andersen ØM, Markham KR, editor. CRC Press, Boca Raton; 2006. Flavonoid functions in plants; pp. 397–441.
4. Davies KM. In: Plant Pigments and their Manipulation. Davies KM, editor. Blackwell Publishing, Oxford; 2004. An introduction to plant pigments in biology and commerce; pp. 1–22.
5. Strack D, Vogt T, Schliemann W. Recent advances in betalain research. Phytochemistry. 2003;62:247–269. doi: 10.1016/S0031-9422(02)00564-2.
6. Zrÿd J-P, Christinet L. In: Plant Pigments and their Manipulation. Davies KM, editor. Oxford: Blackwell Publishing; 2004. Betalains; pp. 185–213.
7. Brockington SF, Walker RH, Glover BJ, Soltis PS, Soltis DE. Complex pigment evolution in the Caryophyllales. New Phytol. 2011;190:854–864. doi: 10.1111/j.1469-8137.2011.03687.x.
8. Mabry T. The betacyanins, a new class of red violet pigments, and their phylogenetic significance.New York: Roland Press; 1964.
9. Gandía-Herrero F, García-Carmona F, Escribano J. Fluorescent pigments: new perspectives in betalain research and applications. Food Res Intl. 2005;38:879–884. doi: 10.1016/j.foodres.2005.01.012.
10. Gandía-Herrero F, Jiménez-Atiénzar M, Cabanes J, Escribano J, García-Carmona F. Fluorescence detection of tyrosinase activity on dopamine-betaxanthin purified from *Portulaca oleracea* (Common Purslane) flowers. J Agric Food Chem. 2009;57:2523–2528. doi: 10.1021/jf803608x.
11. Christinet L, Burdet F, Zaiko M, Hinz U, Zrÿd JP. Characterization and functional identification of a novel plant 4,5-extradiol dioxygenase involved in betalain pigment biosynthesis in *Portulaca grandiflora*. Plant Physiol. 2004;134:265–274. doi: 10.1104/pp.103.031914.
12. Gandía-Herrero F, Escribano J, García-Carmona F. Betaxanthins as substrates for tyrosinase An approach to the role of tyrosinase in the biosynthetic pathway of betalains. Plant Physiol.2005;138:421–432. doi: 10.1104/pp.104.057992.
13. Steiner U, Schliemann W, Böhm H, Strack D. Tyrosinase involved in betalain biosynthesis of higher plants. Planta. 1999;208:114–124. doi: 10.1007/s004250050541.
14. Schliemann W, Kobayashi N, Strack D. The decisive step in betaxanthin biosynthesis is a spontaneous reaction. Plant Physiol. 1999;119:1217–1232. doi: 10.1104/pp.119.4.1217.
15. Sasaki N, Adachi T, Koda T, Ozeki Y. Detection of UDP-glucose *cyclo*-DOPA 5-*O*-glucosyltransferase activity in four o'clocks (*Mirabilis jalapa* L.) FEBS Lett. 2004;568:159–162. doi: 10.1016/j.febslet.2004.04.097.
16. Sasaki N, Wada K, Koda T, Kasahara K, Adachi T. Isolation and characterization of cDNAs encoding an enzyme with glucosyltransferase activity for *cyclo*-DOPA from Four O'clocks and Feather Cockscombs. Plant Cell Physiol. 2005;46:666–670. doi: 10.1093/pcp/pci064.
17. Mueller LA, Hinz U, Uze M, Sautter C, Zrÿd JP. Biochemical complementation of the betalain biosynthetic pathway in *Portulaca grandiflora* by a fungal 3,4-dihydroxyphenylalanine dioxygenase. Planta. 1997;203:260–263. doi: 10.1007/s004250050190.
18. Mueller LA, Hinz U, Zrÿd JP. The formation of betalamic acid and muscaflavin by recombinant DOPA-dioxygenase from *Amanita*. Phytochemistry. 1997;44:567–569. doi: 10.1016/S0031-9422(96)00625-5.
19. Vogt T, Grimm R, Strack D. Cloning and expression of a cDNA encoding betanidin 5-O-glucosyltransferase, a betanidin- and flavonoid-specific enzyme with high homology to inducible glucosyltransferases from the Solanaceae. Plant J. 1999;19:509–519. doi: 10.1046/j.1365-313X.1999.00540.x.
20. Vogt T, Zimmermann E, Grimm R, Meyer M, Strack D. Are the characteristics of betanidin glucosyltransferases from cell-suspension cultures of *Dorotheanthus bellidiformis* indicative of their phylogenetic relationship with flavonoid glucosyltransferases? Planta. 1997;203:349–361. doi: 10.1007/s004250050201.
21. Shimada S, Inoue YT, Sakuta M. Anthocyanidin synthase in non-anthocyanin-producing Caryophyllales species. Plant J. 2005;58:950–959.
22. Shimada S, Otsuki H, Sakuta M. Transcriptional control of anthocyanin biosynthetic genes in the Caryophyllales. J Exp Bot. 2007;58:957–967. doi: 10.1093/jxb/erl256.
23. Shimada S, Takahashi K, Sato Y, Sakuta M. Dihydroflavonol 4-reductase cDNA from non-anthocyanin producing species in the Caryophyllales. Plant Cell Physiol. 2004;45:1290–1298. doi: 10.1093/pcp/pch156.
24. Stafford HA. Anthocyanins and betalains: evolution of the mutually exclusive pathways. Plant Sci.1994;101:91–98. doi: 10.1016/0168-9452(94)90244-5.
25. Kugler F, Stintzing FC, Carle R. Identification of betalains from petioles of differently colored Swiss Chard (*Beta vulgaris* L. ssp. cicla L. Alef. Cv. Bright Lights) by High-Performance Liquid Chromatography-Electrospray Ionization Mass Spectrometry. J Agr Food Chem. 2004;52:2975–2981. doi: 10.1021/jf035491w.
26. Svenson J, Smallfield BM, Joyce NI, Sansom CE, Perry NB. Betalains in red and yellow varieties of the andean tuber crop ulluco (*Ullucus tuberosus*) J Agr Food Chem. 2008;56:7730–7737. doi: 10.1021/jf8012053.
27. Schwinn K, Venail J, Shang Y, Mackay S, Alm V, Butelli E, Oyama R, Bailey P, Davies K, Martin C. A small family of *MYB*-regulatory genes controls floral pigmentation intensity and patterning in the genus *Antirrhinum*. Plant Cell. 2006;18:831–851. doi: 10.1105/tpc.105.039255.
28. Murashige T, Skoog F. A revised medium for rapid

growth and bioassays with tobacco tissue cultures. Physiol Plantarum. 1962;15:743–747.

29. Gleave AP. A versatile binary vector system with a T-DNA organizational-structure conducive to efficient integration of cloned DNA into the plant genome. Plant Mol Biol. 1992;20:1203–1207. doi: 10.1007/BF00028910.

30. Shang Y, Schwinn KE, Hunter DA, Waugh TL, Bennett MJ, Pathirana NN, Brummell DA, Jameson PE, Davies KM. Methods for transient assay of gene function in floral tissues. Plant Methods.2007;3:1. doi: 10.1186/1746-4811-3-1.

31. Clough SJ, Bent AF. Floral dip: a simplified method for *Agrobacterium*-mediated transformation of*Arabidopsis thaliana*. Plant J. 1998;16:735–743. doi: 10.1046/j.1365-313x.1998.00343.x.

CHAPTER 4

PHYLLOBILINS – THE ABUNDANT BILIN-TYPE TETRAPYRROLIC CATABOLITES OF THE GREEN PLANT PIGMENT CHLOROPHYLL

Bernhard Kräutler

Institute of Organic Chemistry and Centre of Molecular Biosciences, University of Innsbruck, Innrain 80/82, A-6020 Innsbruck, Austria.

INTRODUCTION

Have you never wondered what happens to chlorophyll, when the leaves of deciduous plants de-green and display their fall colours, or when fruit ripen and turn to an appealing yellow, red or blue (Fig. 1)? The quest of finding remains of the green plant pigment has, indeed, engaged natural scientists' interest for a considerable time. Generally, their search was guided by the idea that the breakdown products were assumed to be coloured.[1] Their failure in actually identifying chlorophyll degradation products was puzzling, as chlorophyll could not seriously be considered to disappear 'without leaving a trace'. In fact, the global formation of chlorophyll breakdown products has been estimated to amount to about 1000 million tons, each year,[2] and the seasonal disappearance of chlorophyll can be observed (and can be studied) from outer space.[3]

Figure 1: Characteristic colour changes observed in fall leaves and ripening fruit are a visual sign of chlorophyll breakdown (reproduced from *Chem. Biol.*, 2008, **3**, B79).

About twenty-five years ago, in de-greening leaves, a colourless tetrapyrrole was identified as the first non-green catabolite of chlorophyll (Chl).[4] This colourless compound, now named*Hv*-NCC-1, was called a 'rusty pigment', originally, as it easily oxidized with formation of coloured product mixtures.[4,5] When its structure could be established

Citation: Kräutler, B. 2014. Phyllobilins – the abundant bilin-type tetrapyrrolic catabolites of the green plant pigment chlorophyll. Chem. Soc. Rev., 2014, 43, 6227-6238. DOI: 10.1039/C4CS00079J (Tutorial Review).

as a genuine Chl-breakdown product, it was classified as a 'nonfluorescent' Chl-catabolite (NCC), as it displayed no fluorescence (or other apparent photo-activity, see Fig. 2).[4] Later studies have confirmed the broad relevance of the tetrapyrrolic NCCs as products of Chl-breakdown in higher plants.[3,6] The original source of NCCs in specific plants was incorporated into their provisional names, *e.g. Hv*-NCC-1 for the most polar NCC from barley (*Hordeum vulgare*, see Fig. 2). Nowadays, over a dozen NCCs with different chemical structures are known, which are linear tetrapyrroles derived from Chl (for a list, see ESI,‡ Table S1).[6] As will be delineated further below, NCCs are now classified as 1-formyl-19-oxo-phyllobilanes, as they are remarkably similar to bilanes,[6]a well known class of natural linear tetrapyrroles.[7]

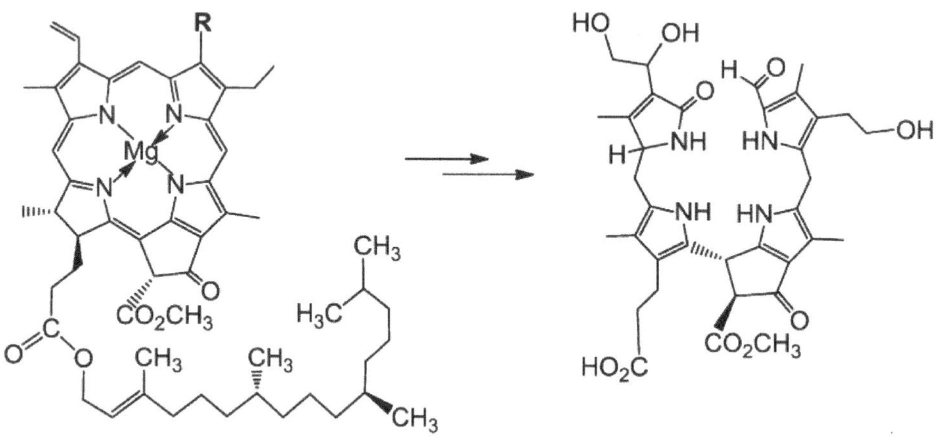

chlorophyll *a* (Chl *a*, R = CH_3)
chlorophyll *b* (Chl *b*, R = HC=O)

Hv-NCC-1

Figure 2: 'Rusty pigment' from senescent leaves of barley (*Hordeum vulgare*), later named *Hv*-NCC-1, was identified as the first non-green Chl-catabolite.[4,6]

Breakdown of Chl is the visible sign of senescence and cell death in leaves and vegetables, as well as of ripening of fruit.[8] As we now know, this process degrades Chl to a variety of linear tetrapyrroles, classified as phyllobilins.[6] These Chl-catabolites, in turn, are 'biomarkers' for senescence in higher plants. In this review, most representative formulas of the phyllobilins are drawn in a pseudo-cyclic fashion (see Fig. 2 and 3), which helps to make visual structure–correlations with the macro-cyclic precursors (Chls, pheophorbide *a*). However, the saturated linkages between the 5-membered heterocyclic rings show high conformational flexibility: there are three saturated linkages in phyllobilanes (see Fig. 3), two in phyllobilenes,*etc.* In the cases of unsaturated linkages the *E*/*Z*-geometry is defined in the formulas of the corresponding Chl-catabolites, as shown below (see *e.g.*Fig. 8 and ESI,‡ Fig. S2).

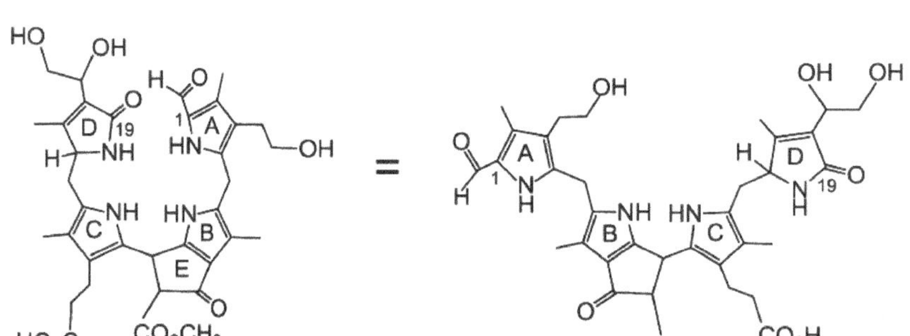

Hv-NCC-1

Figure 3: Constitutional formula of Hv-NCC-1 in 'pseudo-cyclic' (left) and 'extended' versions (right).

Considering the massive amount of Chl broken down each year on Earth, phyllobilins are also an interesting class of natural products from ecological and phytochemical points of view.[9]Systematic isolation from a variety of plant sources, and subsequent structural work, have revealed the basic chemical nature of phyllobilins.[6] They are structurally related to the much better known and biologically important bilins,[7] which originate from heme-breakdown.[10] In spite of their importance in higher plants, the linear tetrapyrroles derived from Chl-breakdown are remarkable newcomers to the area of the heterocyclic natural products.[9]

PHYLLOBILINS – A LONG OVERLOOKED CLASS OF BILIN-TYPE TETRAPYRROLIC HETEROCYCLES

Phyllobilins and key Enzymes in the Common 'Early' Phase of Chlorophyll Breakdown

Structure elucidation of phyllobilins has provided a solid basis for detailed studies of the biochemical processes that constitute the Chl-breakdown pathway, which undergoes its 'early' phase in senescent chloroplasts.[11] In these plastids, the Chls are degreened rapidly in a strictly regulated process to colourless, blue fluorescent Chl-catabolites (FCCs), which do not accumulate in senescent leaves but exist only fleetingly (see below). Along this path, the green plant pigments, Chls *a* and *b*, are first degraded to pheophorbide *a* (Pheo *a*). The macrocycle of the green Pheo *a* is cut open at the 'northern' *meso*-position by pheophorbide a oxygenase (PaO), an oxygen-dependent monooxygenase, which is highly active in senescent leaves.[12]PaO furnishes (an enzyme bound form of) the red Chl-catabolite (RCC), a linear tetrapyrrole. This 1-formyl-19-oxophyllobilin is the progenitor of all phyllobilins formed 'later' during Chl-breakdown. Thus, PaO is considered the key enzyme of the largely common path of Chl-breakdown in higher plants,[12] named the 'phyllobilin/PaO' pathway.[6] However, since RCC, the product of the PaO reaction, remained bound to PaO, its detection in senescent plants was futile, at first. In fact, RCC was first prepared by chemical synthesis from Pheo *a*, and then it was also identified as breakdown intermediate in senescent leaves.[11]

In a type of metabolic channeling PaO-bound RCC is reduced directly at its C15-*meso*-position by RCC-reductase (RCCR), yielding a fluorescent Chl-catabolite (FCC). RCCR of*Arabidopsis thaliana* was expressed in a functional form[13] and its crystal structure (with RCC bound) was elucidated.[14] RCCR occurs in two stereo-divergent lines, which produce either the 'primary' FCC (pFCC) or its C16-epimer, *epi*-pFCC (see Fig. 4). It is a ferredoxin-dependent enzyme, homologous to the 'ferredoxin-dependent bilin reductases', which catalyze a range of reductive modifications of heme-derived bilins.[15] RCCR is presumed to reduce RCC *via* single electron/proton transfer steps,[14] for which electrochemical investigations with RCC provided a mechanistic model.[16] pFCC was first identified from an enzyme extract from senescent leaves of oil seed rape,[17]*epi*-pFCC from a related system using leaves of sweet pepper.[6] So far, the critical absolute configuration at C16 of FCCs remains unassigned. pFCC and related FCCs are colourless tetrapyrroles that luminesce blue (at 450 nm) when excited with UV-light (at about 360 nm), making the detection of minute amounts of FCCs possible.[3]

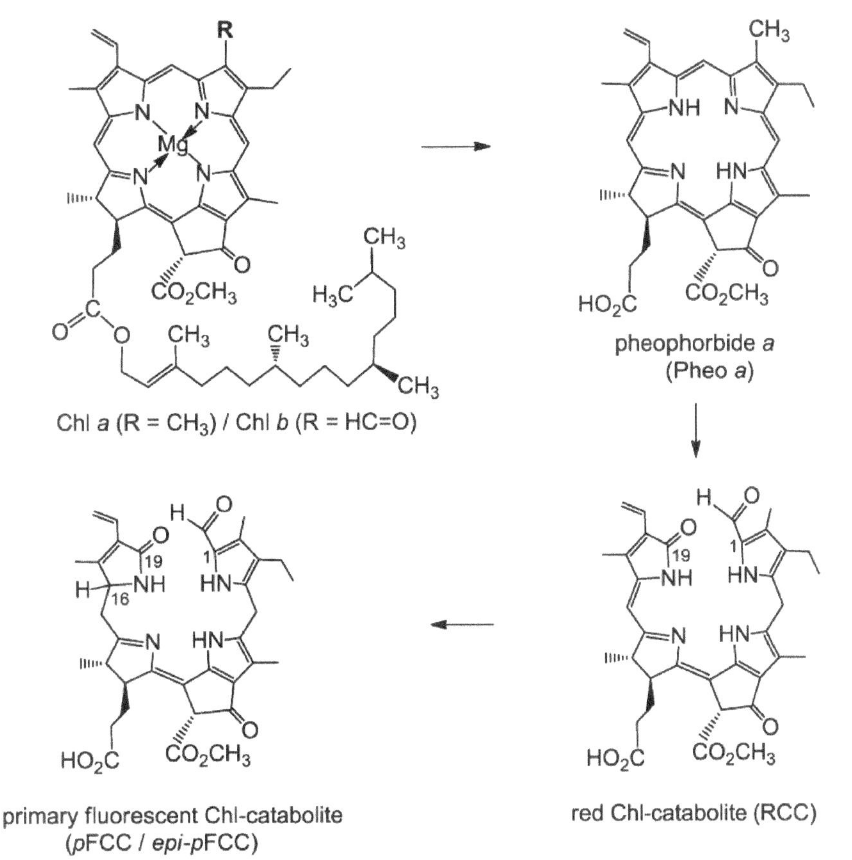

Figure 4: Chlorophyll *a*/*b* is degraded, in senescent chloroplasts, *via* pheophorbide *a* (Pheo *a*) and red Chl-catabolite (RCC) to the 'primary' fluorescent Chl-catabolite (pFCC/*epi*-pFCC), following the 'phyllobilin/PaO' path.[6]

Phyllobilins and key Enzymes in Divergent 'Later' Phases of Chlorophyll Breakdown

Typical FCCs are only short-lived metabolic precursors of the colourless, 'nonfluorescent' Chl-catabolites (NCCs). Only the 'hypermodified' FCCs (hmFCCs) are persistent, in which the propionic acid side chain carries an ester function.[16] hmFCCs are generated by still uncharacterized enzymes and accumulate in specific plant tissues (*e.g.* in ripe bananas, see below).[18]FCCs and NCCs are 1-formyl-19-oxophyllobilins, which are also classified here as type-I phyllobilins. FCCs are also precursors of dioxobilin-type NCCs (DNCCs),[19] which lack the 1-formyl group.[20,21] DNCCs (and their possible descendants, see below) are dioxobilin-type Chl-catabolites, or type-II phyllobilins (see Fig. 5).[6] Interestingly, DNCCs carry the same type of oxo-groups at their 1- and 19-positions as the heme-derived (dioxo)-bilins.

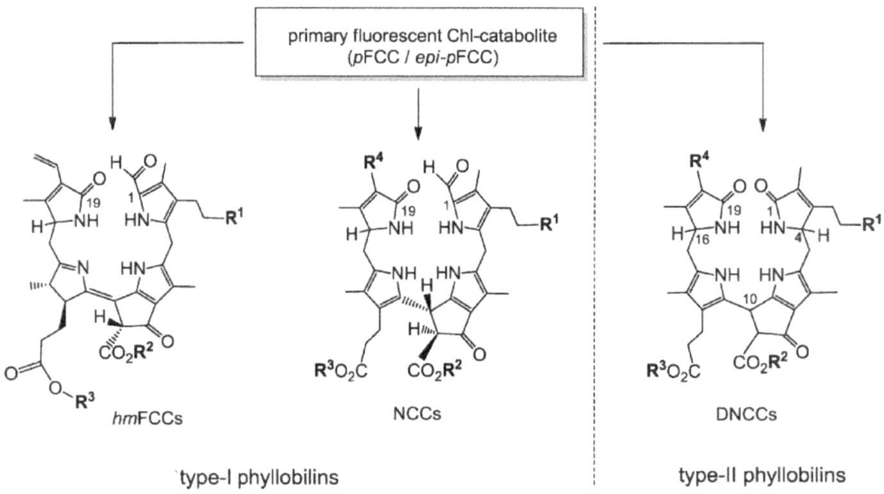

Figure 5: pFCC/*epi*-pFCC represent a three-way branching point of Chl-breakdown (see Fig. 4for formula of pFCC/*epi*-pFCC). Further transformations lead to other type-I phyllobilins, such as NCCs (*e.g. Hv*-NCC-1: R^1 = OH, R^2 = CH_3, R^3 = H, R^4 = CH(OH)–CH_2OH) and 'hyper-modified' FCCs (hmFCCs, *e.g. Mc*-FCC-56: R^1 = OH, R^2 = CH_3, R^3 = daucic acid), or to type-II phyllobilins, such as DNCCs (*e.g.*: *Ap*-DNCC: R^1 = OH, R^2 = CH_3, R^3 = H, R^4 = CH(OH)–CH_2OH).

NCCs were found to accumulate in a variety of leaves, where they localized to the vacuoles of senescent cells.[3] Spontaneous, acid-induced isomerization of FCCs to the corresponding NCCs has been proposed to take place in these acidic 'storage' vessels. The free carboxylic acid-function at the propionate side chain of typical FCCs proved to be the amazing prerequisite for achieving a significant rate and high stereo-selectivity of this isomerization at the pH typically found in vacuoles.[22] Based on a detailed stereo-chemical analysis of the FCC/NCC-isomerization, the newly formed asymmetric centre at C10 was deduced to have (*R*)-configuration. Interestingly, all structurally characterized NCCs were inferred to have the same configuration at this position, based on their similar circular dichroism (CD)-spectra.[9,16]

Over a dozen constitutionally different (colourless) NCCs have been identified, so far (see ESI,‡ Table S1).[6] They were decorated with one or several new functional groups (R^1 to R^4, see Fig. 5), when compared to the structures of the 'earlier' phyllobilins (RCC or 'primary' FCCs): a hydroxyl group at $C3^2$, which may carry a glucopyranosyl or a malonyl group, a free carboxylic acid function at $C8^2$, a 1,2-dihydroxyethyl group at C18, and malonylation at the sugar unit bound to $O3^3$.

In general, the new polar functions of the NCCs appear to increase their hydrophilicity. However, to be consistent with the general relevance of the previously delineated FCC/NCC-isomerization step, these NCCs require the existence of the correspondingly 'modified' FCCs (mFCCs). This, in turn, demands the activities of enzymes catalyzing the suggested modifications at the three FCC-side chains concerned (see Fig. 6 and Table S2, ESI‡). Thus, the structures of NCCs identified, so far, indicate the broad relevance of an FCC-hydroxylation at $C3^2$, and a possible subsequent glycosylation, or attachment of a malonyl group at the newly introduced $O3^3$, the hydrolysis of the methyl ester function at $C8^2$, a dihydroxylation of the vinyl group at C18, and malonylation at a sugar unit bound to $O3^3$. A few such enzymes are now known:[11] a methyl esterase was found in *Arabidopsis thaliana* and was shown to be localized to the cytosol, which hydrolyzes the methyl ester function bound at $C8^2$ of pFCC. A malonyl transferase attaches the malonyl unit to $O3^3$ (as was actually shown with several NCCs).[11]

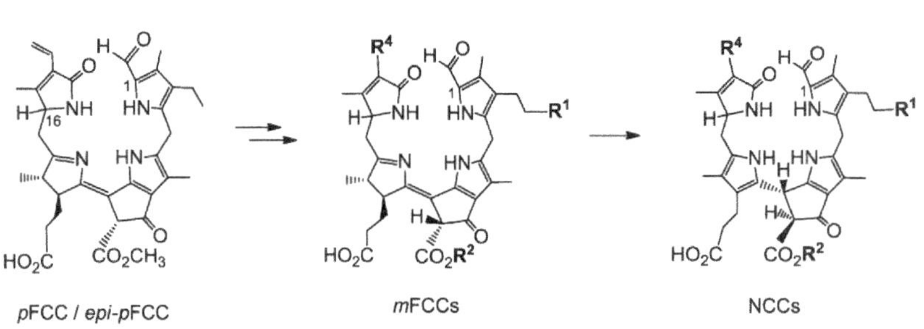

Figure 6: pFCC/*epi*-pFCC is transformed to the non-fluorescent NCCs, *via* modification of peripheral groups to give (often yet elusive) modified FCCs (mFCCs), followed by acid catalyzed isomerization of the latter to corresponding NCCs.

When dioxobilin-type NCCs (DNCCs) were discovered in barley leaves, they were considered to be formed from NCCs by an oxidative deformylation reaction.[6,20] However, the structure of *Ap*-DNCC from senescent leaves of Norway maple (see Fig. 5) suggested an alternative path for the formation of DNCCs.[21] Indeed, in studies with *A. thaliana*, a cytochrome P450 enzyme (cyp89A9) was identified that catalyzed the oxidative deformylation of pFCC to the corresponding DpFCC (two stereo-isomers).[19] In weakly acidic solution, the latter then isomerized to DNCCs of the type also found in senescent leaves of *A. thaliana* (see Fig. 7). Cyp89A9 was localized in the endoplasmatic reticulum, consistent with inferred transport of the FCC–substrate from the chloroplast into the cytosol.[19] In line with this finding and with the observation of NCCs in the vacuole, transport of FCCs through the chloroplast envelope into the cytosol, and (eventually) from this compartment into the vacuoles is a relevant and integral part of Chl-breakdown.[11] Modified DFCCs would, likewise, be expected to be formed in the cytosol, and to be transported into the vacuole, with subsequent rapid isomerization to the corresponding DNCCs.[19]

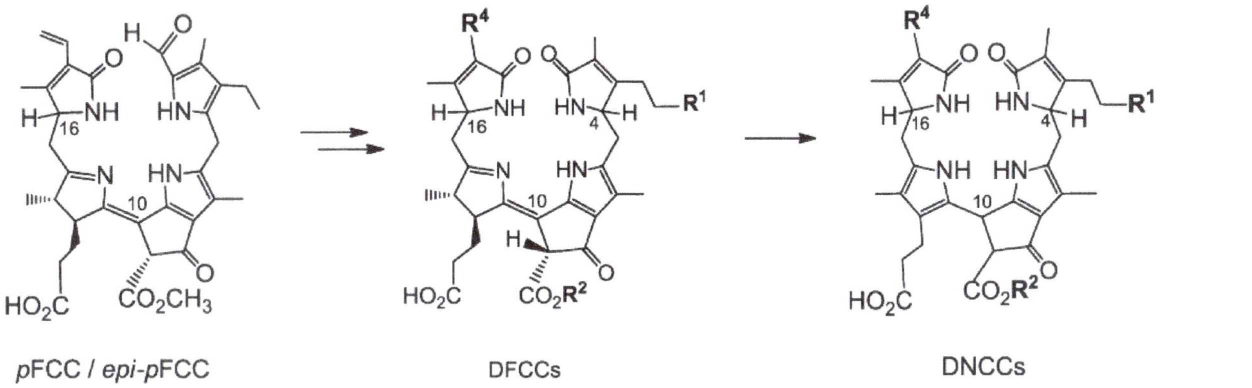

Figure 7: pFCC/*epi*-pFCC has been suggested to loose its formyl group, yielding the still fluorescent DpFCC (R^1 = H, R^2 = CH_3, R^4 = CH=CH_2). Subsequently, DpFCC is further modified to other DFCCs, which then are transformed, *via* acid catalyzed isomerization, to corresponding non-fluorescent DNCCs.

Colourless phyllobilins, such as NCCs and DNCCs, accumulate only temporarily in senescent leaves. Indeed, besides NCCs a (small) variety of DNCCs have meanwhile been identified in extracts of senescent leaves and characterized (see ESI,‡ Table S3). Together with the remarkably persistent 'hypermodified' FCCs (hmFCCs), three major lines of bilin-type Chl-catabolites have thus been discovered, which arise from a diversity of the breakdown pathway in its 'later' phases (see Fig. 5).

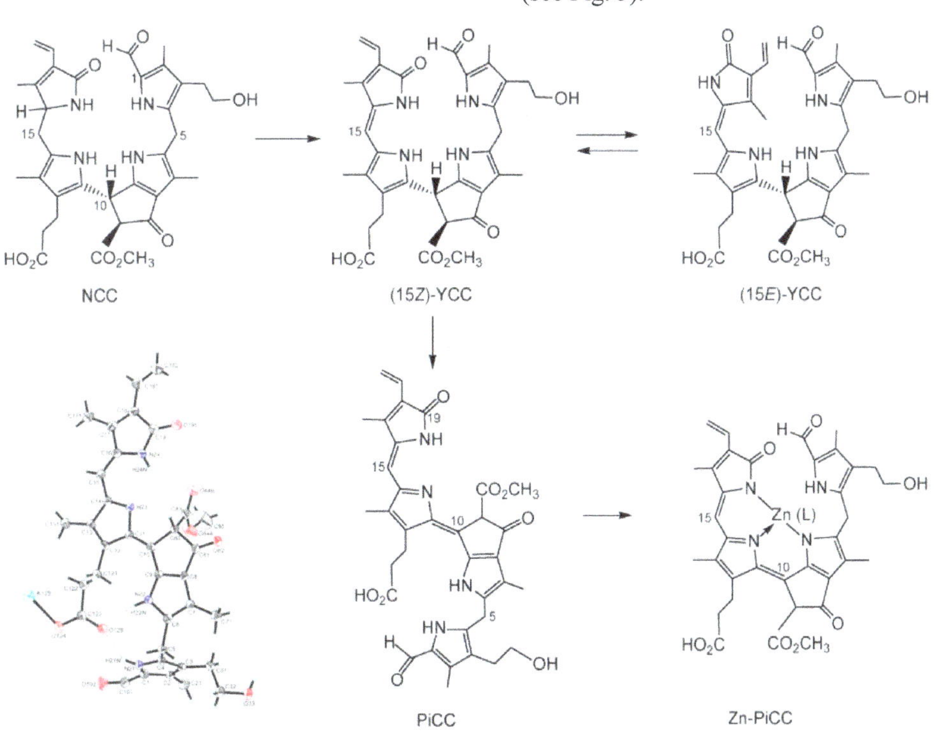

Figure 8: Oxidation of the NCC 3^2-hydroxy-1-formyl-19-oxophyllobilane leads to the corresponding (15*Z*)-YCC, which isomerizes to its 15*E*-isomer photo-chemically. Oxidation of the (15*Z*)-YCC (15*Z*)-3^2-hydroxy-1-formyl-19-oxophyllobilene-*c* occurs readily and yields the corresponding PiCC, whose C10=C11 double bond displays *E*-configuration (see formula and structural model taken from a X-ray crystal structure, bottom left). Binding of divalent transition metal ions by PiCC gives metallo-PiCCs, such as Zn-PiCC (bottom, right), in which both of the double bonds, at C10/C11 and at C15/C16, have been derived to have *Z*-configuration.[24]

In addition, yellow and pink coloured phyllobilins, named yellow Chl-catabolites (YCCs) and pink Chl-catabolites (PiCCs), were found in some fully senescent, yellow leaves as oxidation products of NCCs.[6,23] Part of the yellow and red colours of fall leaves may thus be due to these coloured pigments derived from Chl. YCCs and PiCCs are also available from NCCs by chemical oxidation and they are formed, as well, when NCCs turn 'rusty'. The first PiCC to be characterized by NMR-spectroscopy and mass spectrometry, (10*E*,15*Z*)-1-formyl-3^2-hydroxy-19-oxophyllobiladiene-*b*,*c*, turned out to possess an extended conjugated system, including a new C10═C11 double bond in *E*-configuration (see Fig. 8).[23] In spite of this structural feature (which was confirmed by crystallography), this PiCC has an excellent capacity for binding (divalent) transition metal ions in a three-dentate fashion, requiring the C10═C11 and C15═C16 double bonds to be *Z* in the resulting metal complexes.[24]

It is not known, at this stage, where in the senescent plant cell these coloured phyllobilins are formed and where they are accumulating, or whether the types of metal complexes of the PiCC, as prepared by chemical synthesis, would also exist in the plants. Likewise, the role of these coloured phyllobilins in Chl-breakdown and their further fate are still puzzling.

Chemical Properties and Spectroscopy of Phyllobilins

Phyllobilins are amphiphilic heterocyclic compounds that differ in their polarity by the nature of their peripheral modifications. High performance liquid chromatography (HPLC) is an efficient method of their analysis and separation in small scale. Phyllobilins can be classified on the basis of their UV/Vis-spectra (and of their associated optical properties, see Fig. 9 and 10).

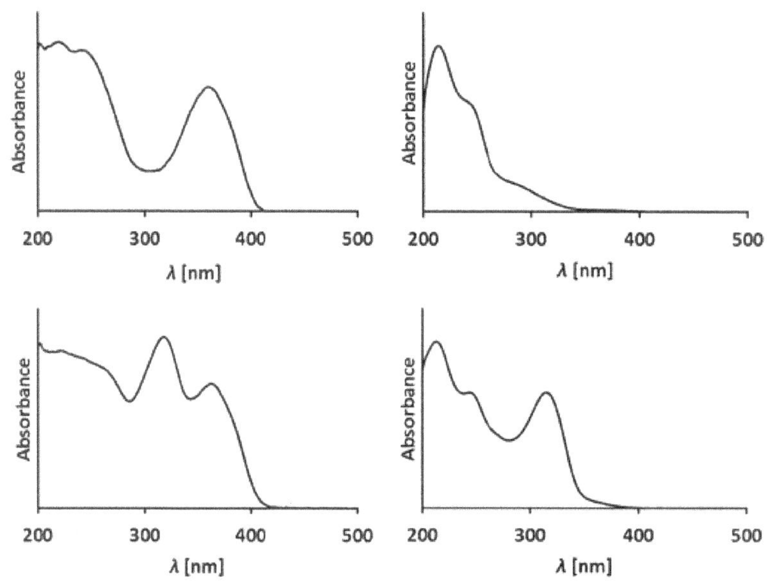

Figure 9: Representative UV-spectra of phyllobilins from *A. thaliana*. Top: UV-spectra of a DFCC and a DNCC, two type-II phyllobilins;[25] bottom: UV-spectra of an FCC and an NCC show an (additional) absorption maximum near 320 nm, due to the α-formyl pyrrole of type-I phyllobilins.[26]

Type-I phyllobilins feature an absorbance maximum near 315 to 320 nm (of the 1-formyl-pyrrole unit at ring A), as well as other absorption bands of their chromophore(s) at rings B to D, which depend upon the eventual conjugation between these ring sections (see Fig. 10). Type-II phyllobilins lack the formyl group and, thus, the corresponding band near 320 nm (see Fig. 9). As a consequence, UV-spectra of DNCCs lack absorption maxima helpful for the (on-line) detection of these phyllobilins, which may therefore have escaped being described in the original analysis of some plant extracts, *e.g.* of barley[20] and of *A. thaliana*.[19] Mass spectrometry offers additional methods to detect phyllobilins routinely, and that allow the provisional identification of phyllobilins ('on-line'). With the help of high-resolution mass spectrometry, the molecular formulas of phyllobilins may be delineated. In simple cases, the chemical constitution could also be derived from a detailed mass spectrometric analysis.[27]

The original classification of the (bilane-type) NCCs as 'rusty pigments'[1,4] reflected the ease, with which these colourless tetrapyrroles underwent apparently unspecific oxidation reactions to coloured materials. Indeed, NCCs (and YCCs) were shown to be excellent antioxidants.[6,28] NCCs were found to undergo chemical oxidation reactions at the C15-(*meso*)-position, leading to yellow bilene-type YCCs (or 1-formyl-19-oxophyllobilenes-*c*), with double bond *Z*-configuration at C15. YCCs feature a chromophore, which is identical to the 'western' half of bilirubin.[7,29] Moreover, YCCs are similarly effective as antioxidants, as bilirubin, and are easily oxidized to PiCCs (or 1-formyl-19-oxophyllobiladienes-*b*,*c*), with further extension of the conjugated chromophore system at C10.[24]

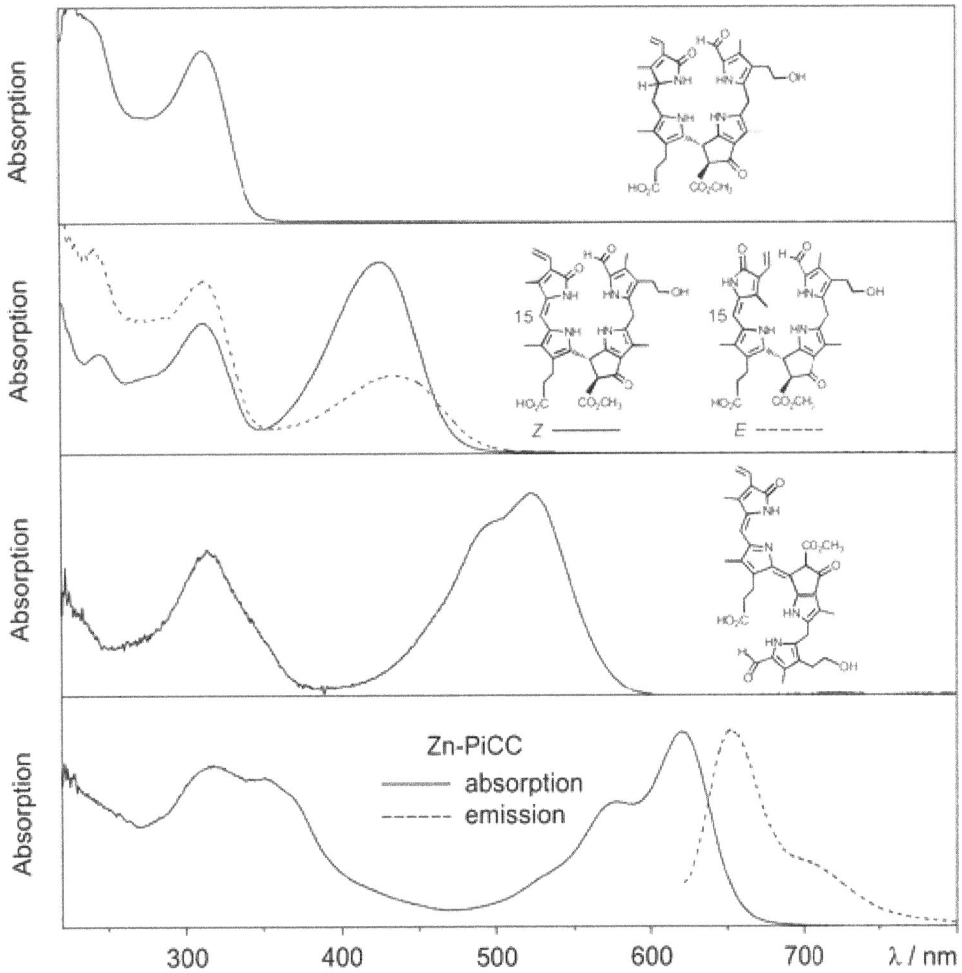

Figure 10: UV/Vis-spectra of an NCC, *E*/*Z*-isomeric YCCs, a PiCC and its Zn(II)-complex Zn-PiCC, together with an emission spectrum of Zn-PiCC (formulas are shown here and/or inFig. 8).

Phyllobilins show diagnostic UV/Vis-spectroscopic properties. With the exception of the colourless and 'nonfluorescent' phyllobilanes (NCCs and DNCCs), phyllobilins also display a correspondingly interesting spectrum of specific photochemical activities. UV-spectra of fluorescent Chl-catabolites (FCCs) exhibit absorption maxima near 317 and 360 nm, and FCCs correspondingly emit blue fluorescence with a maximum near 450 nm.[18] Typical, natural FCCs isomerize readily to NCCs, making some photo-physical studies with them difficult.[22] As FCC esters are persistent, a semi-synthetic FCC methyl ester was prepared and studied in detail. It was found not only to exhibit a relatively high fluorescence quantum yield (Φ_f) of about 0.25 (at room temperature), but, surprisingly, also to sensitize the formation of singlet oxygen with a remarkable quantum yield of about 0.6[30] (for further discussion, see Section 2.7).

UV/Vis-spectra of the (15*Z*)-YCC (15*Z*)-3^2-hydroxy-1-formyl-19-oxo-16,19-dihydro-phyllobilene-*c* and of its isomer (15*E*)-YCC exhibit absorption maxima near 310 and 426 nm, or near 313 and 440 nm, respectively (see Fig. 10).[23] In exploratory photochemical studies this (15*Z*)-YCC was seen to show weak luminescence and to undergo *E*/*Z*-photo-isomerization of its C15–C16 double bond.[23] Absorption spectra of the PiCC (10*E*,15*Z*)-3^2-hydroxy-1-formyl-19-oxo-16,19-dihydro-phyllobiladiene-*b*,*c* displayed absorption maxima near 313 and 523 nm.[24]Binding of Zn(II)-ions (or Cd(II)-ions) by this PiCC gave blue complexes, with absorption maxima near 578 and 620 nm (see Fig. 10). The PiCC was indicated to only exhibit weak luminescence at 620 nm,[24] whereas its blue complexes with Zn(II)- and Cd(II)-ions gave rise to strong red luminescence at 650 nm, allowing the detection of these metal ions at nM concentrations.[24] The photochemistry of phyllobilins is likely to be relevant for their possible physiological functions in plants. Indeed, for most phyllobilins, the detailed photochemical behaviour remains to be studied. Clearly, such knowledge promises to provide important insights into their eventual biological roles (see discussion in Section 2.7).

Naming of Natural Phyllobilins

In the earlier investigations, mostly concerned with the detection, isolation and identification of Chl-catabolites in senescent leaves, a phenomenological nomenclature was typically used. The newly found bilin-type catabolites were thus classified *e.g.* as red, fluorescent and nonfluorescent Chl-catabolites (RCC, FCC, NCC, resp.). Often they were specified further according to their plant source (*e.g. Brassica napus*) and their chromatographic polarity, *i.e.* as *Bn*-FCC-2, or for their position in the Chl-breakdown pathway (*e.g.* as 'primary': pFCC).[17] When the structures of Chl-catabolites became available, semi-systematic names were used for these bilin-type tetrapyrroles, based on the phytoporphyrin core, and supplemented by atom numbering, as established for the Chls:[6,31] *e.g.* pFCC (*Bn*-FCC-2, see Fig. 4) was named a $3^1,3^2$-didehydro-1,4,5,10,17,18,20-(22*H*)-octahydro-13^2-[methoxycarbonyl]-4,5-dioxo-4,5-seco-phytoporphyrinate,[17] and *Hv*-NCC-1 a $3^1,3^2,8^2$-trihydroxy-1,4,5,10,15,20-(22*H*,24*H*)-octahydro-13^2-[methoxycarbonyl]-4,5-dioxo-4,5-seco-phytoporphyrinate[4] (see Fig. 2 and 3). Such long designations eventually became impractical in cases, where the full structural specification of the Chl-catabolites was relevant. Furthermore, with the recent advent of the discovery of the general relevance of 1,19-dioxo-phyllobilins, and their derivatives, the recommended nomenclature of bilins (see ref. 7 and references therein) was adapted for naming the Chl-derived phyllobilins, as described here systematically.

The core structural unit for an NCC is thus represented by the phyllobilane (see Fig. 11), and *Hv*-NCC-1 from barley leaves is a 1-formyl-19-oxo-$3^2,18^1,18^2$-trihydroxy-16,19-dihydrophyllobilane. FCCs are 12,13-dihydro-phyllobilenes-*b*, *i.e.* pFCC is a (10*Z*)-1-formyl-19-oxo-12,13,16,19-tetrahydro-phyllobilene-*b*. Similarly, RCC is a (10*Z*,15*Z*)-1-formyl-19-oxo-12,13,19-(24*H*)-tetrahydro-phyllobiladiene-*b*,*c*. Likewise, *Ap*-DNCC from senescent leaves of Norway maple is named a 1,19-dioxo-$3^2,18^1,18^2$-trihydroxy-1,4,16,19-tetrahydro-phyllobilane (for complete atom numbering see ESI,‡ Fig. S1 and S2).

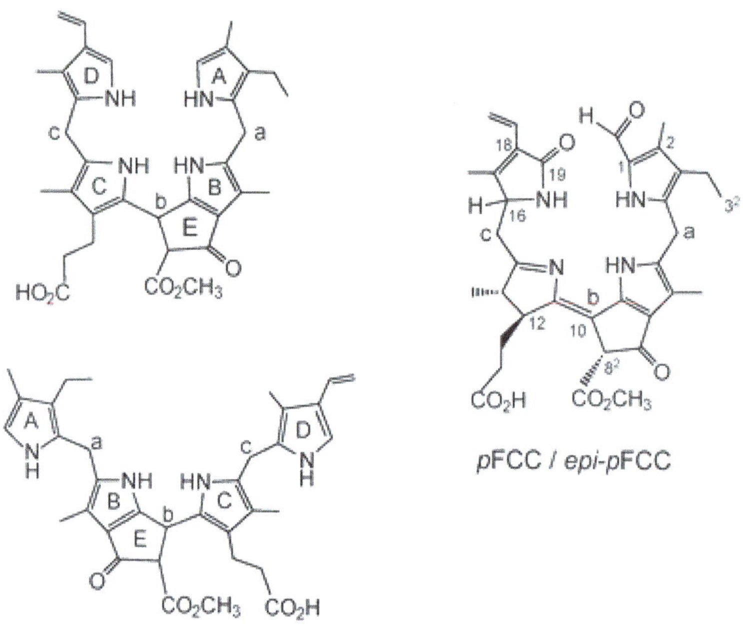

Figure 11: 'Pseudo-cyclic' and 'extended' formulas of phyllobilane with nomenclature of 5-membered rings and *meso*-positions (left), annotated ('pseudo-cyclic') formula of the (10*Z*)-1-formyl-19-oxo-12,13,16,19-tetrahydro-phyllobilene-*b* (pFCC/*epi*-pFCC) as example (right).

An Unexpected Manifestation of Chlorophyll Breakdown – BLUE Luminescent bananas

Typical FCCs are easily detectable by their fluorescence, yet only short lived metabolic precursors of the colourless and essentially photo-inactive NCCs. Strikingly, yellow bananas 'glow' bright blue, when excited by UV-light at wavelengths near 360 nm ('black' light), and when observed in the absence of white light (see Fig. 12).[9,18]

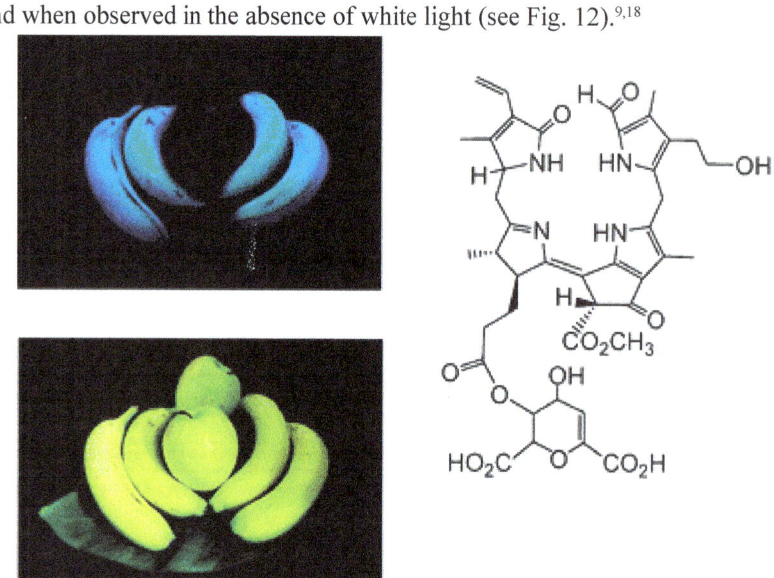

Figure 12: Ripe, yellow bananas show blue fluorescence, when irradiated with 'black light' (*e.g.*at 365 nm) and observed in the dark, whereas ripe, yellow 'Golden Delicious' apples do not fluoresce; formula of the main hmFCC (*Mc*-FCC-56) in peels of ripe bananas.

The peels of other ripe fruit, such as of 'Golden Delicious' apples, which exhibit a similar yellow colour as bananas, show insignificant luminescence, when studied under the same conditions (see Fig. 12). Indeed, only NCCs were found in the peels of ripe apples, and no FCCs.[28] The blue fluorescence of ripe bananas is largely caused by the accumulation of the special 'hypermodified' FCCs (hmFCCs), such as of *Mc*-FCC-56, in the banana peel (see Fig. 12, right).[18,32] In fact, the ripeness of a banana (*Musa acuminata*) could be read from the intensity of its blue fluorescence, caused by temporal accumulation of hmFCCs. Similar hmFCCs, as in banana peels, were also found as major (or sole) tetrapyrrolic Chl-catabolites in some (blue luminescent) senescent leaves, *e.g.* of the banana plant (*Musa acuminata*)[33,34] and of the Peace Lily (*Spathiphyllum wallisii*).[35] The occurrence of blue luminescent, persistent FCCs in bananas (and, potentially, in other ripe fruit) may have a particular role in signaling ripeness of these fruit to frugivores, important in the context of seed distribution.[6,18] Blue luminescence of senescent leaves is less well rationalized, at present. It could be relevant in helping to signal ripe fruit by 'fruit flagging'.[33,34]

Relationships Between Bilins and Phyllobilins

Phyllobilins, the natural, bilin-type breakdown products of chlorophyll (the green 'pigment of leaves'), are ubiquitous linear tetrapyrroles. Their structures are remarkably similar to those of the bilins, linear tetrapyrroles named for their occurrence in bile, as catabolites of heme (the red 'pigment of blood').[7] Phyllobilins are to be kept distinct from 'phytobilins' and 'phycobilins' of plants and other photosynthetic organisms,[7,36] which are linear tetrapyrroles due to further biosynthetic transformations of the 'heme-degradation product' biliverdin.[15] Note, these latter 'plant bilins', represent non-cyclic (tetrapyrrolic) chromophores that play important roles in light capture, when bound to phycobiliproteins, and light sensing in phytochromes.[15,37,38]

Biologically relevant linear tetrapyrroles appear to occur along two main paths in the biosphere: either as early biosynthetic precursors of natural porphyrinoids, or as biosynthetic products, obtained from the porphyrinoids by cleavage of their macroring.[7,37] *Via* the latter paths, natural linear tetrapyrroles are generated by oxygen-dependent cleavages of the porphyrinoid macrocycles of Chl or of heme, at their respective 'northern' *meso*-positions. An iron center is crucial for oxygen activation and catalysis in both cases. In the course of Chl-breakdown, ring opening of metal-free Pheo *a*, is the key step that is achieved by an iron-dependent enzyme, the senescence regulated (Rieske-type) mono-oxygenase PaO. Oxygenolytic Pheo *a* cleavage occurs with retention of the *meso*-carbon atom and its conversion into a formyl group, and it generates RCC as the progenitor of the characteristic type-I phyllobilins.[12] In contrast, the dioxo-bilin biliverdin is formed in the key step during heme-breakdown, in which the particular *meso*-carbon is lost as carbon monoxide. Thus, heme-oxygenases rely on the substrate (heme = iron-protoporphyrin) to present the iron center for activation by the protein (see Fig. 13).[10] These catabolic enzymes are not specific for heme (iron-protoporphyrin IX) and can accept other iron-porphyrins: heme-oxygenase from rat even accepted an iron-complex of Pheo *a* as substrate and converted it into a metal-free dioxobilin-analogue, as a recent exploratory study suggested.[6]

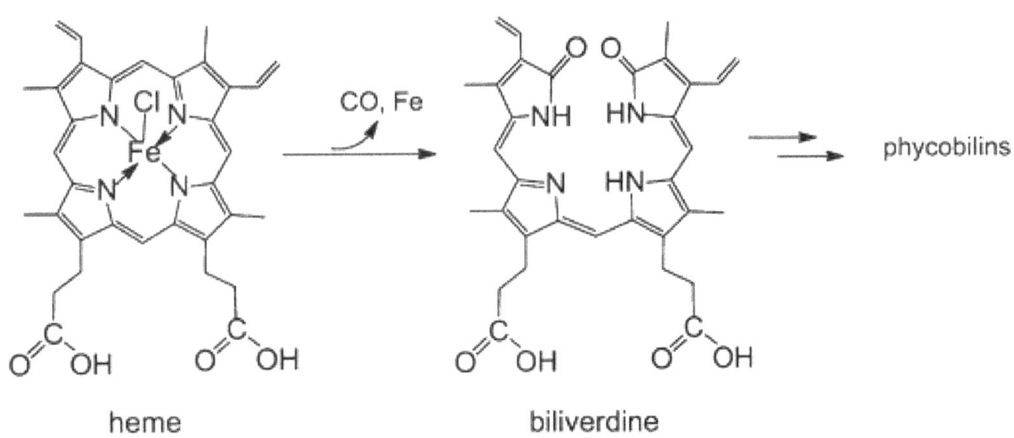

Figure 13: Oxidative opening of the porphyrin macrocycle of heme by heme-oxygenase liberates the iron-ion and CO, and furnishes biliverdin (BV). BV is the biosynthetic precursor of the (dioxo-)bilins, such as bilirubin and phycoviolobilin (see Fig. 14).

Remarkably, in some plants, type-I phyllobilins (1-formyl-19-oxo-phyllobilins), which are generated *via* the 'early' steps of the 'phyllobilin/PaO' pathway, are subsequently deformylated oxidatively to type-II phyllobilins (or 1,19-dioxo-phyllobilins), rendering these Chl-catabolites more similar to the heme-derived (dioxo)-bilins.[19] A puzzling hydroxymethylation at ring A of some type-II phyllobilins was observed recently, which is unprecedented among the known natural bilins,[7] and which remains to be rationalized, both in the context of Chl-breakdown, as well as of other metabolic issues.[25]

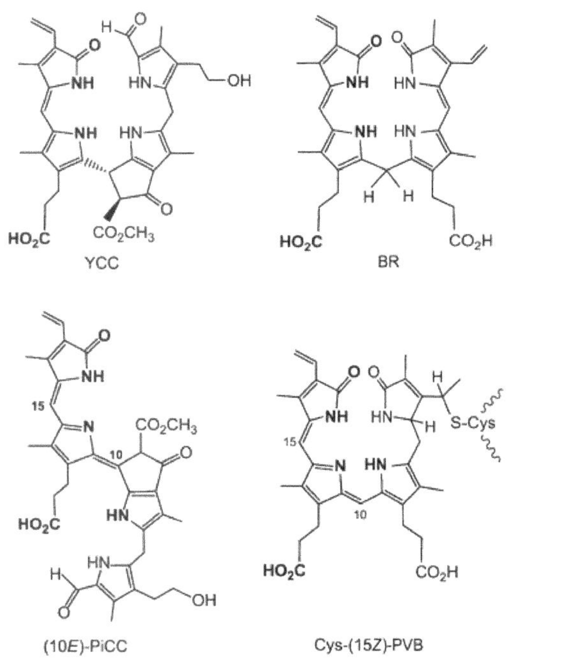

Figure 14: The main chromophores of YCCs and of bilirubin (BR) are strikingly similar, as are those of PiCCs and of (the protein-bound) phycoviolobilin (Cys-(15*Z*)-PVB).

Clearly, a distinct structural hallmark of the phyllobilins is the presence of an 'additional' cyclopentanone-unit ('ring E') annealed with the pyrrole ring B, a structural feature common to the Chls (where 'ring E' is annealed to 'ring C'). When carrying the original methoxy-carbonyl group, or a free carboxylic acid function, the 'extra' ring E is likely to exert steric interactions with ring C, similar to those in the Chls. Interestingly, the colour-giving chromophore of YCCs is structurally related to the one of bilirubin,[7] and the origin of the pink colour of PiCCs derives from a conjugated system similar to the chromophore of a protein-bound heme-catabolite, named phycoviolobilin (Cys-PVB, see Fig. 14).[15,39] In the light-sensing phytochromes and cyanobacteriochromes the protein-bound bilin cofactors undergo a crucial photo-isomerization at their C15=C16 double bond.[38,39]

The Question of Physiological Roles of Phyllobilins

The biological significance of Chl-breakdown in higher plants has mainly been associated with the photo-toxicity of Chl.[1,40] Indeed, the observed rapid degradation of the green plant pigments to colourless linear tetrapyrroles is consistent with this proposal.[3] Remarkably, an entirely different strategy for Chl-detoxification has been observed in aquatic herbivores, such as unicellular and multicellular zooplankton, which feed on algae, and convert Chls from the ingested food, into its non-luminescent and practically photo-inactive pheophorbide *a* (Pheo *a*) derivative, named '$13^2,17^3$-cyclopheophorbide *a* enol'.[6,41]

As a result of Chl-breakdown in higher plants, Chls are withdrawn from Chl-binding proteins (in light-harvesting complexes and in photoreaction centres) of the thylakoids. This, in turn, was found to labilize the protein matrix for controlled proteolytic decomposition, used for the purpose of recuperating protein-derived nitrogen.[1,40] On the other hand, in contrast to earlier views, the four nitrogen atoms of the Chl-molecules appear not to be mobilized as nutrient, since they remain part of the known catabolites.[6]

Phyllobilins were observed as consequences of leaf senescence and of fruit ripening.[8] Chl-breakdown implies reduction of the photosynthetic activity, with stringent consequences for the supply of the plant tissues with energy and crucial metabolites. The vitality of the transforming plant cells is particularly threatened in these developmental phases.[1] Phyllobilins may, therefore, play a particular physiological role as antioxidants.[28] Furthermore, phyllobilins are pigments likely to contribute to the apparent colour of leaves and fruit. Hence, further roles of these tetrapyrrolic Chl-catabolites are possible, either as 'sun-screens', or as (light-emitting) luminophores. To the latter, the exceptional hmFCCs may be counted, which are results of 'biosynthetic Chl-breakdown steps', and which may signal ripeness of bananas and other fruits to frugivores.[6,18] Some phyllobilins, such as FCCs, are unexpectedly effective sensitizers of singlet oxygen, which may act as a stress signal[42,43] or as a toxin for pathogens.[44] In the latter context, the capacity of some phyllobilins to bind and activate transition metals may be relevant.[24,45] On the other hand, this chemical property could also be important in plants for 'heavy metal' detoxification.[46]

The striking structural similarity of linear tetrapyrroles from Chl-breakdown with bilins from 'heme-breakdown' fuels speculations that phyllobilins may also play a variety of other physiological roles in plant cells. Phytobilins, such as phytochromobilin, and phycoviolobilin are chromophores of photoreceptor-proteins, named phytochromes[15,47] or *e.g.* cyanobacteriochromes.[38,39] Phycobilins are important cofactors of phycobiliproteins in photo-system antenna complexes (phycobilisomes).[15] Structurally similar (type-II) phyllobilins could bind well to photoreceptor and antenna proteins. Possibly, they could act as cofactor-mimics, as (competitive) inhibitors of these, or, in general, as 'phycobilin dummies'. Phycobiliproteins have found application as coloured and luminescent additives to food and cosmetics.[15] Furthermore, some phycocyanines, abundant in cyanobacteria, have been characterized as antioxidants, and were shown to have interesting anti-inflammatory effects, to function as phototoxic drugs against cancer, and to have a range of other beneficial health-effects, as deduced from experiments with mammalian cell lines and tissues.[15,38] Likewise, the abundant chlorophyll-derived phyllobilins may also have a variety of physiological effects and find interesting applications, in which their bilin-like structures, as well as their photo-, redox-, and metal chelating chemistry would be important features. Indeed, similar to the situation with various heme-derived bilins, which are now known to arise by biosynthetic transformations of the heme-oxygenase product biliverdin (BV),[38,39] some of the modifications observed in Chl-catabolites may be seen purposeful for their eventual biological function, and they may, likewise, be classified as 'biosynthetic'.[18,25]

RESEARCH OUTLOOK AND BROADER IMPLICATIONS

A variety of bilin-type Chl-catabolites are now identified, their structures have been characterized, and their chemistry has been partially explored.[6] This research has already given a solid basis to biochemical and plant-biological investigations of Chl-breakdown during leaf senescence and during fruit ripening.[11,48] Colourless phyllobilanes typically accumulate in senescent leaves and ripe fruit, and mark an apparently 'final' stage of Chl-breakdown.[40] Nowadays, most earlier questions concerning the basic biochemical processes along the 'phyllobilin/PaO' path appear to be clarified, up to the stage of the phyllobilanes.[6,11,48]

The situation is far less advanced concerning the further whereabouts of Chl-breakdown products. In some freshly senescent leaves, the amounts of the phyllobilanes found came up roughly for the amounts of Chls present in the original green leaf. However, in other cases, and in the typical later phases of leaf senescence, phyllobilanes seem to 'disappear' and their further fate is still a puzzle. The observation of yellow phyllobilenes (YCCs) and pink phyllobiladienes (PiCCs) in a variety of de-greened, yellow leaves may suggest metabolic oxidative processes involving the phyllobilanes. Indeed, a still little characterized oxidative process that leads to highly stereo- and regio-selective hydroxylation of NCCs at the 'western' C15 position appears to be ubiquitous in leaves.[6] The relevance of this process for the decrease of the amounts of phyllobilanes in senescent leaves, and their eventual disappearance, remains to be seen. Clearly, the formation of chemically rather labile, coloured bilin-type Chl-catabolites may point to new physiological roles of phyllobilins, and it may open a path to further breakdown of the Chls in senescent leaves, based on their subsequent transformations. In a related context, the further fate of the known phyllobilins in detached leaves (*e.g.* their further 'breakdown' by microorganisms in the soil) would be another interesting subject, into which, at present, well founded insights are not available.

Biochemical, molecular biological and various plant-biological studies[11] have paved the way, over the last twenty years, to insights important for a deeper understanding of physiological processes and evolutionary factors that contribute to the development, growth and stress resistance of plants,[49] as well as to the ripening of fruit.[50] In this respect, knowledge and control of Chl-breakdown may help *e.g.* to increase the resistance of plants (such as grass) against drought and against lack of nutrients,[49,51] in order to produce better crops and more nutritious crop products (such as barley and soy beans), and to increase the resistance of green vegetables (such as broccoli) against post-harvest deterioration.[53] Along these lines, Chl-breakdown may also be a relevant subject, in a practical sense, in the context of pathogen-induced senescence of higher plants and a possible role in the plants' defense program.[6] Clearly, basic scientific insights into Chl-breakdown may prove to be economically and ecologically beneficial.

In view of these advances, phyllobilins are yet a strikingly little explored class of ubiquitous, plant-derived, amphiphilic heterocyclic natural products, which own a wide range of unusual structural and chemical properties. Remarkably, good evidence for relevant biological roles of phyllobilins in the plants is still lacking.[6] The quest of finding any of the still elusive metabolic functions is an important (and presumably fruitful) challenge. Indeed, little is also known about the further fate of phyllobilins in the tissue of senescent leaves and of ripening fruit. Interestingly, the latter topic and the previous one may be strongly interrelated. Specific enzymatic modifications of the structures of the colourless phyllobilins may suggest further metabolic relevance of these Chl-catabolites and of their descendants, rather than simply be associated with metabolic 'extravagances'.[1] Likewise, as phyllobilins are part of the nutrition of plant-ingesting animals and humans, the effect of these natural products on animal and human metabolism and health is of interest.[6,28] Indeed, the established health benefits of antioxidants in the peels of ripe apples and the overlooked accumulation of NCCs as antioxidants in these and in other fruit, have led to the proposal that NCCs should also be taken into consideration, in that respect (as the old proverb 'An apple a day keeps the doctor away' suggests).[28]

In contrast to the possibly beneficial effects of some phyllobilins, as part of the nutrition, Chls and pheophorbide *a* (Pheo *a*) should be considered to be phototoxic to mammals. In line with this, evidence for active cellular export of Pheo *a* (from plant derived food), as well as for endogenous breakdown to red RCC-like compounds in mice was provided in an animal study.[53] The pathway for the inferred degradation of Pheo *a* is unknown. However, exploratory chemical evidence has been provided, which would suggest that heme-oxygenase could do the job of cutting up and eliminating (iron-complexes of) Chl-derived porphyrinoids in animals and humans.[6]

Chl-breakdown to bilin-type tetrapyrroles is not restricted to higher plants.[6] It has been observed for several lower organisms, notably in the green alga *Auxenochlorella protothecoides*(which release red type-I phyllobilins, related to RCC),[54] as well as for dinoflagellates, marine photo-organism, which generate bilin-type Chl-catabolites *via* opening of the Chl-macroring at the 'western' *meso*-position.[55]

In summary, our increasing insights into the occurrence and chemistry of phyllobilins, and of related Chl-derived bilins, have helped to develop a basic understanding of Chl-breakdown, with broad implications to plant biology and physiology.[1,11,52] Hopefully, this will promote interest in this topic in food and agricultural chemistry, as well as in the nutritional sciences. Last but not least, such knowledge could have practical consequences in agriculture and horticulture, and it could also enlighten school education and botanical excursions.

Acknowledgements

I would like to thank Thomas Müller and the present 'greenish' part of my co-workers in Innsbruck (Chengjie Li, Xiujun Liu, Matthias Roiser, Markus Ruetz, Mathias Scherl, Gerhard Scherzer, Iris Süssenbacher and – not least – Clemens Vergeiner), as well as their previous colleagues, for their excellent research contributions to this topic, Stefan Hörtensteiner at the University of Zürich, as well as Michael Oberhuber (Laimburg, Italy), and their groups, for very fruitful collaborations. I am grateful to the Austrian National Science Foundation for generous financial support (recent FWF projects: P 19596, L-472 and I-563), as well as to an Austrian/Italian/EU interregional project (Interreg-IV ID 5345).

REFERENCES

1. P. Matile, *Exp. Gerontol.*, 2000, **35**, 145 .
2. G. A. F. Hendry, J. D. Houghton and S. B. Brown, *New Phytol.*, 1987, **107**, 255.

3. B. Kräutler and P. Matile, *Acc. Chem. Res.*, 1999, **32**, 35 .
4. B. Kräutler, B. Jaun, K. Bortlik, M. Schellenberg and P. Matile, *Angew. Chem., Int. Ed. Engl.*, 1991, **30**, 1315 .
5. C. Peisker, H. Thomas, F. Keller and P. Matile, *J. Plant Physiol.*, 1990, **136**, 544 .
6. B. Kräutler and S. Hörtensteiner, in *Handbook of Porphyrin Science*, ed. G. C. Ferreira, K. M. Kadish, K. M. Smith and R. Guilard, World Scientific, Singapore, 2014, vol. 28, pp. 117–185 .
7. H. Falk, *Chemistry of Linear Oligopyrroles and Bile Pigments*, Springer, 1989 .
8. B. Kräutler, *Photochem. Photobiol. Sci.*, 2008, **7**, 1114 .
9. S. Moser, T. Müller, M. Oberhuber and B. Kräutler, *Eur. J. Org. Chem.*, 2009, 21 .
10. P. R. Ortiz de Montellano and K. Auclair, Heme Oxygenase Structure and Mechanism, in *The Porphyrin Handbook*, ed. K. M. Kadish, K. M. Smith and R. Guilard, Amsterdam, 2003, vol. 12, pp. 183–210 .
11. S. Hörtensteiner and B. Kräutler, *Biochim. Biophys. Acta, Rev. Bioenerg.*, 2011, **1807**, 977 .
12. S. Hörtensteiner, K. L. Wüthrich, P. Matile, K. H. Ongania and B. Kräutler, *J. Biol. Chem.*, 1998, **273**, 15335 .
13. A. Pružinska, I. Anders, S. Aubry, N. Schenk, E. Tapernoux-Lüthi, T. Müller, B. Kräutler and S. Hörtensteiner, *Plant Cell*, 2007, **19**, 369 .
14. M. Sugishima, Y. Okamoto, M. Noguchi, T. Kohchi, H. Tamiaki and K. Fukuyama, *J. Mol. Biol.*, 2010, **402**, 879.
15. K. E. Overkamp and N. Frankenberg-Dinkel, in *Handbook of Porphyrin Science'*, ed. G. C. Ferreira, K. M. Kadish, K. M. Smith and R. Guilard, World Scientific, Singapore, 2014, vol. 28, pp. 187–226 .
16. M. Oberhuber, J. Berghold and B. Kräutler, *Angew. Chem., Int. Ed.*, 2008, **47**, 3057.
17. W. Mühlecker, K. H. Ongania, B. Kräutler, P. Matile and S. Hörtensteiner, *Angew. Chem., Int. Ed. Engl.*, 1997, **36**, 401 .
18. S. Moser, T. Müller, M.-O. Ebert, S. Jockusch, N. J. Turro and B. Kräutler, *Angew. Chem., Int. Ed.*, 2008, **47**, 8954.
19. B. Christ, I. Süssenbacher, S. Moser, N. Bichsel, A. Egert, T. Müller, B. Kräutler and S. Hörtensteiner, *Plant Cell*, 2013, **25**, 1868.
20. F. G. Losey and N. Engel, *J. Biol. Chem.*, 2001, **276**, 8643.
21. T. Müller, M. Rafelsberger, C. Vergeiner and B. Kräutler, *Angew. Chem., Int. Ed.*, 2011, **50**, 10724 .
22. M. Oberhuber, J. Berghold, K. Breuker, S. Hörtensteiner and B. Kräutler, *Proc. Natl. Acad. Sci. U. S. A.*, 2003, **100**, 6910.
23. M. Ulrich, S. Moser, T. Müller and B. Kräutler, *Chem. – Eur. J.*, 2011, **17**, 2330.
24. C. Li, M. Ulrich, X. Liu, K. Wurst, T. Müller and B. Kräutler, *Chem. Sci.*, 2014 **10.1039/C4SC00348A**.
25. I. Süßenbacher, B. Christ, S. Hörtensteiner and B. Kräutler, *Chem. – Eur. J.*, 2014, **20**, 87 .
26. B. Kräutler, Chlorophyll Breakdown and Chlorophyll Catabolites, in *The Porphyrin Handbook*, ed. K. M. Kadish, K. M. Smith and R. Guilard, Oxford, 2003, vol. 13, pp. 183–209 .
27. T. Müller, S. Vergeiner and B. Kräutler, *Int. J. Mass Spectrom.*, 2014, **48**, 365–366 .
28. T. Müller, M. Ulrich, K.-H. Ongania and B. Kräutler, *Angew. Chem., Int. Ed.*, 2007, **46**, 8699 .
29. D. Lightner and A. F. McDonagh, *Acc. Chem. Res.*, 1984, **17**, 417 .
30. S. Jockusch, N. J. Turro, S. Banala and B. Kräutler, *Photochem. Photobiol. Sci.*, 2014, **13**, 407 .
31. H. Scheer, in *Chlorophylls and Bacteriochlorophylls*, ed. B. Grimm, R. J. Porra, W. Rüdiger and H. Scheer, Dordrecht, The Netherlands, 2006, pp. 1–26 .
32. S. Moser, T. Müller, A. Holzinger, C. Lütz, S. Jockusch, N. J. Turro and B. Kräutler, *Proc. Natl. Acad. Sci. U. S. A.*, 2009, **106**, 15538.
33. S. Banala, S. Moser, T. Müller, C. Kreutz, A. Holzinger, C. Lütz and B. Kräutler, *Angew. Chem., Int. Ed.*, 2010, **49**, 5174.
34. C. Vergeiner, S. Banala and B. Kräutler, *Chem. – Eur. J.*, 2013, **19**, 12294.
35. B. Kräutler, S. Banala, S. Moser, C. Vergeiner, T. Müller, C. Lütz and A. Holzinger, *FEBS Lett.*, 2010, **584**, 4215 .
36. N. Frankenberg and J. C. Lagarias, in *The Porphyrin Handbook*, ed. K. M. Kadish, K. M. Smith and R. Guilard, Oxford, UK, 2003, vol. 13, pp. 211–235 .
37. M. J. Terry and A. C. McCormac, in *Tetrapyrroles: Birth, Life and Death*, ed. M. J. Warren and A. G. Smith, Landes Bioscience, Austin, Texas, 2008, pp. 221–234 .
38. K.-H. Zhao, R. J. Porra and H. Scheer, in *Handbook of Porphyrin Science*, ed. G. C. Ferreira, K. M. Kadish, K. M. Smith and R. Guilard, World Scientific, Singapore, 2012, vol. 22, pp. 1–66 .
39. N. C. Rockwell, S. S. Martin, A. G. Gulevich and J. C. Lagarias, *Biochemistry*, 2012, **51**, 1449.
40. P. Matile, S. Hörtensteiner, H. Thomas and B. Kräutler, *Plant Physiol.*, 1996, **112**, 1403 .
41. Y. Kashiyama, A. Yokoyama, Y. Kinoshita, S. Shoji, H. Miyashiya, T. Shiratori, H. Suga, K. Ishikawa, A. Ishikawa, I. Inouye, K.-i. Ishida, D. Fujinuma, K. Aoki, M. Kobayashi, S. Nomoto, T. Mizoguchi and H. Tamiaki, *Proc. Natl. Acad. Sci. U. S. A.*, 2012, **109**, 17328.
42. K. Apel and H. Hirt, *Annu. Rev. Plant Biol.*, 2004, **55**, 373.
43. C. Triantaphylidès and M. Havaux, *Plant, Cell Environ.*, 2009, **14**, 219 .
44. C. Flors and S. Nonell, *Acc. Chem. Res.*, 2006, **39**, 293.
45. A. Fürstner, *Angew. Chem., Int. Ed.*, 2003, **42**, 3582 .
46. T. Jaffré, Y. Pillon, S. Thomine and S. Merlot, *Plant Sci.*, 2013, **4**, 1 .
47. N. Frankenberg and J. C. Lagarias, *J. Biol. Chem.*, 2003, **278**, 9219.
48. A. Tanaka and R. Tanaka, *Curr. Opin. Plant Biol.*,

2006, **9**, 248.

49. H. Thomas, L. Huang, M. Young and H. Ougham, *BMC Evol. Biol.*, 2009, **9**, 163 .

50. C. S. Barry, *Plant Sci.*, 2009, **176**, 325.

51. B. Christ, A. Egert, I. Süssenbacher, B. Kräutler, D. Bartels, S. Peters and S. Hörtensteiner, *Plant, Cell Environ.*, 2014 DOI:10.1111/pce.12308.

52. G. A. King and E. M. O'Donoghue, *Exp. Gerontol.*, 1995, **6**, 385 .

53. J. W. Jonker, M. Buitelaar, E. Wagenaar, M. A. van der Valk, G. L. Scheffer, R. J. Scheper, T. Plosch, F. Kuipers, R. P. J. O. Elferink, H. Rosing, J. H. Beijnen and A. H. Schinkel,*Proc. Natl. Acad. Sci. U. S. A.*, 2002, **99**, 15649.

54. N. Engel, T. A. Jenny, V. Moser and A. Gossauer, *FEBS Lett.*, 1991, **293**, 131 .

55. H. Nakamura, Y. Kishi, O. Shimomura, D. Morse and J. W. Hastings, *J. Am. Chem. Soc.*, 1989, **111**, 7607 .

CHAPTER 5

THE CO-PIGMENTATION OF ANTHOCYANIN ISOLATED FROM MANGOSTEEN PERICARP (GARCINIA MANGOSTANA L.) AS NATURAL DYE FOR DYE- SENSITIZED SOLAR CELLS (DSSC)

H Munawaroh, G Fadillah, L N M Z Saputri, Q A Hanif, R Hidayat, S Wahyuningsih

Chemistry of Department, Faculty of Mathematics and Natural Sciences, Sebelas Maret University, Jl. Ir. Sutami, Kentingan, Surakarta 57126, Central Java, Indonesia

ABSTRACT

Study of color stability of anthocyanin from extract mangosteen pericarp (*Garcinia mangostana L.*) with co-pigmentation method has been conducted. Malic acid and ascorbic acid used as a co-pigment to stabilize the anthocyanin structure through formation of new binding between anthocyanin. Anthocyanin from mangosteen pericarp were isolated by several steps, including maceration, extraction, and Thin Layer Chromatography (TLC). The anthocyanin separation was conducted by TLC, while the identification of functional groups of those compound, were used FTIR (Fourier Transform Infrared Spectroscopy) for spectra analysis. Ultraviolet- visible absorption spectra have represented differences absorbance and color intensity in various pH. Copigmentation with malic acid and ascorbic acid in many composition and temperature were also well described. Meanwhile, anthocyanin-malic acid and anthocyanin-ascorbic acid have color retention higher than that of pure anthocyanin. Maximum color retention has been achieved at a ratio of 1:3 and 1:5 for ascorbic acid and malic acid, respectively. Therefore, the addition of ascorbic acid and malic acid as a copigment shows the ability to protect color retention of anthocyanin (mangosteen pericarp) from degradation process. The better efficiency of DSSC (η) have been achieved, whereas n of controlled anthocyanin, anthocyanin-ascorbic acid, and anthocyanin-malic acid were 0,1996%, 0,2922%, 0,3029%, respectively.

INTRODUCTION

Based on the manufacture of materials, there are two kinds of solar cells. The first solar cells are made of silicon. However, the fabrication of silicon solar cells is costly. While the new type of solar cells is dye-sensitized solar cells (DSSC), a device for the conversion of visible light into electricity based on the sensitization of wide band gap semiconductors [1]. The first DSSC was pioneered by Grätzel et al. in the early nineties, called Grätzel cells [1,2], these have attracted considerable attention due to their environmental friendliness and low cost [3]. The performance of the cell mainly depends on a dye used as a sensitizer. The important parameters of the efficiency of the cells are determined by the absorption spectrum of the dye and the anchorage of the dye to the surface of TiO_2 [4].

Citation: H Munawaroh, G F adillah, L N M Z Saputri, Q A Hanif, R Hidayat and S Wahyuningsih 2016. The co-pigmentation of anthocyanin isolated from mangosteen pericarp (Garcinia Mangostana L.) as Natural Dye for Dye- Sensitized Solar Cells (DSSC). IOP Conf. Series: Materials Science and Engineering 107 (2016) 012061 doi:10.1088/1757-899X/107/1/012061.

So far, the highest efficiency of DSSC by transition metal coordination compound (ruthenium polypridyl complexes) absorbed on nanocrystalline TiO_2 reached 8.7% [5]. However, ruthenium polypridyl complexes contain heavy metal which is undesirable for environment aspect and a limited amount of ruthenium hence their costly production [6]. Alternatively, natural dyes are being developed due to their availability, low cost [3], easy to prepare, nontoxic, and biodegradable [7].

By far, several of natural dyes that have been used as a sensitizer in DSSC system such as tannin [8], chlorophyll [9], anthocyanin [10], cyanine [11], and carotene [12]. Anthocyanin is very large group of red-blue pigments that can be found in flowers, fruits, fruit juice, leaves, stem, and root of the plants. One of them can be obtained from mangosteen (Garcinia Mangostana L.) pericarp. Anthocyanin in isolated from mangosteen pericarps has been studied before[13], but they have the weakness especially in the color stability and labile pigment due to very easily degraded. The efforts to enhance the stabilization of anthocyanin is co-pigmentation, an interaction between anthocyanin with other molecules such as metal (Al^{3+}, Fe^{3+}, Sn^{2+}, Cu^{2+}) and other organic molecules such as flavonoid compounds (flavonoids, flavanones, flavonols), and more. Their action of co-pigmentation tends to increase the color stability and protect from oxidation [14]. Based on the description above, this research focuses on co-pigmentation anthocyanin with malic acid and ascorbic acid. These extracted dyes were characterized by Thin Layer Chromatography (TLC), Infrared spectroscopy (IR), and UV-Vis absorption spectra. The photovoltaic properties of DSSC by purified product were studied.

EXPERIMENTAL

Sample Preparation

Mangosteen pericarp was dried in an oven with temperature ± 40 °C. And then dried sample changed into a powder.

Sample maceration and extraction

Extraction of mangosteen pericarp powder used ethanol 96% and HCl 1M as a solvent with ratio 1:4 in 24 hours. Residue and filtrate were separated with vacuum filtration. Then filtrate was evaporated to remove the solvent with a rotary evaporator, yielded concentrated extract. Characterization concentrated extract was conducted with UV-vis spectrophotometry, Thin Layer Chromatography (TLC), and Fourier Transform Infrared (FTIR).

Characterization of anthocyanin extracts in various pH

Measurement of the color intensity of anthocyanin extract was performed on six pH conditions (1, 3, 5, 7, 9, 11). Each solution was measured at the maximum wavelength with a spectrophotometer at range 200-800 nm.

The color retention characterization of anthocyanin in co-pigmented with ascorbic acid

In this step, anthocyanin was diluted with ethanol than mixed with ascorbic acid in several ratio; 1:0, 1:1, 1:2, and1:3 (w/w). Then co-pigmented-anthocyanin was conducted in several of temperature (30, 40, 50, and 60 °C) to observe the effect of temperature in color stabilization. Then the result was characterized with UV-visible spectrophotometry in wavelength 520 nm.

The color retention characterization of anthocyanin co-pigmented with malic acid

In this step, anthocyanin was diluted with ethanol than mixed with ascorbic acid in several ratio; 1:5 and 1:10 (w/w). Then co-pigmented-anthocyanin was conducted in various of temperature (60 °C, 80 °C, 100 °C, 120 °C and 140 °C) to observe the effect of temperature in color stabilization. Then the result was characterized with UV-vis spectrophotometry in wavelength 520 nm.

Fabrication of DSSC

A fluorine-doped tin oxide (FTO) glass was used. The conductive glass substrate was first cleaned and coated with a layer of TiO2 paste (0.5 g TiO2 was added 2 mL ethanol and stirred until homogeny) by slip casting and dried at 400 °C for 30 min. After that, the glass substrate coated TiO2 was immersed into anthocyanin dye overnight. To prepare the counter electrode, a platinum paste was deposited on the conductive substrate glass and dried at 400 °C for 30 min. The working electrode and counter electrode were assembled into a cell. A drop of electrolyte solution containing I-/I3-injected in the cell and the hole in the counter electrode was sealed using a sealing spacer. Therefore, DSSC arrangement was FTO|TiO_2|Anthocyanin (or co-pigmented-anthocyanin)| I-/I3- |Pt|FTO.

Characterization and measurement of photo-electrochemical

Solar energy conversion efficiency (the photo current-voltage (I-V) curve) was measured by Keithley multimeter under illumination. Based on I-V curve, the fill factor (FF) is defined as equation (1);

$$FF= (I_{max} \times V_{max})/(I_{sc} \times V_{oc}) \tag{1}$$

Where I_{max} and V_{max} are the photocurrent and photovoltage for maximum power output (P_{max}), I_{sc} and V_{oc} are the short-circuit photocurrent and open-circuit photovoltage, respectively. The overall energy conversion efficiency (η) is defined as equation (2)

$$\eta= (I_{sc} \times V_{oc} \times FF)/P_{in} \tag{2}$$

where P_{in} is the power of incident light.

RESULT AND DISCUSSION

Absorption of Anthocyanin

Figure 1(a) shows the UV-Vis absorption spectrum of mangosteen extract. It was found that the absorption peak of mangosteen extract is about 520 nm. About 320, 340, and 450-520 nm indicate the presence of anthocyanin compounds. The present anthocyanin also depends on pH value. Figure 1(b) shows the absorption spectrum of anthocyanin shift to a greater wavelength in base condition or pH increase. Overall the spectrum shows a peak of anthocyanin at range 400-600 nm.

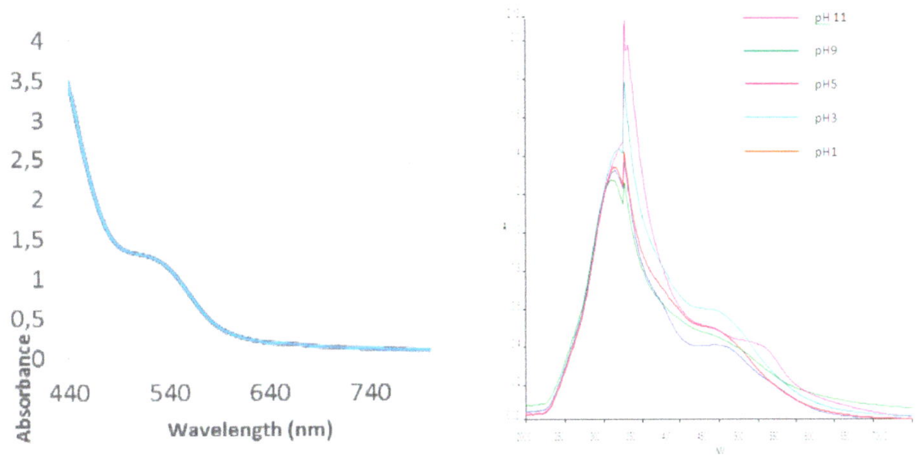

(a) (b)

Figure 1. (a) Spectrum UV-Vis Anthocyanin. (b) Effect of pH.

Besides using UV-Vis, characterization of extract mangosteen was performed also with Thin Layer Chromatography (TLC) and infrared spectroscopy (IR). The purpose of using Thin Layer Chromatography (TLC) is to determine patterns of separation and purity of the compounds. Butanol:acetic acid: water used as eluent with a ratio of 4:1:5. The retention factor (Rf) obtained 0.255, this value related to the previous result that the Rf of anthocyanin approximately 0.1 to 0.4 [15]. While the characterization of anthocyanin using IR spectroscopy to determine the molecular structure of compounds (figure 2).

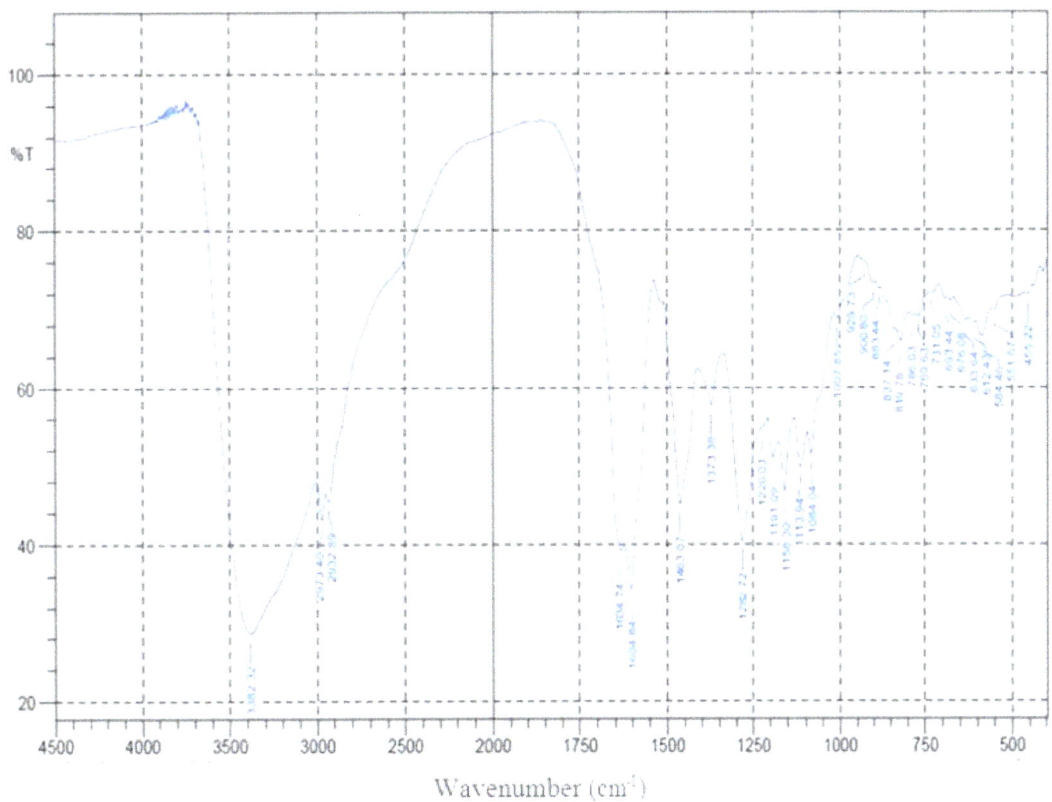

Figure 3. Spectra IR of Anthocyanin.

The FTIR spectra of these compounds show benzene groups vibration peaks between 455.22-929.7 cm^{-1}. The double bond C=C vibrations at 1100-1300 cm^{-1}. Two adjacent peaks at 1604-1634 cm^{-1} indicate carbonyl group (C=O). In addition, to the peak appearing at 2932-2937 cm-1 for C-H stretching. While the wide uptake 3362 cm^{-1} indicate the presence of OH group sand hydrogen bonding that occurs as intermolecular interaction of anthocyanin. All of these function groups establish the structure of anthocyanin that consist of benzene ring, double bond, a carbonyl group, C-H bond, and OH group.

Copigmentation Anthocyanin: Ascorbic Acid

Figure 3 shows the effect of increasing temperature to color stability of anthocyanin-ascorbic acid copigmentation. Increasing absorbance of anthocyanin-ascorbic acid co-pigmentation followed by increasing mass ratio of ascorbic acid were added. The co-pigmentation of anthocyanin-ascorbic acid with ratio 1:3 have the greatest absorbance rather than ratio 1:1 and 1:2. However, there is an irregularity curve at ratio 1:2 caused by influence of heat that make the anthocyanin-ascorbic acid degraded quickly. High heating temperatures relatively can damage the structure of anthocyanin. Heating an anthocyanin-ascorbic acid cause an impact to their color stability due to degradation that may occur. Figure 5 represent that increasing temperature of heating anthocyanin-ascorbic acid was followed by degradation of color retention (color stability).

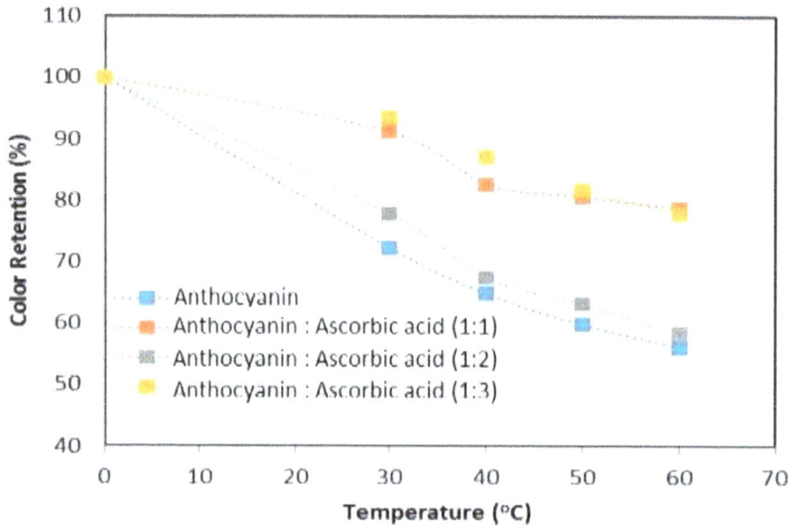

Figure 5. Color Retention of Anthocyanin-Ascorbic Acid.

Copigmentation Anthocyanin: Malic Acid

Figure 4 shows the effect of increasing temperature to color stability of anthocyanin- malic acid copigmentation. The color stabilities anthocyanin-malic acid with ratio 1:0, 1:5, and 1:10 are 74.32; 92.89; and 98.64%, respectively. High heating temperatures relatively can damage the structure of anthocyanin. Heating an anthocyanin-malic acid cause an impact to their color stability due to degradation that may occur. Increasing temperature of heating anthocyanin-malic acid was followed by degradation of color retention (color stability).

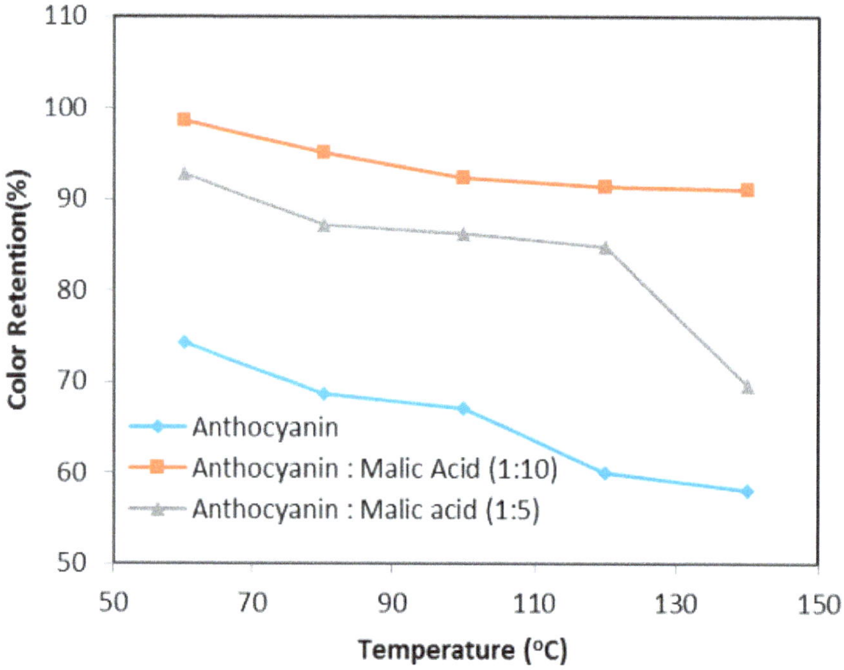

Figure 4. The Color Retention of Anthocyanin-Malic Acid.

Photoelectrochemical Properties of DSSC with Natural Dye

Photovoltaic test of DSSC using co-pigmentation of natural dyes as sensitizers were performed by measuring the current-voltage (I-V) curve by solar simulator. The performance of DSSC was evaluated by energy conversion efficiency (η). The efficiency of anthocyanin, anthocyanin-ascorbic acid, and anthocyanin: malic acid are 0.199; 0.292; 0.303%, respectively. The result shows that copigmentation potentially increase the efficiency of DSSC from 46.40% to 51.75%. The polymerization that may occur between anthocyanin and co-pigment be able to stabilize the color of anthocyanin.

CONCLUSIONS

The anthocyanin was successfully extracted from mangosteen pericarp by maceration method. The characterization using TLC, UV-Vis, and FTIR shows the characteristic of anthocyanin, compared by the standard of anthocyanin. The efficiency of anthocyanin, anthocyanin-ascorbic acid, and anthocyanin-malic acid are 0.199; 0.292; 0.303%, respectively. The co-pigmentation with ascorbic acid and malic acid be able to increase the efficiency of DSSC from 46.40% to 51.75%.

REFERENCES

1. O' Regan B and Gratzel M 1991 A Low-Cost High Efficiency Solar Cell Based on Dye Sensitized Colloidal TiO2 Films Nature 353:737-740
2. Maddu A, Zuhri M and Irmansyah 2010 Penggunaan Ekstrak Antosianin Kol Merah sebagai Foto sentizer pada Sel Surya TiO2 Nanokristal Tersentisasi Dye Bogor Departemen FISIKA FMIPA Institut Pertanian Bogor
3. Zhou H, Liqioing Wu, Yurong Gao and Tingli Ma 2011 Dye-Sensitized Solar Cells Using 20 Natural Dyes as Sensitizer Photochem Photobiol 219:188-194
4. Wongcharee K, Vissanu M and Sumaeth C 2007 Dye-Sensitized Solar Cell Using Natural Dyes Extracted from Rosella Blue Pea Flowers Sol Ener Mat Sol C 91: 566-571
5. Nazeeruddin M K, Klein C, Liska P and Gratzel M 2005 Synthesis of Novel Ruthenium Sensitizers And Their Application In Dye-Sensitized Solar Cells Coord. Chem. Rev 249: 1460-1467
6. Zhang G, Bala H, Cheng Y, Shi D, Lv X, Yu Q and Wang P 2009 High Efficiency and Stable Dye-Sensitized Solar Cells With an Organic Chromophore Featuring a Binary PhiConjugated Spacer Chem Comm 2198-2200
7. Kamiloglu S, Ayca Ayfer P, Beraat O, John Y C and Esra C 2015 Color Retention, Anthocyanin Stability, And Antioxidant Capacity in Black Carrot (Daucuscarota) Jamsand Marmalades: Effect of Processing, Storage Conditions, And In Vitro Gastrointestinal Digestion Roy Soc Ch 13: 1-10
8. Espinosa R, Zumetta I and Santana J L 2005 Nanocrystalline TiO2 Photosensitized With Natural Polymers With Enhanced Efficiency From 400 to 600 Nm Sol Ener Mat Sol C 85: 359-369
9. Kumara G R A, Kaneko S and Okuya M 2006 Shiso Leaf Pigments For Dye Sensitized SolidState Solar Cell Sol Energ Mat Sol C 90: 1220-1226
10. Wongcharee, Meeyoo K V and Chavadej S 2007 Dye-Sensitized Solar Cell Using Natural Dyes Extracted From Rosella and Blue Pea Flowers Teknik Kimia: Universitas Teknologi Mahanakom
11. Furukawa S, Lino H, Iwamoto T and Kukita K 2009 Characteristics Of Dye Sensitized Solar Cells Using Natural Dye Thin solid films 518: 526-529
12. Ortiz, Gomez I A, Maldonado V and Espadas A R P 2010 Dye Sensitized Solar Cells With Natural Dyes Extracted From Achiote Seeds Sol Energ Mat Sol 94: 40-44
13. Tanaka Y, Sasaki N and Ohmiya A 2008 Biosynthesis of plant pigments: anthocyanins betalains and carotenoids The Plant journal 54: 733-749
14. Boulton R 2001 The Copigmentation of Anthocyanins and Its Role In The Color Of Red Wine: A Critical Review Am J Enol Viticul 52 (2): 67-8
15. Harborne J B 1987 Metode Fitokimia: Penuntun Cara Modern Menganalisis Tumbuhan Bandung: Penerbit ITB

CHAPTER 6

TYROSINE HYDROXYLATION IN BETALAIN PIGMENT BIOSYNTHESIS IS PERFORMED BY CYTOCHROME P450 ENZYMES IN BEETS (BETA VULGARIS)

Rasika Sunnadeniya[1] , Alexander Bean[1] , Matthew Brown[1] , Neda Akhavan[1] , Gregory Hatlestad[1] , Antonio Gonzalez[1] , V. Vaughan Symonds[2] , Alan Lloyd[1]

[1] Department of Molecular Biosciences, The University of Texas at Austin, Austin, Texas, United States of America

[2] Institute of Fundamental Sciences, Massey University, Palmerston North, New Zealand

ABSTRACT

Yellow and red-violet betalain plant pigments are restricted to several families in the order Caryophyllales, where betacyanins play analogous biological roles to anthocyanins. The initial step in betalain biosynthesis is the hydroxylation of tyrosine to form L-DOPA. Using gene expression experiments in beets, yeast, and *Arabidopsis*, along with HPLC/MS analysis, the present study shows that two novel cytochrome P450 (CYP450) enzymes, CYP76AD6 and CYP76AD5, and the previously described CYP76AD1 can perform this initial step. Co-expressing these CYP450s with DOPA 4,5-dioxygenase in yeast, and overexpression of these CYP450s in yellow beets show that CYP76AD1 efficiently uses L-DOPA leading to red betacyanins while CYP76AD6 and CYP76AD5 lack this activity. Furthermore, CYP76AD1 can complement yellow beetroots to red while CYP76AD6 and CYP76AD5 cannot. Therefore CYP76AD1 uniquely performs the beet R locus function and beets appear to be genetically redundant for tyrosine hydroxylation. These new functional data and ancestral character state reconstructions indicate that tyrosine hydroxylation alone was the most likely ancestral function of the CYP76AD alpha and beta groups and the ability to convert L-DOPA to cyclo-DOPA evolved later in the alpha group.

INTRODUCTION

Betalains are water-soluble nitrogenous pigments classified as the red-violet betacyanins and yellow betaxanthins. These pigments variously accumulate in flowers, fruits, and vegetative tissues of plants in most core families of the Caryophyllales order, and where they exist, betalains exclusively replace anthocyanins. The betacyanins play important and analogous roles to anthocyanins in attracting pollinators and seed dispersers. While the structural and regulatory genes of anthocyanin biosynthesis are well known, betalain biosynthesis has been poorly understood at the molecular level creating a phylogenetic mystery surrounding their origin(s).

Much of the biosynthetic pathway involved in betalain biosynthesis has recently been described (Fig 1A), yet the

Citation: Sunnadeniya R, Bean A, Brown M, Akhavan N, Hatlestad G, Gonzalez A, et al. (2016) Tyrosine Hydroxylation in Betalain Pigment Biosynthesis Is Performed by Cytochrome P450 Enzymes in Beets (Beta vulgaris). PLoS ONE 11(2): e0149417. doi:10.1371/journal.pone.0149417

enzyme(s) that perform the initial step of the pathway, the hydroxylation of L-tyrosine to L-3,4-dihydroxyphenylalanine (L-DOPA), have largely remained unknown. Once L-DOPA is formed it can either be converted to betalamic acid (BA; step 2) by DOPA 4,5-dioxygenase (DODA) [1–4] or to cyclo-DOPA (step 3) [4,5]. BA is the key intermediate forming the common chromophore of all betalains. BA can spontaneously condense with amino acids or other amine groups to form yellow fluorescent betaxanthin pigments [6] or spontaneously condense with cyclo-DOPA to form red betacyanin pigments [7]. Earlier work identified the single enzyme responsible for converting L-DOPA to cyclo-DOPA in *Beta vulgaris* (beet) as BvCYP76AD1 [4], which is absolutely necessary to make red betalain pigments in this species.

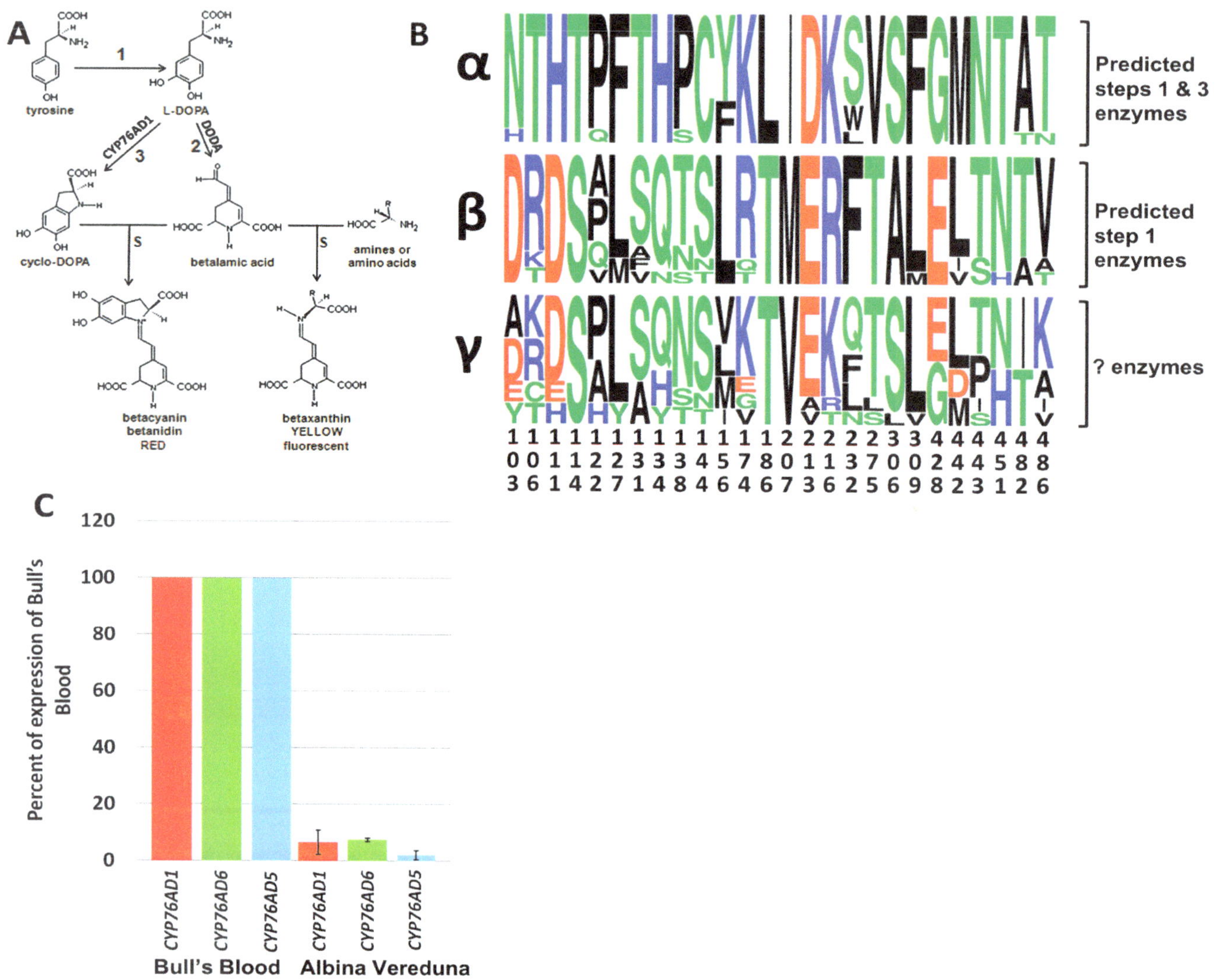

Figure 1. Identification of candidate genes for tyrosine hydroxylation. A. Biosynthetic scheme for betacyanin and betaxanthin production. Tyrosine hydroxylation is step 1. B. LOGO analysis of the three CYP76AD subfamily clades. C. qRT-PCR analysis of the beet CYP76AD1, CYP76AD5, and CYP76AD6 genes in non-transgenic beets. Bull's Blood is a very dark red table beet and Albina Vereduna is a white table beet. Bull's Blood is YY, expressing the dominant, high-pigment Y allele of BvMYB1 while Albina Vereduna is yy.

Investigators have proposed the hydroxylation of L-tyrosine to L-DOPA was performed by the monophenolase activity of a tyrosinase (or a polyphenoloxidase-PPO) [7,8]. There are reports of PPO transcripts that increase with betalain pigment accumulation in pokeweed fruits (*Phytolacca americana*; Phytolaccaceae [9], and PPOs have been purified from beet [10] and red Swiss Chard (same species as beet) [11]. A tyrosinase with both monophenolase and diphenolase [7] activities, and one with just monophenolase activity [12] have been purified from *Portulaca grandiflora* (Portulacaceae), forming L-DOPA from tyrosine and cyclo-DOPA from L-DOPA *in vitro*. Mushroom tyrosinases have been shown to perform the tyrosine hydroxylation *in vitro* and in transgenic plant tissues [13]. Even though these findings suggest PPO involvement in betalain biosynthesis, others point out that there is no direct evidence for PPO involvement in betalain biosynthesis *in vivo* [13–15]. Thus, until recently there was no direct evidence for identification of the genes and enzyme(s) responsible for performing the key first step, tyrosine hydroxylation, in the betalain biosynthesis pathway, representing the only ring biosynthetic step not molecularly identified. DeLoache et al. [5] recently described a yeast synthetic biology assay in which CYP76AD1 was shown to have both tyrosine hydroxylating activity and the ability to produce cyclo-DOPA from L-DOPA [4]. Polturak et al. [16] subsequently showed that CYP76AD6 is also capable of tyrosine hydroxylation.

Here, the search for the genes/enzymes responsible for tyrosine hydroxylation in the beet betalain biosynthetic pathway is described. Two novel beet cytochrome CYP450 genes (*CYP76AD5* and *CYP76AD6*) were identified with mRNA levels positively correlated with high betalain levels. These two genes are in the beta clade of CYP76AD genes described by Brockington et al. [17], while *CYP76AD1* is in the alpha clade. The newly identified CYP76AD5 and CYP76AD6 and the previously characterized CYP76AD1 [4,5] are all able to perform tyrosine hydroxylation leading to betalain biosynthesis not only in beet tissues, but also in yeast and Arabidopsis. However, while CYP76AD1 is also able to produce cyclo-DOPA, CYP76AD5 and CYP76AD6 do not have this activity. These differential clade biochemical activities are verified by showing that alpha and beta clade members, CYP76AD3 and CYP76AD15 respectively, from a representative of a different family,*Mirabilis jalapa* (Nyctaginaceae), have the same activities as alpha and beta clade members from beets (Amaranthaceae). The maintenance of both alpha and beta CYP76AD proteins would allow plants to differentially regulate these genes to produce different ratios of yellow and red pigment, thus allowing yellow, orange, red, or patterned pigmentation.

MATERIALS AND METHODS

Cytochrome P450 and PgDODA Cloning and Sequence

The *CYP76AD1* (Genbank HQ656023; see S1 Table for list of genes) sequence was obtained from Roche/454 pyrosequencing performed on cDNA from 7-day-old *Beta vulgaris*cv. W357B seedling hypocotyls [4]. cDNA from 7-day-old hypocotyls of three white beet cultivars, table beet varieties Blankoma and Albina Vereduna, and sugar beet variety C869, was sequenced by SOLiD (Applied Biosystems) nextgen sequencing. The SOLiD RNAseq reads were mapped separately on to the assembled contigs of the red table beet cultivar W357B Roche/454 reference sequence database [4]. The approximate fold changes of gene expression from red to each white beet variety was calculated as follows. The ratio of the number of reads for a particular contig over the total number of RNAseq reads in that beet variety was determined. Then the fold difference in these ratios was determined in red vs. white.

In order to search for candidate genes for the betalain pathway tyrosine hydroxylation step, specifically CYP450s or PPOs, genes with relatively high expression levels in red beets compared to white beets were identified. Five-fold change in expression level was used as the base line to identify candidate genes. Expression levels of candidates were then tested through qPCR.

Previous work reported the identification and cloning of *CYP76AD1*, which lies at the R locus [4,18]. CYP76AD5 and CYP76AD6 were both found to be differentially expressed in white and red beet RNAseq databases. The cDNA sequences were amplified by PCR from cDNA made from red hypocotyls of 7-day-old W357B seedlings. Primer sequences: CYP76AD5StartF 5'ATGGATAACACTACACTTGCATTGATACTTTCTTC3'; CYP76AD5StopR 5'TTATTTCCTAGGTACGGGAATAACTTGGAGAGGC3'; CYP76AD6StartF 5'ATGGATAACGCAACACTTG3'; CYP76AD6StopR 5'CCAGTTCCCAGAAACTAG3'. Gateway recombination sequences were included in all primers but are not shown. The products were cloned into pDONR222 (Invitrogen) and sequenced: *CYP76AD5* (1506 bp; Genbank KM592961), *CYP76AD6* (1500 bp; Genbank KM592962). These clones were recombined into the plant TDNA vector pB7WG2 [19]. This creates p35S::gene for overexpression of the*CYP76AD5*, and *CYP76AD6* coding regions in plants. These plasmids were transformed into *Agrobacterium rhizogenes* ARqua1 [20] for beet transformation, which produces transformed hairy root tissues, or *Agrobacterium tumefaciens* GV3101 [21] for Arabidopsis transformation. A p35S::GUS construct was used as a control.

Cloning of the *Portulaca grandiflora* DODA cDNA (*PgDODA*) into pDONRZeo and subsequently into the yeast vectors, pVV214 or pVV200 [22] has been described [4]. To express PgDODA in plants, the cDNA was recombined into the plant TDNA vector pK7WG2 [19] to make p35S::PgDODA. This was transformed into *A. tumefaciens* GV3101 [21] for Arabidopsis transformation.

The cDNA sequences for *CYP76AD3* and *CYP76AD15* were amplified from 7-day-old red 4 o'clock seedling cDNA using the following primers: CYP76AD3StartF 5'ATGGATTTCTTAACCCTTGTCATG3';CYP76AD3StopR 5'TCAATATTTAATAAGAGGAATAATCTC3'; CYP76AD15StartF 5'ACGTTCAAGCATAGACTAACC3'; CYP76AD15StopR 5'TGATTATCTCCCAACCATCG3'. Gateway recombination sequences were included in all primers but are not shown. The products were recombined into pDONR222 (Invitrogen) and sequenced: *CYP76AD3* (1680 bp; Genbank HQ656026) *CYP76AD15* (1515 bp; Genbank KM516798). These were recombined into pVV214 for yeast expression.

Over Expression in Beet Hairy Root Cultures

Seeds of yellow beet cultivar Golden Globe (from W. Atlee Burpee & Co.) were sterilized and germinated on MS media with 3% sucrose in petri plates. Overnight liquid LB cultures of the*A. rhizogenes* strains were placed into a 1 ml tuberculin syringe. A 30-gauge needle was used to make wounds on the hypocotyls of 10-day-old seedlings, injecting the *A.rhizogenes* culture into the wounds. Inoculated seedlings were maintained on MS media with 3% sucrose for 2 days and then transferred to the same media with Timentin 200mg/L to kill the bacteria. Plates were incubated at 22°C in continuous light. "Hairy roots" emerged from wound sites on the hypocotyl in about 2–3 weeks. Roots were excised from the wound site and maintained on MS with 3% sucrose and 200 mg/L Timentin. At least ten independent transgenic lines with consistent phenotype were produced for each construct.

Arabidopsis Transformation

Arabidopsis thaliana mutant *ttg1-1* in the L*er* ecotype (from the Arabidopsis Biological Resource Center) was used for all experiments because it does not produce anthocyanin pigments. Arabidopsis plants were transformed by vacuum infiltration with the described constructs in *A. tumefaciens* GV3101 [21].

Quantitative RT-PCR

For gene expression analysis, tissue was collected from

7-day-old untransformed hypocotyls of the beet cultivars Bull's Blood (red; from Johnny's Selected Seeds) and Albina Vereduna (white; from John Scheepers Kitchen Garden Seeds) or of transformed hairy roots generated on Golden Globe by overexpressing *CYP76AD1*, *CYP76AD5*, or*CYP76AD6*. Total RNA was extracted using Qiagen plant RNeasy mini kit and used in 20 μL reverse transcriptase reactions using 1 μg RNA and 1 μg oligo-dT. qRT-PCR was performed as described [23]. Four technical replicates were performed for each target from each biological replicate and for actin controls using 400 nM of the appropriate qRT-PCR primers: BvACT (Genbank HQ656028) 5›TCTATCCTTGCATCTCTCAG3› and 5›ATCATACTCGCCCTTGGAGA3›; CYP76AD1RTF 5› CTTTTCAGTGGAATTAGCCCACC3› and CYP76AD1RTR 5›CCCAATATCTTCCATAATGTTCCA3›; CYP76AD5RTF 5›TCATTTTCATAAGTTCTTCG3› and CYP76AD5RTR 5›CAGTCAACTCATCACTCATC3›; CYP76AD6RTF 5›GCTAACCGAACCATTCCTGA3› and CYP76AD6RTR 5›TTGGACAGCGGAGATTTTTC3›; BvPPO (Genbank KR337592) BvPPORTF 5›AAAACAGTTGAGGCTCCGACC3' and BvPPORTR 5'GGAAGAGCTTTCATTAAGGC3'. Results were analyzed by the comparative cycle threshold method according to the user manual for the ABI PRISM Sequence Detection System. Each experiment was performed with three biological replicates and mean values of all technical replicates are shown.

Expression in Yeast and Feeding Assay

The cDNAs of *CYP76AD1*, *CYP76AD6*, *CYP76AD5*, *CYP76AD3*, and *CYP76AD15* in yeast expression vector pVV200 [22], and pVV200 empty vector control were transformed into the yeast strain WAT11 [24]. The WAT11 yeast strain contains a galactose-inducible nicotinamide adenine dinucleotide phosphate (NADPH)–cytochrome P450 reductase from Arabidopsis. Yeast cultures (pVV200/*CYP76AD1*; pVV200/*CYP76AD5*; pVV200*CYP76AD6*; and pVV200/empty) were grown overnight in minimal medium with galactose and supplemented with 100 mg/L leucine, 20 mg/L histidine, and 40 mg/L adenine. The cultures were pelleted by centrifugation. The yeast were resuspended to an OD_{600} of 1.1 in 3 mL of above media with 10 mM tyrosine and 25 mM ascorbic acid.

The WAT11 yeast strain was also co-transformed with the above vector constructs and*PgDODA* cloned into pVV214 [4,22] to make yeast carrying: pVV200/*CYP76AD1* + pVV214/*PgDODA*; pVV200/*CYP76AD6* + pVV214/*PgDODA*; pVV200/*CYP76AD5* + pVV214/*PgDODA*; pVV200/*CYP76AD3* + pVV214/*PgDODA*; pVV200/*CYP76AD15* + pVV214/*PgDODA*; and pVV200/empty vector + pVV214/*PgDODA*. These yeast co-expression cultures were also grown overnight in minimal medium (with galactose and supplemented with 100 mg/L leucine, 20 mg/L histidine and 40 mg/L adenine) and resuspended in 3 mL as above with three separate substrate feedings: 10 mM L-DOPA + 25 mM ascorbic acid; 10 mM tyrosine + 25 mM ascorbic acid; and no enzyme substrate + 25 mM ascorbic acid, for each independent strain. Cultures were grown for 24 hrs after feeding and pelleted by centrifugation and 500 μL of the supernatant was retained for pigment analysis as described below in the pigment chemical analysis section.

Pigment Chemical Analysis

The preparation of pigment standards for High Performance Liquid Chromatography / Mass Spectrometry (HPLC/MS) analysis was performed by preparative Thin Layer Chromatography (TLC) from red and yellow beet extracts as described [25].

The supernatants from the yeast feeding assays, or plant pigment extracts from transgenic hairy roots overexpressing *CYP76AD1*, *CYP76AD5*, or *CYP76AD6* in beet cultivar Golden Globe, were filtered through a 0.2 μm membrane filter. The filter residue was rinsed with 100% methanol to elute the pigment completely. Eluted pigment was concentrated by vacuum at 30°C and resuspended in 100 μL of 0.1% ascorbic acid. Resuspended pigments were analyzed by HPLC/MS.

Five microliter aliquots of extracts were injected into an Agilent 1260 Infinity HPLC system interfaced with an Agilent 6530 Accurate-Mass QTOF mass spectrometer. The samples were separated on a ZORBAX Eclipse XDB-C18 column (5 μm, 50 x 4.6 mm) using a two-stage LC program. The first stage of the program was a 2 minute isocratic run with a mobile phase of 1% formic acid in water at a flow rate of 0.1 mL/min. The second stage of the program, lasting 6 minutes, included a flow rate gradient from 0.1 mL/min to 0.5 mL/min and a mobile phase gradient of 0–33% B, where mobile phase A is 1% formic acid in water and mobile phase B is 0.1% formic acid in acetonitrile. Mass spec detection of the eluent was performed using positive mode electrospray ionization. Gas temp was 350^0 C, drying gas flow was 12 L/min, nebulizer pressure was 50 psig, VCap was 5000 V, fragmentor voltage was 200 V, the skimmer was set to 65 V. The mass spec data was acquired scanning the m/z range from 100 m/z to 3200 m/z.

The exact mass of L-DOPA is 197.188 g/mol. The observed mass for L-DOPA in Fig 2 is 198.076 for CYP76AD1, CYP76AD5, and CYP76AD6, and 198.083 for the 35S:GUS vector. The observed mass for L-DOPA in Fig 3 is 198.076 for CYP76AD1, CYP76AD5, and CYP76AD6, and 198.075 for the empty vector.

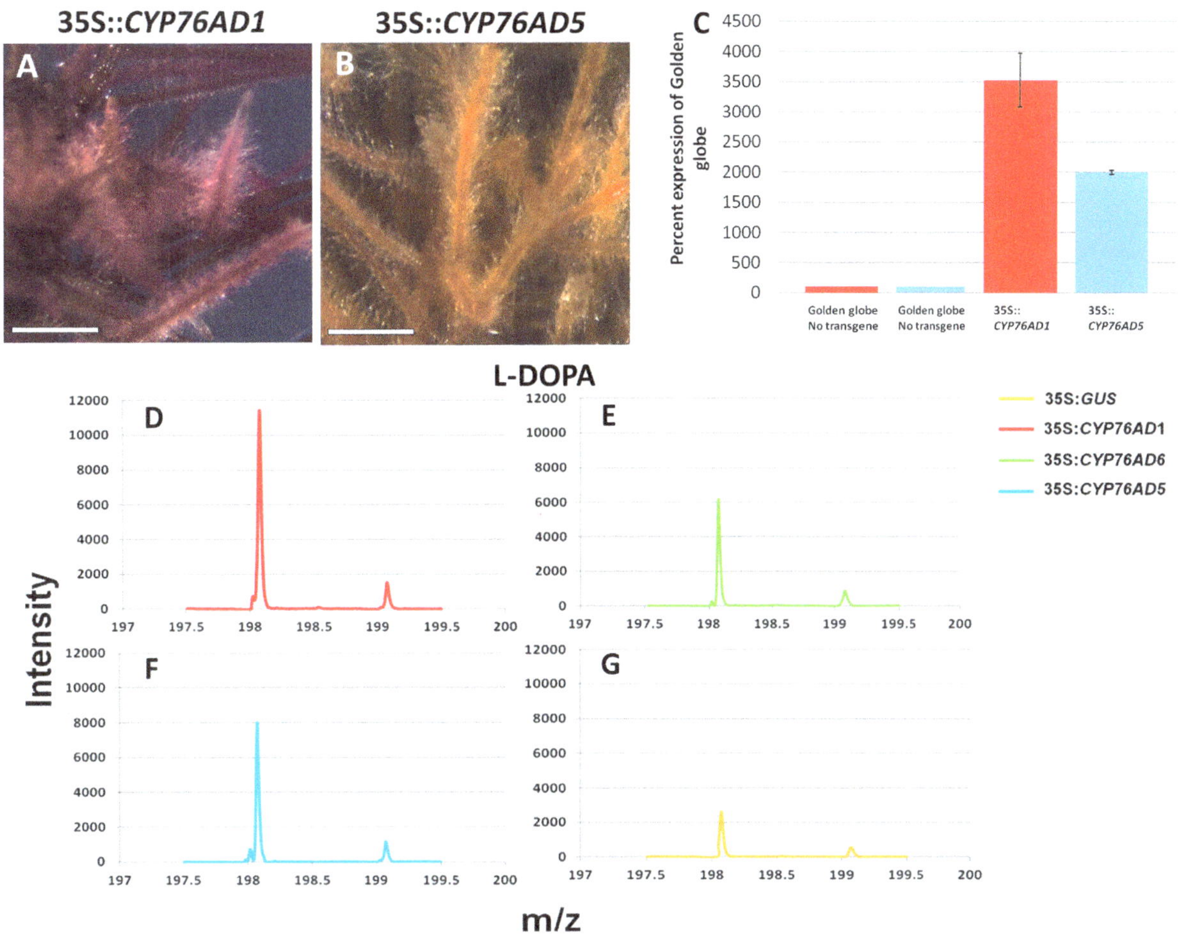

Figure 2. Overexpression of CYP76AD1, CYP76AD6, or CYP76AD5 results in increased L-DOPA content in beet. A. and B. CYP76AD1 complemented, but CYP76AD5 did not complement the *rr*(*CYP76AD1*) beet mutant (Golden Globe) from yellow to red. C. qRT-PCR verified that*CYP76AD1* and *CYP76AD5* were overexpressed in transgenic hairy root cultures. D, E, F, and G. Mass spectrometry analysis of L-DOPA in Golden Globe beet roots overexpressing CYP76AD1, CYP76AD6, CYP76AD5, and GUS, respectively. Error bars are s.d. Each experiment was replicated three times. Scale bars are 3.5 mm.

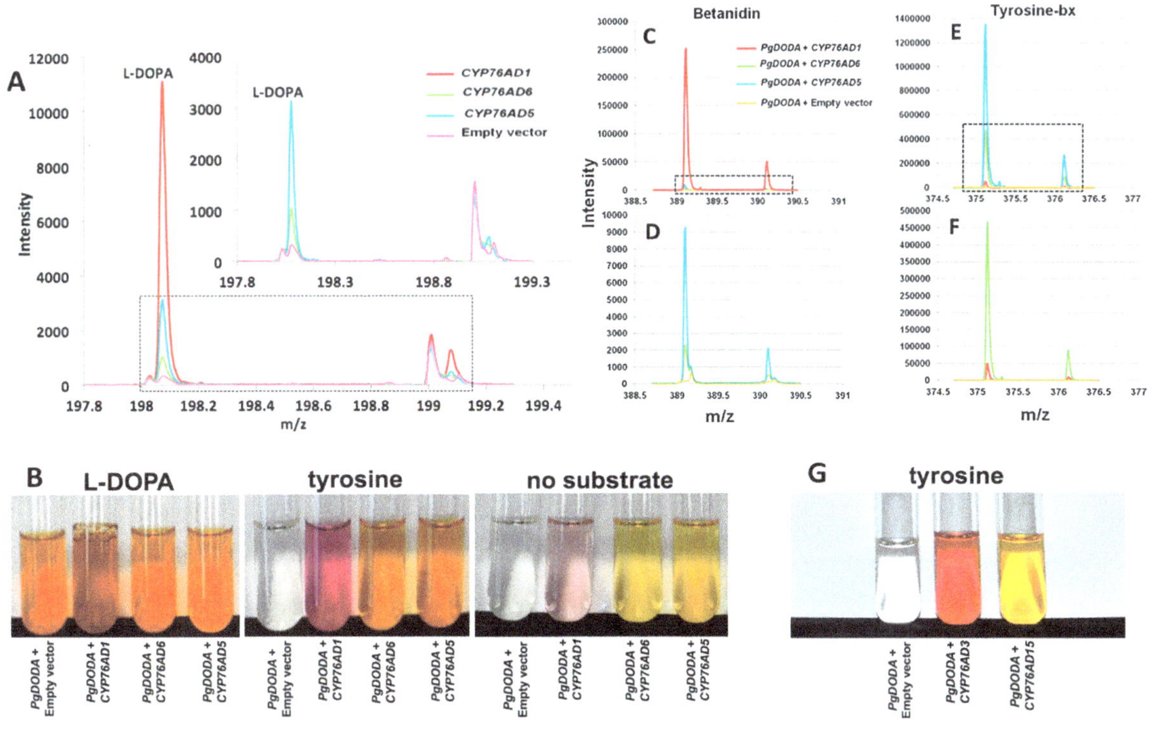

Figure 3. All tested CYP76AD proteins perform tyrosine hydroxylation in yeast. A. HPLC/ms analysis of yeast expressing CYP76AD1, CYP76AD6, and CYP76AD5 individually shows that all will hydroxylate tyrosine to L-DOPA. Inset shows expanded view of lower peaks. Yeast

containing empty vector either have weak tyrosine hydroxylation activity or produce a compound that co-migrates with L-DOPA. B to F. Same yeast strains as in A with the addition of PgDODA. B. images of the cultures fed L-DOPA, tyrosine, or no added substrate with the tyrosine-fed panel corresponding to C through F. C and D. HPLC/MS analysis of the red betanidin pigment with the lower peaks for CYP76AD6, CYP76AD5, and empty vector expanded in D. E. and F. HPLC/MS analysis of the yellow tyrosine-betaxanthin pigment with the lower peaks for CYP76AD6, CYP76AD1, and empty vector expanded in F. Note that CYP76AD1 produces large amounts of betanidin and small amounts of betaxanthin. G. Yeast cultures expressing PgDODA alone and PgDODA with CYP76AD3 or CYP76AD15 from *Mirabilis jalapa* and fed tyrosine. CYP76AD3 has CYP76AD1-like activity producing red pigment and CYP76AD15 has CYP76AD6/5-like activity producing yellow pigment.

LOGO Analysis of Conserved Amino Acid Sequences Indicative of the CYP76AD Clades

The sequences used to create the phylogeny (S1 Fig; S1 Table) were aligned using Clustal Omega (http://www.ebi.ac.uk/Tools/msa/clustalo/). The alignment was analyzed for residues that appeared to diverge according to the phylogenetic clades (highlighted in S2 Fig). These residues were assembled into a shortened sequence alignment and made into a frequency plot via WebLogo (Fig 1B; http://weblogo.berkeley.edu/logo.cgi). The number for each residue corresponds to its position in CYP76AD1 and the height of the residue is proportional to its appearance within the clade members.

Phylogenetic Analysis of the Betalain Pathway CYP450s

The cDNA sequences of *BvCYP76AD1*, *BvCYP76AD5*, and *BvCYP76AD6* were each submitted to BLAST searches at NCBI and OneKP to identify orthologs from across the Caryophyllales and beyond. The resulting set of sequences and selected outgroup sequences (members of two other CYP76 subfamilies) were aligned using the MUSCLE alignment tool in GENEIOUS [26] and edited to match reading frames manually. The resulting alignment was assessed in JMODELTEST 2 [27] to determine the best fitting model of molecular evolution. Phylogeny estimation was performed using Bayesian inference as implemented in MRBAYES 3.2 [28] under GTR+I+G. Following 20 million generations, convergence was complete (standard deviation of split frequencies was <0.001) and a consensus tree generated from the trees sampled every 1000 iterations after discarding the first 25% of samples as burn-in (S1 Fig).

Ancestral amino acid sequences were generated for all internal nodes of the CYP76 gene tree using CODEML as implemented in PAMLX version 1.3.1 [29]. Input for the reconstructions included the same gene tree shown in S1 Fig and an amino acid alignment of the same sequences, produced using the MUSCLE alignment option in GENEIOUS version R9 [26]. New ancestral sequences were aligned with the original sequences analyzed in the gene tree and amino acid frequency logo and compared at sites of interest.

RESULTS

To identify betalain pathway gene/s responsible for tyrosine hydroxylation, RNAseq was performed on three white beet cultivars—Blankoma, Albina Vereduna, and sugar beet cv. C869. Red beets contain a dominant Y betalain regulatory locus allele [18,30] while the white beets are homozygous yy. White beets make betalains but accumulate very low levels of betalain pigmentation compared to red beets.

The reads from the RNAseq data of the white beet varieties were mapped onto a red beet reference sequence database [4] and genes were identified that were both expressed in lesser amounts in white beets as compared to red and might encode tyrosine hydroxylating enzymes, such as CYP450s or PPOs. This candidate gene search revealed two novel CYP450 genes from the same gene family as *CYP76AD1*: *BvCYP76AD5* and *BvCYP76AD6*(Fig 1B, S1 Fig). In the RNAseq database analysis, the expression of CYP76AD5 is down 5, 41 and 9 fold in white beet cultivars Albina Vereduna, Blankoma, and C869, respectively, compared to red beet, while the expression of *CYP76AD6* is down 14, 115 and 6 fold. Meanwhile, known betalain biosynthetic genes, *Beta vulgaris DODA1* (*BvDODA1*) and*CYP76AD1*, are down-regulated many fold in all three white beet varieties: *BvDODA1* is down 71, 207, 57 fold while *CYP76AD1* is down 19, 22, 26 fold in Albina Vereduna, Blankoma, and C869, respectively.

Genes considered to be likely candidates via RNAseq analysis were also subjected to quantitative RT-PCR (qRT-PCR) to confirm the expression differences observed in the RNAseq data. qRT-PCR of *CYP76AD1*, *CYP76AD5* and *CYP76AD6* showed that these genes were expressed at 7%, 2%, and 8% respectively, in Albina Vereduna white beets vs. red beet cultivar Bull's Blood (Fig 1C).

A single *PPO* (*BvPPO*; genbank KR337592) was also identified in the RNAseq database and in the sugar beet genome [31] and its expression was analyzed in white beets vs. red via qRT-PCR. The expression level of *BvPPO* was 2.1 fold higher in Albina Vereduna and 3.8 fold higher in sugar beet C869 than in the dark red cultivar Bull's Blood. This lack of correlation between *BvPPO* expression and betalain production does not support a role for this gene in betalain biosynthesis.

cDNAs for *CYP76AD5* and *CYP76AD6* were amplified from red beet seedlings. *CYP76AD5*and *CYP76AD6* have open reading frames of 1506 bp and 1500 bp respectively; the nucleotide identity is 81% while the amino acid identity is 85%. The amino acid identity between *CYP76AD1* and *CYP76AD6* is 72% while the identity between *CYP76AD1* and*CYP76AD5* is 69%.

Previous work showed that CYP76AD1 performs step 3 (Fig 1A) of the betalain pathway converting L-DOPA to cyclo-DOPA [4]. Golden Globe is the yellow beet commonly sold in grocery stores and it carries a mutant allele of *CYP76AD1*, which lies at the R locus. Ectopic expression of CYP76AD1 in Golden Globe complements it from yellow to red. To test whether *CYP76AD5* or *CYP76AD6* can also perform step 3, these genes were overexpressed in Golden Globe tissue via *Agrobacterium rhizogenes* and pigment phenotypes were observed in transformed hairy roots. Overexpression of *CYP76AD1* in yellow beetroots complemented them to red (Fig 2A). Overexpression of *CYP76AD5* (Fig 2B) and *CYP76AD6* (not shown) in yellow beets resulted in yellow, non-complemented roots.

qRT-PCR was performed to verify that the respective genes were overexpressed in all hairy root cultures (Fig 2C).

Increases in L-DOPA levels in tissues carrying these transgenes were observed using HPLC/MS analysis (Fig 2D to 2G). The L-DOPA mass spec peak intensity levels (m/z 198) were highest for CYP76AD1, followed by CYP76AD5, and CYP76AD6, in tissues overexpressing these genes respectively, vs. 35S::GUS overexpression control lines, indicating that all three of these genes are able to convert tyrosine to L-DOPA in beets.

CYP76AD1, *CYP76AD6*, and *CYP76AD5* were also expressed in yeast to test for tyrosine hydroxylation activity by detecting L-DOPA formation after growing in media containing 10 mM tyrosine (Fig 3A). L-DOPA, the product of tyrosine hydroxylation, was detected with all three CYP450s through HPLC/MS analyses of extracts (Fig 3A). The L-DOPA peak intensity was highest with *CYP76AD1*, followed by *CYP76AD6* and *CYP76AD5* respectively. There is also a very low intensity mass spectrometry peak that co-migrates with L-DOPA in the empty vector control. It is not known if this is L-DOPA or another compound.

The CYP450s were also co-expressed in yeast with the DODA gene (step 2) from *Portulaca grandiflora*, *PgDODA*, to analyze red and yellow betalain pigment formation (Fig 3B to 3F). These were grown in media for 24 hours with: 10 mM tyrosine, the substrate for step 1; or 10 mM L-DOPA, the substrate for both steps 2 and 3 (Fig 1A); or no supplemented feeding. Of the cultures fed L-DOPA: *PgDODA* alone (*PgDODA* + empty vector); *PgDODA* + *CYP76AD6*;*PgDODA* + *CYP76AD5* produced visible yellow to orange pigmentation; while *PgDODA* +*CYP76AD1* produced red pigmentation. Of the cultures fed tyrosine, *PgDODA* + empty vector produced no color; *PgDODA* + *CYP76AD6*; *PgDODA* + *CYP76AD5* produced yellow to orange pigmentation; while *PgDODA* + *CYP76AD1* produced red pigmentation. In the absence of added substrate feeding all four cultures produced colors similar to tyrosine feeding but with lower intensity. Pigments produced without supplemental feeding were undoubtedly through the conversion of native yeast-produced tyrosine to betalain pigments.

The pigments extracted from the four yeast cultures co-expressing DODA and fed tyrosine were analyzed by HPLC/MS to detect the presence of betanidin, the undecorated red betalain pigment, and tyrosine-betaxanthin (tyrosine-bx) as a representative of yellow betaxanthins (Fig 3C to 3F). Betanidin (m/z 389) was detected at very high mass spec intensity in *PgDODA* + *CYP76AD1* as previously reported [4] and interestingly it was also present at very low intensities in *PgDODA* + *CYP76AD6* and *PgDODA* + *CYP76AD5* at about 27 and 100 fold less intensity, respectively (Fig 3C with lower peaks expanded in Fig 3D). These data suggest that CYP76AD6 and CYP76AD5 both have very weak ability to produce cyclo-DOPA, step 3, which is required to produce the betanidin pigment (Fig 3D). Betanidin was absent in the *PgDODA* + empty vector sample, but note that this sample has a small peak that runs at a slightly higher m/z ratio than betanidin. Tyrosine-bx (m/z 375) was detected at high intensities in both *PgDODA* + *CYP76AD6* and *PgDODA* + *CYP76AD5*samples. The tyrosine-bx level was very low in *PgDODA* + *CYP76AD1* and absent in*PgDODA* + empty vector (Fig 3E with lower peaks expanded in Fig 3F).

While expression in yeast clearly indicates the activities of the divergent beet CYP76ADs, these activities were verified by expression in *Arabidopsis thaliana*. Expression of*35S*::*PgDODA* alone results in yellow betaxanthin production only when the plants are fed L-DOPA (data not shown), the same as seen in yeast, and the same that has been previously demonstrated in Arabidopsis [32].

Similarly, *35S*::*CYP76AD5* expression alone in Arabidopsis results in no pigmentation in most of the tissues of the living plant (Fig 4A, 4B and 4C left; exception discussed below). When *PgDODA* and *CYP76AD5* are co-expressed in Arabidopsis, the plants develop intense yellow betaxanthin pigmentation (Fig 4A, 4B and 4C right) without any substrate feeding, again repeating the yeast result. This yellow pigment is easily visible in all non-photosynthetic tissues, where chlorophyll does not mask the yellow color. This betaxanthin pigment is also highly fluorescent [33] and can be seen via fluorescence microscopy (Fig 4C and 4D). In addition, it is noted that expression of CYP76AD5 alone causes the seeds to develop black pigmentation and become small and shriveled (Fig 4E), and upon drying at maturity, many of the plants develop this same gray or blackish pigment, mostly on siliques and stems (not shown). This is probably a form of melanin pigment produced from L-DOPA, but the pigment has not yet been analyzed.

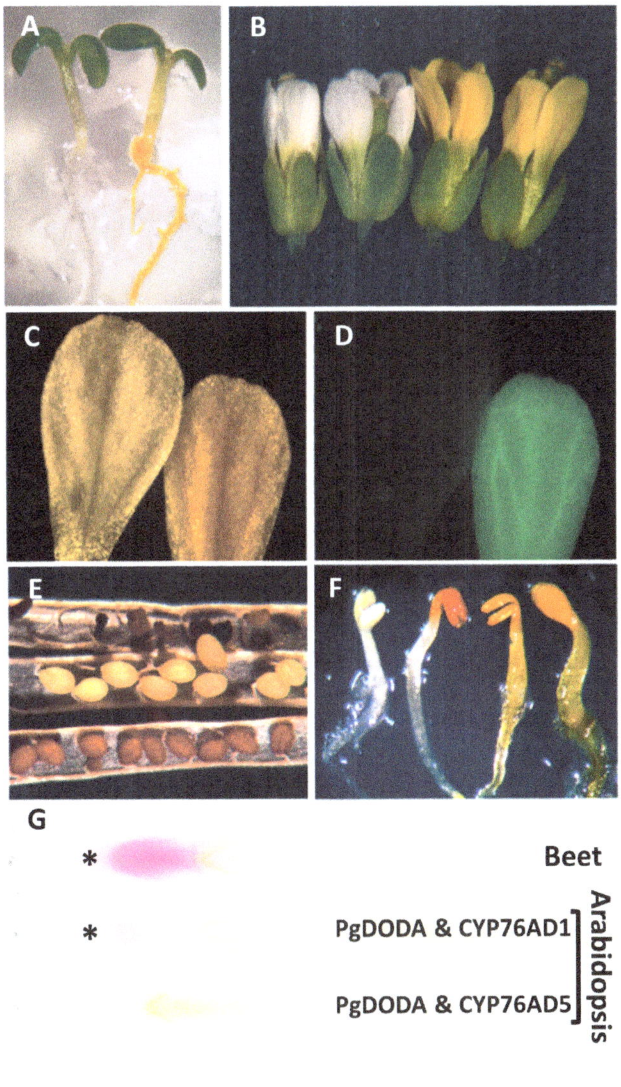

Figure 4. Arabidopsis expressing betalain biosynthetic genes produce visible pigment phenotypes. All Arabidopsis are *ttg1-1* mutant except the wild type L*er* ecotype in panel E. A. left,*35S:CYP76AD5*

alone, right *35S:CYP76AD5* and *35S:PgDODA*. B. left two flowers and right two flowers the same as in A. C and D are images of the same two petals. Left and right, the same as in A and B. C. white light, D. excitation with UV irradiation showing fluorescence of betaxanthins. E. mature siliques with seeds; top-*35S:CYP76AD5* in *ttg1-1* alone, middle-*35S:PgDODA* in *ttg1-1* alone, bottom-wild type L*er*, no transgene. F. seedlings, left to right- *35S:CYP76AD1* and *35S:PgDODA*unfed; *35S:CYP76AD1* and *35S:PgDODA* fed 1.5 mM tyrosine; *35S:CYP76AD5* and*35S:PgDODA* fed 1.5 mM tyrosine; *35S:CYP76AD5* and *35S:PgDODA* unfed. G. Thin Layer Chromatography of betalain pigments: top-untransformed red beet; middle-Arabidopsis transformed with *35S:CYP76AD1* and *35S:PgDODA* fed 1.5 mM tyrosine; bottom-Arabidopsis transformed with *35S:CYP76AD5* and *35S:PgDODA* fed 1.5 mM tyrosine. Asterisks mark the presence of red betacyanins.

Expression of 35S::CYP76AD1 alone does not produce pigmentation in any plant tissues, not even in the seeds (not shown). When PgDODA and CYP76AD1 are co-expressed in *A.thaliana*, no pigmentation is visible without feeding, despite the predicted presence of red and/or yellow coloration. However, when 3-day-old seedlings were transferred from germination media to the same media with 1.5 mM tyrosine prior to greening, red pigmentation was observed in the cotyledons and upper hypocotyl (Fig 4F second from left). If they are transferred after the cotyledons have expanded and turned green, the red pigment is not produced. Thin Layer Chromatography (TLC) was used to verify that this pigment was betacyanin, where it can be seen that the red pigment co-migrates with beet betacyanin (Fig 4G), and that similarly treated PgDODA/CYP76AD5 co-expressing plants do not make betacyanin but only make yellow betaxanthins under these conditions (Fig 4Fthird from left, Fig 4G).

Phylogenetic analysis by Brockington et al. [17] indicates that CYP76AD1 is a member of a distinct clade (alpha) from CYP76AD5 and CYP76AD6 (beta). To test whether the two divergent CYP76AD1-like and CYP76AD5/6-like clades do indeed correspond to the divergent functions of CYP76AD1 and CYP76AD5/6, the activities of another member from each group was analyzed from a betalain-producing species that is not closely related to *B.vulgaris* (Amaranthaceae family). The sequence of *Mirabilis jalapa* (4 o'clocks; Nyctaginaceae family) MjCYP76AD3 is BvCYP76AD1-like (alpha clade) while MjCYP76AD15 is BvCYP76AD5/6-like (beta clade). Co-expression of MjCYP76AD3 and PgDODA in yeast produced red pigment (agrees with the findings of Deloache et al. [5]) and expression of MjCYP76AD15 and PgDODA produced only yellow pigment (Fig 3G), consistent with the hypothesis that the CYP76AD1-like alpha clade members are able to perform steps 1 and 3 while the CYP76AD5/6-like beta clade members just perform step 1 (Fig 1A).

DISCUSSION

Betalain pigments are particularly interesting from an evolutionary perspective. Questions remain about how and why they arose within a single order of flowering plants, the Caryophyllales, and why they are mutually exclusive with the almost ubiquitous anthocyanins. Given the recently identified biosynthetic steps and pathway regulatory mechanisms described here and in previous publications [1,4,5,16,30] these questions are now being addressed. Recent molecular phylogenetic analysis [17] indicates that betalains arose a single time following gene duplication and neofunctionalization of precursors of the modern *CYP76AD* and *DODA* genes.

Betalain pigments are also of interest for their many applications in the food industry, for the molecular breeding of flower and fruit colors, and, with recent discoveries, for the metabolic engineering of plant and non-plant organisms to produce these pigments. Betalains used as added or engineered food colorants may provide protection against certain oxidative stress-related disorders in humans [34–37].

Three main enzymatic steps convert tyrosine to the betalain pigment backbone. The DOPA 4,5-dioxygenase responsible for producing betalamic acid from L-DOPA (step 2; Christinet et al., 2004) and the CYP76AD1 enzyme responsible for conversion of L-DOPA to cyclo-DOPA in beet (step 3) [4] have been identified. Tyrosine hydroxylation (step 1) is the last of the three main betalain ring structure biosynthetic steps to be described at the molecular level. Deloache et al. [5] first showed that the CYP76AD1 enzyme was also able to perform tyrosine hydroxylation (thus performing both steps 1 and 3) which was confirmed by Polturak et al. [16] and these results are confirmed here. The present work demonstrates that other members of the alpha clade also have steps 1 and 3 activities and that three members of the beta clade only have step 1 activity (tyrosine hydroxylation). This confirms and extends Polturak et al.'s work on *CYP76AD6* [16]. It has been shown that yellow or golden beets are the result of mutations in *CYP76AD1* and that *CYP76AD1* lies at the beet *R*locus [4,18]. While transgenic expression of CYP76AD1 will complement yellow roots to red, performing step 3 (Fig 2A), expression of beta clade members, CYP76AD5, CYP76AD6, and CYP76AD15 will not. Thus beets appear to have a single enzyme that performs step 3 but they are at least threefold redundant for step 1. The functional divergence of the similar CYP450 genes presented here would allow betalain producing plants to selectively produce yellow or red pigment by differential regulation of the CYP450 that produces cyclo-DOPA. It has been shown that mutating the phenylalanine at position 309 to leucine in CYP76AD1 ablates its ability to produce cyclo-DOPA but leaves its ability to hydroxylate tyrosine intact [5]. It is noted here that the vast majority of beta clade members have leucine at this position with at least one member having methionine, a conservative change from leucine. It is also noted that the vast majority of the gamma clade members have leucine at this position, which may indicate that they are also tyrosine hydroxylases.

If CYP76AD1, CYP76AD5 and CYP76AD6 are the only enzymes responsible for tyrosine hydroxylation in beet, then knocking them all out should prevent any betalain pigment production. It has been shown that VIGS against CYP76AD1 and CYP76AD6 leads to loss of betalain pigmentation [16]. Based on the high sequence identity between CYP76AD5 and CYP76AD6, it's quite possible that they were both silenced in those experiments. This triple redundancy for step one function, and possibly more redundancy, would explain the lack of a step 1 mutant in beet, and may explain that same absence in any known species. Within beet, CYP76AD5 and CYP76AD6 are the most similar paralogs to each other (~85% identity) and to CYP76AD1 (~70% identity), but there are several other paralogs in the beet genome, from the CYP76AD gamma clade, in the 50–60% amino acid identity range and all of these have leucine at the critical F309 position. This may indicate that the gamma clade members can also redundantly perform tyrosine hydroxylation, but they have not been tested. The fact that Arabidopsis expressing DODA do not produce betaxanthins indicates that there is no native tyrosine hydroxylating activity in *A. thaliana* [32]. However, in support of the existence of non-PPO/non-CYP76AD plant enzymes able to perform the L-DOPA to cyclo-DOPA step is the fact that *Antirrhinum majus* (snapdragon) cells expressing the*PgDODA* gene and fed L-DOPA produce not only yellow betaxanthin but also the red betacyanin pigments [32]. The arguments are that if a PPO were producing cyclo-DOPA, the same enzyme would produce LDOPA from tyrosine leading to betalamic acid, and, CYP76AD subfamily enzymes are not found outside the Caryophyllales order [17] so snapdragon would not have these enzymes. This is in contrast to beet where mutations in CYP76AD1 alone appear to knock out the ability to produce cyclo-DOPA (step 3, Fig 1) or betanin; i.e. there is no enzyme in beets other than CYP76AD1 that performs this function.

Recent phylogenetic analyses indicate that both the CYP76AD and DODA alpha and beta groups have arisen via gene duplication and subsequent diversification and that these events may underlie the origin of betalains [17]. At this point the minimum number of changes required for the functional divergence of CYP76AD1-like and CYP76AD5/6-like enzymes is not known, but polymorphisms identified from protein alignments of presumed orthologs and the mutagenesis work of DeLoache et al. [5] point to where these mutations probably lie.

It will be interesting to determine whether the CYP76AD1 (step 1 and 3 performing enzymes), or CYP76AD5/6 (step 1 only enzymes) CYP450s better represent the ancestral type. If the CYP76AD5/6 type predates the CYP76AD1 type, it would indicate that those first betalain-producing plants only had yellow betalains, which would be unlikely to supplant anthocyanins through analogous biological function. The latter could not have occurred until the new, step 3 activity evolved, providing betacyanins with hues similar to anthocyanins. If the CYP76AD1 type evolved first, this would indicate that red betalains predominated, allowing anthocyanins to be functionally replaced in many or all contexts. It would also mean that evolving CYP450s that limit the production of red while leaving yellow intact was an important subsequent evolutionary event providing a selective advantage. Of the polymorphisms that differentiate the CYP76AD1 and CYP76AD5/6 groups, the next closest orthologs (gamma group) are more similar to the CYP76AD5/6 genes, suggesting that tyrosine hydroxylation (step 1) was the ancestral function and cyclo-DOPA formation (step 3) evolved later in the CYP76AD1 lineage. Further, ancestral sequence reconstructions indicate that the ancestral copy of the alpha and beta groups would have had a leucine at the pivotal amino acid position 309, which likely would have limited its activity to step 1, thus producing only yellow pigments. This is confounded by the finding here that CYP76AD5/6 appear to have weak step 3 activity when overexpressed in yeast; however, in the beet *r* mutant, AD5/6 cannot replace CYP76AD1's step 3 activity.

The current taxonomic distribution of betalains has been a perplexing feature of angiosperm evolution. The most recent phylogenetic analyses suggest that betalains have had a single origin in the Caryophyllales, with subsequent betalain losses and reversions to anthocyanin production. What remains unclear is why betalains may have arisen in the first place and how they supplanted the well-entrenched and successful anthocyanins. As an alternative to outright anthocyanin replacement by betalains, the betalain pathway may have evolved in a lineage that lost the ability to synthesize anthocyanins. Indeed, some extant species in the Caryophyllales reportedly make neither betalains nor anthocyanins [38]. In this scenario, betalains would not have replaced anthocyanins outright but instead evolved to compensate for the loss of red/violet pigments. Such a scenario would likely require long-term persistence of a lineage that made no or little anthocyanins. This scenario is also consistent with the absence of species that possess both pigment pathways but would also require some groups to regain the anthocyanin pathway, such as the Molluginaceae and Caryophyllaceae. Signatures of such pathway recoveries might be expected to be evident in the sequences of the anthocyanin synthesis or regulatory networks.

SUPPORTING INFORMATION

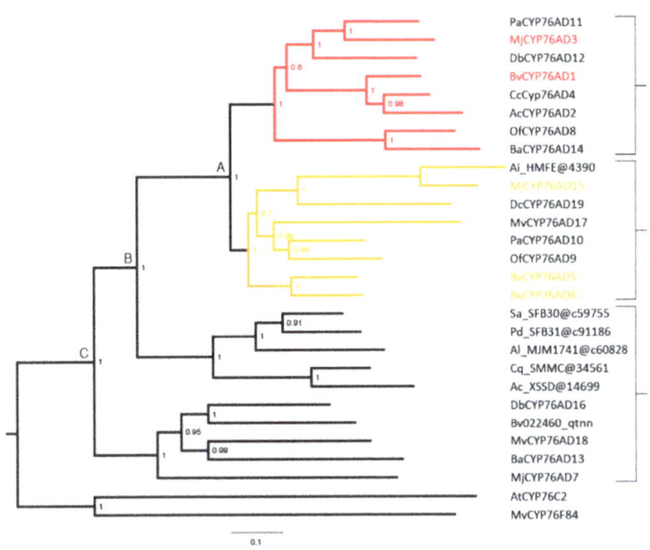

S1 Fig. Phylogenetic analysis of the CYP76AD subfamily members used to generate the LOGO shown in Fig 1B. The tree presented here is very similar to that presented by Brockington et al. [16], except that the gamma group is polyphyletic; this may be a product of sampling as only complete or near complete coding sequences were used here. The previously defined alpha and beta groups are recovered here with strong support and the activities of those two groups are indicated by color-coded branches; the alpha group is shown in red to indicate their step 1 and step 3 roles leading to betacyanin production and the beta clade is shown in yellow to reflect their step 1 only activity, leading to yellow betaxanthins. The genes that were functionally characterized in this paper are highlighted in color as well. Node labels A, B, and C indicate where ancestral sequences were reconstructed. Each node is also labeled with calculated probability.

doi:10.1371/journal.pone.0149417.s001

BvCYP76AD1	---------------MDHA-TLAMI-LAIWFIS----------FHFIKLLF----SQQTT
OfCYP76AD8	---------------MDTP-TLSYF-ISAITFYYIAFQIVKLGFNVIMTSK----KTKRR
BaCYP76AD14	---------------MDNT-TLAIL-LS-TSYFIIYLIITKLGFHYKVLP-----KNPKQ
AcCYP76AD2	---------------MDNA-TLAMI-LTIWFIS----------INFIKMFF----YHQNT
CcCYP76AD4	---------------MDNA-TLAML-LAIWFIS----------FHFIKMLF----TNQST
MjCYP76AD3	---------------MDFL-TLVMI-LSIIFFF----------YNLLKMKF----TTHSD
PaCYP76AD11	---------------MDHT-TLAMI-LSVIFLL----------YNLVKAIF----SQ-SN
DbCYP76AD12	---------------MDYT-TLVMI-LSIVFFC----------YNLFNLLF----TR-KN
MvCYP76AD17	---------------MDQF-TLGMI-LSFLFVW----------FYLVKTIF----GPFGN
MjCYP76AD15	---------------MENT-MLGVI-LATIFLT----------FHIMKMLF----SP--S
Ai_HMFE@4390	---S--N
DcCYP76AD19	--MF----FHHPK
PaCYP76AD10	---------------MDHT-TIAML-VSIVFLC----------FHLIKSFS----SHSKN
OfCYP76AD9	---------------MDHA-ILATL-VCILFVC----------FHLLWR------RSYKN
BvCYP76AD5	---------------MDNT-TLALI-LSSLFVC----------FQLIRSFI----NHAKK
BvCYP76AD6	---------------MDNA-TLAVI-LSILFVF----------YHIFKSFF----TNS-S
MjCYP76AD7	---------------MNHVLTIPVIFL---LII----------LIIIH------KLIFSS
MvCYP76AD18	---------------MEYHT-FPLILIPLISTI----------IYLLK------ISTKPS
BaCYP76AD13	---------MGNQTTMELYTTLTLFI--SIVSF----------LFLHH------FLKNSP
Bv022460_qtnn	---------------MEYFT-TLLLLLS--III----------LTILL------STKLFT
DbCYP76AD16	MLSFNNIQELCNNYQMENYTITSIVLFPIIFIL----------LPLLY------LSLKSR
Cq_SMMC@34561	---------------------LLLL-FSFVWAF----------LYLFK-HFNSSKLTSNN
Ac_XSSD@14699	--------------------IILIL-IPFIWVW----------KL----------KSSSK
Al_MJM1741@c60828	--------------MDGSILLALI-VSSIWVI----------FFLLKPKLSKHA---SV
Pd_SFB31@c91186	--------------MDYS-MLAAL-LSLLWAC----------FYFLKIKSNSSKLTSIN
Sa_SFB30@c59755	--------------MDYS-MLAAL-VSFLCAC----------FYFLKLNLSSSKLTS--

BvCYP76AD1	KL-LPPGPKPLPIIGNILEVGKKPHRSFANLAKIHGPLISLRLGSVTTIVVSSADVAKEM
OfCYP76AD8	RLPLPPGPKPLPIIGNVFELGPKPHRSFASLAKVYGPLMSLRLGSVTTIIVSSSDVAKEM
BaCYP76AD14	RLRLPPGPKPLPIIGNVLELGSKPHRSFTNLAKVHGPLISLRLGSVTTIIVSSSHVAKEM
AcCYP76AD2	KLSLPPGPKPLPIIGNILEVGKKPHRSFANLAKIHGPLISLRLGSVTTIVVSSAEVAKEM
CcCYP76AD4	KL-LPPGPKPLPIIGNILEVGKKPHRSFANLAKIHGPLISLKLGSVTTIVVSSAEVAKEM
MjCYP76AD3	-AQLPPGPKPMPIFGNIFELGEKPHRSFANLAKTHGPLMSLRLGSVTTIVVSSAEVAKEM
PaCYP76AD11	-TKLPPGPKPVPIFGNIFELGDKPHRSFANLAKIHGPLITLKLGSVTTIVVSSAEVAKEM
DbCYP76AD12	-TKLPPGPKTIPIFGNIFELGKKPHQSFANLAKIHGPLMSLKLGSVTTIVVSSAEVAREM
MvCYP76AD17	ASKIPPGPTPLPIVGNIFQLGKLPHRSFSNLSKIYGSMMTLKLGMVTTIVVSSAEVAKEM
MjCYP76AD15	KVKLPPGPRPLPIIGNILELGDKPHRSFANLAKIHGPLVTLKLGSVTTIVVSSSEVAKEM
Ai_HMFE@4390	KARLPPGPKPLPIIGNILELGEKPHRAFANLAKIHGPLITLKLGSVTTIVVSSSEVAKEM
DcCYP76AD19	NNKLPPGPKQIPIFGNIFDLGEKPHRSFTNLAKIYGPLISLKLGSVTTIVVSSSDVAREM
PaCYP76AD10	PSQLPPGPKPLPIFGNIFELGEKPHRSFARLAKIHGPLISLKLGSVTTIVVSSSEVAKEM
OfCYP76AD9	SSKLPPGPKPVPIFGNIFELGEKPHRSFADLAKIHGPLISLKLGSVTTIVVSSSDVAKEM

BvCYP76AD5 SNKLPPGPKRMPIFGNIFDLGEKPHRSFANLAKIHGPLVSLQLGSVTTVVVSSADVAKEM
BvCYP76AD6 SRRLPPGPKPVPIFGNIFDLGEKPHRSFANLSKIHGPLISLKLGSVTTIVVSSASVAEEM
MjCYP76AD7 NHKLPPGPKPWPIIGNIHLLGNKPHRSLSKLAKKYGPLISLKLGTITTIVISSPDIAKEM
MvCYP76AD18 KSNLPPGPKPWPIIGNIHLLGHKPHRSVSELSKKYGPIMTLKLGSVTTIVISSPKVAEEM
BaCYP76AD13 KSKLPPGPKPWPIIGNIHLLGSKPHCSVANLSKIYGPVMSLKLGSIVTIVISSPKVAKEM
Bv022460_qtnn KSNLPPGPKPWPIIGNILELGKLPHQAVDKLSKTYGPILSLKLGSITTIVISSPEIVKEM
DbCYP76AD16 QSKLPPGPKPWPIIGNILLLGDKPHQSVAKLSKIYGPLMSLKLGSITTIVISSPTIAKEM
Cq_SMMC@34561 NSRLPPGPKPVPILGNLPHLGDSPHRYLANLAKTYGPLISVKFGSITTIVVSSSTVAKEM
Ac_XSSD@14699 LSRLPPGPKPVPILGNLGHLGDKPHRYLASLVKTYGPLIHLKFGSITTIVVSSSAVAKEM
Al_MJM1741@c60828
YSRLPPGPKSLPIIGNLHQLGERPHRSLAELARTYGPLMYLKFGTINTIVVSSSKVAKEM
Pd_SFB31@c91186 GSRLPPGPRPMPIIGNLHQLGDTPHRTLAELSNKYGPLISLRFGSITTIVVSSSLVAKEL
Sa_SFB30@c59755
NGRLPPGPKPLPVIGNLHQLGETPHRSLAELARTYGPLICLKFGSITTIVVSSSVVAKEM

OfCYP76AD8 FLKNDQPLSSTRTIPNSVTAGDHHKLTMSWLPVSPKWRSFRKITTFHLLSPQRLDACSGL
BaCYP76AD14 FLKNDQSLSSNRTIPHSVTAGDHHKLTMSWLPVSPKWRSFRKITTFHLLSPQRLDACCSL
AcCYP76AD2 FLKKDQPLS-NRNVPNSVTAGDHHKLTMSWLPVSPKWRNFRKITAVHLLSPLRLDACQSL
CcCYP76AD4 FLKKDQPLS-NRTVPNSVTAGDHHKLTMSWLPVSPKWRNFRKITAVHLLSPLRLDACQSL
MjCYP76AD3 FLKNDQSLA-DRSVPNSVTAGDHHKLTMSWLPVSPKWKNFRKITAVHLLSPQRLDACHAL
PaCYP76AD11 FLTNDQLLA-NRNVPNSVTAGDHHKLTMSWLPVSPKWKTFRKITAVHLLSPQRLDACQAL
DbCYP76AD12 FLKNDQLLS-NRTVPNSVTAGDHHKTTMSWLPVSQKWRNFRKITAVHLLSPQRLDSCQAL
MvCYP76AD17 FLQHDLAFS-NRMIPDSVTGGKHDKLSVTWIPVSPKWRHLRKIFALNLLSNQKLDETQSQ
MjCYP76AD15 FLKNDQPLA-NRTIPDSVRAGNHDKLSMSWLPVSPKWRNLRKISAVQLLSTQRLDASQAH
Ai_HMFE@4390 FLKNDQALA-NRTI-----QGNHDKLSMSWLPVSPKWRNLRKISAVQLLSSQRLDSSQSH
DcCYP76AD19 FLKSDHVLS-NRTIPDSVKAGNHDKLSMSWLPVSQKWRHLRKIAATQLLSTQRLDSSESL
PaCYP76AD10 FLKHDQVLA-NRTIPDSVRAGNHDKLSMSWLPVSVKWRSLRKIVVVQLFSTQRLDVSQSL
OfCYP76AD9 FLKHDQVFA-NRTIPDSVRAGNHDKLSMSWLPVSAKWRNMRKISAVQLLSTQRLDSNQGL
BvCYP76AD5 FLKNDQALA-NRTIPDSVRAGDHDKLSMSWLPVSAKWRNLRKISAVQLLSTQRLDASQAH
BvCYP76AD6 FLKNDQALA-NRTIPDSVRAGDHDKLSMSWLPVSQKWRNMRKISAVQLLSNQKLDASQPL
MjCYP76AD7 FTKHDLALS-ARQVPEAAKIADHHLYSIIFLPICPKWRGYRKISTVHLFTNQILDTSQHL
MvCYP76AD18 FLKHDLALS-GRKTPYAICVQDHDKYSMAFIPVSPKWRNLRKIATVQLLTNHRLDASQGL
BaCYP76AD13 FLKHDSACS-SRRVPDCFKIIGHEQVSMVWLPANHRWRNLRKISATQLFTNLRLDANQGL
Bv022460_qtnn FLEHDLALS-SRPSPDASRVGNHNKFSIVWLPVSPKWRDLRKIATIQLFTTQRLDSSQEL
DbCYP76AD16 FLKHDLAFS-SRNVPDVRTIDNHDKLSIVWLPVCPKWRDLRKIATIQLFTNQRLDATRDL
Cq_SMMC@34561 FQKHDLTLS-TRHVSAAVRSNGHDKFSVAWLPVCAKWRTLRKISALQLFSTQRLDASQAL
Ac_XSSD@14699 FQKHDLTVS-SRQVCAALRANGHDQFSVVWLPVCAKWRSLRKISAIYLFSSQRLDASQAL
Al_MJM1741@c60828 FQKHDLSLA-NRKVPAAVRANGHDMYSIAWLPICPKWRTLRKISAVHSLSNQRLDA
SQSF
Pd_SFB31@c91186 FQKHDLALA-NRKVPAAVRSNEHDKYSIAWLPVCPKWRNLRKISAVHSLSTQRLDA
SQSL
Sa_SFB30@c59755 FQKHDLSLA-NRKVPAAVRANDHDKFSIAWLPVCPKWRNLRKISAIHSLSNQRLDD
SQSL

BvCYP76AD1 RHAKVQQLYEYVQECAQKGQAVDIGKAAFTTSLNLLSKLFFSVELAHH-KSHTSQEFKEL
OfCYP76AD8 RQAKVQQLYEYVLECSRTGQAVDIGKAAFTTSLNLLSKLFFSLELANH-TSDKSQEFKEL
BaCYP76AD14 RQAKVQQLFEYVLQCSRTGQPVDIGKAAFTTSLNLLSKLFFSLELAHH-RSTKSQEFKDL
AcCYP76AD2 RHAKVQQLYQYVQECALKGQSVDIGKAAFTTSLNLLSKLFFSKELACH-KSHESQELKQL
CcCYP76AD4 RHAKVQQLFQYVQECAQKGQAVDIGKAAFTTSLNLLSKLFFSKELASH-KSRESQEFKQL
MjCYP76AD3 RHAKVKQLYEYVQECALKGEAVDIGKAAFTTSLNLLSNLFFSVELANH-TSNTSQEFKQL
PaCYP76AD11 RHTKVKQLYEYVQECAKRGEAVDIGKAAFTTSLNLLSNLFFSVELANH-TSSSSQEFKEL

DbCYP76AD12 RQAKVKQLFNYIHECAQKGEAVDIGKAAFTTSLNLLSNLFFSVELANH-KSSSSQEFKQL
MvCYP76AD17 RQAKVQQLLEYVQECAKKGCAVDIGRATFTTTLNLLSNTVFSMELAHY-SSNASQEFKRL
MjCYP76AD15 RQAKIKQLIEYVKKCSKIGQYVDIGQVAFTTSLNLLSNTFFSKELASF-DSDNAQEFKQL
Ai_HMFE@4390 RQSKVKQLVDYVAECSKLGQSVDIGRVAFTTSLNLLSNTFFSKELASF-DSNNAQEFKNL
DcCYP76AD19 RQAKVEKLLEYVQECCNSGLSVDIGRAVFTTTLSLLSNTFFSKELANY-SSSKSQEFKQL
PaCYP76AD10 RHAKVQQLHEYITECSKKGEPVDIGRAAFTTSLNLLSNTFFSMELANH-SSSASQEFKQL
OfCYP76AD9 RQAKVQQLLEYVQECCKKGQPVDIGRAAFTTSLNLLSNTFFSMELAQH-SSSASQEFKQL
BvCYP76AD5 RQSKVQQLLEYVHDCSKKGQPVDIGRAAFTTSLNLLSNTFFSVELASH-ESSASQEFKQL
BvCYP76AD6 RQAKVKQLLSYVQVCSEKMQPVDIGRAAFTTSLNLLSNTFFSIELASH-ESSASQEFKQL
MjCYP76AD7 RRKKVHELVEYIKCCCENNQVVEIGKAAFTTSLNLLSNTFFSFDLASDIDFGYSIKFKNI
MvCYP76AD18 RREKVHELVQYARDCCEKGIAMDIGKAGFITSLNLLSNTFFSMNLTSY-DSSFSGEFRDI
BaCYP76AD13 RREKINDLIQFVKDSCGSGLAIDIGKAAFTTSLNLLSNTFFSVELASY-DSSISHEFKEL
Bv022460_qtnn RQIKVNELVDYVRQCCEKGLPVDVGKAGFTTTLNMLSNTFFSMDLASH-ASSNSQEFKDL
DbCYP76AD16 RRKKVEELVGYARQCCEQGLALDIGKAAFTTSLNLLSNTFFSMNLGSH-DSSWSQEFKEL
Cq_SMMC@34561 RQEKVSKLMEYVKECSKSGEAIDVGGVAFTTSLNLLSNTFFSFDLASY-NSSDTGEFKEL
Ac_XSSD@14699 RQEKVSKLMEYVNECCKKGEALDIGVAAFTTSLNLLSNTFFSFDLAGY-SSRDSGEFKEL
Al_MJM1741@c60828 RQAKVADLLEYVKGCSVTGEVVDIGKAAFTTSLNLLSNTFFSFDLASH-
RSSDSGQFKDL
Pd_SFB31@c91186 RQAKVMELLEFVRGCCKAGEAVDVGKAAFTTSLNLLSNTFFSFDLASH-
SSSDSGEFKEL
Sa_SFB30@c59755 RQAKVLELLDYVRQCCKVGEAVDVGKAAFTTSLNLLSNTFFSLDLASH-
SSTDSGEFKEL
BvCYP76AD1 IWNIMEDIGKPNYADYFPILGCVDPSGIRRRLACSFDKLIAVFQGIICERLAPDSS--TT
OfCYP76AD8 IWNIMEDIGKPNYADYFPCLKYFDPSGIRRRLACSFEKLIEVFQVIIRQRLSLSSS---G
BaCYP76AD14 IWNIMEDIGKPNIADHFPCLKYFDPSGIRRRLASSFERLIEVFQDIIRQRMSLSFG----
AcCYP76AD2 IWNIMEDIGKPNYADYFPILGCIDPLGIRRRLAANFDKLISVFQTIISERLENDIN----
CcCYP76AD4 IWNIMEDIGKPNYADYFPILGCVDPSGIRRRLASNFDKLIEVFQCIIRQRLERNP-----
MjCYP76AD3 IWDIMEDIGKPNYADYFPLLKYVDPSGIRRRLAANFDKLIDVFQSFISKRLSSAY----S
PaCYP76AD11 IWDIMEDIGKPNYADYFPVLKCVDPWGIRRRLESNFDKLIEVFQSFIRKRLSTEPF--SA
DbCYP76AD12 IWNIMEDIGKPNYADYFPVLKYVDPSGIRRRLASNFNKLIDVFQGFIRLRMSTNS---SC
MvCYP76AD17 MWCIMEEIGRPNYADYFPFLRHFDPFGNRRRLTAAYAKLIGFFENIVHERLNARSAKL--
MjCYP76AD15 MWCIMEEIGRPNYADYFPILGYVDPFGARRRLSRYFDQLIEVFQVIIRERLTHDNNIV--
Ai_HMFE@4390 MWCIMEEIGRPNYADYFPILGYVDPFGVRRRLSRYFDQLIEVFQE----RLAKNSNNV--
DcCYP76AD19 MWDIMVEIGRPNYADYFPMLGFMDPFSIRRRLSSYFDKLIGVFQEIISERLSLRKK--SC
PaCYP76AD10 MWCIMEEIGRPNYADFFPILGYVDPFGIRRRLAVYFDKLIAVFQEIIRERQKARFTNL--
OfCYP76AD9 MWCIMEEIGRPNYADYFPILGYFDPFGIRRRLTAYFDKLIAIFQDIIHERLKARSTG---
BvCYP76AD5 MWNIMEEIGRPNYADFFPILGYLDPFGIRRRLAGYFDQLIAVFQDIIGERQKIRSANLSG
BvCYP76AD6 MWNIMEEIGRPNYADFFPILGYIDPFGIRRRLAGYFDKLIDVFQDIIRERQKLRSSNSSG
MjCYP76AD7 VGELSEALAKPNLSDFFPIIKSLDLQGVKKKMEVLFTDMWDVFRKVVEERLSDNSK----
MvCYP76AD18 VWNMLEEAGRPNLSDFFPVLKGLDLQGMRKRYSVYYYKMIAIFDEIINQRLKDS-T----
BaCYP76AD13 VWRVLEVGATPNISDFFPLIRTFDLQGTRRRGKYYLLKLRGIFEKIIDERLSCQ-T----
Bv022460_qtnn VWSLLEEGAKPNVSDFFPIVRELDLQGVSKNRRVHMKKLMGIFEEIIDGRLTKL-K----
DbCYP76AD16 VWNLLEEGAKPNVSDYFPILKRFDLQGAVRRVSRYVDKLMAVFEDIIDERLKNS-T----
Cq_SMMC@34561 VWKIMEEIGKPNLVDCFPMLRFLSVFSVNCNLMGYGNKLNEVFVDIINKRLESFGV---S
Ac_XSSD@14699 VWKIMEEIGKPNWVDCFPMLRFFSKFSVNRKLMGYGNKLNEIFEDIIQKRLNDCDV---C
Al_MJM1741@c60828
VWKIMEEIGKPNLADCFPMLRFFSSISVNQQLMGYGNKLNEVFADIINERLKNGVSN---
Pd_SFB31@c91186
VWKIMEEIGKPNLADCFPVLRFVSAVSVNCQLMGYGNKLNEVFAGIIEERLRSGASSD-S

Sa_SFB30@c59755 VWRIMEEIGKPNLADCFPVLRFISAVSVNSQLMGHSNKLNKVFAEIINERLKTPS--
S-A

BvCYP76AD1 TTTTTDDVLDVLLQLFKQN--ELTMGEINHLLVDIFDAGTDTTSSTFEWVMTELIRNPEM
OfCYP76AD8 TNDHNNDVLDVLLDLYQQK--ELSMEEINHLLVDIFDAGTDTTSSTFEWAMAELIKNPRM
BaCYP76AD14 -SSHNNDVLDVLLGLYNQK--ELTMDEINHLLVDIFDAGTDTTSSTFEWSMAELMKNRRI
AcCYP76AD2 SNATTNDVLDVLLQLYKQK--ELSMGEINHLLVDIFDAGTDTTSSTFEWVMAELIRNPKM
CcCYP76AD4 STPPTNDVLDVLLELYKQN--ELSMGEINHLLVDIFDAGTDTTSSTFEWVMAELIRNPEM
MjCYP76AD3 SATSLDDVLDVLLKLLKEK--ELNMGEINHLLVDIFDAGTDTTSNTFEWAMAELMRNPIM
PaCYP76AD11 SAKTPNDVLDVLLNLLKEE--ELNMGEINHLLVDIFDAGTDTTSSTFEWAMAELVRNPEM
DbCYP76AD12 GATNPNDVLDVLLNLYKGD--DLNMDEINHLLVDIFDAGTDTTSSTFEWAMAELVKNPKM
MvCYP76AD17 --STTNDVLDTLLNLHQDN--QLTMDEIYHILVDIFDAGTDTTANTLEFAMAQLVKNPET
MjCYP76AD15 --GNNNDVLATLLDLYKQN--ELTMDEINHLLVDIFDAGTDTTASTLEWAMSELIKNPHI
Ai_HMFE@4390 --DIKSDILATLLDLYKQN--ELSMDEINHLLVDIFDAGTDTTASTLEWAMAELIKNPHI
DcCYP76AD19 GKDNTNDVLSTLLNLYEEN--ELSMDEVNHLLVDIFNAGTDTIASTLEWAMSELLKNPNI
PaCYP76AD10 --STTNDVLDTLLNLYQEN--ELSMDEINHLLVDIFDAGTDTTASTMEWAMAELVKNPDI
OfCYP76AD9 -SSSTNDILDTLLNLYQEN--ELSMDEINHLLVDIFDAGTDTTASTLEWAMAELVKNPEI
BvCYP76AD5 GKQTTNDILDTLLNLYDEK--ELSMGEINHLLVDIFDAGTDTTASTLEWAMAELVKNPDM
BvCYP76AD6 AK-QTNDILDTLLKLHEDN--ELSMPEINHLLVDIFDAGTDTTASTLEWAMAELVKNPEM
MjCYP76AD7 --GLKDDVLDTLLKLVDEQ--EISLDEVVHFIMDLFSAGTETTLITLEWAMTELLRCPDK
MvCYP76AD18 --SDKDDVLGSLLKLVKED--ELSLDDVRHLLLDLFIAGTDTTSATLEWAMTELLVNPQK
BaCYP76AD13 --SGKDDVLDTLLKLVKQN--ELSLHEVQHFLIDLFVAGTDTTSSVLEWAMTELLRNPEK
Bv022460_qtnn --DVKDDVLSTLLKLVKDE--ELNLDDVKHMLMDLFLAGTDTTSITLEWAMTELLRNPEK
DbCYP76AD16 --EAKDDVLNLLINLVKEN--ELSLHNVKHMLFDLFLAGTDTTSSTVEWAMAELQRHPEK
Cq_SMMC@34561 GNNDGGDVLDTLIKIMKENESELSLDDIKHLLMDFFTAGTDTTSSTLEWAMTELLHSPEK
Ac_XSSD@14699 E--NYGDVLDTLLRLMKEKESGLTLADIKHLLMDFFTAGTDTTSSTLEWALTELLHNPEK
Al_MJM1741@c60828 CADNDADVLDTLLRLMK-NDSELCLDDIKHLLMDFFTAGTDTTSSTLEWAMTELL
HDPEK
Pd_SFB31@c91186
SGVGDGDVLDTLLRLMKENDSELSLDDIKHLLMDFFTAGTDTTSSTLEWAMTELLRNPEK
Sa_SFB30@c59755
SAITGGDVLDTLLRLMEENDSELSFDDINHLLMDFFTAGTDTTSSTLEWAMTELLHNPEK

BvCYP76AD1 MEKAQEEIKQVL---GKDK--QIQESDIINLPYLQAIIKETLRLHPPTVFLLPRKADTDV
OfCYP76AD8 METAQAEIKLIL---GKDL--HIQESDIPKLPYLRAIIKETLRLHPPTVFLLPRKADADV
BaCYP76AD14 MEKAQAEILHVL---GKNS--YIQESDISKLPYLRAIIKETLRLHPPTVFLLPRKADADV
AcCYP76AD2 MEKAQQEIHEVL---GKDR--QIQESDIIKLPYLQALIKETLRLHPPTVFLLPRKADMDV
CcCYP76AD4 MAKAQDEIEQVL---GKDR--QIQESDIIKLPYLQAIIKETLRLHPPTVFLLPRKADTDV
MjCYP76AD3 MKRAQNEIALVL---GKDN-ATIQESDIANMPYLQAIIKETLRLHPPTVFLLPRKAITNV
PaCYP76AD11 MKKAQDEIEQVL---GKDA--IIQESDIPKMPYLQAIIKETLRLHPPTVFLLPRKASSNV
DbCYP76AD12 MKKAQAEIQQVL---GKDS--IIRESDIPNMPYLQAIIKETLRLHPPTVFLLPRKADADV
MvCYP76AD17 LVKAQTEIEQAL---GKSS-SIIQESDISKLPYLQAIIKETFRLHPPAVFLLPRKAETDI
MjCYP76AD15 MAKAQEEVRRATMSHGGATVAEIQESDINNLPYIQSIIKETLRLHPPTVFLLPRKADVDV
Ai_HMFE@4390 MAKAKAEVRQVTMSRGESTLAQIEESDISNLPYIQAIVKEALRLHPPTVFLLPRKADMDV
DcCYP76AD19 MTKAQIEIKNAL---ENNSS-KIQESDIPKLPYLQAIIKETLRLHPPTVFLLPRKADEDV
PaCYP76AD10 MLKARIEIKQAL---GNDSSLIIQESDIAKLPYLQAIVKETLRLHPPTVFLLPRKAEANV
OfCYP76AD9 MIKAQDEINAAL---GKDCS-VIHESDIVKLPYLQAIVKETLRLHPPTVFLLPRKADLDV
BvCYP76AD5 MVKVQDEIEQAI---GKGCS-MVQESDISKLPYLQAIIKETLRLHPPTVFLLPRKADADV

BvCYP76AD6 MTKVQIEIEQAL---GKDCL-DIQESDISKLPYLQAIIKETLRLHPPTVFLLPRKADNDV

MjCYP76AD7 MAKAQAEIDQVI---DMDG--SVQESDIPKLPYIQAIVKEILRLHPATPFLVPRMAEQEV

MvCYP76AD18 IEIAQKQISQVF---AKDQ--VIHESDIPKLPYIHAVVKETFRMHPSVPFLLPHWAENDV

BaCYP76AD13 MGKLQNEINNAF---GNDQSKSIQESDISKLPYLQGVVKETLRLHPPAPFLLPRKVEKDI

Bv022460_qtnn MEKVQIELDKVL---GKDS--SLQESMISKLPYIQAIVKETLRLHPPTPFLIPHKAEKDV

DbCYP76AD16 MLKAQAEIDQVI---PNDG--FVQEMDISKLAYIQAIVKETLRLHPPAPFLIPHKTVKDV

Cq_SMMC@34561 MAKAQAELKQVL---GENR--IVGESDISKLPYLQAIVKETLRMHPPTVFLLPRKATNDV

Ac_XSSD@14699 MANVQAELDEVI---GKNK--VVGESDISKLPYLQATVKETLRMHPPTVFLLPRKANNEV

Al_MJM1741@c60828 MIKAQAELKQVL---GNDNL-MVLESDIPSLPYLQAIVKETLRMHPPTVFLLPRQAD
ADV

Pd_SFB31@c91186 MAKAQAELEQVL---GNDNV-AVQESDTLKLPYLEAIVKETLRMHPPTVFLLPRKAN
ADV

Sa_SFB30@c59755 MAKAQAELEQVL---GNDNV-LVQESDISKLPYLQSIVKETLRMHPPTVFLLPRKAD
ADV

BvCYP76AD1 ELYGYIVPKDAQILVNLWAIGRDPNAWQNADIFSPERFIG---CEIDVKGRDFGLLPFGA

OfCYP76AD8 ELYGYTVPKNAQILVNLWALGRDPKVWENPDVFSPERFLG---CDIDVKGRNFGLLPFGA

BaCYP76AD14 ELYGYVVPKDAQILVNLWALGRDPAVWENPDEFSPDRFMG---SEIDVKGRDFGLLPFGA

AcCYP76AD2 ELYGYVVPKDAQILVNLWAIGRDSQVWEKPNVFLPERFLG---SDVDVKGRDFGLLPFGA

CcCYP76AD4 ELYGYIVPKDAQILVNLWAIGRDSQAWENPKVFSPDRFLG---CEIDVKGRDFGLLPFGA

MjCYP76AD3 KLYGYIVPKNAQILVNLWAIGRDPKVWKNPNEFLPDRFLN---SDIDVKGRDFGLLPFGA

PaCYP76AD11 ELYGYVVPKNAQILVNLWAIGRDPTVWDNPNMFSPERFLN---SDIDVKGRDFGLLPFGA

DbCYP76AD12 ELYGYVVPKNAQILVNLWALGRDPLVWKSPNVFKPERFLG---SEIDFKGRDFGLLPFGA

MvCYP76AD17 TLCGYVVPKNAQIMLNLWSIGRDPKVWQNPEVFSPERFLN---CDIDVRGRHFELLPFGA

MjCYP76AD15 QLFGYVVPKNAQVLVNLWAIGRDPNVWPDPEVFSPERFMD---CEIDVKGRDFELLPFGA

Ai_HMFE@4390 QLYGYVVPKNAQVLVNLWAIGRDPNVWSNPELFRPERFMD---CEIDVKGRDFELLPFGA

DcCYP76AD19 ELFGYIVPKNAQVLVNLWAIARDEKAWENPELFNPERFLG---RNIDVKGRNFELLPFGA

PaCYP76AD10 QLYGYLIPKNAQLLVNLWAIGRDPNVWNDPLVFSPERFLASSSCEIDVKGRHFELLPFGA

OfCYP76AD9 ELYGYVVPKNAQILVNLWAIGRDPKVWSNPEVFSPERFLD---STIDVKGRDFELLPFGA

BvCYP76AD5 ELYGYVVPKNAQVLVNLWAIGRDPKVWKNPEVFSPERFLE---SNIDYKGRDFELLPFGA

BvCYP76AD6 ELYGYVVPKNAQVLVNLWAIGRDPKVWKNPEVFSPERFLD---CNIDYKGRDFELLPFGA

MjCYP76AD7 QLCDYYVPKNAQILVNVWLIGRDPSVWSNPETFIPERFLG---RDIDVKGQDFELIPFGS

MvCYP76AD18 QLSSYYVPKDAQIWVNVWSMGRDPSVWLDPNSFIPERFLD---KDIDVKGRDFELLPFGA

BaCYP76AD13 DLVGYHVPKNSTIWVNVWSMGRDPSNWSNPEVFMPERFLD---SEIDVKSRHSELIPFGA

Bv022460_qtnn LLCNYLVPKNSIIWVNLWSIARSPSVWPNPESFSPERFLE---MEIDIKGRDFKLIPFGS

DbCYP76AD16 QLCDYTVPKNALVWVNVWSIGRDPSVWTDPDSFVPERFLE---REIDFKGRNFELIPFGA

Cq_SMMC@34561 ELYGYVVPKNAQVFVNLWAISRDPNHWVNPDLFSPERFLE---REIDMKGQDFGLIPFGA

Ac_XSSD@14699 ELYGYVVPKDAQIFVNLWAISRDPNHWTDPDLFSPERFLE---REIDMKGQDFGLIPFGA

Al_MJM1741@c60828 ELYGYLVPKNAQIFVNLWAISRDPNAWENPDVFIPERFLG---
RDIDMKGQDFGLIPFGS

Pd_SFB31@c91186 DLYSYLVPKNAQVLVNLWAISRDPNTWENPNVFSPERFLD---
RDIDMKGQDFGLIPFGA

Sa_SFB30@c59755 ELYGYVVPKNAQVLVNLWAISRDPNAWENPDVFSPERFLD---
RDIDMKGQDFGLIPFGA

BvCYP76AD1 GRRICPGMNLAIRMLTLMLATLLQFFNWKLEG-----DISPKDLDM---DEKFG-IALQK

OfCYP76AD8 GRRICPGMNLAYRMLTLMLATLLQSFDWKLPN-----EMNPQNLDM---DEKFG-IALQK

BaCYP76AD14 GRRICPGMNLAIRMLTLMLATLLRSFDWKLPE-----GEAPAQLDM---DEKFG-IALQK

AcCYP76AD2 GKRICPGMNLAIRMLTLMLATLLQFFNWKLED-----GMNPQDLDM---DEKFG-IALQK

CcCYP76AD4 GKRICPGMNLAIRMLTLMLATLLQFFNWKLQD-----GMSLEDLDM---EEKFG-IALQK

MjCYP76AD3 GRRICPGMNLAYRMLTLMLATLLQSFDWKLPH-----RNSPLDLDM---DEKFG-IALQK

PaCYP76AD11	GRRICPGMNLAYRMLTLMLATLLQSFDWKLGD-----GVNPKDLDM---DEKFG-IALQK
DbCYP76AD12	GRRICPGMNLAYRMLTLMLATLLQSFDWKVAD-----GTNPQDMDM---DEKFG-IALQK
MvCYP76AD17	GRRICPGITLSYRMLHLMLAALIRSFDWKLED-----GTNPKDLDV---TEKFG-IALQK
MjCYP76AD15	GRRICPGLSLAYRMLNLMLANMVHSFDWKLPGVENGSGSEMDSLDM---DEKFG-IALQK
Ai_HMFE@4390	GRRICPGLSLAYRMLNLMLANLIHSFDWKLPDI----GSDVGELDM---DEKFG-ITLQK
DcCYP76AD19	GRRICPGVTLANRMMNLMLANLIHSFDWKLEE-----GVNPSDLDM---NEKFG-ITLQK
PaCYP76AD10	GRRICPGLTLAYRMLNLMLVTLVNSFDWKLED-----VTSAKDLDM---DEKFG-ITLQK
OfCYP76AD9	GRRICPGLTLASRMLNLMLATLVHNFNWKLED-----GMIPKDLDM---TEKFG-ITLQK
BvCYP76AD5	GRRICPGLTLAYRMLNLMMANFLHSYDWKLED-----GMHPKDLDM---DEKFG-ITLQK
BvCYP76AD6	GRRICPGLTLAYRMLNLMLATLLQNYNWKLED-----GINPKDLDM---DEKFG-ITLQK
MjCYP76AD7	GRRICPGMSLGYRMVHLMLANLLHSFDWSLPN-----GLDPKDLDM---EDTFG-LTLRK
MvCYP76AD18	GRRICPGLPLAYRMVHLTLATLLHSFDWKLGN-----GMSQVWDHVAKISAAYGGPTLQM
BaCYP76AD13	GRRICPGLPLAYRMVHLMLANLVQFFNWKVDR-----WSNPEEIDM---EEKFG-ITLQK
Bv022460_qtnn	GRRMCPGMPLAYRMTHMLLATLLHSFNWKYGE------ASPKDIDM---KEKFG-LTLQK
DbCYP76AD16	GRRTCPGMPLAYRMTHLMLATLLHSFDWKLSD-----GVNPENLDM---EEKFG-ITVQK
Cq_SMMC@34561	GRRICPGDTLAFRMLNLMLGNLLHAFDWKVGD-----GLIPEDLDM---TDKFG-ITIQK
Ac_XSSD@14699	GRRICPGDTLAIRMLNLMLGNLLHGFNWKVGD-----GIRPEDLDM---SDKFG-ITIQK
Al_MJM1741@c60828	GRRICPGLTLAYRMLNLMLGTLIHAFHWELAN-----GLSPKDLDM---MDKFG-ITIQK
Pd_SFB31@c91186	GRRICPGLALAHRMLNLMLGTLIHAFNWKLGD-----GLNPEDLDM---TDKFG-ITIQK
Sa_SFB30@c59755	GRRICPGLTLAYRMLNLMLGTLIHSFNWKLGD-----GLSPEDLDM---TDKFG-ITIQK
BvCYP76AD1	TKPLKLIPIPRY----
OfCYP76AD8	TKPLQIIPLSKD----
BaCYP76AD14	TTPLKIIPIFKNSI--
AcCYP76AD2	NKPLEIIPSLRH----
CcCYP76AD4	TKPLRIIPVSRY----
MjCYP76AD3	TKPLEIIPLIKY----
PaCYP76AD11	TKPLQVIPVLKY----
DbCYP76AD12	TTPLQIIPVYKY----
MvCYP76AD17	AEPLEIVPVPKTSLIA
MjCYP76AD15	TKP-------------
Ai_HMFE@4390	VKPLQIIP--------
DcCYP76AD19	VIPLKLTPIAKL----
PaCYP76AD10	VKPLQLIPIPKK----
OfCYP76AD9	VNPLQVIPIQKN----
BvCYP76AD5	VKPLQVIPVPRK----
BvCYP76AD6	VKPLQVIPVPRN----
MjCYP76AD7	AQPLDAVAHYRS----
MvCYP76AD18	IR--------------
BaCYP76AD13	AKPLEVVPILR-----
Bv022460_qtnn	AQPLQAIPIPR-----
DbCYP76AD16	VQPLQLIPAVRGSRK-
Cq_SMMC@34561	AIPLRAIP--------
Ac_XSSD@14699	ALPLRAIP--------
Al_MJM1741@c60828	EKPLLAIPISK-----
Pd_SFB31@c91186	AKPLRAIPIPK-----
Sa_SFB30@c59755	AKPLRAIPVAK-----

S2 Fig. Amino acid alignment of the CYP76AD subfamily members used to generate the phylogenetic tree and LOGO. Amino acids that appear to differentiate the three primary groups resolved in the gene tree are highlighted. The top (alpha) group consists of the CYP76AD1-like homologs (functional steps 1 and 3); center (beta) group, CYP76AD5/6-like (functional step 1 only); bottom (gamma) group, unknown function. Red: conserved among CYP76AD1-like. Yellow: conserved among CYP76AD5/6-like. Blue: conserved among CYP76AD1, 5, and 6. Green: conserved among the unknown CYP76AD members. The numbers below the alignment are the amino acid positions in CYP76AD1 and correspond to the amino acids used in the LOGO analysis.

S1 Table. Sequences discussed in the text and used to produce the LOGO analysis, sequence alignment, and phylogenetic tree. Shown are the species name, family, gene name, and accession number for all genes referenced in the manuscript.

doi:10.1371/journal.pone.0149417.s003

Species	Family	Gene	GenBank Accession Number
Beta vulgaris	Amaranthaceae	PPO	KR337592
		ACTIN	HQ656028
		CYP76AD1	HQ656024
		CYP76AD5	KM592961
		CYP76AD6	KM592962
		Bv022460_qtnn	KQ090199.1
Amaranthus cruentus	Amaranthaceae	CYP76AD2	HQ656025
		Ac_XSSD@14699	KR376441.1
Celosia cristata	Amaranthaceae	CYP76AD4	KC857455
Chenopodium quinoa	Amaranthaceae	Cq_SMMC@34561	KR376438.1
Mirabilis jalapa	Nyctaginaceae	CYP76AD3	HQ656026
		CYP76AD7	KM516790
		CYP76AD15	KM516798
Opuntia ficus-indica	Cactaceae	CYP76AD8	KM516791
		CYP76AD9	KM516792
Phytolacca americana	Phytolaccaceae	CYP76AD10	KM516793
		CYP76AD11	KM516794
Phytolacca dioica	Phytolaccaceae	Pd_SFB31@c91186	KR376427.1
Seguieria aculeata	Phytolaccaceae	Sa_SFB30@c59755	KR376422.1
Dorotheanthus bellidiformis	Aizoaceae	CYP76AD12	KM516795
		CYP76AD16	KM516799
Basella alba	Basellaceae	CYP76AD13	KM516796
		CYP76AD14	KM516797
Mollugo verticillata	Molluginaceae	CYP76AD17	KR779855
		CYP76AD18	KR779856
		CYP76F84	KR779857
Dianthus caryophyllus	Caryophyllaceae	CYP76AD19	KR779858
Arabidopsis thaliana	Brassicaceae	CYP76C2	NM_130119.3

Acknowledgments

This research was supported by funds from the US National Science Foundation.

Author Contributions

Conceived and designed the experiments: RS AB MB NA GH AG VVS AL. Performed the experiments: RS AB MB NA GH AG VVS AL. Analyzed the data: RS AB VVS AL. Contributed reagents/materials/analysis tools: VVS AL. Wrote the paper: RS AB VVS AL.

REFERENCES

1. Christinet L, Burdet FX, Zaiko M, Hinz U, Zryd JP. Characterization and functional identification of a novel plant 4,5-extradiol dioxygenase involved in betalain pigment biosynthesis in *Portulaca grandiflora*. Plant Physiol. 2004:134: 265–274. pmid:14730069 doi: 10.1104/pp.103.031914
2. Sasaki N, Abe Y, Goda Y, Adachi T, Kasahara K, Ozeki Y. Detection of DOPA 4,5-dioxygenase (DOD) activity using recombinant protein prepared from *Escherichia coli* cells harboring cDNA encoding *DOD* from *Mirabilis jalapa*. Plant Cell Physiol. 2009:50: 1012–1016. doi: 10.1093/pcp/pcp053. pmid:19366710
3. Gandía-Herrero F, García-Carmona F. Characterization of recombinant *Beta vulgaris*4,5-extradiol dioxygenase active in the biosynthesis of betalains. Planta 2012:236: 91–100. doi: 10.1007/s00425-012-1593-2. pmid:22270561
4. Hatlestad GJ, Sunnadeniya RM, Akhavan NA, Gonzalez A, Goldman IL, McGrath JM, et al. The beet R locus encodes a new cytochrome P450 required for red betalain production. Nature Genetics 2012:44: 816–820. doi: 10.1038/ng.2297. pmid:22660548
5. DeLoache WC, Russ ZN, Narcross L, Gonzales AM, Martin VJJ, Dueber JE. An enzyme-coupled biosensor enables (S)-reticuline production in yeast from glucose. Nature Chemical Biology 2015:11: 465–471. doi: 10.1038/nchembio.1816. pmid:25984720
6. Schliemann W, Kobayashi N, Strack D. The decisive step in betaxanthin biosynthesis is a spontaneous reaction. Plant Physiology 1999:119: 1217–1232. pmid:10198080 doi: 10.1104/pp.119.4.1217
7. Steiner U, Schliemann W, Böhm H, Strack D. Tyrosinase involved in betalain biosynthesis of higher plants. Planta 1999:208: 114–124. doi: 10.1007/s004250050541
8. Strack D. Schliemann W. Bifunctional Polyphenol Oxidases: novel functions in plant pigment biosynthesis. Angew. Chem. Int. Ed. 2001:40: 3791–3794. doi: 10.1002/1521-3773(20011015)40:20<3791::aid-anie3791>3.0.co;2-t
9. Joy RWIV, Sugiyama M, Fukuda H Komamine A. Cloning and characterization of polyphenol oxidase cDNAs of *Phytolacca americana*. Plant Physiol. 1995:107: 1083–1089. pmid:7539531 doi: 10.1104/pp.107.4.1083
10. Gandía-Herrero F, García-Carmona F, Escribano J. Purification and characterization of a latent polyphenol oxidase from beet root (*Beta vulgaris* L.). J. Agric. Food Chem. 2004:52: 609–15. pmid:14759157 doi: 10.1021/jf034381m
11. Gao Z.-J, Han X-H, Xiao X-G. Purification and characterization of polyphenol oxidase from red Swiss chard (*Beta vulgaris* subspecies *cicla*) leaves. Food Chem. 2009:117: 342–348. doi: 10.1016/j.foodchem.2009.04.013
12. Yamamoto K, Kobayashi N, Yoshitama K, Teramoto S, Komamine A. Isolation and purification of tyrosine hydroxylase from callus cultures of *Portulaca grandiflora*. Plant Cell Physiol. 2001:42: 969–975. pmid:11577191 doi: 10.1093/pcp/pce125
13. Nakatsuka T, Yamada E, Takahashi H, Imamura T, Suzuki M, Ozeki Y, et al. Genetic engineering of yellow betalain pigments beyond the species barrier. Scientific Reports 2013:3: 1970. doi: 10.1038/srep01970. pmid:23760173
14. Mayer AM. Polyphenol oxidases in plants and fungi: Going places? A review. Phytochemistry 2006:67, 2318–231. pmid:16973188 doi: 10.1016/j.phytochem.2006.08.006
15. Sullivan M. Beyond brown: polyphenol oxidases as enzymes of plant specialized metabolism. Front. Plant Sci. 2015:5: 783. doi: 10.3389/fpls.2014.00783. pmid:25642234
16. Polturak G, Breitel D, Grossman N, Sarrion-Perdigones A, Weithorn E, Pliner M, et al. Elucidation of the first committed step in betalain biosynthesis enables the heterologous engineering of betalain pigments in plants. New Phytologist 2015: doi: 10.1111/nph.13796.
17. Brockington SF, Yang Y, Gandia-Herrero F, Covshoff S, Hibberd JM, Sage RF, et al. Lineage-specific gene radiations underlie the evolution of novel betalain pigmentation in Caryophyllales. New Phytologist 2015:207: 1170–1180. doi: 10.1111/nph.13441. pmid:25966996
18. Keller W. Inheritance of some major color types in beets. J. Agric. Res. 1936:52: 27–38.
19. Karimi M, Inze D, Depicker A. GATEWAY vectors for *Agrobacterium*-mediated plant transformation. Trends Plant Sci. 2002:7: 193–195. pmid:11992820 doi: 10.1016/s1360-1385(02)02251-3
20. Quandt HJ, Pühler A, Broer I. Transgenic root nodules of *Vicia hirsuta*: a fast and efficient system for the study of gene expression in indeterminate-type nodules. Molecular Plant-Microbe Interactions 1993:6: 699–706. doi: 10.1094/mpmi-6-699
21. Koncz C, Schell J. The promoter Tl-DNA gene 5 controls the tissue-specific expression of chimaeric genes carried by a novel type of *Agrobacterium* vector. Mol. Gen. Genet. 1986:204: 383–396. doi: 10.1007/bf00331014
22. an Mullem V, Wery M, De Bolle X, Vandenhaute J. Construction of a set of*Saccharomyces cerevisiae* vectors designed for recombinational cloning. Yeast 2003:20: 739–746. pmid:12794935 doi: 10.1002/yea.999
23. Zhao M, Morohashi K, Hatlestad G, Grotewold E, Lloyd AM. The TTG1-bHLH-MYB complex controls trichome cell fate and patterning through direct targeting of regulatory loci. Development 2008:135: 1991–1999. doi: 10.1242/dev.016873. pmid:18434419
24. Urban P, Mignotte C, Kazmaier M, Delorme F, Pompon D. Cloning, yeast expression, and characterization of the coupling of two distantly related *Arabidopsis thaliana*NADPH–cytochrome P450 reductases with P450 CYP73A5. J. Biol. Chem. 1997:272: 19176–19186. pmid:9235908 doi: 10.1074/jbc.272.31.19176
25. Bilyk A. Thin layer chromatographic separation of beet pigments. J. Food Sci. 1981:46: 298–299. doi: 10.1111/j.1365-2621.1981.tb14588.x

26. Kearse M, Moir R, Wilson A, Stones-Havas S, Cheung M, Sturrock S, et al. Geneious Basic: an integrated and extendable desktop software platform for the organization and analysis of sequence data. Bioinformatics 2012:28: 1647–1649. doi: 10.1093/bioinformatics/bts199. pmid:22543367

27. Darriba D, Taboada GL, Doallo R, Posada D. jModelTest 2: more models, new heuristics and parallel computing. Nature Methods 2012:9: 772. doi: 10.1038/nmeth.2109

28. Ronquist F, Teslenko M, van der Mark P, Ayres DL, Darling A, Höhna S, et al. MrBayes 3.2: efficient Bayesian phylogenetic inference and model choice across a large model space. Syst. Biol. 2012:61: 539–542. doi: 10.1093/sysbio/sys029. pmid:22357727

29. u B, Yang Z. PAMLX: a graphical user interface for PAML. Mol. Biol. Evol. 2013:30: 2723–2724. doi: 10.1093/molbev/mst179. pmid:24105918

30. Hatlestad G, Akhavan NA, Sunnadeniya RM, Elam L, Cargile S, Hembd A, et al. The beet Y locus encodes an anthocyanin-MYB-like protein that activates the betalain red pigment pathway. Nature Genetics 2015:47: 92–96 doi: 10.1038/ng.3163. pmid:25436858

31. Dohm JC, Minoche AE, Holtgrawe D, Capella-Gutierrez S, Zakrzewski F, Tafer H, et al. The genome of the recently domesticated crop plant sugar beet (*Beta vulgaris*). Nature 2014:505: 546–549 doi: 10.1038/nature12817. pmid:24352233

32. Harris NN, Javellana J, Davies KM, Lewis DH, Jameson PE, Deroles SC, et al. Betalain production is possible in anthocyanin-producing plant species given the presence of DOPA-dioxygenase and L-DOPA. BMC Plant Biology 2012:12: 34. doi: 10.1186/1471-2229-12-34. pmid:22409631

33. Gandía-Herrero F, García-Carmona F, Escribano J. Botany: Floral fluorescence effect. Nature 2005:437: 334. pmid:16163341 doi: 10.1038/437334a

34. Kapadia GJ. Azuine MA. Sridhar R. Okuda Y. Tsuruta A. Ichiishi E, et al. Chemoprevention of DMBA-induced UV-B promoted, NOR-1-induced TPA promoted skin carcinogenesis, and DEN- induced phenobarbital promoted liver tumors in mice by extract of beetroot. Pharmacol. Res. 2003:47: 141–148. pmid:12543062 doi: 10.1016/s1043-6618(02)00285-2

35. Stintzing FC, Carle R. Functional properties of anthocyanins and betalains in plants, food, and in human nutrition. Trends in Food Science & Technology 2004:15: 19–38. doi: 10.1016/j.tifs.2003.07.004

36. han MI, Harsha PSCS, Giridhar P, Ravishankar GA. Pigment identification, nutritional composition, bioactivity, and in vitro cancer cell cytotoxicity of *Rivina humilis* L. berries, potential source of betalains. LWT–Food Sci. Technol. 2012:47: 315–323. doi: 10.1016/j.lwt.2012.01.025

37. Gandia-Herrero F, Cabanes J, Escribano J, Garcia-Carmona F, Jimenez-Atienzar M. Encapsulation of the most potent antioxidant betalains in edible matrixes as powders of different colors. J. Agric. Food Chem. 2013:61: 4294–4302. doi: 10.1021/jf400337g. pmid:23521423

38. Brockington SF, Walker RH, Glover BJ, Soltis PS, Soltis DE. Complex pigment evolution in the Caryophyllales. New Phytologist 2011:190: 854–864. pmid:21714182 doi: 10.1111/j.1469-8137.2011.03687.x

CHAPTER 7

EFFECTS OF GIBBERELLIC ACID AND CALCIUM CHLORIDE ON COLOUR, PHENOLIC COMPOUNDS, CAROTENOIDS AND QUALITY ATTRIBUTES OF WHITE CABBAGE (BRASSICA OLERACEA) DURING STORAGE UNDER REFRIGERATION

Williamson D.D[1], Benkeblia N[2]

[1] Department of Chemistry, University of the West Indies, Mona Campus, Kingston 7, Jamaica.

[2] Department of Life Sciences, University of the West Indies, Mona Campus, Kingston 7, Jamaica.

ABSTRACT

The effects of gibberellic acid (150 and 300 mg/L) and calcium chloride (40 g/L) on chlorophylls a and b, carotenoids, total phenolic compounds, colour and physiological weight losses of white cabbage (*Brassica oleracea* L. var. capitata) during storage at 10 ± 1°C were investigated. Results showed that chlorophyll a and total phenolic compounds decreased in all samples and no significant difference was noted between treated heads and control. Chlorophyll b and carotenoids did not vary much during storage and a steady state was noted in all samples. Treatments maintained well the total colour difference compared to control, and GA treated cabbages maintained better the colour after the fourteen days storage. The chemical treatments also reduced significantly the physiological weight loss which was much greater in controls than in GA treated heads after the storage duration.

INTRODUCTION

White Cabbage (Brassica oleracea L. var. capitata) is one of the most commonly cultivated vegetables across the entire island of Jamaica. This crop is cultivated mainly in rural parishes and is a staple food for many Jamaican families. In this context it is customary for the immature leaves within the cabbage head to be consumed while the mature outer leaves are usually rejected. Cabbage as a vegetable contains chlorophylls, carotenoids, vitamins and phenolic compounds that are linked with dietary activities, colour, characteristic flavour and therapeutic value.

Chlorophylls, carotenoids and phenolic compounds are important biological compounds found in all vegetables and contribute to the overall visual quality attributes of cabbage. Degradation of these compounds will lead to quality losses of cabbage heads including yellowing of outer leaves, core elongation, internal yellowing in the apex region, and sometimes in rootlet development at the core-end. These quality defects may affect sales as they are obvious to the consumer when the cabbage has reached market [1]. Gibberellic acid (GA3) has been widely used in agriculture. It was reported that GA_3 plays a role in the development of colours and metabolic activities

Citation: Williamson D.D, Benkeblia N (2015) Effects of Gibberellic Acid and Calcium Chloride on Colour, Phenolic Compounds, Carotenoids and Quality Attributes of White Cabbage (Brassica oleracea) during Storage under Refrigeration. Int J Food Sci Nutr Diet. 04(7), 239-245. doi: dx.doi.org/10.19070/2326-3350-1500043.

in the ripening process of fruits and vegetables, and delays shrivelling and senescence of some fruits [2]. It has also been reported that gibberellic acid reduces physiological weight loss of treated mangoes [3] and lemon [4] during storage. Calcium chloride is also used to extend the shelf-life of crop commodities [5]. The application of calcium chloride maintains the rigidity of cell walls; and delays membrane lipid catabolism and senescence [6]. Thus, tissues remain firm longer due to the inhibition of many enzymes such as polygalacuturonase which is involved in softening, therefore, ripening is delayed and shelf-life extended [2, 7]. Many investigations have been conducted on the postharvest application of calcium chloride to extend the shelflife of crops. These have shown that calcium chloride reduces physiological weight loss and delays softening of mangoes [8] and rotting of apples [9] during storage.

Although interesting data exist on the application of gibberellic acid and/or calcium chloride during the pre and the postharvest stages of some commodities, no work reported the application of these two chemicals on cabbage and their potential use to extend the shelf-life of this crop. Therefore, the goal of this study is to investigate the impact of gibberellic acid and calcium chloride on the colour, chlorophylls and carotenoids of cabbage heads during storage under refrigerated temperatures.

MATERIALS AND METHODS

Plant Material

Freshly harvested cabbage heads of uniform size, and free from damage, disease or spoilage, were obtained from the local fresh crops market in Kingston. Cabbages were transported immediately to the laboratory and packed in perforated polyethylene bags. From the total cabbages, fifty were selected and 10 heads per bag were packed and kept few hours at 10 ± 1°C prior to chemical treatments.

Chemicals Treatments and Storage

Prior to packaging and storage, cabbages were sprayed six times using spray bottles with the following solutions: gibberellic acid of 150 mg/L (GA1) and 300 mg/L (GA2), and calcium chloride (40 g/L) (CC). Control samples were sprayed with distilled water. After spraying, samples were left to dry at room temperature, then stored at 10 ± 1°C and 90-95 % RH during 14 days.

Chlorophyll a, b and Carotenoids Assessments

Chlorophylls a, b and carotenoids are determined according to the method described by Costache et al[10]. One gram of cabbage leaf is homogenised in 50 mL of 90% methanol using a mortar and pestle. The homogenate was filtered through two layers of cheesecloth, then centrifuged 1,000 × g to obtain a clear extract prior to being analysed. The absorbances of the solutions are read at 653 nm and 666 nm against 90 % methanol blank.

Determination of Total Phenolic Compounds (TPC)

Total phenolic compounds were extracted and assayed as described by Singleton and Rossi [11] with some modifications. Samples (5 g) were homogenized in 70% ethanol containing Nametabisulfite ($Na_2S_2O_5$, 20 g/L) and extracted usingan ultra sound assisted extraction (UAE) with an ultra sound sonicator at room temperature. The extracts were centrifuged at 1,000 × g for 10 min and the supernatant collected for TPC assay. Total phenolic compounds (TPC) of extracts were quantified colorimetrically using Folin-Ciocalteu reagent and chlorogenic acid as standard. Five millilitres of Folin-Ciocalteu (diluted ten-fold in distilled water), 2 mL of sodium bicarbonate (200 g/L) and 2 mL of distilled water were added to 1 mL of extract. After 15 min incubation at room temperature, the absorbance was read at 730 nm using Thermo Scientific™ GENESYS 10S UV-VIS spectrophotometer, and results expressed in chlorogenic acid equivalents(mg CAE/g fresh weight).

Colour Analysis

The colour of the arils was measured on the fruit surface using a Konica & Minolta CR-400 chromameter (Konica & Minolta Sensing Inc., Osaka, Japan) and the data were expressed in, L^*, a^* andb^* values (CIE). The colour reading was taken in triplicate at the equatorial region of each fruit and averaged to give a value for each cabbage head. The values a^* and b^* were used to calculate chroma$(C^* = [a^*2 + b^*2]1/2)$, which indicates the intensity or colour saturation, and hue angle (H^* = arctangent [b*/a*]), where 0° = red-purple, 90° = yellow, 180° = bluish-green, and 270° = blue [12].

Physiological Weight Loss (PWL)

Weight loss was calculated as a percentage (%) of the initial mass, with an electronic scale. The physiological weight loss was estimated from the difference between the initial and the final weight after each storage period.

Statistical Analysis

Experimental work was duplicated and experiments were run in triplicate. Experimental results were averaged and reported as mean ± standard deviation (SD). A single factor analysis of variance (ANOVA) was performed using GraphPad Prism 4.03 (GraphPad Software, Inc., 2236 La Jolla, CA, USA). Treatment means were compared by least significant difference (*LSD*) at $P \leq 0.05$.

RESULTS AND DISCUSSION

Variation of Chlorophylls a, b and Carotenoids

Chlorophyll a varied significantly during storage as shown in Figure 1A. Chlorophyll a decreased significantly in control, CC and GA1 after 11 days storage, while in GA2, the decrease was significant after 14 days storage. After 4 days, chlorophyll a decreased by 46%, 27% and 50% in control, CC and GA1, respectively, and the decrease reached 81%, 76% and 81% after 14 days storage, respectively. In GA2 treated cabbage, chlorophyll a decreased by 12% after four days, but the decrease was similar to that observed in other treated cabbage and reached 80%.

Surprisingly, chlorophyll b in cabbage did not show significant variation during storage (Figure 1B). Chlorophyll b varied from an initial value of 2.49 mg/g DW and averaged 2.53, 2.56 and 2.63 mg/gDW after four, eleven and fourteen days storage, respectively. Similarly, total carotenoids decreased slightly but did not vary significantly in cabbage during storage (Figure 1C). Total carotenoids yielded 2.71mg/g DW in fresh cabbage heads, and decreased by 14%, 3%, 15% and 4% in control, CC, GA1 and GA2 after fourteen days storage, respectively. On the other hand, carotenoids in cabbage averaged 2.62, 2.55 and 2.46 mg/g DW after four, eleven and fourteen days, respectively.

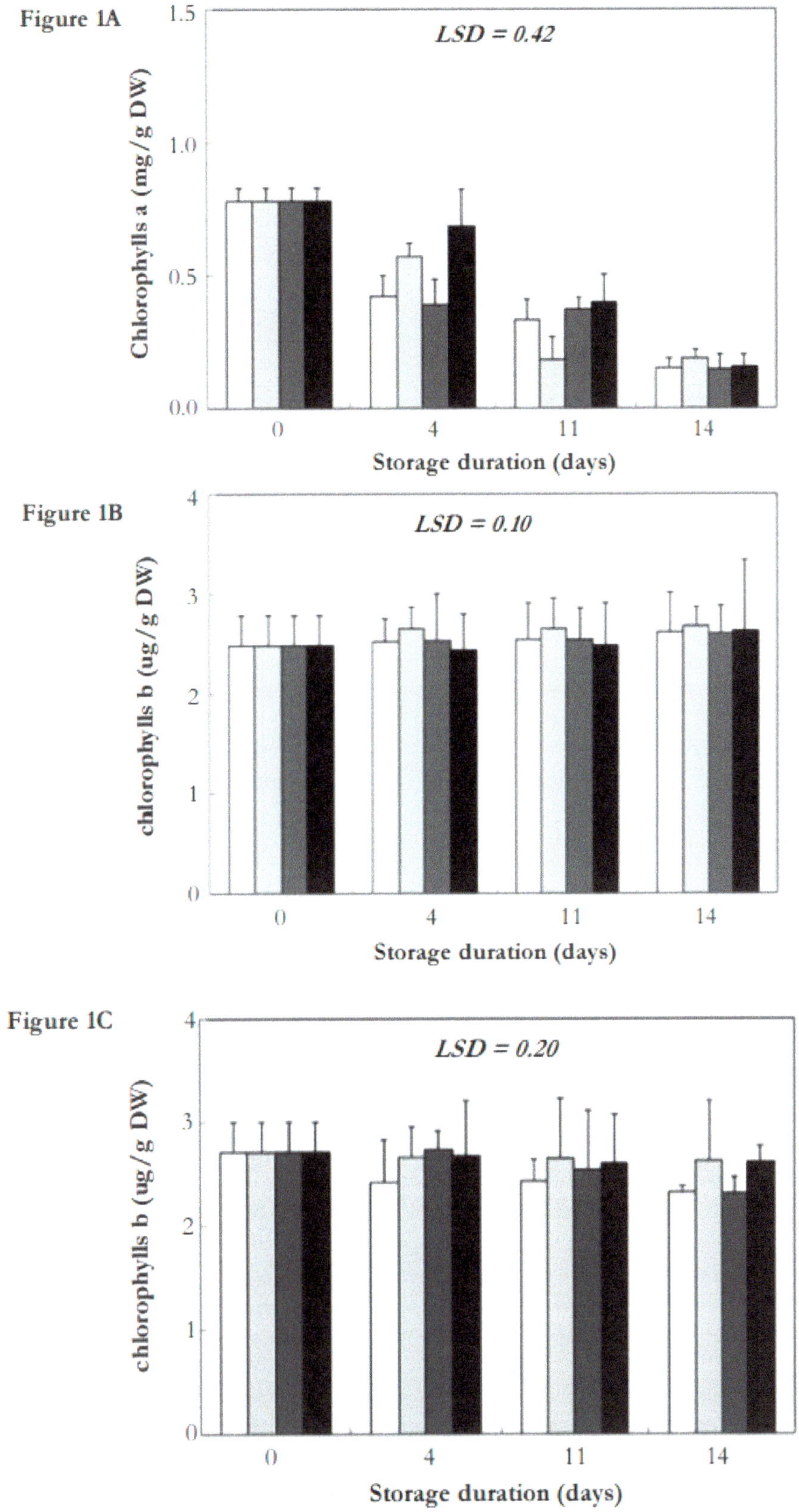

Figure 1. Chlorophylls a(A), chlorophylls b(B) and carotenoids (C) concentration in cabbage heads stored at 10 ± 1°C during 14 days. **□Control, Calcium Chloride, GA1 ■ GA2**

Comparatively, chlorophylls a/b ratio was estimated and data showed the ratio decreased progressively from 0.31 to 0.17, 0.22, 0.15 and 0.28 after four days, to 0.13, 0.07, 0.15 and 0.16 after eleven days, and to 0.06, 0.07, 0.06 and 0.06 after fourteen days in control, CC, GA1 and GA2, respectively (Figure 2A). On the other hand, the chlorophylls [a+b]/carotenoids ratio varied very slightly and decreased from 1.20 to 1.21, 1.21, 1.07 and 1.17 after four days, to 1.18, 1.07, 1.15 and 1.11 after eleven days, and to 1.19, 1.09, 1.19 and 1.07 after fourteen days in control, CC, GA1 and GA2, respectively (Figure 2B).

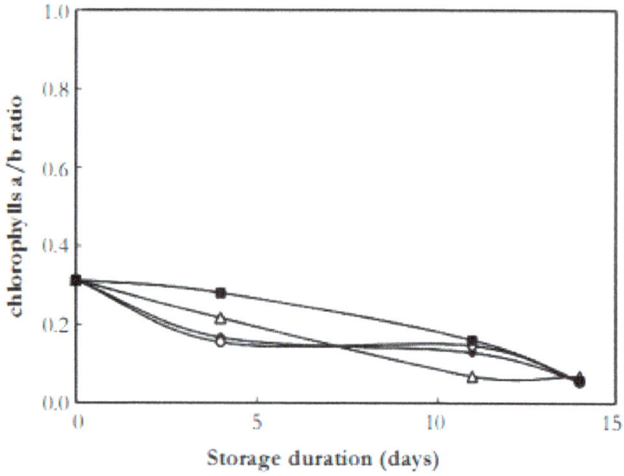

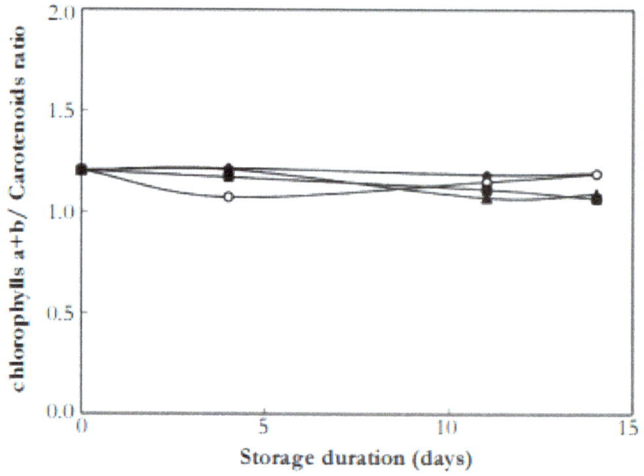

Figure 2. Chlorophylls a/b (A) and Chlorophylls a+b/carotenoids (B) ratios in cabbage heads stored at 10 ± 1°C during 14 days.

The use of gibberellic acid as a preharvest treatment is extensively documented. While the postharvest treatment reported is mainly on the use of GA to extend the shelf-life and longevity of ornamental flowers, few reports are available on fruits and vegetables. However, some discrepancies have been reported on the effect of postharvest application of GA on chlorophyll content. Siddiqui et al [13] reported a decrease of c.a. 50% in chlorophylls of mango fruits stored twelve days at room temperature of 28 ± 2°C, and no difference in effect was noted between the different GA concentrations used.

On the contrary, many studies reported that GA retained green colour of fruits and chlorophylls. Gross et al [14] reported that GA treatments delayed the senescence and retained chlorophyll in persimmon for long periods. Pre and postharvest applications of GA also retained the green fruit colour of citrus fruits. Mahajan et al [7] used different concentrations of GA3 ranging from 25 to 75 ppm on guava fruit and no significant difference in effect was noted among these concentrations. Singh and Dankhar [15]reported that postharvest application of GA retained chlorophylls in okra during storage and retention was higher even with visible senescence. They also noted that chlorophyll a decreased more rapidly than chlorophyll b which surprisingly exceeded the content of chlorophyll a in some fruits.

Janowska and Jerzy [16] investigated the effects of GA on the longevity of *Zantedeschia elliottiana* leaves and noted that GA reduced the degradation of chlorophylls, but no significance was noted between the concentrations and exposure time. Indeed, the mechanisms of GA on chlorophylls are not clearly understood, but it is likely that GA delays the activity of chlorophyllase during ripening of GA treated strawberry fruit as reported by Martínez et al [17], and/or promotes chlorophyll synthesis [14]. Indeed, the visible colour of plants results from the predominant pigments, even though other pigments are present but are not visible unless the predominant is degraded. Consequently, the degradation of chlorophylls causes colours change in the tissues from green to many other colours such as yellow and orange observed in ripe and senescent tissues [18].

Calcium chloride is widely used in agriculture and food preservation as well. Investigations showed that postharvest use of calcium chloride maintains firmness and visual quality, therefore, extending the shelf life of the fresh produce. It was reported that dipping fruits in calcium chloride was effective in delaying senescence and extended figs [19] and jujube [20] shelf-life during storage. On the other hand, calcium chloride treatments have also been reported as effective in reducing chlorophyll [21]. Calcium chloride has also been used in combination with gibberellic acid and other organic acids. Bhanja and Lenka [22] combined gibberellic acid and calcium chloride treatments of sapota (*Achraszapota*) fruit, and they noted that treatments reduced the physiological weight losses and percentage rotting of fruits during storage, and the shelf-life was extended to 36 days compared to control (8 days).

Indeed, these results are not in contradiction with our results, as we noted that chlorophyll a content is much lower than chlorophyll b. Even though chlorophyll a decreased chlorophyll b did not vary much, therefore, making the change in total chlorophylls (a+b) not significant as shown in Figure 2.

Variation of Total Phenolics Content (TPC)

The total phenolics content decreased progressively during storage of cabbage heads, however, the decrease was significant in the control after four days, while in GA2 TPC decreased significantly after fourteen days (Figure 3). After four days, TPC decreased from 50.91 to 40.20, 39.97, 45.13 and 47.26

mg/g DW in control, CC, GA1 and GA2, respectively. After fourteen days storage, TPC decreased significantly to 26.17, 26.38, 28.91 and 27.10 mg/g DW in control, CC, GA1 and GA2, respectively, and no significant difference was noted between control and treatments.

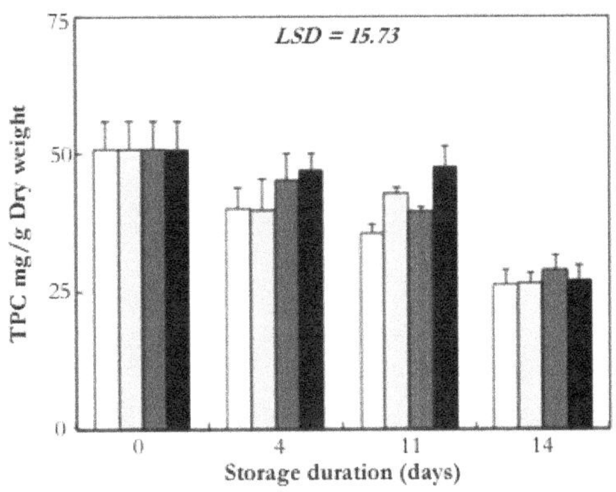

Figure 3. Total phenolic compounds concentration in cabbage heads stored at 10 ± 1°C during 14 days. For legends see Figure 1.

Many investigations have reported on TPC of white cabbage heads, and different TPC has been determined. For example and among literature cited, Leja et al [23] reported values ranging between 12.58 and 47.6 mg GAE/100 g FW. The large variations of TPC in white cabbages vary with different factors such as the botanical cabbage group [24], variety [25], extraction solvents used [26], and growing region [27]. Many works reported on the total phenolics of cabbage, but few exist on the variation of TPC during storage, and no work reports on the effects of GA and calcium chloride on TPC in stored cabbage. Hounsome et al [28] studied the variation of different phenolic compounds during storage and they noted that longterm storage significantly reduced the content of many phenolic compounds such as caffeic, syringic, gallic and trihyroxybenzoic acids, as well as flavonoids, while other phenolics such as cinnamic, dimethoxybenzoic acid, chalcone and flavone were less affected by storage. Hagen et al [29]stored another Brassica variety called curly kale (*Brassica oleracea L. var. acephala*) and noted that total phenolics increased slightly after six weeks storage at 1°C. Nevertheless, in stored lettuce, escarole and rocket salad tissues, TPC varied differently. TPC increased after one day but decreased after three days storage at 4°C. In escarole, TPC increased significantly after three days, while in rocket tissues TPC decreased after three days at the same temperature [30].

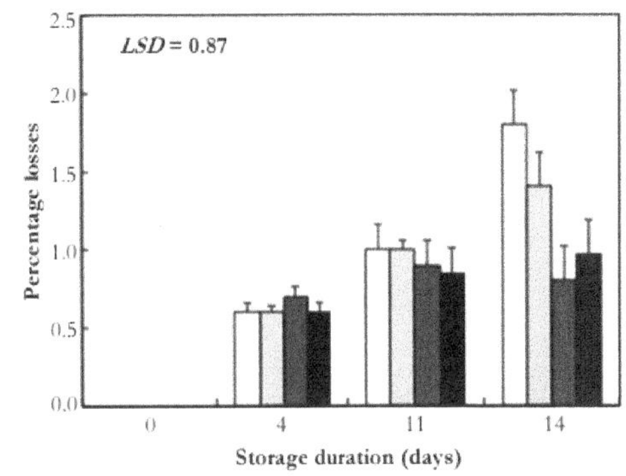

Figure 4. Weight losses in cabbage heads stored at 10 ± 1°C during 14 days. For legends see Figure 1.

Colour Variation

During storage, colour of cabbage heads showed different variations (Table 1). Data of a^* are indicating that cabbage heads are less after eleven days storage, and the lightness decreased after the same storage period as indicated by the decrease of L^*. The b^* decreased from positive to negative values and the H^* increased after fourteen days storage indicating that the yellowish colour of cabbages is decreasing and the heads are turning darker and did not retain the yellow colour (as shown in Figure 5). However, no significant difference was noted between control and treated cabbages during the storage period.

Moreover, the total colour difference between all three coordinates was determined by the following equation:

$$\Delta E = \sqrt{(\Delta L^2 + \Delta a^2 + \Delta b^2)}$$

Table 1. Variation of colour parameters of cabbage heads during storage. Values with different superscript letters within the same columns are significantly different.

	a^*	b^*	L^*	C^*	H^*
	Freshly harvested				
	70.60 ± 2.90	10.46 ± 0.30	31.94 ± 1.97	33.62 ± 1.90	108.23 ± 1.02
	After 4 days storage at 10 ± 1°C				
Control	70.60 ± 2.90^{a}	10.46 ± 0.30^{a}	31.94 ± 1.97^{a}	33.62 ± 1.90^{a}	108.23 ± 1.02^{a}
CC	69.66 ± 6.54^{a}	11.14 ± 0.86^{b}	25.98 ± 1.35^{b}	28.30 ± 1.33^{b}	113.26 ± 9.78^{a}
GA1	70.28 ± 3.94^{a}	14.25 ± 0.74^{b}	30.70 ± 2.07^{a}	33.85 ± 2.17^{a}	114.96 ± 0.60^{a}
GA2	72.44 ± 2.68^{a}	9.63 ± 4.85^{a}	28.61 ± 3.87^{ab}	35.25 ± 2.27^{a}	123.59 ± 6.96^{b}
	After 11 days storage at 10 ± 1°C				
Control	58.16 ± 7.75^{b}	-10.63 ± 1.16	18.94 ± 2.01^{c}	24.26 ± 2.12^{b}	113.05 ± 2.22^{a}
CC	61.79 ± 5.96^{b}	-13.61 ± 1.11	30.96 ± 1.67^{a}	33.83 ± 1.98^{b}	113.67 ± 8.55^{a}
GA1	57.47 ± 8.04^{b}	-9.96 ±1.07	22.90 ± 1.94^{b}	24.99 ± 2.10^{b}	113.47± 7.10^{a}
GA2	49.76 ± 7.53^{c}	-13.75 ± 3.44	27.17 ± 6.05^{ab}	30.46 ± 6.94^{ab}	126.42 ± 11.4^{b}
	After 14 days storage at 10 ± 1°C				
Control	70.60 ± 2.90^{a}	10.46 ± 0.30	19.94 ± 1.97^{a}	23.62 ± 1.90^{b}	116.23 ± 1.02^{a}
CC	63.36 ± 3.02^{b}	-12.85 ± 2.28	24.37 ± 1.66^{b}	27.61 ± 2.50^{ab}	117.29 ± 2.86
GA1	71.82 ± 6.08^{a}	-13.20 ± 1.53	24.99 ± 3.57^{b}	28.26 ± 5.72^{ab}	117.84 ± 9.72
GA2	72.72 ± 2.54^{a}	-10.63 ± 1.20	23.21 ± 1.72^{b}	25.55 ± 2.01^{b}	124.50 ± 13.5^{b}

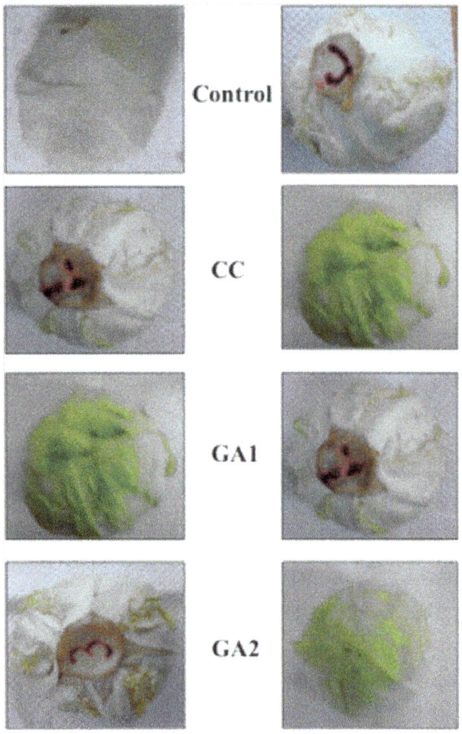

Figure 5. Cabbage heads stored at 10 ± 1°C during 14 days.

Data are showing that control, CC, GA1 and GA2 averaged 12.00, 25.56, 24.69 and 22.92 after four, eleven and fourteen days storage indicating that the colour of control and GA2 are lighter than CC and GA1. Few works have reported on the variation on colour of cabbage during storage, while no reference is readily available on the effects of gibberillic acid and calcium chloride on stored cabbage heads. Manolopoulou and Varzakas [31] reported similar results and noted that lightness L^* and H^* values of fresh cabbage stored at 0°C decreased after 15 days storage, but L^* decreased after 20 days storage, and these data are in agreement with our results indicating that the reduction was limited regardless of the chemical treatments and the initial colour was to some extent preserved. The increase of H* value observed after fourteen days might be due to browning caused by the senescence [32]. On the other hand, it was suggested that the total colour difference (ΔE) could be used as a discoloration index[33], and our data are suggesting that all treatments did not preserve the colour of cabbage.

Physiological Weight Loss (PWL)

During storage, PWL increased progressively during storage, however, control and CC treated cabbages showed higher PWL than GA treated ones (Figure 4). After four days storage, PWL averaged 0.6% to 0.7% in control and treated cabbages, and after eleven days PWL averaged 1.00%, 1.00%, 0.90% and 0.85% in control, CC, GA1 and GA2, respectively, with no significant difference between control and all treated cabbages. Nevertheless, after fourteen days storage, PWL averaged 1.80%, 1.40%, 0.80% and 0.97% in control, CC, GA1 and GA2, respectively. By fitting linear regression lines to weight losses vs time, results showed average losses of 0.13%, 0.11%, 0.06% and 0.07% per day in control, CC, GA1 and GA2, respectively, indicating that GA treatment was more efficient in reducing PWL than CC treatment.

Extensive literature exists on PWL of fresh crops during storage. Weigh losses of fresh crops is the consequence of respiration and oxidative reactions, mainly sugars. However, degradation of energy-source compounds is correlated to many factors such as the physiological stage, temperatures and storage duration, as well as other factors such as gas composition (eg., modified atmosphere packaging), and even the microbial charge of the stored crops [34]. Manolopoulou and Varzakas [31]reported that weight losses of cabbage averaged 3% after twenty-three days storage at 4°C. In minimally processed cabbage, weight losses were higher and averaged 2.81% and 3.32% in control, and 2.28% and 3.35% in calcium chloride treated heads stored during twenty-two days at 0°C and 5°C, respectively [35].

CONCLUSION

The application of gibberellic acid and calcium chloride to extend the shelf-life of cabbage heads stored under refrigeration showed that these chemical treatments did not reduce chlorophyll a and total phenolic compounds losses but maintained chlorophyll b and carotenoid levels which contributed to better colour preservation although no statistical difference was noted. However, all the chemical treatments reduced significantly the physiological weight losses which were much higher in the control. We can conclude that gibberellic acid and calcium chloride could be used to maintain the quality attributes of fresh cabbage heads stored under refrigerated temperature, however, the optimal storage period should not exceed two weeks.

Acknowledgements

This work was supported by a New Research Initiative Research Grant from the Office of the Graduate Studies & Research, UWI, Mona.

REFERENCES

1. Kramchote S, Srilaong V, Wongs-Aree C, Kanlayanarat S (2012) Low temperature storage maintains postharvest quality of cabbage (*Brassica oleraceae* var. capitata L.) in supply chain. Int Food Res J 19(2): 759-763.
2. Pila N, Gol NB, Rao TVR (2010) Effect of post-harvest treatments on physicochemical characteristics and shelf life of tomato (*Lycopersicon esculentum* Mill.) fruits during storage. Am Eurasian J Agric Environ Sci 9(5): 470-479.
3. Ahmed MS, Sanjay S (1999) Effect of various post-harvest treatments on shelf-life of amrapali mango. Orrisa J Hortic 27: 29-33.
4. Sindhu SS, Singhrot RS (1993) Effect of Pre-harvest spray of growth regulator and fungicides on the shelf life of lemon cv. Baramasi-A note. Haryana J Hort Sci 22: 204-204.
5. Martín-Diana AB, Rico D, Frías JM, Barat JM, Henehan G, et al. (2007) Calcium for extending the shelf life of fresh whole and minimally processed fruits and vegetables: a review. Trends Food Sci Technol 18(4): 210-218.

6. Ferguson IB (1984) Calcium in plant senescence and fruit ripening. Plant Cell Environ 7(6): 477-489.
7. Mahajan BVC, Ghuman BS, Bons HK (2011) Effect of postharvest treatment of calcium chloride and gibberellic acid on storage behaviour and quality of guava fruits. J Hort Sci Ornamen Plants 3: 38-42.
8. Mootoo A (1991) Effect of post-harvest calcium chloride dips on ripening changes in 'Julie' mangoes. Trop Sci 31(3): 243-248.
9. Nayital RK, Chopra SK, Sharma YP (1990) Effect of post-harvest application of calcium chloride, fungicides and wax emulsion on the storage behaviour of apple cv. red delecious. Indian Food Packer 44: 31-38.
10. Costache MA, Campeanu G, Neata G (2012) Studies concerning the extraction of chlorophyll and total Carotenoids from vegetables. Romanian Biotechnol Lett 17(5): 7702-7708.
11. Singleton VL, Rossi Jr JA (1965) Colorimetry of total phenolics with phosphomolybdic– phosphotungstic acid reagents. Am J Enol Vitic 16(3): 144-158.
12. McGuire RG (1992) Reporting of objective color measurements. Hort Science 27(12): 1254-1255.
13. Siddiqui MW, Dutta P, Dhua RS, Dey A (2014) Changes in biochemical composition of mango in response to pre-harvest gibberellic acid spray. Agric Consp Sci 78(4): 331-335.
14. Gross J, Bazak H, Blumenfeld A, Ben-Arie R (1984) Changes in chlorophyll and carotenoid pigments in the peel of Triumph persimmon (*Diospyros kaki* L.) induced by pre-harvest gibberellin (GA3) treatment. Sci Hort 24(3-4):305-314.
15. Singh BP, Dhankhar BS (1980) Effect of growth regulators and pre-packaging on the storage life of okra. Haryana Agric Univ J Res 10(3): 398-402.
16. Janowska B, Jerzy M (2003) Effect of gibberellic acid on the post-harvest leaf longevity of *Zantedeschia elliottiana* (W. Wats.) Engl. J Fruit Ornam Plant Res 11: 69-76.
17. Martínez GA, Chaves AR, Añón MC (1996) Effect of exogenous application of gibberellic acid on color change and phenylalanine ammonia-lyase, chlorophyllase, and peroxidase activities during ripening of strawberry fruit (*Fragaria x ananassa* Duch.). J Plant Growth Regul 15: 139-146.
18. Schoefs B (2002) Chlorophyll and carotenoid analysis in food products. Properties of the pigments and methods of analysis. Trends Food Sci Technol 13(11): 361-371.
19. Irfan PK, Vanjakshi V, Prakash KMN, Ravi R, Kudachikar VB (2013) Calcium chloride extends the keeping quality of fig fruit (*Ficus carica* L.) during storage and shelf-life. Postharvest Biol Technol 82: 70-75.
20. Li L, Ban Z, Li X, Xue T (2014) Effect of 1-methylcyclopropene and calcium chloride treatments on quality maintenance of 'Lingwu Long' jujube fruit. J Food Sci Technol 51(4): 700-707.
21. Smout C, Sila DN, Vu TS, Van Loey A, Hendrickx M (2005) Effect of preheating and calcium pre-treatment on pectin structure and thermal texture degradation: a case study on carrots. J Food Eng 67(4): 419-425.
22. Bhanja PK, Lenka PC (1994) Effect of pre- and postharvest treatments on storage life of sapota fruits cv. Oval. Orissa J Hort 22(1): 54-57.
23. Leja M, Kamińska I, Kołton A (2010) Phenolic compounds as the major antioxidants in red cabbage. Folia Hort 22(1): 19-24.
24. Cartea ME, Francisco M, Soengas P, Velasco P (2011) Phenolic compounds in brassica vegetables. Molecules 16(1): 251-280.
25. Singh J, Upadhyay AK, Bahadur A, Singh B, Singh KP, et al. (2006) Antioxidant phytochemicals in cabbage (*Brassica oleracea* L. var. capitata). Sci Hort 108(3): 233-237.
26. Podsedek A (2007) Natural antioxidants and antioxidant capacity of Brassica vegetables: A review. LWT/Food Sci Technol 40(1): 1-11.
27. Kusznierewicz B, Bartoszek A, Wolska L, Drzewiecki J, Gorinstein S, et al. (2008) Partial characterization of white cabbages (Brassica oleracea var. capitata f. alba) from different regions by glucosinolates, bioactive compounds, total antioxidant activities and proteins. LWT/Food Sci Technol 41(1): 1-9.
28. Hounsome N, Hounsome B, Tomos D, Edwards-Jones G (2009) Changes in antioxidant compounds in white cabbage during winter storage. Postharvest Biol Technol 52(2): 173-179.
29. Hagen SF, Borge GIA, Solhaug KA, Bengtsson GB (2009) Effect of cold storage and harvest date on bioactive compounds in curly kale (*Brassica oleracea* L. var. acephala). Postharvest Biol Technol 51: 36-42.
30. Degli'innoocenti E, Pardossi A, Tattini M, Guidi L (2008) Phenolic compounds and antioxidant power in minimally processed salad. J food Biochem 32(5): 642-653.
31. Manolopoulou E, Varzakas TH (2013) Effect of modified atmosphere packaging (MAP) on the quality of 'ready-to-eat' shredded cabbage. Int J Agric Food Res 2(3): 30-43.
32. Kaji H, Ueno M, Osajima Y (1993) Storage of shredded cabbage under a dynamically controlled atmosphere of high oxygen and high carbon dioxide. Biosci Biotechnol Biochem 57(7): 1049-1052.
33. Hu W, Jiang A, Qi H, Pang K, Fan S (2007) Effects of initial low oxygen and perforated film package on quality of fresh-cut cabbage. J Sci Food Agric 87(11): 2019-2025.
34. Paull RE (1999) Effect of temperature and relative humidity on fresh commodity quality. Postharvest Biol Technol 15(3): 263-277.
35. Manolopoulou E, Varzakas T (2011) Effect of storage conditions on the sensory quality, colour and texture of fresh-cut minimally processed cabbage with the addition of ascorbic acid, citric acid and calcium chloride. Food Nutri Sci 2(9): 956-963.

CHAPTER 8

MARENNINE, PROMISING BLUE PIGMENTS FROM A WIDESPREAD HASLEA DIATOM SPECIES COMPLEX

Romain Gastineau [1], François Turcotte [2], Jean-Bernard Pouvreau [3], Michèle Morançais [4], Joël Fleurence [4], Eko Windarto [1], Fiddy Semba Prasetiya [1], Sulastri Arsad [1], Pascal Jaouen [5], Mathieu Babin [2], Laurence Coiffard [4], Céline Couteau [4], Jean-François Bardeau [6], Boris Jacquette [1], Vincent Leignel [1], Yann Hardivillier [1], Isabelle Marcotte [7], Nathalie Bourgougnon [8], Réjean Tremblay [2], Jean-Sébastien Deschênes [2], Hope Badawy [6], Pamela Pasetto [6], Nikolai Davidovich [9], Gert Hansen [10], Jens Dittmer [6] and Jean-Luc Mouget [1]

[1]FR CNRS 3473 IUML, Mer-Molécules-Santé (MMS), Université du Maine, Ave O. Messiaen, 72085 Le Mans cedex 9, France

[2]Institut des Sciences de la mer de Rimouski, Université du Québec à Rimouski, 310 des Ursulines, Rimouski, QC G5L 3A1, Canada

[3]EA 1157, Laboratoire de Biologie et Pathologie Végétales (LBPV), Université de Nantes, SFR 4207 QUASAV, 44322 Nantes, France

[4]FR CNRS 3473 IUML, Mer-Molécules-Santé (MMS), Université de Nantes, 44322 Nantes, France

[5]FR CNRS 3473 IUML, UMR-CNRS 6144 (GEPEA), Université de Nantes, CRTT 37 Bd de l'Université, F-44602 Saint-Nazaire, France

[6]UMR CNRS 6283, Institut des Molécules et Matériaux du Mans (IMMM), Université du Maine, Ave O. Messiaen, 72085 Le Mans cedex 9, France

[7]Department of Chemistry, Université du Québec à Montréal, P.O. Box 8888, Downtown Station, Montréal, QC H3C 3P8, Canada

[8]Université de Bretagne-Sud, Laboratoire de Biotechnologie et Chimie Marines, Campus de Tohannic, F-56017 Vannes, France

[9]Karadag Nature Reserve of the National Academy of Sciences, p/o Kurortnoe, Feodosiya 98188, Crimea

[10]Department of Biology, Marine Biological Section, University of Copenhagen, Øster Farimagsgade 2D, 1353 Copenhagen, Denmark

ABSTRACT

In diatoms, the main photosynthetic pigments are chlorophylls *a* and *c*, fucoxanthin, diadinoxanthin and diatoxanthin. The marine pennate diatom *Haslea ostrearia* has long been known for producing, in addition to these generic pigments, a water-soluble blue pigment, marennine. This pigment, responsible for the greening of oysters in western France, presents different biological activities: allelopathic, antioxidant, antibacterial, antiviral, and growth-inhibiting. A method to

Citation: Gastineau R, Turcotte F, Pouvreau J-B, et al. Marennine, Promising Blue Pigments from a Widespread Haslea Diatom Species Complex. Marine Drugs. 2014;12(6):3161-3189. doi:10.3390/md12063161.

extract and purify marennine has been developed, but its chemical structure could hitherto not be resolved. For decades, *H. ostrearia* **was the only organism known to produce marennine, and can be found worldwide. Our knowledge about** *H. ostrearia***-like diatom biodiversity has recently been extended with the discovery of several new species of blue diatoms, the recently described***H. karadagensis***,** *H. silbo* **sp. inedit. and** *H. provincialis* **sp. inedit. These blue diatoms produce different marennine-like pigments, which belong to the same chemical family and present similar biological activities. Aside from being a potential source of natural blue pigments,** *H. ostrearia***-like diatoms thus present a commercial potential for aquaculture, cosmetics, food and health industries.**

Keywords: aquaculture, biological activities, cosmetics and food industry, *Haslea ostrearia*-like diatoms, marennine-like blue pigments

INTRODUCTION

Seas and oceans cover more than 70% of the Earth's surface; water mainly transmits and scatters blue wavelengths while absorbing the red part of the solar light spectrum, thus leading astronauts to say our planet is blue. In or under the sea, however, the blue color is scarcely distributed, as it can be observed only in a few organisms, such as the blue jellyfish *Cyanea lamarckii*, the blue coral *Heliopora coerulea*, the blue sea star *Linckia laevigata*, the giant clam *Tridacna maxima*—the mantle of which is commonly bright blue—and some species of surgeonfish from the genus *Acanthurus*, all of which being emblematic species. Micro-organisms appear to be better providers of the blue color, as blue pigments have been evidenced long ago among bacteria, albeit first in terrestrial species. For instance, in Proteobacteria, *Pseudomonas aeruginosa* produces pyocyanin [1,2], a blue pigment that possesses antibiotic activities [3], and *Pantoea agglomerans* has recently been shown to produce a novel "deep blue" pigment [4]. The study of marine bacteria started and expanded later, but a few species are known today to synthesize blue pigments [5]. For example, glaukothalin is produced by different species from the genus*Rheinheimera* [6,7], and indigoidine is encountered in a strain of the marine bacteria genus *Phaeobacter*, in which it plays a role in the colonization of surfaces [8]. It is worth noting that indigoidine is also encountered in the terrestrial plant pathogen *Erwinia chrysanthemi* [9], in which this blue pigment seems to be partly responsible for its pathogenicity [10].

Regarding photosynthetic organisms, blue pigments can be observed in prokaryote as well as in eukaryote species. Formerly known as "blue algae", cyanobacteria have specific accessory protein–pigment complexes, the phycobiliproteins. Some of these phycobiliproteins, like phycocyanin and allophycocyanin, have a blue color due to their absorption of orange and/or red light [11]. Phycobiliproteins were discovered in the 19th century, phycocyanin being first described in a strain of *Oscillaria* sp. [12], and allophycocyanin in the red macroalga *Porphyra vulgaris* [13]. The partial protein nature of these two molecules was suggested by Mölisch [14], in his work on phycoerythrin [15]. Kylin [16] further demonstrated that each molecule was a complex of a chromophore (phycobilin) covalently bound to a protein. Phycocyanin and allophycocyanin are not restricted to cyanobacteria, as they have been evidenced in two groups of photosynthetic eukaryotes, Rhodophyta and Glaucophyta. Members of another group, the Cryptophyta, contain only one of these two blue pigments, phycocyanin (e.g., [17]). Until very recently, only two other photosynthetic eukaryotes, both members of the Heterokontophyta, were known to produce blue pigments, the recently discovered *Aurearena cruciata*(Aurearenophyceae) in its senescent stage [18], and the long-known pennate diatom*Haslea ostrearia* (Bacillariophyceae), during its exponential phase of growth and aging [19]. *H. ostrearia*, the "blue navicula," produces the specific pigment marennine, responsible for the "greening" of oysters.

In this review, we focus on the diatom *H. ostrearia*, its blue pigment marennine, and their many interactions with oysters. We present an historical perspective of our knowledge of marennine complemented by some new insights, regarding its chemical nature, characterization, and biological properties, particularly considering biophysical processes required to absorb light. Indeed, an explanation why the color blue is relatively seldom in organisms and organic molecules could be related to the physical processes required to absorb red or yellow light, that is, light with relatively low energy. If the color is not stemming from metal centers, it is usually the excitation of delocalized electrons from their molecular orbital ground state π to the excited state π^*, which makes a molecule appear colored. One double bond alone is usually not sufficient as it absorbs light in the UV range. Only the conjugation of several double bonds broadens the energy range of the different π orbitals, reducing the energy gap between the highest π and the lowest π^* orbital. More than ten conjugated double bonds are necessary to reduce the energy gap to absorb red light [20] so that the molecule appears blue, and this is a rather rare constellation. Similar considerations count for aromatic rings, where usually adjacent groups can reduce the absorption energy, but hardly into the red region.

We also present data about *H. karadagensis*, the second species of blue diatoms recently discovered in the Black Sea, and *H. provincialis* sp. inedit. and *H. silbo* sp. inedit., two new blue diatom species discovered in the Mediterranean Sea and the Canary Islands, respectively [21]. At least one of the blue pigments produced by these new species (*H. karadagensis*) is different from marennine, but all belong to the same chemical family. Finally, we address the unexpected biodiversity of blue diatoms, and the added-value potential of marennine-like pigments, in aquaculture, cosmetics, food and health industry.

THE MARENNINE–OYSTER CONNECTION

In the following, we will refer to blue pigments as "marennine" when produced by *H. ostrearia*, and as "marennine-like pigments" when produced by another species of blue diatoms (Figure 1). Indeed, no other name can be provided in accordance with the IUPAC standards, as the chemical structures of these pigments are as of yet unresolved. Marennine is the anglicized version of "marennin," proposed by Lankester [22], in reference to Marennes-Oléron (Marennes, 45°49′25.8″ N 1°06′16.3″ W), a region of western France where oysters have been farmed for centuries, especially renowned for producing "green oysters". In the Marennes-Oléron Bay, as well as in Bourgneuf Bay (Bouin, 46°58′27.4″ N 1°59′54.9″ W), oysters are matured and fattened in old salt marshes, the so-called "claires". In these shallow and nutrient-rich ponds, phytoplankton proliferates and may feed oysters whose size rapidly increases and whose organoleptic properties improve. The duration of the fattening period and the density of the oysters in the ponds are regulated according to specific standards for the bivalves to be commercialized as "fines de claires." In these ponds, *H. ostrearia* may become dominant year after year, producing large amounts of marennine, which are released into the seawater. It adheres to oyster gills, a phenomenon that increases culinary attractiveness and therefore the market value of the bivalves (red label "fines de claires vertes").

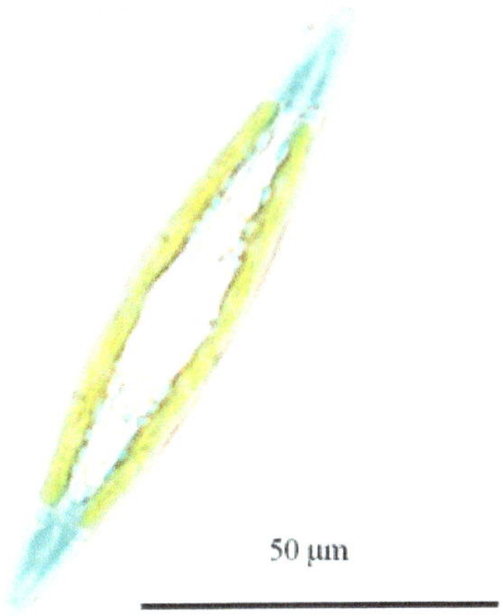

Figure 1: Light micrograph of *Haslea ostrearia*.

A Brief History of the Greening of Oysters, Signature of the Diatom Haslea ostrearia

As a biological phenomenon, green oysters provoked questions and wonderment for centuries. The earliest indication known is "The history of the generation and ordering green oysters, commonly called Colchester-oysters", by Thomas Sprat, at the end of the 17th century [23]. Thomas Sprat was an Anglican bishop of Rochester, famous for his*History of the Royal Society of London*, in which he included an article on the green oysters of Colchester. He mentioned that oysters from the Tolesbury ponds could present a green color during summer, a phenomenon he suspected resulted from the actions of the sun and the Earth, leading to a green coloration of pond bottom, before oyster gills turn green. The same environmental causal factors, along with the brackishness of waters, were addressed 150 years later [24]. Other hypotheses, not referring to environmental issues, ascribed the greening to a disease similar to obesity [25], or to "liver malfunction" [26]. The role of grass, plants and mosses grown on the shores of the ponds and possibly ingested by oysters or some kind of Priestley's green matter was also hypothesized [27,28,29]. In line with these speculations, some authors suspected that the water of ponds, and consequently oysters, could be colored by pigments originating from green macroalgae [30]. Finally, some authors estimated that greening could result from the presence of specific metallic ions in pond sediments [31], or from unusual interactions with copper [32,33], a metal long known for inducing changes in color in oysters [34]. Indeed, such unusual color may be perceived as unappetizing, possibly reflecting the presence of pollutants as zinc or copper [35,36,37]. In France, however, green oysters are gastronomically famous and more expensive than ordinary oysters. Moreover, from an historical perspective, green oysters have always been celebrated in this country as a delicacy [38,39,40], even being considered a dish fit for a king—at the very least for the Sun King; it was one of Louis XIV's favorite meals [41].

The first experimental work on green oysters was reported by Benjamin Gaillon [42], an officer of French Customs during the Restoration and the early July Monarchy. Gaillon worked in the town of Dieppe (Normandy) and dedicated his free time to the life sciences, with a special interest in green oysters, common at that time in the Dieppe area. Gaillon sampled green oysters, scraped their shells and made microscopic examinations. He observed small motile organisms with a "green" color, not blue like *H. ostrearia* cells, possibly because of the limits of his microscope device. Gaillon hypothesized that these motile organisms could be responsible for the greening of oysters, and considered them to be animals, which he called *Vibrio ostrearius*, based on the classification of worms by Bruguière [43]. Controversially but not impartially, Goubeau de la Bilennerie [24] argued that the greening exclusively depended on environmental factors and oyster farmers' know-how, a phenomenon thus appearing more acceptable for oyster consumers. Indeed, a possible impact of Gaillon's theory on consumers and green oyster sales was of some concern for Goubeau de la Bilennerie, president of the court in Marennes (for a full study, see [44]).

Haslea ostrearia, a Very Uncommon Diatom

The topic of green oysters involved Gaillon in a second controversy, with Jean-Baptiste Bory de Saint-Vincent, botanist, explorer, and Dragoons cavalry officer during the Napoleonic wars. Whereas Gaillon advocated for an animal nature of *V. ostrearius*, Bory de Saint Vincent classified it alongside diatoms into his "psychodiaire" kingdom, which contained all organisms whose position between animal and plant was unclear. Bory [45] thus proposed the name *Navicula ostrearia*, which remained for 150 years. Taking advantage of scanning electron microscopy facilities, Simonsen [46] transferred the "blue navicula" from the genus *Navicula* to *Haslea*, a new genus he created for this purpose, based on specific morphological features of the frustule, and he used *H. ostrearia* as the type-species. By doing so, Simonsen acknowledged G.R. Hasle for her considerable work on phytoplankton in general, and on diatoms in particular.

Haslea ostrearia is a tychopelagic diatom [47]—an organism that can be benthic or epiphyte—but also planktonic [48]. *H. ostrearia* is euryhaline [49,50], and can develop in high light environments [51]. Thus this diatom seems well adapted to oyster ponds, characterized by shallow and nutrient-rich water, where it mainly proliferates in autumn/spring, and can outcompete other microalgae [47,52]. Marennine produced during*H. ostrearia* blooms is released into the seawater, and the ponds turn green. In such ponds, oysters can become green in a few days, by exhibiting light to dark-green gills (Figure 2a).

This phenomenon is not restricted to oyster ponds in western France, as it can happen spontaneously elsewhere, in Great Britain [23], Denmark [53], the United States [36,54],*etc.* It is worth noting that the greening happens naturally in peculiar environments, usually protected bays with quiet waters and fresh water inlets. The greening is not limited to oysters, as it also occurs in other invertebrates, polychaetes, crabs, littorina, mussels [55], sea anemones [42], scallops, clams and cockles (Figure 2b–d). If greening invertebrates is not the blue diatoms' *raison d'être*, it is as least their signature.

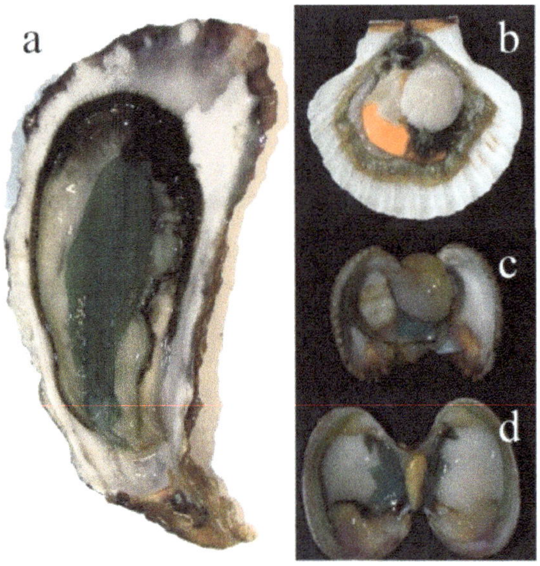

Figure 2: Greening effect of marennine on bivalves. (**a**) Pacific oyster; (**b**) Scallop; (**c**) Cockle; (**d**) Clam.

An Expanding Species—Complex of Blue Diatoms from the Genus Haslea

Over the past two centuries, diatoms with blue tips, or marine invertebrates with greened gills have been reported from almost all seas and oceans, which made *Navicula* or *Haslea ostrearia* a unique and cosmopolitan species [56,57], and marennine a curiosity. Indeed, all blue diatoms observed were ascribed to *N.* or *H. ostrearia*, with the possible exception of a sample collected in Honduras, identified as *N. fusiformis* var. *ostrearia* by Grunow [58]. The biodiversity of blue diatoms has recently been unraveled, with the collection of phytoplankton samples in different countries and continents, and their examination using various and complementary methods, *i.e.*, biometry of natural populations, morphometry of the frustules, molecular markers, reproductive compatibility, UV-visible spectrophotometry. The first novelty came from the shores of the Black Sea (Karadag Biological Station, Crimea) with a second species of blue diatom characterized and named*Haslea karadagensis* [56]. Despite a similarity with *H. ostrearia* when observed in light microscopy, striae density of the frustule [56], *rbc*L (RubisCO large sub-unit) and ITS (Internal Transcribed Spacers) markers [56], and the inability to interbreed [59], constitute evidence that they are two different species. Another major difference that allows distinguishing between the two species concerns their pigments, as blue apices in *H. karadagensis* appear darker than in *H. ostrearia* (Figure 3). Both pigments exhibit a comparable and reversible bathochromic shift when pH increases, but different λ_{max} [21]. Furthermore, UV-visible spectrophotometry shows that the pigment produced by *H. karadagensis* presents two isobestic points when pH varies from 2 to 12 [21], in comparison with only one in marennine [21]. Regarding biological properties, both pigments demonstrated a greening effect on bivalves, and as detailed below, antibacterial, antiviral and antifungal activities [57,60].

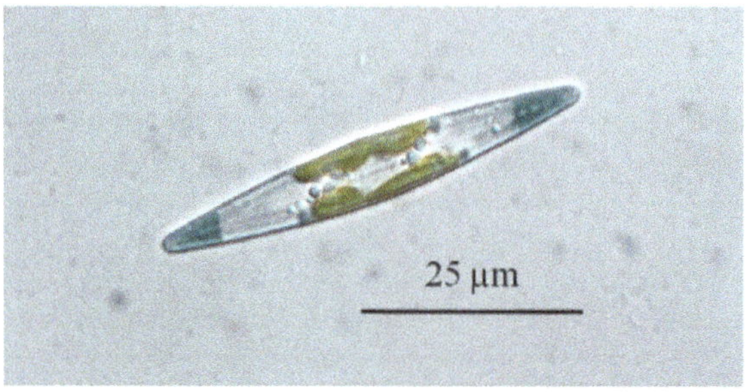

Figure 3: Light micrograph of *Haslea karadagensis*.

More samples of blue diatoms were obtained from the Canary Islands (La Gomera), and the Mediterranean Sea (French coast, between Toulon and Nice). Investigated by the same complementary approaches, these samples proved to be two new species of blue diatoms. They were named *Haslea silbo* sp. inedit. and *Haslea provincialis* sp. inedit., respectively [21]. They are currently being characterized. Both species produce a blue pigment, which is highly similar if not identical to marennine, as evidenced by UV-visible spectrophotometry (not shown), and Raman spectrometry (Figure 4). Both techniques did not allow discriminating between intracellular and extracellular forms of these pigments, in contrast to marennine. However, these two techniques could be less suited for going further in the study of these pigments. For instance, Raman spectroscopy showed that the pigment purified from a supernatant (extracellular form) of *H. provincialis* sp. inedit. displays a spectral signature similar to IMn, but the NMR spectra appear to be more distinctive. Figure 5 compares the ^{1}H NMR spectra of the extracellular form of the pigment from *H. provincialis* sp. inedit. with EMn and IMn from *H. ostrearia*. All three spectra show a general similarity in the represented signal groups, indicating that the molecules belong to the same substance class. There is essentially an important signal group at 3.4 to 5.4 ppm, assigned to sugar ring hydrogens, and the signals in the aliphatic region between 0.8 and 2.4 ppm (for a more detailed description of the NMR studies see paragraph 4.3). There are only weak signals in the aromatic region. However, there are differences in the details. Most strikingly, both extracellular forms contain a strong CH_2signal at 1.22 ppm. Furthermore, the intracellular form of marennine appears to contain less anomeric protons in alpha conformation (4.9–5.7 ppm). Overall, from what can be seen in simple ^{1}H NMR spectra, the difference between the two extracellular forms appears smaller than the difference between EMn and IMn of *H. ostrearia*. These preliminary results appeal for a more thorough study of the marennine-like pigments produced by the different species of blue diatoms, which constitute an original family of natural blue pigments.

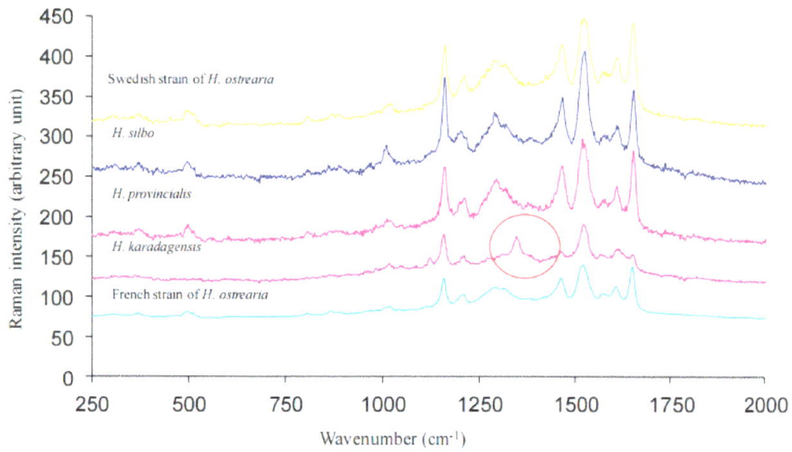

Figure 4: Raman spectra obtained *in vivo* on the blue pigments contained in the apices of different strains of blue diatoms. Noticeable differences can be seen between the pigment of *H. karadagensis* and the others in the 1240 cm^{-1} to 1420 cm^{-1} region.

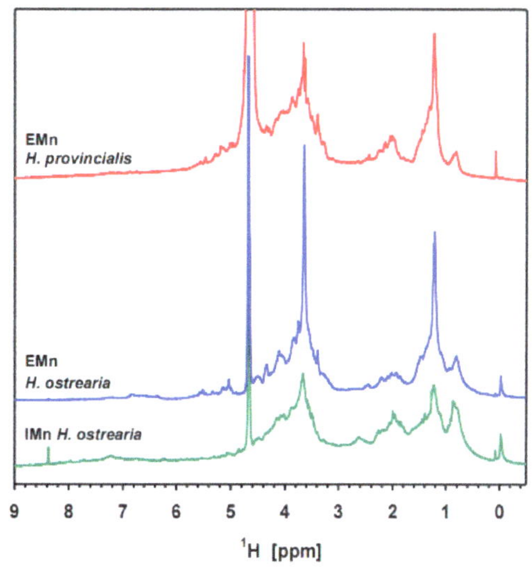

Figure 5: [1H NMR]. ^{1}H 1D NMR spectra of the intracellular form (IMn) of marennine from *Haslea ostrearia*(bottom, green), the extracellular form (EMn) of marennine (middle, blue), and the extracellular form of the marennine-like pigment from *Haslea provincialis* sp. inedit. (top, red), all dissolved in D_2O. The signal at 4.7 ppm stems from residual HDO, the high signal at 3.64 ppm in the extracellular marennine partly stems from an impurity.

MASS PRODUCTION AND METHODS OF PURIFICATION

Overview of Former Attempts and State of the Art

A series of engineering- and process-oriented studies were conducted at the end of the 1990s dealing with biotechnology issues of *H. ostrearia* production, in addition to marennine release and extraction with photobioreactors and membrane separations [61,62]. Vandanjon *et al.* [63] proposed a method to produce a large amount of extracellular marennine (EMn) by simultaneous concentration and desalting of the pigment released into the culture medium (Blue Water) using a 3 kDa cut-off ultrafiltration membrane. The resulting permeate contained EMn and a mixture of several components in a large range of molecular weights, as only a 3 kDa cut-off membrane was used. This experimental approach therefore did not allow the obtaining of suitably pure pigment. On the other hand, the aim of this work was mainly to concentrate a large amount of marennine for developing potential industrial applications, which did not require an absolutely pure product.

Subsequently, in order to increase the recovery of the intracellular marennine (IMn), a continuous-flow-high-pressure homogenizer was evaluated [64]. Cells were partly broken from 30 MPa, but a pressure of 100 MPa (1 cycle) was required to obtain optimal pigment release. The latter was directly linked to the physical cell breakage dependent upon the applied pressure and the number of disintegration cycles. Granulometric analyses by laser diffraction (0.04–2000 μm) revealed a size reduction of cell fragments when increasing these two operating parameters.

In view of optimizing both production rates in photobioreactors and extraction yields of marennine, Vandanjon *et al.* [63] studied the effects of shear stress on *H. ostrearia* cells due to circulation in pumps and valves of the production or harvesting systems. For the pumps, it was shown that shear stress was dependent on the type of pump, but that mechanical shear could have different effects even if the pumps and the number or frequency of loops were the same.

In throttling valves, the aim was to correlate the effect of shear to a parameter related to the inner geometry of the valve and to operating conditions. An overall parameter was then evaluated, *i.e.*, the pressure drop coefficient *Kv*that integrates both the type of valve and its opening degree. As a consequence, the modeling of the shear effects was conceivable; basic descriptive data used so far (type of pump, geometry or opening degree of the valve, *etc.*) were completed and partially substituted by quantitative parameters (rotating velocity, capacity, or internal leakage for the pumps, *Kv* coefficient for the valves).

A new photobioreactor coupled with an ultrafiltration system (immersed membranes) was investigated for the continuous culture of *H. ostrearia* in order to improve marennine production and recovery [65]. The system, with a simple design, was particularly interesting, because energetic costs were minimized, and the cells were not submitted to any shear stress due to pumping or circulation. The photobioreactor was of the cylindrical type; a membrane module was placed at the bottom of the reactor and the hydrostatic pressure was used as a driving force both for the permeation and periodical backflushing steps. The production of biomass and marennine was stable for a three-week period, with marennine concentration three times higher than in a conventional batch photobioreactor.

A final study dedicated to bioprocessing aimed to compare the pigment productivity obtained with two types of photobioreactors [66]. In the first process, cells were free and recycled in a photobioreactor combined with membrane ultrafiltration equipment (external loop). In the second system, cells were entrapped in a tubular agar gel layer in a photobioreactor of original design. The influence of nitrate concentration and renewal rate was examined. Experiments, conducted over long-term periods (up to 40 days) without any external contamination revealed that marennine productivities of more than 5–7 mg 10^9 $cell^{-1}$ day^{-1} could be reached with both bioreactors. The advantages and drawbacks of each process design were also discussed.

Marennine is insoluble in all organic solvents, and first attempts to extract it from algal biomass were made using various aqueous solvents, e.g., pure distilled or tap water, or bicarbonate or phosphate buffers. Regarding purification, Robert *et al.* [19] published a purification procedure in which intracellular marennine (IMn) was extracted using organic solvent and water mixtures and further solubilized in 500 mmol L^{-1} K_2SO_4 at 80 °C. This method obviously increased the possibility of obtaining denatured pigment and was not convenient to produce large amounts of IMn. More recently, Pouvreau *et al.* [67] developed a new method for the extraction and purification of both forms of marennine (Figure 6). EMn and IMn were selectively extracted from Blue Water (blue-colored culture medium) and algal pellet, respectively, and were then purified by a three-step semi-preparative procedure using two ultrafiltration steps and one anion-exchange chromatography step. This method was easily applicable to a large production system. Finally, after dialysis, UV-visible-photodiode array analysis showed that EMn and IMn were not contaminated, suggesting that these compounds reached the degree of purity required for further biochemical investigations.

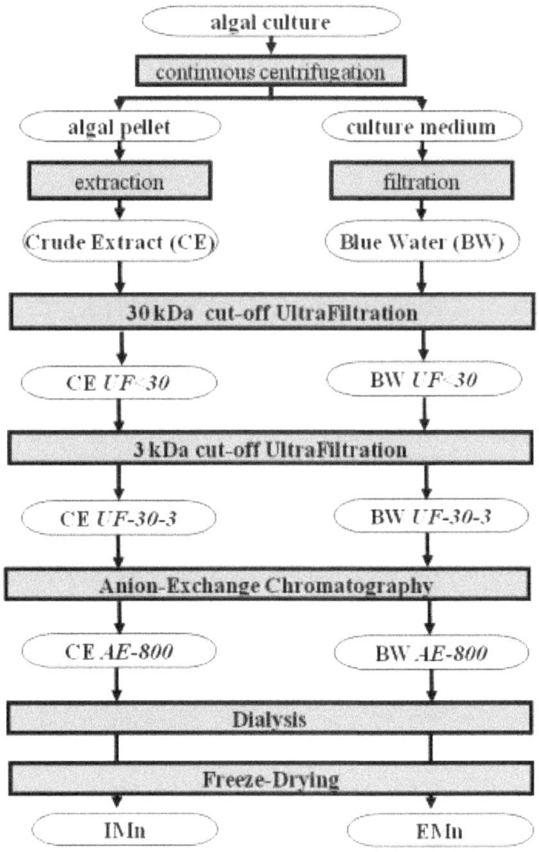

Figure 6: Purification process of marennine and marennine-like pigments.

Recent Transfers to a Pre-Industrial Scale

Fundamental research regarding the characterization of marennine-like pigments can be sustained by algal production at the laboratory scale, but mass production of *H. ostrearia*is a prerequisite before achieving any added value at the industrial scale. Therefore different attempts were made to grow this peculiar diatom. According to the literature, *H. ostrearia* has been cultivated mostly at the laboratory scale (<100 L), but also at a semi-pilot scale (up to 10 m^3) [68], or using photobioreactors (<7 L) [65]. At a larger scale, *H. ostrearia* was grown industrially for several years under indoor controlled conditions to produce marennine, mainly for the intensive greening of oysters (knowledge transfer from U. de Nantes to SOPROMA, Bouin, France); however, this production unit ceased its activity about seven years ago due to lack of funding. In fact, the only structures for the mass production of *H. ostrearia* that have ever been fully operational are oyster ponds. Unfortunately, blooms of *H. ostrearia* in these ponds (and as a consequence the greening of oysters) remain erratic and non-controllable. Production units of *H. ostrearia* still need to be developed, taking into account the specificities of benthic microalgae, that use tank surface to grow and generally tend to form biofilm, which makes them difficult to grow in conventional culture systems designed for suspended microalgae. Since the technology to produce massive algal biofilms is not readily available, a simple photobioreactor could be designed in order to maximize *H. ostrearia* productivity and, mostly, excretion of marennine. Some relevant traits of *H. ostrearia* autecology must be considered to design the culture system: growth is characterized by a settlement on the marine sediment, where maximal biomass (as far as 350,000 cells L^{-1}) is obtained in conditions of shallow waters (as in the oyster ponds) and high light intensity [47]. Marennine is then released in the water, with concentration ranging between 2 and 5 mg L^{-1} [68].

With this goal—but also these constraints in mind—a photobioreactor (PBR) dedicated to the production of *H. ostrearia* and marennine has been devised. This simple PBR was designed in order to obtain large quantities of marennine to further study its biological activities and chemical structure. Also, it was designed in a simple manner so that it could be easily setup by end-users, in laboratories as well as in hatcheries. Flat-bottom polyethylene circular tanks of 200 L were used (diameter: 122 cm), covered by acrylic glass with light supplied from above by T5-5000K high-output fluorescent tubes (General Electric, Mississauga, Canada) at about 25 cm from the cover. A drain was installed at the lowest point of the tank, allowing water sampling and harvesting. Pre-filtered air (0.22 μm) was supplied to the dead space inside the tank through the cover to keep a positive pressure inside the tank, preventing contamination from the outside and purging air heated by the proximity of the light source. The proximity of the air source relative to the water surface also generated a slight agitation at the surface, thus ensuring a slow but constant water movement in the tank and providing nutrient renewal to the cells.

The study was conducted at the Université du Québec à Rimouski (UQAR) Station Aquicole at Pointe-au-Père (Québec, Canada), with the use of two PBRs. Algal production was carried out using the NCC-136 strain of *H. ostrearia* isolated from Bourgneuf Bay (Bouin, France) and provided by the Nantes Culture Collection (NCC). Cells were grown in a semi-continuous mode in sterilized seawater enriched with F/2 [69] and 30 mg L^{-1} silicates. Culture was initiated in 500 mL Erlenmeyer flasks filled with 200 mL seawater, and then transferred to 2.8 L Erlenmeyer flasks filled with 2 L seawater. Growth irradiance was 125 μmol photons m^{-2} s^{-1} and room temperature was maintained at 20 °C. Two of these 2 L flasks were then used to inoculate one PBR with approximately 2000 cells mL^{-1}. To minimize light attenuation by the water column (<10 cm), tanks were half filled with 100 L filtered (1 μm) natural seawater (salinity 28) supplemented with commercial nutrients (f/2 and silicate from Fritz) and ultrafiltered at 50 kDa (Romicon, KOCH Membrane, Wilmington, MA, USA). The irradiance level was 180 μmol photons m^{-2} s^{-1} (PAR Radiometer, Q201, Macam photometrics LTD., Livingston, Scotland, UK), measured under the light source at the bottom of the empty tank. Light intensity decreased in a linear way to reach 100 μmol photons m^{-2} s^{-1} at the border of the PBR (Figure 7a). Room temperature was kept at 16 °C by an air-conditioner, for a maximal water temperature of 19.5 °C. Both rooms were kept in a 14/10 h light/dark cycle. Two cultures were run in separate PBRs, under the same conditions. Marennine concentration was determined on the cell-free culture water (syringe-filtered on 0.22 μm) by optical absorption according to the Beer-Lambert law. Optical density (OD) was measured at 677 nm in a 10 cm cell by means of a Cary 100 Bio UV-Visible spectrophotometer (Agilent Technologies, Mississauga, Ontario, Canada), using the specific extinction coefficient for EMn following Pouvreau *et al.* [70]. Optimal algal biomasses in the tanks, based on cells and biofilm appearance, were obtained after 12 days of culture growth, as with pre-culture realized in smaller volumes (confirmed with Nageotte counting chambers). Marennine release into the medium started after 10 days of growth (Figure 7b), which corresponds to the end of the exponential phase. Marennine production was relatively constant between day 10 and day 33, with a production rate of about 0.3 mg L^{-1} day^{-1}. Maximal extracellular concentration obtained was about 6.2 mg L^{-1} in both PBRs.

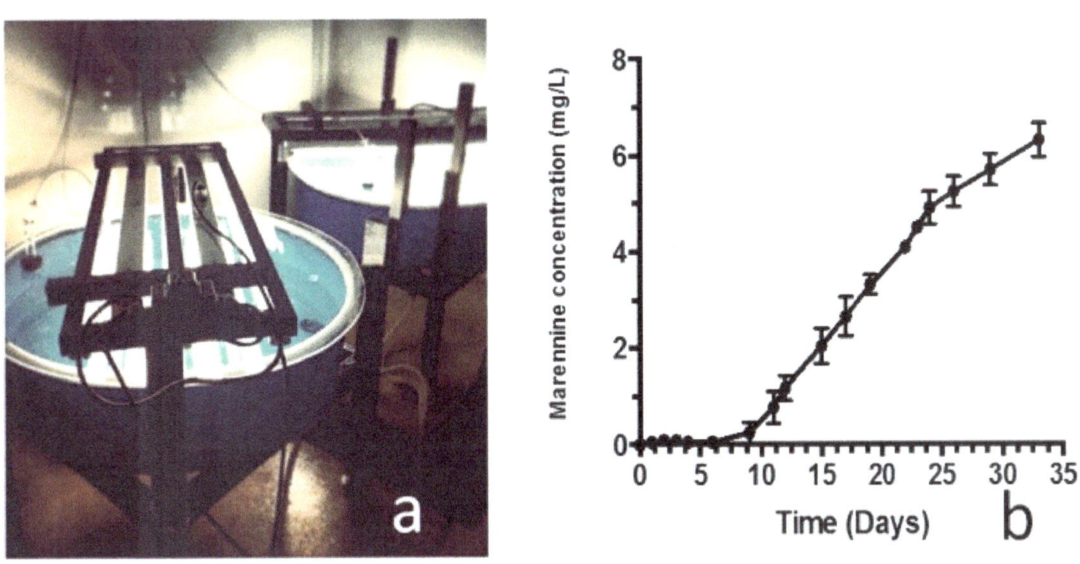

Figure 7: (**a**) Photography of the two PBRs used in this experiment; (**b**) Marennine concentration (mg L^{-1}) in the PBR, measured spectrophotometrically on the culture medium. Values are means of concentrations obtained in each PBR (n = 2).

Different culture methods were tried in the past by different authors, with estimated marennine productivities ranging from 3 to 100 mg L^{-1} [65,66,68,71,72,73,74], but these values were obtained using a cytophotometric method and a calibration curve established with unpurified marennine, and not referring to a published extinction coefficient [75], which is more an estimation of marennine production in a culture rather than a true quantification of the pigment [70]. In the literature, there is no quantification of marennine production in natural environments or in a culture that uses the molar extinction coefficient determined on purified marennine by Pouvreau *et al.* [70]. Therefore, a direct comparison of our results with other results in the literature is difficult.

HISTORICAL PERSPECTIVE ON MARENNINE STUDIES

Biosynthesis of Marennine: When, Where, Why, and How It Happens

Although the work of Pouvreau *et al.* [70] represents the most advanced investigation on marennine characterization thus far, the exact structure of marennine is still unknown, and the description of its biosynthesis pathway still a distant goal. It is also unclear when marennine synthesis is triggered in the cell. There is consensus on the stimulating effect of high light conditions and long photoperiods [49,76,77] with just a few authors advocating the opposite [78]. An impact of the light quality, with a positive effect of blue wavelength, has been documented [19,76,77]. Some authors suggested that marennine production could be influenced by organic substances like glucosamines [78,79], or result from salt-induced stress [32,49,80,81], or nutrient deficiencies, e.g., iron, vitamins [47,49]. In the line of the "nutrient deficiency hypothesis," it has also been suggested that marennine synthesis could result from, or be enhanced by, the senescence of cells [82]. It is worth noting that this could be the result of an apparent and transient accumulation inside the cells, due to a reduced division rate, rather than an increase in production yield. Furthermore, it has been shown that *H. ostrearia* cultures in the exponential growth phase release marennine continuously in the medium [19], but the mechanisms responsible for the release of marennine out of the cell and the factors that control this release are not yet determined. However, according to Nassiri *et al.* [83], marennine is accumulated in vesicles at the apex of the cells and excreted by exocytosis. A type of secretory vesicles containing marennine (a few μm in diameter) can be observed outside the plasmalemma in contact with cells [83]. These vesicles attach to the cells during their erratic movements, then they detach from the cells, still float for a few minutes, before they finally collapse and release marennine in the medium.

There is no consensus on a possible relation between marennine production and photosynthetic pigment content, some authors having observed a negative correlation [32,49,78,84,85], others no correlation [19,47,86,87]. A striking feature associated with marennine production and accumulation in the cell apices is the reduction in size of chloroplasts [32,49,55,78,88,89,90,91,92,93], possibly related with a decrease in photosynthetic activity [86]. However, the plastid size change could also reflect the decrease of the cellular content of major photosynthetic pigments (chlorophylls *a* and *c*, fucoxanthin), when growth irradiance increases [51], a common photoacclimation response in phytoplankton. Moreover, the impact of this size modification on the organization and stability of thylakoids remains uncertain, as observations were contradictory [83,85].

Regarding the autecology of *H. ostrearia* in oyster ponds, a few authors hypothesized a correlation between the accumulation of marennine in the cells and a change in algal behavior, switching from the planktonic to the benthic stage, together with changes in its metabolism [47,78,79], blue cells of *H. ostrearia* becoming able to assimilate reduced forms of nitrogen and carbon [52]. Moreover, the many biological activities displayed by marennine and marennine-like pigments (see below) could explain why these specific pigments represent a competitive advantage for the blue diatoms.

The Various Hypotheses Regarding the Chemical Nature of Marennine

For more than hundred years, the presence of such intriguing and unusual blue pigment inside unicellular algae has been a fruitful matter for contradictory theories and inconclusive experimentation. The hypothesis of a metallic salt was proposed early and rapidly rejected [22,26,35,36,94]. Then there was the "pigment connection", with hypotheses successively relating marennine to carotenoids [79], or to chlorophylls [32,47,55,80,81,85], possibly resulting from their degradation. Other authors linked marennine with cyanobacterial pigments, thus suggesting a protein nature for this water soluble molecule [54,75,95] or proposed it could be an anthocyanin, with respect to stress and pigment accumulation in cells [96]. Finally, recent advances related marennine to a mixture of different macromolecules [97], or to a polymer possibly of polyphenolic nature [70]. It has been shown that marennine exists in two slightly different forms, intracellular and extracellular [67], and that for both forms, the color changes with pH from blue (acidic pH) to green (basic pH) [70].

Recent Discoveries on Marennine's Structure

Following the preliminary characterization by Pouvreau *et al.* [70], a series of experiments are currently being conducted to extend our knowledge about the chemical structure of marennine, mainly using nuclear magnetic resonance (NMR) techniques. All current NMR experiments have been carried out on a Bruker Avance 400 MHz spectrometer equipped with a 5 mm BBFO$^+$ probehead. Samples of EMn, purified as described in Pouvreau *et al.* [67] were dissolved in 0.5 mL of 25 mM phosphate buffer in D_2O with 40 mM NaN_3 at pD 6.6 (corresponding to pH 7.0) to a concentration of 2.4 mM. A ^{1}H–^{13}C HSQC was acquired with 64 repetitions and 800 increments in the indirect dimension, applying the echo–antiecho scheme, until a maximal t_1 of 24.8 ms. A ^{1}H TOCSY with 90 ms DIPSI-2 mixing was acquired using 16 repetitions and 800 increments in the indirect dimension until a maximal t_1 of 106 ms.

The standard one-dimensional ^{1}H spectrum of EMn shows signal groups without any resolved individual signals (Figure 5 and top of Figure 8a). Marennine is, therefore, a macromolecule with a relatively complex structure. ^{1}H diffusion spectra ("DOSY") reveal a diffusion coefficient of about 10^{-10} m^2/s for both EMn and IMn, consistent with the mass in the 10 kDa range determined by mass spectrometry [70]. The high signal at 3.58 ppm is identified as impurity.

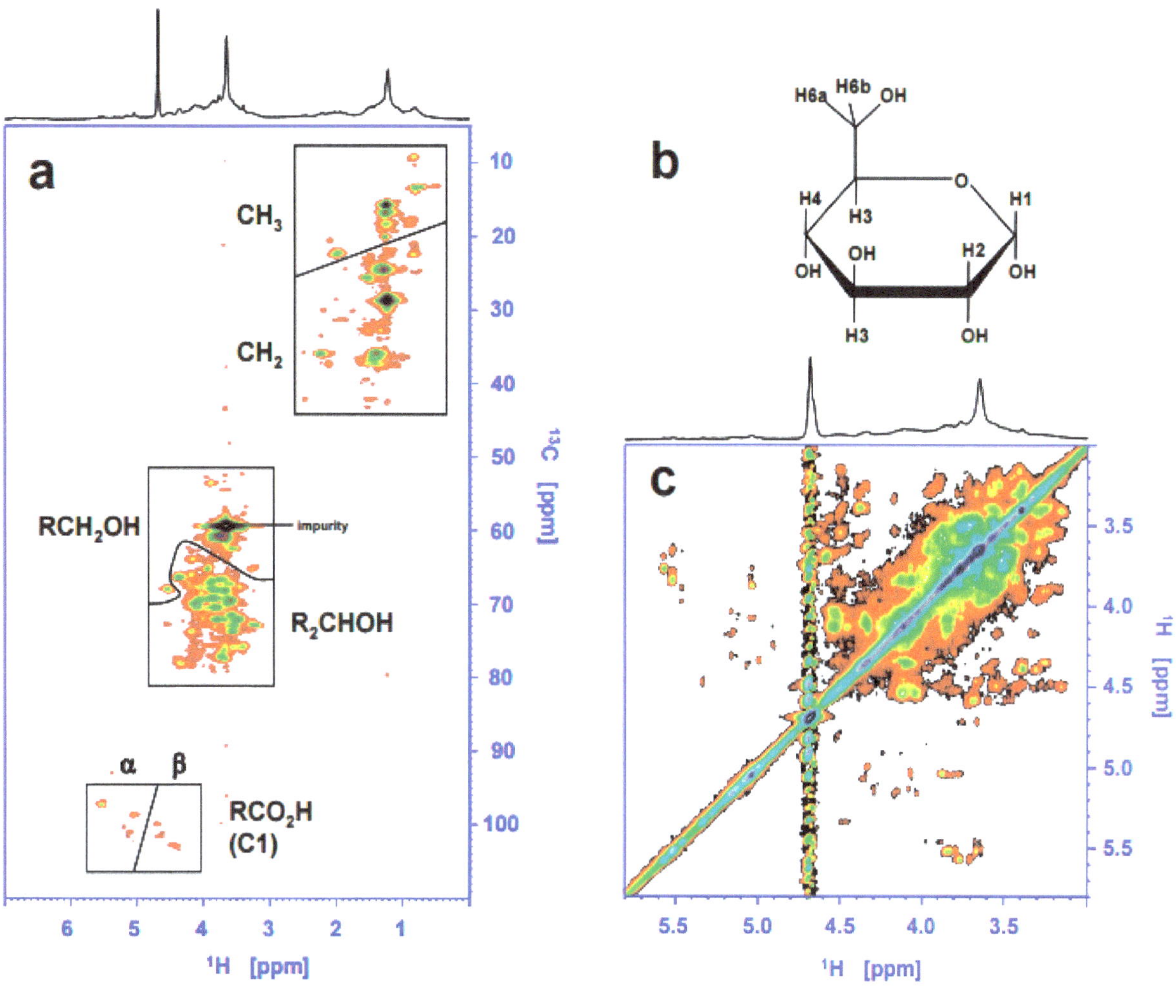

Figure 8: [2DNMR]. (**a**) ^{1}H–^{13}C HSQC of the extracellular form of marennine (EMn) in deuterated phosphate buffer. On top the corresponding 1D ^{1}H spectrum. The lines separate areas of carbons with—among others—two hydrogen atoms and with one or three hydrogens, as determined by an additional edited HSQC with lower resolution and signal-to-noise ratio; (b) Generic pyranose; (c) Ring region of a ^{1}H– ^{1}H TOCSY of EMn with 90 ms mixing time. The strong ^{1}H signal at 3.65 ppm stems from an impurity.

A (partial) resolution of signals can only be achieved by two-dimensional [2D] NMR. The correlation signals in a ^{1}H–^{13}C HSQC (Heteronuclear Single Quantum Coherence) spectrum identify chemical groups by a combination of the hydrogen and carbon chemical shifts. An edited HSQC enables, in addition, the determining of the number of hydrogen atoms attached to a carbon by the sign of the signal. The sensitivity is, however, lower. We have acquired an HSQC (Figure 8a) and an edited HSQC of EMn (not shown). Three groups of signals can be seen in the HSQC spectrum, the most striking being in the chemical shift ranges of (60–75 ppm) × (3.5–4.5 ppm). This chemical shift distribution matches—among others—the ring carbons (C2–C5 in the case of a pyranose) and hydrogens of saccharides. This is supported by a small group of signals in the relatively unusual chemical shift region of (95–105 ppm) × (4.3–5.8 ppm), characteristic of anomeric carbons of aldoses (C1) and their hydrogens. According to the edited HSQC, these signals belong to a carbon bound to—among others—one hydrogen atom. Only some signals with relatively small carbon shifts stem from carbons bound to two hydrogens, corresponding to a CH_2OH group (C6; also C1 in some furanoses). According to the chemical shifts of both hydrogen and carbon atoms, the anomeric signals can further be divided into a group attributable to carbohydrates in the α (97–102 ppm) × (4.9–5.7 ppm) and the β form (usually 103–106 ppm, here lower) × (4.3–4.8 ppm). The interpretation of the HSQC spectrum is confirmed by ^{1}H–^{1}H correlation experiments. Figure 8b displays the ring region of the TOCSY (Total Correlation SpectroscopY) spectrum. Typical for polysaccharides, there is much spectral overlap in the region of cross peaks between the non-anomeric ring protons by a high density of signals with little dispersion (Figure 8c). Cross-peaks of H2 to α-anomeric hydrogen signals H1 are faint even after relatively long mixing time (90ms), because the equatorial–axial *J*-coupling is weak [98].

The fact that marennine consists to a large extent of glycosidic elements suggests a connection to the main energy storage compound of diatoms, chrysolaminarin, in particular under consideration of the large amount of marennine produced by *H. ostrearia*. The spectra are, however, more complex than those of chrysolaminarin [99]. Moreover, there is a large number of signals in the aliphatic region at about (10–40 ppm) × (1.0–2.5 ppm). The CH_3 and CH_2 areas are separated by their chemical shifts and the edited version of the HSQC. A particularly high $^{13}CH_2$ signal has a chemical shift of 28 ppm; it is already clearly visible in the 1D ^{1}H spectrum at 1.22 ppm. Diffusion experiments (Diffusion Ordered SpectroscopY, DOSY) confirm that this signal comes from the macromolecule and not from an impurity. It could indicate the presence of longer CH_2chains. Such a unit would, however, conflict with the general hydrophilic character of marennine. Steroids are an aglycone substance group that could explain the general dispersion of aliphatic signals.

In a previous study using primarily biochemical tests on various potential substance groups [70], a glycosidic digestion test on sugars had been negative. It is possible, however, that the other chemical groups compromised the detection by this method. On the other hand, it was concluded that marennine was a substance with polyphenolic or similar rings, but signals from the aromatic region in the HSQC spectrum are too weak to detect any correlation under the present experimental conditions (see also Section 5). The Folin-Ciocalteu and the Prussian Blue tests on polyphenols might have been positive due to these few groups, while NMR identifies the scaffold of the macromolecule. This also counts for the chromophore. Whatever its nature is, polyphenolic or of a different type, it is very possible that it constitutes only a small part of the molecule, which does not become apparent from the first impression from the principal NMR signals of this molecule. Pouvreau *et al.* [70] also report the elemental composition of marennine, where, remarkably, oxygen constitutes about 50% of the molecular mass. This result—since then confirmed using different samples of purified marennine—is consistent with a polyphenol but also with a glycoside.

MARENNINE POSSIBLE FUNCTIONS AND BIOLOGICAL ACTIVITIES

During a bloom of *H. ostrearia* in an oyster pond, the quantity of marennine released into the environment should represent a high "cost of production", which raises a question about the advantages that the species may derive from this peculiar biosynthesis pathway. More generally, the ecological significance of marennine-like pigments remains to be considered, especially in regards to the amount of pigment produced during the diatom life cycle. As presented above, historically, a wealth of information is available on marennine, but concerning the diatom *H. ostrearia* itself, information about the biological function of this specific blue pigment is scarce and inconclusive. The longest known and most obvious biological effect of marennine is the greening of oysters. Experimentally, oysters placed in either a suspension of *H. ostrearia* or a green supernatant of a culture, turn green within a few hours [55,57,82,89,100,101], but little is known about the mechanism. Some authors suggested there might be an interaction between marennine and some proteins in the gills [36], especially in some specialized secreting cells [22,36]. In oyster gills *in vivo*, it is possible that marennine binds and precipitates proteins like tannins [60]. A tannin-binding protein effect would be in agreement with the suggested polyphenolic nature of marennine [70]. Regarding *H. ostrearia* itself, Schubert *et al.* [102] showed that marennine did not play any role in light capture and photochemical activity; however, it could indirectly influence photosynthesis by absorbing in the red part of the spectrum, with peaks at 672 and 677 nm for the intracellular and the extracellular forms of the purified pigment (neutral pH), respectively [70], and at 669 nm for raw supernatants of *H. ostrearia* cultures [19]. Marennine could thus be considered a photoprotective molecule at high irradiance levels [103], or a factor able to modify the light spectrum in the water column when accumulated in the medium [70,77].

Apart from this shading effect, it was shown that marennine has antioxidant activity [103], that it can afford some protection against metals such as copper [104], and act directly as an allelochemical by inhibiting the growth of some algal species encountered in oyster ponds and modifying inter-specific competition among phytoplankton [49,105]. These results have been reinforced by co-cultivation experiments of *H. ostrearia* with other species, which underlined the sensitivity of centric species like *Skeletonema costatum*,*Chaetoceros calcitrans*, *C. gracilis*—all species abundantly used in aquaculture—as well as the insensitivity of others like *Pavlova lutheri* (not shown). This could explain the occasional dominance of *H. ostrearia* in oyster ponds, concomitant with an almost complete elimination of other diatom species [47,68,79], thus revealing the importance of chemical ecology in marine phytoplankton and environments [106]. A few authors hypothesized that *H. ostrearia* could itself be affected by marennine, a sort of autotoxin associated with pathological processes [32,49,79]. However, it was further demonstrated that *H. ostrearia* was rather tolerant to marennine [70].

Last but not least, preliminary works conducted on *H. ostrearia* aqueous extracts, thus containing marennine, displayed antiviral, anticoagulant [107] or growth-inhibiting properties [108]. These activities have been recently confirmed using purified marennine, which exhibited antibacteria, antivirus, and antiproliferative activites [60], or using the pigment synthesized by *H. karadagensis*, the pigment of which demonstrated antifungi, antibacteria, and antiviral activities [57]. Hence, aside from the greening action, marennine and marennine-like pigments could be especially useful in prophylaxis in the context of oyster farming, due to their antibacterial and antiviral activities.

In the last decades, it has been observed that the cultivated Pacific oyster *Crassostrea gigas* presents massive and recurrent summer mortalities, this being of great concern for the oyster industry all over the world. In France, for instance, bacteria such as *Vibrio aestuarianus* [109], *V. splendidus* [110], and viruses belonging to the Malacoherpesviridae, like the ostreid herpesvirus OsHV-1, distantly related to other members of the Herpesviridae [111] are frequently associated to, if not responsible for, these severe summer mortality events.

Marennine-like pigments displayed antibacterial activities against three marine bacteria, *V. aestuarianus*, *Pseudoalteromonas elyakowii*, *Polaribacter irgensii* [57,60]. In a first attempt to explain the mechanism of action of the antibacterial activity exhibited by marennine-like pigments, Tardy-Laporte *et al.* [112] demonstrated, using ^2H solid-state NMR on intact *Escherichia coli*, that the extracellular form of the pigment produced by *H. provincialis* sp. inedit. disturbs the bacteria membranes, unlike the intracellular form. More specifically, their results suggest that the pigment released in the medium exerts its antibiotic action by interacting with the lipopolysaccharides on the bacterium's surface, thus rigidifying the outer membrane.

Along the lines of this membrane-mediated inhibition effect, a series of experiments were conducted with *V.*

splendidus cells incubated for 3 h with different concentrations of EMn (0.1 μg mL^{-1} to 1.0 mg mL^{-1}), then washed and resuspended in fresh marine medium, for 48 h. The conclusion was that the higher the marennine concentration, the higher the inhibition of the growth would be (Figure 9). The effective concentration reducing bacteria growth rate by 50%, EC_{50}, was 2.89 μg mL^{-1}, a value in the range of many anti-bacterial marine compounds described thus far [113].

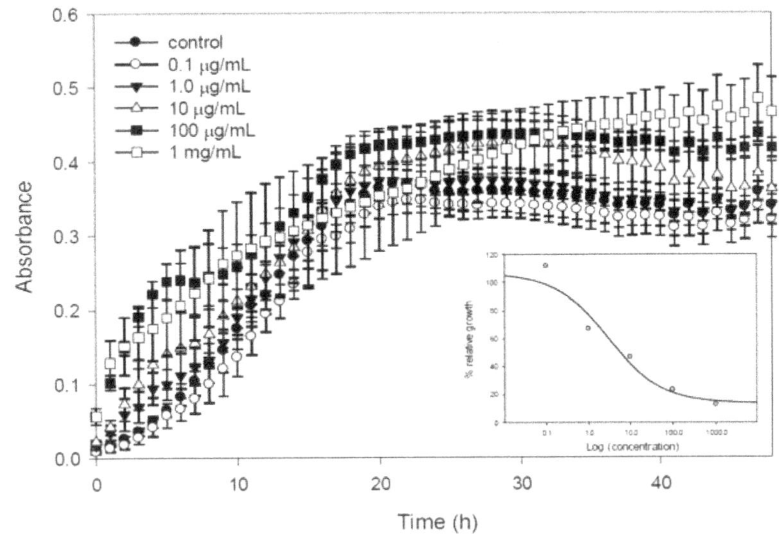

Figure 9: Growth of *Vibrio splendidus* after 3 h in contact with marennine. *V. splendidus* was grown in modified marine media overnight. Cells were washed and then incubated for 3 h in 0, 0.1, 1.0, 10, 100, or 1000 μg mL^{-1} marennine. Cells were washed and then incubated for 3 h in 0, 0.1, 1.0, 10, 100, or 1000 μg mL^{-1} marennine. Cells were washed in sterile water and brought to an optical density of 0.5 before the 3 h incubation. Cells were then added to fresh marine media in a 96 well plate and growth kinetics were done for 48 h with measurements every 30 min. Inset: maximum growth rates expressed in function of the control (100%) with marennine concentrations.

Regarding antiviral activities, due to a lack of a suitable bivalve cell line for culturing oyster herpes virus, a heterologous model using fibroblastic Vero cells and human HSV-1 was chosen in some studies, to investigate the effect of antiviral activity [114]. It has been shown that marennine intracellular and extracellular forms exhibited antiviral activity against the HSV-1 *herpes* virus [60], with 50% antiviral effective concentration (EC_{50}) values of 24.0 and 27.0 μg mL^{-1}, respectively (for a multiplicity of infection (MOI) of 0.001 ID_{50}/cells). The blue-grey pigment produced by *H. karadagensis* presents similar antiviral activity [98], with the extracellular form being more active than the intracellular form (EC_{50} of 23 μg mL^{-1} and 62 μg mL^{-1}, respectively). Both intracellular forms of the pigments present a relative cytotoxicity against the fibroblastic Vero cells, not observed with their extracellular counterparts. This underlines the need for further experiments regarding a possible toxicity, and the mode of action of marennine-like pigments. A sulfated polysaccharide, naviculan, was isolated from *Navicula directa*, a diatom collected from deep-sea water in Toyama Bay, Japan [115]. This compound was shown to inhibit HSV-1 and HSV-2 (half maximal inhibitory concentration, IC_{50} = 7–14 μg mL^{-1}) by interfering with the early stages of viral replication, most likely affecting viral adhesion and penetration into host cells. Very few other biologically active secondary metabolites have been reported from diatoms [116,117], and except for a few studies [57,60,107], little is known about the antiviral activity of diatom extracts or compounds. The selection of antiviral marine compounds for aquaculture environments being a possible solution to control viral disease transmission [118], a possible valorization of the biological activities of marennine-like pigments, would constitute a new and promising field of research. The fact that marennine-like pigments seem to have the potential to inhibit growth of pathogenic marine bacteria and virus replication, is of special interest for oyster producers, as bivalves lack antibody-mediated humoral immunity, and possess instead an innate immune system. Experiments are in progress to test the hypothesis that marennine in solution or present on gills could protect oysters from these pathogens in realistic *in vivo*conditions.

THE PUTATIVE COMMERCIAL POTENTIAL OF MARENNINE-LIKE PIGMENTS

At the industrial scale, marennine is solely exploited in aquaculture for the greening of oysters, thus as a natural blue-green colorant. The main source of natural blue colorants is indigo dye, extracted from different species of higher plants. For the food industry, however, there is a lack of available natural blue pigment alternatives compared to red, orange, and yellow natural colorants [119]. More simple to produce and less expensive to use, some of the synthetic blue colorants used in food chemistry have to face alleged or assessed reputations of harmfulness, some of them possibly encountering commercial restrictions or banishment (e.g., Patent Blue V E131). Presently, the use of marine blue dye in the food industry is restricted to phycocyanin, mainly produced by cyanobacteria but also some Rhodophyta [120]. Phycocyanin is a food colorant, which is known in Japan under the name of "Lina-Blue", and it is mainly used in ice-creams or drink preparations. In this context, marennine is also a natural blue pigment, which could complement the source of marine blue dyes available for the food industry, and it presents some merits but also a few flaws. People have consumed green oysters for centuries, without any disease or anaphylactic reactions recorded, which should suggest that this pigment is non-toxic, when considering a standard food intake. Marennine is produced by a marine microalga, which augurs well for an environmental-friendly production system. It is water soluble, and its extraction process does not require massive use of solvents. However, the structure of marennine is still unknown, and because marennine exhibits a wide range of biological activities, its possible cytotoxicity and its stability as a pigment have to be carefully tested before considering it for possible use in cosmetics. Indeed, before turning into a reality in Europe, and in France especially, this type of application will need to be submitted for approval. For example, the use of "pure marennine" as a food colorant would indeed require nutritional and toxicity studies on rats, and cellular tests to demonstrate that the pure molecule has no mutagenic properties. These necessary steps represent an immense obstacle to using marennine as a food additive. An alternative method, already applied for the upgrading of phycocyanin and phycoerythrin, would be to promote the use of aqueous extracts enriched in pigment for their colorant properties. Generally, this application is authorized, at least for phycocyanin, and appears to be realistic from an economic point of view regarding cosmetics; however, this would become an even more crippling problem for any development in the food and health industry.

Illustrating the added-value potential of some microalgae in cosmetic industry, a few companies such as LVMH group and Daniel Jouvance have invested in their own microalgal production units. Indeed, some microalgae (e.g., *Chlorella*, *Odontella*,*Tetraselmis*, *Dunaliella*, *Emiliania*, *Noctiluca*) are established in the skincare market. Microalgal extracts are mainly found as skincare products, e.g., anti-aging, emollient or moisturizing, and also sunscreen products. Preliminary experiments were conducted to study the possible photo-protective and anti-inflammatory effects of marennine. The photo-protective potential was thus studied, to determine sun protection factor (SFP) and UVA protection factor (PF-UVA) of marennine, using an *in vitro* method. An O/W emulsion placebo was prepared in the laboratory as previously described [121]. Marennine was incorporated into the formulation components at 10% (w/w) in order to study the potential biological properties. Thirty milligrams of product exactly weighed were spread on polymethylmetacrylate (PMMA) plates over the whole surface (25 cm^2) using a cot-coated finger (15 mg remain on the finger cot). SPF and PF-UVA of the creams were measured *in vitro* [122,123]. Three plates were prepared for each product to be tested and nine measurements were performed on each plate. Transmission measurements between 320 and 400 nm were carried out using a spectrophotometer equipped with an integrating sphere (UV Transmittance Analyzer UV1000S, Labsphere, North Sutton, NH, USA). Emulsions containing 10% (w/w) marennine exhibited SPF (1.28 ± 0.05) and PF-UVA (1.24 ± 0.04) values (means ± SD, $n = 3$), which demonstrate that marennine could not be considered a molecule interesting enough in the domain of the topical photo-protection.

The determination of a possible anti-inflammatory effect of marennine incorporated into a cream formulated in the laboratory was carried out using a test with Phorbol-12-Myristate-13-Acetate (PMA). Introduction of mouse ear edema was based on the method of Carlson*et al.* with some modifications [124,125,126]. Firstly, the thickness of the mouse ears was measured using a model micrometer gauge (Oditest®, Kroeplin, Schlüchtern, Germany). Ten mL of preparation with marennine or 0.1% (w/w) butyrate hydrocortisone-based lotion (Locoïd®, Astellas Pharma—Levallois-Perret, France) were applied using a ripette genix electro dispenser (Fisher Scientific, Illkirch, France), on the mice's right ears, twice at 5 min intervals. Ten mL of placebo emulsion were applied according to the same protocol, on the mice's left ears. Thirty minutes later, 10 mL of a hydro-alcoholic solution of Phorbol-12-Myristate-13-Acetate (250 mg/mL) were then applied on each ear, in order to cause an edema. After 3.5 h, the thickness of the ears was once again determined using the Oditest®. Five mice were used for each product tested. The cream containing 10% (w/w) marennine demonstrated a moderate anti-inflammatory effect, with an edema inhibition of 62.5%, as compared to 100% for the control butyrate d'hydrocortisone (0.1% w/w). These preliminary experiments show that marennine can hardly be considered as a potential UV filter, but it could be valorized in soothing creams. It might be interesting to explore potential applications in the management of atopic skin or rosacea, for example.

CONCLUSIONS

Aquaculture still represents the most immediate and "natural" valorization approach of marennine and marennine-like pigments, due to the greening of bivalves. Beside this coloring action, it is hypothesized that marennine-like pigments could act as natural prophylactic agents in hatchery and oyster farming, due to their antibacterial and antiviral activities, which were demonstrated *in vitro* at the laboratory scale. Alongside possible usage as natural blue colorants for cosmetics—provided they prove to be inexpensive to produce, stable in the formulations, and safe to use—the many biological activities evidenced thus far represent a new research avenue, and a great potential of valorization for the marennine-like pigments.

Acknowledgments

This publication benefited from funding from the European Commission under the Community's Seventh Framework Programme BIOVADIA (contract No. FP7-PEOPLE-2010-IRSES-269294, "Biodiversity and Valorisation of Blue Diatoms"). I.M. wishes to thank the Natural Sciences and Engineering Research Council of Canada for funding.

Conflicts of Interest

The authors declare no conflict of interest.

REFERENCES

1. Wrede F., Strack E. Über das Pyocyanin, den blauen Farbstoff des *Bacillus pyocyaneus*. Z. Physiol. Chem. 1929;181:58–76. doi: 10.1515/bchm2.1929.181.1-3.58.
2. Frank L.H., DeMoss R.D. On the biosynthesis of pyocyanine. J. Bacteriol.1959;77:776–782.
3. Hassan H.M., Fridovich I. Mechanism of the antibiotic action of pyocyanine. J. Bacteriol. 1980;141:156–163.
4. Fujikawa H., Akimoto R. New Blue Pigment Produced by *Pantoea agglomerans* and Its Production

Characteristics at Various Temperatures. Appl. Environ. Microbiol.2011;77:172–178. doi: 10.1128/AEM.00264-10.

5. Fenical W. Chemical Studies of Marine Bacteria: Developing a New Resource. Chem. Rev. 1999;93:1673–1683. doi: 10.1021/cr00021a001.

6. Brettar I., Christen R., Höfle M.G. *Rheinheimera baltica* gen. nov., sp. nov., a blue coloured bacterium isolated from the central Baltic Sea. Int. J. Syst. Evol. Microbiol.2002;52:1851–1857. doi: 10.1099/ijs.0.02151-0.

7. Grossart H.P., Thorwest M., Plitzko I., Brinkhoff T., Simon M., Zeeck A. Production of a Blue Pigment (Glaukothalin) by Marine *Rheinheimera* spp. Int. J. Microbiol. 2009;2009doi: 10.1155/2009/701735.

8. Cude W.N., Mooney J., Tavanaei A.A., Hadden M.K., Frank A.M., Gulvik C.A., May A.L., Buchan A. Production of the Antimicrobial Secondary Metabolite Indigoidine Contributes to Competitive Surface Colonization by the Marine Roseobacter *Phaeobacter*sp. Strain Y4I. Appl. Environ. Microbiol. 2012;78:4771–4780. doi: 10.1128/AEM.00297-12.

9. Starr M.P., Cosens G., Knackmuss H.J. Formation of the Blue Pigment Indigoidine by Phytopathogenic *Erwinia*. Appl. Microbiol. 1966;14:870–872.

10. Reverchon S., Rouanet C., Expert D., Nasser W. Characterization of Indigoidine Biosynthetic Genes in *Erwinia chrysanthemi* and Role of This Blue Pigment in Pathogenicity. J. Bacteriol. 2002;184:654–665.

11. Tandeau de Marsac N. Phycobiliproteins and phycobilisomes: The early observations.Photosynth. Res. 2003;76:197–205. doi: 10.1023/A:1024954911473.

12. Esenbeck N. Ueber einen blau-rothen Farbstoff, der sich bei der Zersetzung von Oscillatorien bildet. Liebigs Ann. Chem. (Ann. Pharm.) 1836;18:75–82.

13. Sorby H.C. On the characteristic colouring-matters of the red groups of algae. J. Linnean Soc. Bot. 1877;15:34–40. doi: 10.1111/j.1095-8339.1875.tb00214.x.

14. Mölisch H. Das Phycocyan, ein krystallisirbarer Eiweisskörper. Bot. Z. 1895;53:131–135.

15. Mölisch H. Das phycoerythrin, seine Krystallisirbarkeit und chemische. Natur. Bot. Z.1894;52:177–189.

16. Kylin H. Über die roten und blauen Farbstoffe der Algen. Z. Physiol. Chem.1912;76:396–425. doi: 10.1515/bchm2.1912.76.5-6.396.

17. Apt K.E., Collier J.L., Grossman A.R. Evolution of the Phycobiliproteins. J. Mol. Biol. 1995;248:79–96. doi: 10.1006/jmbi.1995.0203.

18. Kai A., Yoshii Y., Nakayama T., Inouye I. Aurearenophyceae *classis nova*, a New Class of Heterokontophyta Based on a New Marine Unicellular Alga *Aurearenacruciata*gen. et sp. nov. Inhabiting Sandy Beaches. Protist. 2008;159:435–457. doi: 10.1016/j.protis.2007.12.003.

19. Robert J.M., Morançais M., Pradier E., Mouget J.L., Tremblin G. Extraction and quantitative analysis of the blue-green pigment "marennine" synthesized by the diatom*Haslea ostrearia*. J. Appl. Phycol. 2002;14:299–305. doi: 10.1023/A:1021184532581.

20. Brill T.B. Light: Its Interaction with Art and Antiquities. Plenum Press; New York, NY, USA: 1980. The origin of color in organic molecules; pp. 119–154.

21. Gastineau R. Ph.D. Thesis. University of Le Mans; Le Mans, France: Sep, 2011. Biodiversité, Reproduction et Phylogénie des Diatomées Bleues du Genre Haslea et Valorisation de Leurs Pigments de Type Marennine.

22. Lankester E.R. On green oysters. Q. J. Microsc. Sci. 1886;26:71–94.

23. Sprat T. History of the Royal Society. Martyn; London, UK: 1667.

24. Goubeau de la Bilennerie J.F. Dissertation sur les huîtres vertes de Marennes, avec des observations critiques de plusieurs naturalistes touchant la reproduction des huîtres en général, et des causes de la couleur verte que ces animaux peuvent acquérir. Goulard; Rochefort, France: 1821.

25. Pasquier A. Essai Médical sur les Huîtres. Didot le Jeune; Paris, France: 1818.

26. Valenciennes A. Sur les causes de la coloration en vert de certaines huîtres. Comptes Rendus Hebdomadaires des sEances de l'Académie des Sciences. 1841;12:345–346.

27. Valmont de Bomare J.C. Huître in Dictionnaire Raisonné d'Histoire Naturelle.Lacombe; Paris, France: 1768.

28. Cadet de Gassicourt C.-L. Cours Gastronomique, ou les Diners de Manant-Ville.Capelle et Renand; Paris, France: 1809.

29. Bory de Saint-Vincent J.-B. Dictionnaire Classique d'Histoire Naturelle. Tome Dixième, Rey et Gravier/Baudouin; Paris, France: 1826.

30. Bosc G., Lair P.A. Nouveau Dictionnaire d'Histoire Naturelle Appliquée aux Arts, à l'agriculture, à l'économie Rurale et Domestique, à la Médecine, etc. Tome XV.Deterville; Paris, France: 1817. Huître.

31. Coste V. Voyage d'Exploration sur le Littoral de la France et de l'Italie. Imprimerie Impériale; Paris, France: 1861.

32. Bachrach E. Le bleuissement des Diatomées et le verdissement des huîtres. Revue critique. Revue des Travaux de l'Institut des Pêches Maritimes. 1935;8:112–123.

33. Bizio B. Memorie dell'Imperial Regio Instituto Veneto di Scienze Lettere ed Arti.Volume 2 Antonelli; Venice, Italy: 1845. Ricerche sopra il coloramento in verde delle branchie delle ostriche.

34. Boon D.D. Coloration in bivalves: A Review. J. Food Sci. 1977;42:1008–1012. doi: 10.1111/j.1365-2621.1977.tb12655.x.

35. Boyce R.W., Herdman W.A. The possible infectivity of the oyster, and upon the green disease in oysters. Comm. Br. Assoc. Adv. Sci. 1896;66:663–669.

36. Mitchell P.H., Barney R.L. The occurrence in Virginia of green-gilled oysters similar to those of Marennes. Bull. U. S. Bur. Fish. 1918;35:135–149.

37. Wang W.X., Yang Y.B., Guo X.Y., He M., Guo F., Ke C.H. Copper and zinc contamination in oysters: Subcellular distribution and detoxification. Environ. Toxicol. Chem.

2011;30:1767–1774. doi: 10.1002/etc.571.

38. Gayot de Pitaval F. Bibliothèque des gens de cours, ou Mélange Curieux de bons mots d'Henri IV, de Louis XIV, de Plusieurs Princes et Seigneurs de la Cour, et autres Personnes Illustres Tome V. Le Gras; Paris, France: 1732.
39. Compain de Saint-Martin A. Nouvelle Méthode Géographique. Volume 2 Mérigot; Paris, France: 1771.
40. Prudhomme L.M. Dictionnaire Géographique Universel Tome IV. Kilian/Picquet; Paris, France: 1828.
41. Grelon M. Saintonge, Pays des Huîtres Vertes. Rupella; La Rochelle, France: 1978.
42. Gaillon B. Des huîtres vertes, et des causes de cette coloration. J. de Physique, de Chimie, d'histoire Naturelle et des Arts. 1820;91:222–225.
43. Bruguière J.G. Encyclopédie Méthodique: Histoire Naturelle des vers. Volume 1 Chez la veuve Agasse; Paris, France: 1792.
44. Briée C. Ph.D. Thesis. University of Nantes; Nantes, France: Oct, 2010. Le verdissement des huîtres: Deux siècles de transformation d'un problème biologique.
45. Bory de Saint-Vincent J.B. Dictionnaire Classique d'Histoire Naturelle. Rey et Gravier/Baudouin; Paris, France: 1824. Navicules; pp. 562–565.
46. Simonsen R. The diatom plankton of the Indian Ocean expedition of RV "Meteor"Hrsg. Dtsch. Forschungsgemeinschaft. 1974;19:46–49.
47. Robert J.M. Ph.D. Thesis. University of Nantes; Nantes, France: Apr, 1983. Fertilité des eaux des claires ostréicoles et verdissement: Utilisation de l'azote par les diatomées dominantes.
48. Round F.E., Crawford R.M., Mann D.G. The Diatoms. Biology AndMorphology of the Genera. Cambridge University Press; Cambridge, UK: 1990. p. 747.
49. Neuville D., Daste P. Recherches sur le *déterminisme de la production de marennine*par la Diatomée marine *Navicula ostrearia* (Gaillon) Bory en culture *in vitro*. Rev. Gen. Bot. 1978;85:255–303.
50. Wraige E.J., Belt S.T., Lewis C.A., Cooke D.A., Robert J.M., Massé G., Rowland S.J. Variations in structures and distributions of C25 highly branched isoprenoid (HBI) alkenes in cultures of the diatom, *Haslea ostrearia* (Simonsen) Org. Geochem.1997;27:497–505. doi: 10.1016/S0146-6380(97)00086-7.
51. Mouget J.L., Tremblin G., Morant-Manceau A., Morançais M., Robert J.M. Long-term photoacclimation of *Haslea ostrearia* (Bacillariophyta): Effect of irradiance ongrowth rates, pigment content and photosynthesis. Eur. J. Phycol. 1999;34:109–115. doi: 10.1080/09670269910001736162.
52. Maestrini S.Y., Robert J.-M. La production micro-algale des claires à huîtres: Particularités nutritionnelles; importance de l'azote organique dissous. Production et relations trophiques dans les écosystèmes marins. IFREMER Act. Coll. 1984;5:185–214.
53. Petersen C.G.J. "Grüne Austern" in Dänemark in 1911–1912. Int. Rev. Hydrobiol. Hydrogr. 1916;7:39–41. doi: 10.1002/iroh.19140070103.
54. Ryder J.A. Report of Commission of Fish and Fisheries for 1882. Governement Printing Office; Washington, DC, USA: 1884. On the cause of the greening of oysters. With a supplementary note on the coloration of the blood corpuscles of the oysters.
55. Ranson G. L'absorption de matières organiques dissoutes par la surface extérieure du corps chez les animaux aquatiques. Annales de l'Institut Océanographique. 1927;4:174.
56. Gastineau R., Davidovich N.A., Bardeau J.F., Caruso A., Leignel V., Hardivillier Y., Jacquette B., Davidovich O.I., Rincé Y., Gaudin P., et al. *Haslea karadagensis*(Bacillariophyta): A second blue diatom, recorded from the Black Sea and producing a novel blue pigment. Eur. J. Phycol. 2012;47:469–479.
57. Gastineau R., Hardivillier H., Leignel V., Tekaya N., Morançais M., Fleurence J., Davidovich N.A., Jacquette B., Gaudin P., Hellio C., et al. Greening effect on oysters and biological activities of the blue pigments produced by the diatom *Haslea karadagensis*(Naviculaceae) Aquaculture. 2013;368–369:61–67.
58. Grünow A. New diatoms from Honduras. Month. Microscop. J. 1877;18:165–186. doi: 10.1111/j.1365-2818.1877.tb00123.x.
59. Davidovich N.A., Gastineau R., Gaudin P., Davidovich O.I., Mouget J.L. Sexual reproduction in the second described blue diatom, *Haslea karadagensis*. Fottea.2012;12:219–229.
60. Gastineau R., Pouvreau J.B., Hellio C., Morançais M., Fleurence J., Gaudin P., Bourgougnon N., Mouget J.L. Biological activities of purified marennine, the blue pigment produced by the diatom *Haslea ostrearia* and responsible for the greening of oysters. J. Agric. Food Chem. 2012;60:3599–3605. doi: 10.1021/jf205004x.
61. Vandanjon L. Ph.D. Thesis. University of Nantes; Nantes, France: Jun, 1997. Etude d'un procédé de valorisation d'une microalgue marine: Concentration et purification par techniques à membranes d'un pigment naturel produit par la diatomée Haslea ostrearia.
62. Rossignol N. Ph.D. Thesis. University of Nantes; Nantes, France: Nov, 1999. Procédés d'extraction et de séparation par membranes appliqués à la production du pigment endo- et exocellulaire synthétisé par la diatomée Haslea ostrearia Simonsen. Mise en oeuvre d'un photobioréacteur à membrane à fonctionnement continu.
63. Vandanjon L., Jaouen P., Rossignol N., Quéméneur F., Robert J.-M. Concentration and desalting by membrane processes of a natural pigment produced by the marine diatom*Haslea ostrearia* Simonsen. J. Biotechnol. 1999;70:393–402. doi: 10.1016/S0168-1656(99)00092-9.
64. Rossignol N., Moan R., Jaouen P., Robert J.-M., Quéméneur F. Continuous high-pressure disruption of marine diatom *Haslea ostrearia*. Assessment by laser diffraction particle sizer. Biotechnol. Technol. 1999;13:909–913. doi: 10.1023/A:1008975527835.
65. Rossignol N., Jaouen P., Robert J.-M., Quéméneur F. Production of exocellular pigment by the marine diatom *Haslea ostrearia* Simonsen in a photobioreactor

equipped with immersed ultrafiltration membranes. Bioresour. Technol. 2000;73:197–200. doi: 10.1016/S0960-8524(99)00171-6.

66. Rossignol N., Lebeau T., Jaouen P., Robert J.-M. Comparison of two membrane-photobioreactors, with free or immobilized cells, for the production of pigments by a marine diatom. Bioprocess Eng. 2000;23:495–501. doi: 10.1007/s004499900186.

67. Pouvreau J.-B., Morançais M., Massé G., Rosa P., Robert J.-M., Florence J., Pondaven P. Purification of the blue-green pigment "marennine" from the marine tychopelagic diatom *Haslea ostrearia* (Gaillon/Bory) Simonsen. J. Appl. Phycol.2006;18:769–781. doi: 10.1007/s10811-006-9088-9.

68. Turpin V., Robert J., Goulletquer P. Limiting nutrients of oyster pond seawaters in the Marennes-Oléron region for *Haslea ostrearia*: Applications to the mass production of the diatom in mesocosm experiments. Aquat. Living Resour. 1999;12:335–342. doi: 10.1016/S0990-7440(99)00114-X.

69. Guillard R.R.L. Culture of phytoplankton for feeding marine invertebrates. In: Smith W.L., Chanley M.H., editors. Culture of Marine Invertebrate Animals. Plenum Press Inc.; New York, NY, USA: 1975. pp. 26–60.

70. Pouvreau J.-B., Morançais M., Fleury F., Rosa P., Thion L., Cahingt B., Zal F., Fleurence J., Pondaven P. Preliminary characterization of the blue-green pigment "marennine" from the marine tychopelagic diatom *Haslea ostrearia* (Gaillon/Bory) Simonsen. J. Appl. Phycol. 2006;18:757–767. doi: 10.1007/s10811-006-9087-x.

71. Lebeau T., Junter G.A., Jouenne T., Robert J.-M. Marennine production by agar-entrapped *Haslea ostrearia* Simonsen. Bioresour. Technol. 1999;67:13–17. doi: 10.1016/S0960-8524(99)00096-6.

72. Lebeau T., Gaudin P., Junter G.A., Mignot L., Robert J.-M. Continuous marennine production by agar-entrapped *Haslea ostrearia* using a tubular photobioreactor with internal illumination. Appl. Microbiol. Biotechnol. 2000;54:634–640. doi: 10.1007/s002530000380.

73. Lebeau T., Gaudin P., Moan R., Robert J.-M. A new photobioreactor for continuous marennine production with a marine diatom: Influence of the light intensity and the immobilised-cell matrix (alginate beads or agar layer) Appl. Microbiol. Biotechnol.2002;59:153–159. doi: 10.1007/s00253-002-0993-9.

74. Rosa P., Turpin V., Robert J., Goulletquer P., Masse G., Charente A.B., Bay A. Oyster greening by outdoor mass culture of the diatom *Haslea ostrearia* Simonsen in enriched seawater. Aquac. Res. 2001;32:801–809. doi: 10.1046/j.1365-2109.2001.00615.x.

75. Robert J.M., Hallet J.N. Absorption Spectrum *in vivo* of the blue pigment "marennine" of the pennate diatom *Navicula ostrearia* Bory. J. Exp. Bot. 1981;32:341–345. doi: 10.1093/jxb/32.2.341.

76. Tremblin G., Cannuel R., Mouget J.L., Rech M., Robert J.M. Change in light quality due to a blue-green pigment, marennine, released in oyster-ponds: Effect on growth and photosynthesis in two diatoms, *Haslea ostrearia* and *Skeletonema costatum*. J. Appl. Phycol. 2000;12:557–566. doi: 10.1023/A:1026502713075.

77. Mouget J.L., Rosa P., Vachoux C., Tremblin G. Enhancement of marennine production by blue light in the diatom *Haslea ostrearia*. J. Appl. Phycol. 2005;17:437–445. doi: 10.1007/s10811-005-0561-7.

78. Moreau J. Contribution aux recherches écologiques sur les claires à huîtres du bassin de Marennes-Oléron. Rev. Trav. Inst. Pêches. Marit. 1970;34:380–462.

79. Ranson G. Le verdissement des huîtres. Sciences. 1937;8:13–24.

80. Bachrach E., Jouvent A. Sur la pigmentation bleue de certaines diatomées. Bulletin Mensuel de la Société Linnéenne de Lyon. 1933;2:36–37.

81. Bachrach E., Simonet M. Le phénomène du bleuissement chez les diatomées. Revue des Travaux de l'Institut des Pêches Maritimes. 1936;9:113–116.

82. Bornet E., Chatin A.D. Le verdissement des huîtres. Bulletin des Séances de laSociété Royale et Centrale d'agriculture. 1895;55:429–439.

83. Nassiri Y., Robert J.M., Rincé Y., Ginsburger-Vogel T. The cytoplasmic finestructure of the diatom *Haslea ostrearia* (Bacillariophyceae) in relation to marennine production.Phycologia. 1998;37:84–91. doi: 10.2216/i0031-8884-37-2-84.1.

84. Moreau J. Recherches préliminaires sur le verdissement en claires: L'évolution de leurs divers pigments liée au complexe pigmentaire de *Navicula ostrearia* Bory. Rev. Trav. Inst. Pêches Marit. 1967;31:372–382.

85. Genevès L., Choussy M., Barbier M., Neuville D., Daste P. Ultrastructure et composition pigmentaire comparées des chromatophores de la Diatomée *Navicula ostrearia* (Gaillon) Bory normale et bleue. C. R. Acad. Sci. Sér. D Sci. Nat.1976;282:449–452.

86. Tremblin G., Robert J.M. Comportement photosynthétique de *Haslea ostrearia* en relation avec sa pigmentation bleue. C. R. Acad. Sci. Ser. III Sci. de la Vie.1996;319:939–944.

87. Tremblin G., Robert J.M. Carbon fixation by the peculiar marine diatom *Haslea ostrearia*. Photosynthetica. 2001;39:215–220. doi: 10.1023/A:1013779125330.

88. Sauvageau C. A propos de la présence de la diatomée bleue dans la Méditerranée.Travaux des Lab. (Arcachon) 1906;6:46–59.

89. Sauvageau C. Le verdissement des huîtres par la diatomée bleue. Travaux des Lab. (Arcachon) 1907;2:1–128.

90. Neuville D., Daste P. Observations concernant la production de pigment bleu par la Diatomée *Navicula ostrearia* (Gaillon) Bory, maintenue en culture unialgale. C. R. Acad. Sci. Sér. D Sci. Nat. 1970;271:2389–2391.

91. Neuville D., Daste P. Variations structurales présentées par la Diatomée *Navicula ostrearia* (Gaillon) Bory, soumise à diverses conditions de culture. Comptes Rendus de la Société de Biol. 1971;165:1723–1726.

92. Robert J.-M. Etude préliminaire des relations entre l'état de verdissement d'une claire ostréicole et l'examen biométrique de populations planctoniques et benthiques du*Navicula ostrearia* Bory, pigmenté; applications. C.

R. Acad. Sci. Sér. D Sci. Nat.1974;278:1405–1408.

93. Robert J.-M. Le verdissement des huîtres dans les claires de la baie de Bourgneuf.Bull. de la Soc. des Sci. Nat. de l'Ouest de la France. 1975;73:123–129.

94. Sullivan W.K. Appendix H, Report of the commission appointed to inquire into the methods of oyster culture in the United Kingdom and France, with a view to the introduction of improved methods of cultivation of oysters in Ireland. Alexander Thom; Dublin, Ireland: 1870. Report on the composition of the soils of oyster grounds; and on qualities which exert most influence on oyster cultivation.

95. Bocat L. Sur la marennine de la diatomée bleue; comparaison avec la phycocyanine.Comptes Rendus des Seances de la Société de Biologie et de ses Filiales. 1907;62:1073–1075.

96. Neuville D., Daste P. Production de pigment bleu par la Diatomée *Navicula ostrearia*(Gaillon) Bory, maintenue en culture unialgale sur un milieu synthétique carencé enazote nitrique. C. R. Acad. Sci. Sér. D Sci. Nat. 1972;274:2030–2033.

97. Hardouin V., Vandanjon L., Jaouen P., Robert J.-M. Procédé combiné extraction-membranes pour l'isolement et la purification de pigments naturels. Actes Colloque Interfiltra Intermembr. 1994:175–181.

98. Bubb W.A. NMR spectroscopy in the study of carbohydrates: Characterizing the structural complexity. Concept Magnetic Res. A. 2003;19:1–19. doi: 10.1002/cmr.a.10080.

99. Størseth T.R., Kirkvold S., Skjermo J., Reitan K.I. A branched β-d-(1→3,1→6)-glucan from the marine diatom *Chaetoceros debilis* (Bacillariophyceae) characterized by NMR. Carbohydr. Res. 2006;341:2108–2134. doi: 10.1016/j.carres.2006.05.005.

100. Dyer W.T.T. Greening of oysters. Nature. 1877;16:397. doi: 10.1038/016397a0.

101. Puységur M. Notice sur la cause du verdissement des huîtres. Revue Maritime et Coloniale. 1880;64:248–256.

102. Schubert H., Tremblin G., Robert J.M., Sagert S., Rincé Y. *In-vivo* fluorescence measurement of photosynthesis of *Haslea ostrearia* Simonsen in relation to marennine content. Diatom Res. 1995;10:341–349. doi: 10.1080/0269249X.1995.9705354.

103. Pouvreau J.B., Morançais M., Taran F., Rosa P., Dufossé L., Guérard F., Pin S., Fleurence J., Pondaven P. Antioxidant and free radical scavenging properties of marennine, a blue-green polyphenolic pigment from the diatom *Haslea ostrearia*(Gaillon/Bory) Simonsen responsible for the natural greening of cultured oysters. J. Agric. Food Chem. 2008;56:6278–6286. doi: 10.1021/jf073187n.

104. Minier C., Tutundjian R., Galgani F., Robert J.M. Copper tolerance in *Haslea ostrearia* assessed by measurements of *in vivo* esterase activity. Mar. Environ. Res.1998;46:579–582. doi: 10.1016/S0141-1136(97)00077-9.

105. Pouvreau J.B., Housson E., le Tallec L., Morançais M., Rincé Y., Fleurence J., Pondaven P. Growth inhibition of several marine diatom species induced by the shading effect and allelopathic activity of marennine, a blue-green polyphenolic pigment of the diatom *Hasleaostrearia* (Gaillon/Bory) Simonsen. J. Exp. Mar. Biol. Ecol. 2007;352:212–225. doi: 10.1016/j.jembe.2007.07.011.

106. Ianora A., Bentley M.G., Caldwell G.S., Casotti R., Cembella A.D., Engström-Öst J., Halsband C., Sonnenschein E., Legrand C., Llewellyn C.A., et al. The Relevance of Marine Chemical Ecology to Plankton and Ecosystem Function: An Emerging Field. Mar. Drugs. 2011;9:1625–1648. doi: 10.3390/md9091625.

107. Bergé J.P., Bourgougnon N., Alban S., Pojer F., Billaudel S., Chermann J.C., Robert J.M., Franz G. Antiviral and anticoagulant activities of a water-soluble fraction of the marine diatom *Haslea ostrearia*. Planta Medica. 1999;65:604–609. doi: 10.1055/s-1999-14032.

108. Carbonnelle D., Pondaven P., Morançais M., Massé G., Bosch S., Jacquot C., Briand G., Robert J.M., Roussakis C. Antitumor and antiproliferative effects of an aqueous extract from the marine diatom *Haslea ostrearia* (Simonsen) against solid tumors: Lung carcinoma (NSCLC-N6), kidney carcinoma (E39) and melanoma (M96) cell lines.Anticancer Res. 1999;19:621–624.

109. Garnier M., Labreuche Y., Nicolas J.L. Molecular and phenotypiccharacterization of*Vibrio aestuarianus* subsp. francensis subsp. nov., a pathogen of the oyster *Crassostrea gigas*. Syst. Appl. Microbiol. 2008;31:358–365. doi: 10.1016/j.syapm.2008.06.003.

110. Denis S., de Decker S., Philippe H., Laetitia C., Maeva R., Garcia C. A Large-Scale Epidemiological Study to Identify Bacteria Pathogenic to Pacific Oyster *Crassostrea gigas*and Correlation Between Virulence and Metalloprotease-like Activity. Microb. Ecol.2010;59:787–798. doi: 10.1007/s00248-009-9620-y.

111. Segarra A., Pépin J.F., Arzul I., Morga B., Faury N., Renault T. Detection and description of a particular ostreid herpesvirus 1 genotype associated with massive mortality outbreaks of Pacific oysters, *Crassostrea gigas*, in France in 2008. Virus Res.2010;153:92–99.

112. Tardy-Laporte C., Arnold A.A., Génard B., Gastineau R., Morançais M., Mouget J.L., Tremblay R., Marcotte I. A 2H solid-state NMR study of the effect of antimicrobials on intact non-mutated *Escherichia coli*. BBA-Biomembranes. 2013;1828:614–622.

113. Mayer A.M.S., Hamann M.T. Marine pharmacology in 2001–2002 marine compounds with anthelmintic, antibacterial, anticoagulant, antidiabetic, antifungal, antiinflammatory, antimalarial, antiplatelet, antiprotozoal, antituberculosis, and antiviral activities; affecting the cardiovascular, immune and nervous systems and other miscellaneous mechanisms of action. Comp. Biochem. Physiol. C. 2005;140:265–286.

114. Olicard C., Renault T., Torhy C., Benmansour A., Bourgougnon N. Putative antiviral activity in haemolymph from adult Pacific oysters. Crassostrea gigas. Antivir. Res.2005;66:147–152. doi: 10.1016/j.antiviral.2005.03.003.

115. Lee J.B., Hayashi K., Hirata M., Kuroda E., Suzuki E.,

Kubo Y., Hayashi T. Antiviral sulfated polysaccharide from *Navicula directa*, a diatom collected from deep-sea water in Toyama Bay. Biol. Pharm. Bull. 2006;29:2135–2139.

116. Shimizu Y. In: Drugs from the Sea. Fusetani N., editor. Karger; Basel, Switzerland: 2000. pp. 30–45.

117. Scholz B., Liebezeit G. Screening for biological activities and toxicological effects of 63 phytoplankton species isolated from freshwater, marine and brackish water habitats.Harmful Algae. 2012;20:58–70. doi: 10.1016/j.hal.2012.07.007.

118. Yasuhara-Bell J., Lu Y. Marine compounds and their antiviral activities. Antivir. Res.2010;86:231–240. doi: 10.1016/j.antiviral.2010.03.009.

119. Newsome A.G., Murphy B.T., van Breemen R.B. Isolation and Characterization of Natural Blue Pigments from Underexplored Sources Physical Methods in Food Analysis.American Chemical Society; Washington, DC, USA: 2013. pp. 105–125.

120. Kuddus M., Singh P., Thomas G., Al-Hazimi A. Recent Developments in Production and Biotechnological Applications of *C*-Phycocyanin. BioMed Res. Int.2013;2013 doi: 10.1155/2013/742859.

121. Couteau C., Pommier M., Paparis E., Coiffard L.J.M. Study of the efficacy of 18 sun filters authorized in European Union tested *in vitro*. Pharmazie. 2007;62:449–452.

122. Groves G.A., Agin P.P., Sayre R.M. *In vitro* and *in vivo* methods to define sunscreen protection. Australas. J. Dermatol. 1979;20:112–119. doi: 10.1111/j.1440-0960.1979.tb00217.x.

123. Couteau C., Chammas R., Alami S., Choquenet B., Paparis E., Coiffard L.J.M. Combination of UVA-filters and UVB-filters or inorganic UV-filters—Influence on the Sun Protection Factor (SPF) and the PF-UVA determined by *in vitro* method. J. Dermatol. Sci. 2008;50:159–161.

124. Brétéché A., Duflos M., Dassonville A., Nourrisson M.R., Brelet J., le Baut G., Grimaud N., Petit J.Y. New *N*-pyridinyl(methyl)-indole-2- and 3-(Alkyl) carboxamides and Derivatives Acting as systemic and topical Inflammation Inhibitors. J. Enzym. Inhib. Med. Chem. 2002;17:415–424. doi: 10.1080/1475636021000005695.

125. Carlson R.P., O'Neil-Davis L., Chang J., Lewis A.J. Modulation of mouse ear oedema by clyclooxygenase and lipooxygenase inhibitors and other pharmacologic agents. Agents Actions. 1985;17:197–204. doi: 10.1007/BF01966592.

126. Collin X., Robert J.M., Duflos M., Wielgosz G., le Baut G., Bobin-Dubigeon C., Grimaud N., Lang F., Petit J.Y. Synthesis of *N*-Pyridinyl(methyl)-1,2-dihydro-4-hydroxy-2-oxoquinoline-3-carboxamides and analogues and their anti-inflammatory activity in mice and rats. J. Pharm. Pharmacol. 2001;53:417–423. doi: 10.1211/0022357011775505.

CHAPTER 9

SEASONAL VARIATION IN THE PIGMENT CONTENT OF DOMINANT MACROPHYTES FROM WULAR LAKE, KASHMIR HIMALAYA, INDIA

Dar NA, Pandit AK and Ganai BA

Department of Environmental Science, University of Kashmir, Hazratbal, Srinagar, India

ABSTRACT

The pigment content (Chlorophyll-a, Chlorophyll-b and Total Chlorophyll) of dominant macrophytes was investigated in this present study along with subsequent analysis of relevant physico-chemical variables. In this study, the relationship between the pigment content of dominant species and the ambient environmental parameters was established. In general, Chlorophyll-a, Chlorophyll-b and Total Chlorophyll contents varied from 1.5 ± 0.1 µg/ml to 3.7 ± 0.5 µg/ml, 0.3 ± 0.2 µg/ml to 1.7 ± 0.5 µg/ml and 2.1 ± 0.9 µg/ml to 4.1 ± 0.5 µg/ml respectively with highest values during summer coinciding with increasing water temperatures. Chlorophyll-a and Chlorophyll-b and Total Chlorophyll levels were positively correlated ($P<0.05$) with water temperature and negatively correlated with nutrient salts such as nitrate, ammonia and total phosphorus. Maximum pigment content occurred in submerged macrophytes, whereas minimum levels were obtained in rooted- floating macrophytes. The analysis of Chlorophyll A, B and Total in the macrophytes indicated the presence of species specificity as well as of seasonal dynamics of these pigments. Statistical analysis computed among environmental parameters and pigment content suggests the potential role played by the abiotic parameters on the pigment content of macrophytes.

INTRODUCTION

Aquatic macrophytes which are considered to be the photosynthetic organisms of freshwater habitats are central to the function of fresh water ecosystems [1]. Macrophytes are a potential source of food and fodder for humans and bovine population, besides serving as a base of aquatic food chain [2-6]. In order to evaluate the food potential of macrophytes and estimation of their nutritional value, the knowledge of their chemical composition is of paramount importance [5,7,8]. It is well known that under the influence of abiotic factors such as light, temperatute, sediment composition and water chemistry, macrophytes can modify their metabolic responses (photosynthesis and growth rates) and levels of proximate constituents [9-12] thus necessitating the determination of biochemical ingredients of the tissues produced under each ecological condition. Light and temperature act together in affecting growth, morphology, photosynthesis, chlorophyll composition and reproduction of macrophytes [13-18]. The biological functions such as photosynthesis, physiology, etc.

Citation: Dar NA, Pandit AK, Ganai BA (2013) Seasonal Variation in the Pigment Content of Dominant Macrophytes from Wular Lake, Kashmir Himalaya, India. Biochem Pharmacol 2:124. doi:10.4172/2167-0501.1000124.

in macrophytes are dependent on chlorophyll a (chl a), a key photosynthetic pigment. Consequently, changes in chl a can indicate plant growth [19]. or disturbances from stressors [20]. It is believed that the distinct variation in pigment content as well as specific features of the pigment complex ensure adaptation of macrophytes to light conditions in water stratum [17,18,21].

The present study aims to analyze the seasonal variations and comparative analysis in the pigment content of aquatic macrophytes characterized by different degree of submergence (emergent, rooted floating, free floating, and submerged leaves) with respect to relevant hydrological parameters (like surface water temperature, transparency, nutrients (ammonia, nitrate and phosphate).

MATERIALS AND METHODS

Study Site

Geographical the lake is situated at an altitude of 1580 m (a.m.s.l) lying between 34°16′24.67″ - 34° 20′26.26″ N latitudes and 74° 33′41.42″ - 74o 44′02.90″E longitudes of Kashmir Valley. The oxbow type lake is mono- basined, elliptical in shape and is of fluviatile in origin, formed by the meandering of River Jhelum. Various rivers and streams which contribute their might draining to the Wular lake are River Jhelum, Erin Nallah, Madhumati Nallah, Ferozepur Nallah, Sukhnag Nallah, Ningal Nallah, Ajus Nallah, etc. and the only exit (outlet) of the lake is River Jhelum. The macrophyte dominant lake has well developed littoral zone. Extended littorals in general, are dominated by (i) reeds like Phragmites australis, Typha angustata and Sparganium ramosum, (ii) rooted floating-leaf types like Nymphoides peltatum, Trapa natans and Nymphaea mexicana, and (iii) freefloating types dominated by Salvinia natans, Azolla sp. and Lemnids, all growing together intermixed and forming a complex physiognomy. The slightly deeper zone has abundant growth of submergeds like Ceratophyllum demersum, Potamogeton crispus and Potamogeton lucens forming meadows at various places. The littorals towards the north-eastern side are densely populated with Salix plantation which is eating the vitals of the lake.

Sampling and Analysis of Pigment Content

Seasonal sampling of aquatic plants was carried out between March, 2011 to November, 2011 spelling three seasons. The aquatic plants were collected according to their availability from nine sampling locations within the Wular lake, in order to determine the chlorophyll content of these macrophytes (Figure 1). These aquatic plants were harvested in a lush - green condition using quadrants of 0.25 m2 in an area [22,23] from different areas within each stand to form three representative samples for each species. The harvested materials were placed in polyethylene bags and transported to the laboratory in cool boxes (temperature 4°C) where they were incised to include only leaves. These were then washed thoroughly and moisture was drained before being analyzed for chlorophyll content.

Chlorophyll content (Chl a, Chl b and Total Chlorophyll) of macrophytes was determined Spectrophoto-metrically, in acetone extracts prepared from fresh samples using standard methods [24,25]. Acetone extract of the macrophytes was prepared by macerating 1 g fresh weight of plant leaves with 80% aqueous acetone using a mortar and pestle. The decanted suspension was centrifuged for 3 minutes at 1320 r.p.m. After centrifugation, the upper green clear solution was decanted from the colorless residue and then made up to 10ml with 80% acetone in 10 ml test tubes. The solution was then subjected to centrifugation at 10,000 r.p.m for 10 minutes. The absorbance of the solution was determined using a spectrophotometer (Systronics-116) at wavelengths of 665 and 649 nm respectively. The results were expressed as μg/ml.

Chlorophyll 'a' (μg/ml)=11.63 (A_{665})-2.39 (A_{649})

Chlorophyll 'b' (μg/ml)=20.11 (A_{649})-5.18 (A_{665})

Total Chlorophyll (μg/ml)=6.45 (A_{665}) +17.72 (A_{649})

Analysis of Hydrological Parameters

Surface water samples were collected from all the selected stations in clean TARSON bottles for analysis of hydrological parameters like surface water temperature, transparency, nitrate, ammonia and phosphate on seasonal basis. The temperature of each water sample were measured at the field by means of a mercury thermometer, while as transparency in water was measured with Secchi disc. Other water quality parameters such as nitrate, ammonia and phosphate, which require photometric measurements, were done according to the standard methodology outlined in APHA [26] by using model spectrophotometer (Systronics-116) on the same day in Aquatic Ecology Laboratory, Department of Environmental Science, University of Kashmir-190006, India.

Statistical Analysis

All the data were expressed in terms of mean ± S.D (standard deviation) and range. In addition, Pearson correlation coefficient (r) was also computed between the pigment content of the macrophytes and environmental parameters. Descriptive statistics were calculated using StatSoft Statistica 7.0 statistics software.

RESULTS

Seasonal Variation in Hydrological Parameters

Table 1. Seasonal variations (Mean±S.D) of some physico- chemical parameters in surface waters from Wular lake

Parameter (units)	Spring±S.D	Summer±S.D	Autumn±S.D	Mean±S.D
Water temperature (°C)	15.3 ± 3.7	25.3 ± 7.9	14.4 ± 7.3	18.3 ± 4.9
Nitrate(µg/l)	482.6 ± 214.8	299.8 ± 179.7	431.85 ± 167	404.8 ± 77
Ammonia(µg/l)	99 ± 31.3	69 ± 32.7	123 ± 46.7	97 ± 22.1
Total phosphorus(µg/l)	173.7 ± 13.4	124 ± 53.9	194 ± 61.1	163.9 ± 29.4
Transparency (m)	0.5 ± 0.1	0.7 ± 0.2	0.9 ± 0.3	0.7 ± 0.2

Maximum temperature was recorded in summer (25.3 ± 7.9°C) and minimum in autumn (14.4 ± 7.3°C). Ammonia and Total Phosphorus exhibit same seasonal patterns with autumn depicting higher values (123 ± 46.7 and 194 ± 61.1) and summer depicting lower values (69 ± 32.7 and 124 ± 53.9), respectively. The seasonal mean values of nitrate in the present investigation varied between 299.8 ± 179.7 µg/l and 482.6 ± 214.8 µg/l in summer and spring respectively. In contrast maximum transparency values (0.9 ± 0.3 m) were obtained in autumn against a minimum (0.5 ± 0.1 m) being recorded in spring (Table 1).

Seasonal Variation in Pigment Content

Perusal of data on the pigment content of macrophytes revealed significant difference for Chlorophyll-a, Chlorophyll-b and Total Chlorophyll in the different groups of macrophytes collected from Wular lake. The levels of Chlorophyll-a, the primary photosynthetic pigment in macrophytes in the present study varied from 1.5 ± 0.1 µg/ml to 3.7 ± 0.5 µg/ml. The values were highest in Ceratophyllum demersum (3.7 ± 0.5 µg/ml) and lowest in Azolla sp. (1.5 ± 0.1µg/ml). The Chlorophyll-b content recorded its highest value in Ceratophyllum demersum where the data observed (1.7 ± 0.5 µg/ml) was succeeded by Potamogeton crispus with a value of (1.6 ± 0.5 µg/ml). The lowest Chlorophyll-b values were obtained in Azolla sp. having a value of (0.3 ± 0.2 µg/ml).The Total Chlorophyll content also recorded its highest value in Ceratophyllum demersum (4.1 ± 0.5 µg/ml) followed by Lemna minor with a value of 4.0 ± 0.7 µg/ml, decreasing to a lowest of 2.1 ± 0.9 µg/ml in Azolla sp. (Table 2).

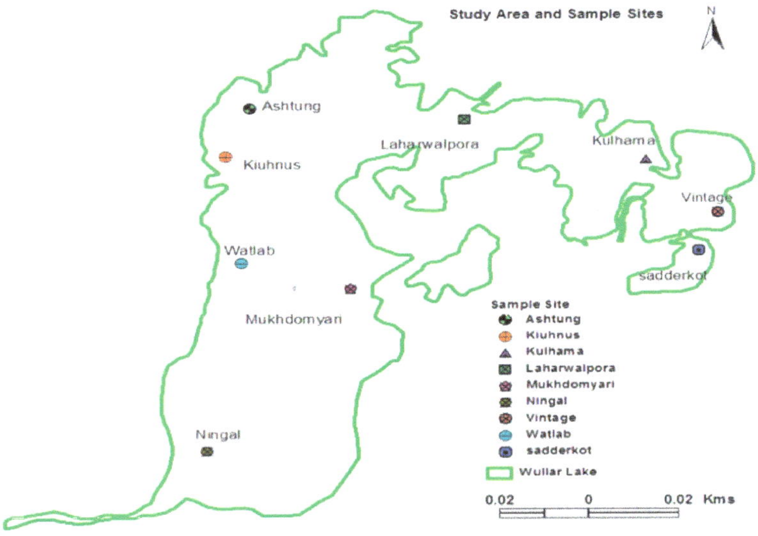

Figure 1: Map showing study area and sampling stations.

Table 2. Mean values of pigment content found in macrophytes harvested from Wular lake Note: Chl.a = Chlorophyll a, Chl.b = Chlorophyll b, Total Chl. = Total Chlorophyll and S.D = standard deviation of the mean

Plant	Species	Chl.a (Mean ± S.D)	Chl.b (Mean±S.D)	Total Chl. (Mean±S.D)
Emergent	*Phragmites australis*	2.2 ± 0.3	0.7 ± 0.3	2.4 ± 0.8
	Polygonum hydropiper	2.0 ± 0.6	0.4 ± 0.2	2.2 ± 0.6
Rootedfloating	*Nymphaea mexicana*	2.8 ± 0.4	1.2 ± 0.5	3.7 ± 0.8
	Nymphoides peltatum	2.4 ± 0.5	0.6 ± 0.4	2.6 ± 0.5
Submerged	*Potamogeton crispus*	3.1 ± 0.6	1.6 ± 0.5	3.5 ± 0.6
	Ceratophyllum demersum	3.7 ± 0.5	1.7 ± 0.5	4.1 ± 0.5
Free floating	*Lemna minor*	2.8 ± 0.4	1.4 ± 0.6	4.0 ± 0.7
	Azolla sp.	1.5 ± 0.1	0.3 ± 0.2	2.1 ± 0.9

The macrophytes collected during different seasons exhibited distinct seasonal profile in the pigment content. Significant changes were seen in all the macrophytes except that of Ceratophyllum demersum and Potamogeton crispus where slight fluctuations were noticed. In the present work, maximum pigment content was observed during summer season, the period of active growth characterized by optimum nutrient load and temperature. The lowest levels of Chlorophyll-a and Chlorophyll-b were obtained during the period of less vigorous growth in autumn, and they increased considerably during the period of sprouting and early growth stages of macrophytes in spring and were highest in summer. However, relative content of Total Chlorophyll in the macrophytes under study was notably lower in spring season than in autumn (Figures 1-4).

In general submerged macrophytes represented in the present study by Ceratophyllum demersum and Potamogeton crispus were characterised by high pigment content. While as lower pigment levels were obtained in emergents and rooted- floatings. Chlorophyll-a and Chlorophyll -b and Total Chlorophyll showed highly significant positive correlation (r=0.96, 1.00 and 0.95, P<0.05) with water temperature and negatively correlated with nutrient salts such as nitrate, ammonia and total phosphorus. However, transparency showed a non-significant positive correlation (r=0.23, P=0.05) with Total Chlorophyll and and negative correlation with Chlorophyll-a and Chlorophyll-b (Table 3).

Table 3. Correlation coefficients between environmental variables and pigment content of macrophytes of Wular lake *Correlation at 0.05

	Water Temperature	Transparency	Nitrate	Ammonia	Total Phosphorus
Water Temperature	1.00				
Transparency	-0.07	1.00			
Nitrate	-0.94	-0.27	1.00		
Ammonia	-0.93	0.44	0.74	1.00	
Total Phosphorus	-0.98	0.28	0.85	0.98	1.00
Total Chlorophyll	0.95*	0.23	-1.00*	-0.77*	-0.67*
Chlorophyll a	0.96*	-0.35	-0.81*	-1.00*	-1.00*
Chlorophyll b	1.00*	-0.06	-0.95*	-0.92*	-0.97*

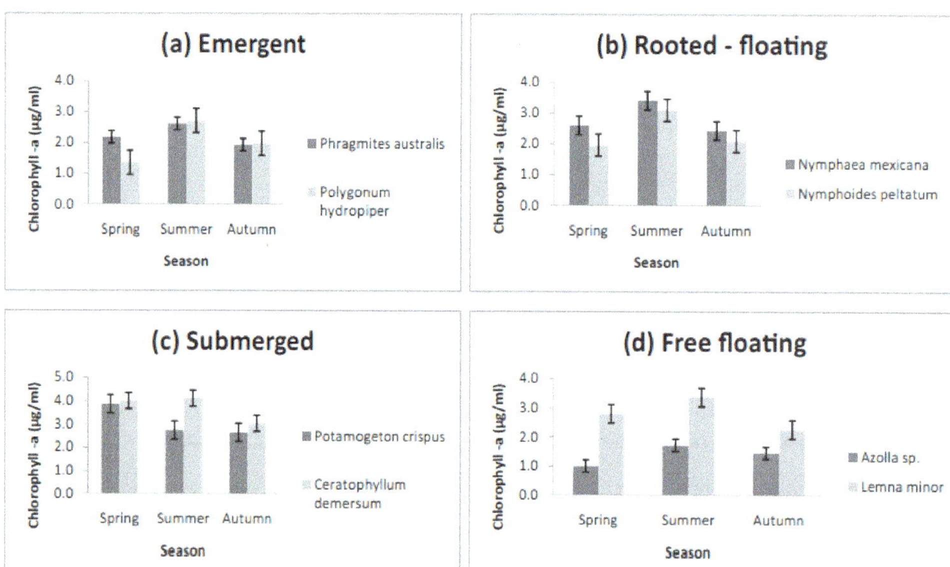

Figure 2: Seasonal variation of Chlorophyll – a (µg/ml) in (a) emergent, (b) rooted – floating, (c) submerged, and free floating aquatic macrophytes.

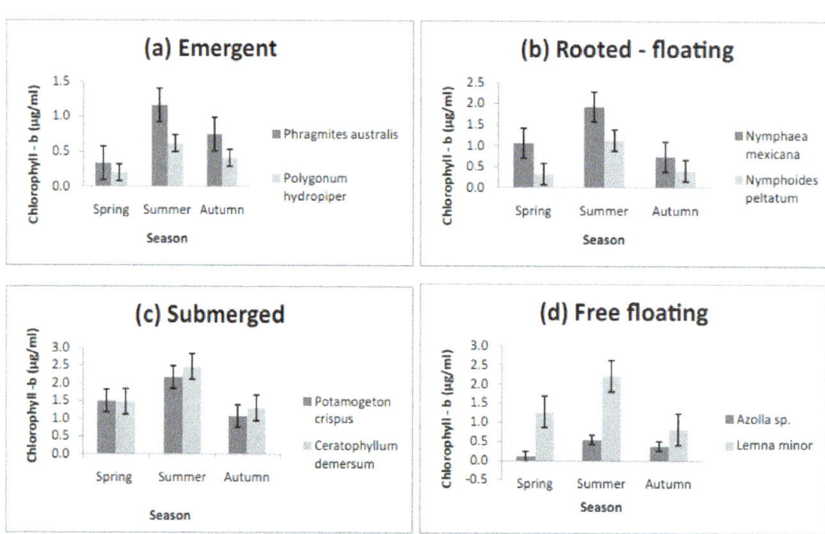

Figure 3: Seasonal variation of Chlorophyll – b (µg/ml) in (a) emergent, (b) rooted – floating, (c) submerged, and frees floating aquatic macrophytes

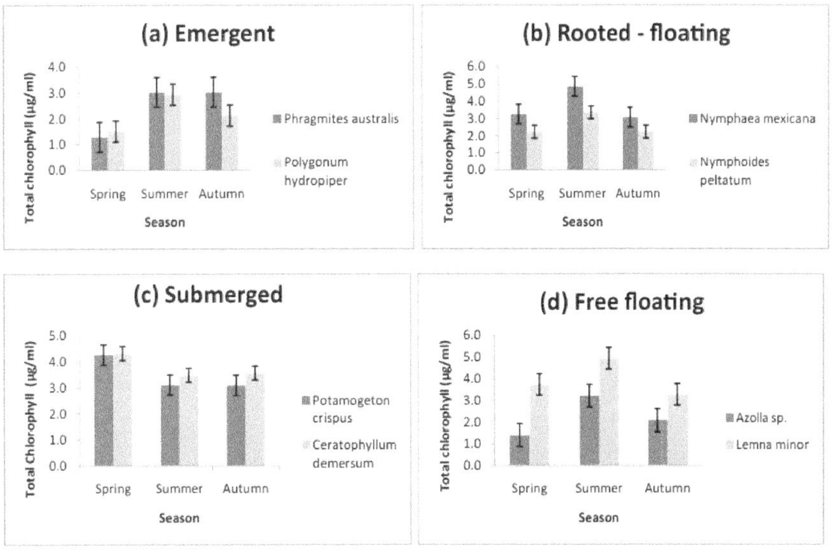

Figure 4: Seasonal variation of Total Chlorophyll (µg/ml) in (a) emergent, (b) rooted – floating, (c) submerged, and frees floating aquatic macrophytes.

DISCUSSION

The analysis of our data revealed some distinct trends in the pigment content in the fresh water macrophytes sampled during different seasons. In the present work, maximum pigment content of the macrophytes was observed during summer season; the period characterized by optimum nutrient load and water temperature. Chlorophyll-a and Chlorophyll-b contents were significantly lower in autumn, Total chlorophyll was the only exception. The trend may be attributed to the positive role of water temperature in promoting greater chlorophyll concentration and photosynthesis [13-16]. The summer fall in the amounts of nutrients (nitrate, ammonia and total phosphorus) is attributed to the greater photosynthetic activity of macrophytes especially submergeds leaving very small quantities of such nutrients in water [27]. This is confirmed from the significant positive correlation between Chlorophyll-a, Chlorophyll -b and Total Chlorophyll and water temperature and negative correlation with nutrients (nitrate, ammonia and total phosphorus).The lower quantity of chlorophyll pigment in macrophytes in other seasons reflects their adaptation to decreased water temperature [28]. Our results are in accordance with the published data [29,30]. Thus the active period of pigment synthesis in macrophytes coincides with the increase in water temperature and decrease in nutrient load.

The macrophytes as a whole were characterized by low chlorophyll pigment content per fresh weight basis in comparison to the terrestrial plants of middle latitudes [31]. The comparative analysis of data (Table 2) showed appreciable variations in the pigment content among different groups of macrophytes with different degree of leaf submergence (emergent, floating, free floating and submerged). In emergent and rooted- floating macrophytes, the low chlorophyll pigment content might be attributed to their existence at high irradiance [17]. In rooted- floating macrophytes the leaves are perpendicular to incident radiation. As a consequence of which they contain lower pigment content compared to submerged aquatic plants. In submerged macrophytes, the leaves are almost vertically oriented [17]. As a result, they do not shade each other resulting in higher pigment content in them. In the submerged leaves, in contrast to the emergent and floating ones, an increase in the photosynthetic rate of a single chloroplast as well as their existence in water under low light conditions results in an increase in the pigment content in their chloroplast. Our results further gain support from the findings of [17,32] who also reported higher pigment content in submerged aquatic plants. The free floating macrophytes represented in our study by Azolla sp. and Lemna minor contain higher pigment levels in contrast to emergent and rootedfloating macrophytes is due to their presence of homogenous type of mesophyll structure [14,17] suggested that within individual plants, total chlorophyll can vary significantly in response to light, with greatest chlorophyll concentration at the basal (darkest) portion of the plant. The normal life activity is ensured by low pigment content in plants adapted to high radiation. One of the advantages in having the low pigment content is that it prevents cell damage caused by photooxidation [17]. This regularity was observed for desert and highmountain plants existing under conditions of high light intensity [31].

CONCLUSION

It is evident from the study that different groups of macrophytes showed distinct profile in their seasonal pigment content. The results obtained in our study establish that physico-chemical parameters have profound influence on the pigment content of aquatic plants. Significant correlation values computed among the environmental parameters and pigment content suggest that the abiotic parameters seem to have potential role in photosynthetic pathways of macrophytes.

Acknowledgements

The authors are highly grateful to Head, Department of

Environmental Science, University of Kashmir, Hazratbal, Srinagar-190006 for providing fund and facilities for smoothly conducting of research work.

REFERENCES

1. Wetzel RG (2001) Limnology: Lake and River Ecosystems. Third edition, Academic Press Elsevier, San Diego, USA.
2. Tardío J, Pascual H, Morales R (2005) Wild food plants traditionally used in the Province of Madrid, central Spain. Econ Bot 59: 122-136.
3. Rahman AH, Rafiul Islam AK, Naderuzzaman AK, Hossain MD, Rowshatul A (2007) Studies on the aquatic angiosperms of the Rajshahi University campus. Res J Agric Biol Sci 3: 474-480.
4. Hasan MR, Chakrabarti R (2009) Use of algae and aquatic macrophytes as feed in smallscale aquaculture: A review. Fisheries and Aquaculture Technical Paper 531, Rome, p 123 .
5. Swapna MM, Prakashkumar R, Anoop KP, Manju CN, Rajith NP (2011) A review on the medicinal and edible aspects of aquatic and wetland plants of India. J Med Plants Res 5: 7163-7176.
6. Smith JE (2011) Algae. Encyclopedia of biological invasions (Simberloff D, Rejmanek M, editors), University of California, Los Angeles, CA, USA Press, pp. 11-15.
7. Chapman VJ (1950) Seaweeds and their uses. Taylor & Francis Publishers, USA.
8. Hawkins SJ, Hartnoll RG (1983) Grazing of intertidal algae by marine invertebrates. Oceanography and marine biology 21: 195-282.
9. Koskimies SK, Nyberg H (1987) Effects of temperature and light on the lipids of Sphagnum magellanicum. Phytochemistry 26: 2213-2221.
10. Roslin AS (2001) Seasonal variations in the lipid content of some marine algae in relation to environmental parameters in the Arockiapuram coast. Seaweed Res Utiln 23:119- 127.
11. Orduna- Rojas J, Robledo D, Dawes CJ (2002) Studies on the Tropical Agarophyte Gracilaria cornea J. Agardh (Rhodophyte, Gracilariales) from Yucatan, Mexico.1. Seasonal Physiological and Biochemical Responses. Bot Mar 45: 453-458.
12. Rajasulochana N, Baluswami M, Parthasarathy MDV, Krishnamurthy V (2002) Chemical analysis of Grateloupia lithophila Boergesen. Seaweed Res Utiln 24: 79-82.
13. Barko JW, Hardin DG, Mathews MS (1982) Growth and morphology of submersed freshwater macrophytes in relation to light and temperature. Canadian Journal of Botany 60: 877-887.
14. Barko JW, Adams MS, Clesceri NL (1986) Environmental factors and their consideration in the management of submerged aquatic vegetation: A review. J Aquat Plant Manage 24: 1-10.
15. Chambers PA (1982) Light, temperature and the induction of dormancy in Potamogeton crispus and Potamogeton obtusifolius. Ph.D. thesis. University of St. Andrews.
16. Nekrasova GF, Ronzhina DA, Maleva MG, P'yankov VI (2003) Photosynthetic Metabolism and Activity of Carboxylating Enzymes in Emergent, Floating, and Submerged Leaves of Hydrophytes. Russ J Plant Physiol 50: 57-67
17. Ronzhina DA, Nekrasova GF, P'yankov VI (2004) Comparative Characterization of the Pigment Complex in Emergent, Floating, and Submerged Leaves of Hydrophytes. Russian J of Plant Physiology 51: 21–27.
18. Robledo D, Freile-pelegrin Y (2005) Seasonal variation in photosynthesis and biochemical composition of Caulerpa spp. (Bryopsidales, Chlorophyta) from the Gulf of Mexico. Phycologia 44 : 312–319.
19. Raven PH, Evert RF, Eichhorn SE (1992) Biology of plants. W. H. Freemann and Company Press, New York, NY
20. Blackburn GA (2007) Hyperspectral remote sensing of plant pigments. J Exp Bot 58: 855-867.
21. Terrados J, Ros JD (1992) the influence of temperature on seasonal variation of Caulerpa prolifera (Forsskal) Lamouroux photosynthesis and respiration. J Experimental Marine Biol Ecol 162: 199-212.
22. Westlake DF (1965) Some basic data for investigations of the productivity of aquatic macrophytes. Memories of Institute of Italian Idrobiologia 18: 229-248.
23. Westlake DF (1971) Macrophytes. In: Vollenweider RA. (Ed) A manual of methods for measuring primary production in aquatic environments. Oxford: Blackwell Scientific Publication 25-32.
24. Strain HH, Cope BT, Svec WA (1971) Analytical procedures for the isolation, identification, estimation and investigation of the chlorophylls. Methods Enzymol 23: 452-476.
25. Lichtenthaler HK, Wellburn AR (1983) Determinations of total carotenoids and chlorophylls a and b of leaf extracts in different solvents. Biochem Soc Transactions 603: 591-592.
26. APHA (2005) Standard methods for the examination of water and waste waters. American Public Health Association ed. Washington DC: 1217 pp.
27. Rich PV, Wetzel RG, Thuy NV (1971) Distribution, production and role of aquatic macrophytes in a southern Michigam marl lake. Freshwat Biol 1: 3-21.
28. Dembitsky VM (1996) Betaine ether-linked glycerolipids: chemistry and biology. Prog Lipid Res 35: 1-51.
29. Kizevetter IV, Sukhoveeva MV, Shmel'kova LP (1981) Promyslovye morskie vodorosli i travy Dal'nevostochnykh morei (Industrial Marine Algae and Grasses of Far East Seas), Moscow: Legkaya i pishchevaya promyshlennost.
30. Goncharova SN, Kostetsky EY, Sanina NM (2004) The Effect of Seasonal Shifts in Temperature on the Lipid Composition of Marine Macrophytes. Russ J Plant Physiol 51: 169-175.
31. Popova OF, Slemnev NN, Popova IA, Maslova TG

(1984) The Content of Plastid Pigments in Plants of the Gobi Desert and the Karakums, Bot Zh 69: 334-344.

32. Nikolic LS, Pajevi CB, Ljevnai C (2009) Primary production dynamics of dominant hydrophytes in Lake Provala (Serbia). Central Eur J Biol 4: 250-257.

CHAPTER 10

STRAWBERRY FLAVOR: DIVERSE CHEMICAL COMPOSITIONS, A SEASONAL INFLUENCE, AND EFFECTS ON SENSORY PERCEPTION

Michael L. Schwieterman[1,8], Thomas A. Colquhoun[1,2,8], Elizabeth A. Jaworski[2,8], Linda M. Bartoshuk[3,8], Jessica L. Gilbert[4,8], Denise M. Tieman[4,8], Asli Z. Odabasi[5,8], Howard R. Moskowitz[6] , Kevin M. Folta[1,4,8], Harry J. Klee[1,4,8], Charles A. Sims[5,8], Vance M. Whitaker[5,7,8], David G. Clark[1,2,8]

[1] Plant Molecular and Cellular Biology Program, University of Florida, Gainesville, Florida, United States of America

[2] Department of Environmental Horticulture, University of Florida, Gainesville, Florida, United States of America

[3] College of Dentistry, University of Florida, Gainesville, Florida, United States of America

[4] Horticultural Sciences Department, University of Florida, Gainesville, Florida, United States of America

[5] Food Science and Human Nutrition Department, University of Florida, Gainesville, Florida, United States of America

[6] Moskowitz Jacobs Inc., White Plains, New York, United States of America

[7] Gulf Coast Research and Education Center, University of Florida, Wimauma, Florida, United States of America

[8] Plant Innovation Program, University of Florida, Gainesville, Florida, United States of America

ABSTRACT

Fresh strawberries (*Fragaria x ananassa*) are valued for their characteristic red color, juicy texture, distinct aroma, and sweet fruity flavor. In this study, genetic and environmentally induced variation is exploited to capture biochemically diverse strawberry fruit for metabolite profiling and consumer rating. Analyses identify fruit attributes influencing hedonics and sensory perception of strawberry fruit using a psychophysics approach. Sweetness intensity, flavor intensity, and texture liking are dependent on sugar concentrations, specific volatile compounds, and fruit firmness, respectively. Overall liking is most greatly influenced by sweetness and strawberry flavor intensity, which are undermined by environmental pressures that reduce sucrose and total volatile content. The volatile profiles among commercial strawberry varieties are complex and distinct, but a list of perceptually impactful compounds from the larger mixture is better defined. Particular esters, terpenes, and furans have the most significant fits to strawberry flavor intensity. In total, thirty-one volatile compounds are found to be significantly correlated to strawberry flavor intensity, only one of them negatively. Further analysis identifies individual volatile compounds that have an enhancing effect on perceived sweetness intensity of fruit independent of sugar content. These findings allow for consumer influence in the breeding of more desirable fruits and vegetables. Also, this approach garners insights into fruit metabolomics, flavor chemistry, and a paradigm for enhancing liking of natural or processed products.

Citation: Schwieterman ML, Colquhoun TA, Jaworski EA, Bartoshuk LM, Gilbert JL, Tieman DM, et al. (2014) Strawberry Flavor: Diverse Chemical Compositions, a Seasonal Influence, and Effects on Sensory Perception. PLoS ONE 9(2): e88446. doi:10.1371/journal.pone.0088446.

INTRODUCTION

Modern fully ripe strawberry (*Fragaria x ananassa)* fruit is characterized by its large size [1], vibrant red color [2], reduced firmness [3], distinct aroma [4], and sweet fruity flavor [5]. The flesh of the strawberry is a swollen receptacle, a false fruit, and the seeds or achenes are the true fruit [6], which will be collectively referred to as strawberry fruit. The three stages of non-climacteric, auxin dependent strawberry fruit development; division, expansion and ripening, involve gains in diameter and fresh weight; during which color shifts from green to white to dark red in roughly forty days following anthesis [7]. Ripening of strawberry fruit results in the accumulation of multiple sugars and organic acids, culminating with peak volatile emission [8].

Flavor is the perceptual and hedonic response to the synthesis of sensory signals of taste, odor, and tactile sensation [9]. In the case of strawberry and other fruits, sensory elicitation is the result of multiple direct interactions between plant and human: sugars and acids, pigments, turgor and structure, and volatile compounds, which elicit the senses of taste, vision, tactile sensation, and olfaction, respectively, in the development of flavor [10]–[13]. A consumer based survey indicated sweetness and complex flavor as consistent favorable attributes of the "ideal" strawberry experience [14]. Much emphasis is placed on sugars, acids, and volatile compounds as these metabolites are primary sensory elicitors of taste and olfaction which attenuate the perception and hedonics of sweetness and flavor. Thus a ripe strawberry is metabolically poised to elicit the greatest sensory and hedonic responses from consumers.

During strawberry fruit development sucrose is continually imported from photosynthetic tissue. A consistently high sucrose invertase activity contributes to carbon sink strength in all developmental stages of fruit [15]. Delivered sucrose is hydrolyzed into glucose and fructose, and these three carbohydrates constitute the major soluble sugars of ripe strawberries, a result of their continual accumulation during fruit development [16]. In fact, an approximately 150% increase in their sum during ripening has been observed [8], [15]. The influx of carbon initiates a complex network of primary and secondary metabolism specific to ripening strawberry fruit [16]. For example, the metabolic activity of ripening strawberry is visualized by the late accumulation of the predominant red pigment, pelargonidin 3-glucoside [17], an anthocyanin derived from the primary metabolite phenylalanine [16].

The dynamics of fruit development are genetically driven. Microarray analysis determined nearly 15% of probed expressed genes exhibit significant differential expression (60% up, 40% down) in red compared to green fruit [18]. One up regulated gene, *Polygalacturonase 1* (*FaPG1*), contributes to fruit softening [19] by aiding in catalytic cell wall disassembly [20]. Reduction of firmness is also attributed to dissolution of middle lamella, a pectin rich cell wall layer that functions in cell-to-cell adhesion [3]. Active shifts in transcription throughout ripening result in metabolic network reconfiguration altering the chemical and physical properties.

Metabolic profiling indicates an accumulation of sugars, organic acids, and fatty acids as well as the consumption of amino acids during fruit development. Subsequently alkanes, alcohols, aldehydes, anthocyanins, ketones, esters, and furanones increase during fruit ripening [7]. Many of these chemical classes serve as precursors to volatile synthesis [21], thus facilitating a metabolic flux through biosynthetic pathways for increased and diverse volatile emissions in ripe strawberry fruit, predominantly furans, acids, esters, lactones, and terpenes [8]. Over 350 volatile compounds have been identified across *Fragaria* [22], however within a single fruit, far fewer compounds are detectable and even less contribute to aroma perception.

A cross comparison of five previous studies which analyze strawberry volatiles depicts the lack of agreement in defining chemical constituents of strawberry aroma. Each source considers a highly variable subset of volatiles, which are determined by signal intensity and/or human perception of separated compounds [4], [5], [23]–[25]. Mutual volatiles across studies include butanoic acid, methyl ester; butanoic acid, ethyl ester; hexanoic acid, methyl ester; hexanoic acid, ethyl ester; 1,6-octadien-3-ol, 3,7-dimethyl- (linalool); butanoic acid, 2-methyl-; and 3(2*H*)-furanone, 4-methoxy-2,5-dimethyl-, the current consensus of integral strawberry aroma compounds. Comparisons of consumer preference among a variety of fresh strawberries and their volatile profiles describes less preferable varieties as possessing less esters, more decalactones and hexanoic acid [4]. The breadth of volatile phenotypes previously reported highlights the diversity across strawberry genotypes and underscores the complexity of the aggregate traits of aroma and flavor.

Florida strawberry production is concentrated on ten thousand acres near the Tampa Bay. Mild winters allow for annual horticulture which requires continual harvest of ripe fruit from late November through March. Environmental effects on fruit quality are partially attributed to gradually increasing temperatures beginning in mid-January. One result is a late season decline of soluble solids content (SSC) [26], [27]. In fact, increasing temperature is known to be responsible for increasing fruit maturation rate and decreasing SSC independent of flowering date [26]. Previous work also identifies variability of SSC, as well as titratable acidity (TA) and multiple classes of volatile compounds across harvest dates [28]. The complex fruit biochemistry, which is variably affected by genetic, environmental, and developmental factors, coupled with individuals' perceptional biases has made defining strawberry flavor cumbersome.

Here we exploit the genetic and within-season variability of fruit to provide as many unique strawberry experiences as possible to a large sample of consumers. To enhance the range and diversity of flavors and chemical constituents 35 genetic backgrounds were included: public and private cultivars representing a large proportion of commercial strawberry acreage in North America, University of Florida advanced breeding selections, and European cultivars (Fig. 1). Parallel assays of ripe strawberry samples quantify fruit traits of TA, pH, and fruit firmness, as well as the content of malic acid, citric acid, glucose, fructose, sucrose, and 81 volatile compounds of diverse chemical classes. The contributions of these attributes to fruit quality is determined by simultaneously evaluating samples for perceived sensory intensities of sourness, sweetness, and strawberry flavor, as well as the hedonic responses of texture liking and overall liking by consumer panelists. Data analyses determine significant biochemical and consumer response differences between early and late season fruit, gross variation of strawberry experiences, and factors influencing hedonics and sensory perception of strawberry fruit consumption using a psychophysics approach. Ultimately, an effect of particular volatile constituents to enhance sweetness intensity independent of sugar content of fruit was found. These findings have great implications in the breeding of more desirable fruits and vegetables, as well as for the food industry as a whole.

Figure 1. Photographs of strawberry production field, plants, and harvested fruits. Photographs characterizing the commercial style production and harvest standards employed in this study. Annual plasticulture of strawberry (A) is common practice in Florida production fields. Winterstar™ strawberry plants (B) bearing flowers and fruit of varying developmental stages and ripeness. Harvested fruit of cultivar 'Winter Dawn' (C) demonstrating ripeness used in study, 90–100% red.

METHODS

Ethics Statement

All human consumer panels are conducted at the Food Science and Human Nutrition Department at the University of Florida in Gainesville, FL. The University of Florida Institutional Review Board 2 (IRB2) chaired by Ira S. Fischler approved the protocol and written consent form (case 2003-U-0491), which participants are required to complete.

Plant Material

Thirty-five strawberry cultivars and selections were grown at or in the near vicinity of the Gulf Coast Research and Education Center (14625 County Road 672, Wimauma, FL) during the 2010–2011 (season 1) and 2011–2012 (season 2) winter seasons. Fruit are cultivated according to current commercial practices for annual strawberry plasticulture in Florida [1],[29](Fig. 1A). The cultivars are chosen to represent a large proportion of commercial strawberry acreage in North America from both public and private breeding programs. Additional breeding selections and European cultivars are added to enhance the range of diversity for flavors and chemical constituents. Weekly cultivar representation is determined by fruit availability during a particular harvest week and attempting to maximize genetic diversity, except for the highly replicated cultivar 'Festival'. Fully-ripe fruit by commercial standards, 90–100% red compared to white [30] (Fig. 1B-C), is harvested from three to five cultivars on Monday mornings, delivered to the respective laboratories, and stored at 4°C in the dark overnight for simultaneous analysis of fresh strawberry fruit volatiles, firmness, and sensory analysis on Tuesdays; as well as sample preparation for later sugar and acid measurements. Six harvests in both seasons allows for the complete analysis of 54 samples. Weather data is obtained from the Balm, FL station of the Florida Automated Weather Network (http://fawn.ifas.ufl.edu/data/reports) for date ranges January 3, 2011 through February 28, 2011 and December 26, 2011 through March 13, 2012. Daily maximum and minimum temperature recording height is 60 cm, and daily average relative humidity, rainfall, and solar radiation are recorded at 2 m.

Volatile Analysis

At least 100 g or seven berries of each sample are removed from 4°C dark overnight storage prior to volatile collection. Samples are homogenized in a blender prior to splitting into three 15 g replicates for immediate capturing of volatile emissions. The remainder is frozen in N_2 (l) and stored at –80°C for later sugar and acid quantification. A two hour collection in a dynamic headspace volatile collection system [31] allows for concentration of emitted volatiles on HaySep 80–100 porous polymer adsorbent (Hayes Separations Inc., Bandera, TX, USA). Elution from polymer is described by Schmelz [32].

Quantification of volatiles in an elution is performed on an Agilent 7890A Series gas chromatograph (GC) (carrier gas; He at 3.99 ml min^{-1}; splitless injector, temperature 220°C, injection volume 2 µl) equipped with a DB-5 column ((5%-Phenyl)-methylpolysiloxane, 30 m length ×250 µm i.d. × 1 µm film thickness; Agilent Technologies, Santa Clara, CA, USA). Oven temperature is programmed from 40°C (0.5 min hold) at 5°C min^{-1} to 250°C (4 min hold). Signals are captured with a flame ionization detector (FID) at 280°C. Peaks from FID signal are integrated manually with Chemstation B.04.01 software (Agilent Technologies, Santa Clara, CA). Volatile emissions (ng^1 gFW^{-1} h^{-1}) are calculated based on individual peak area relative to sample elution standard peak area. GC-mass spectrometry (MS) analysis of elutions are performed on an Agilent 6890N GC in tandem with an Agilent 5975 MS (Agilent Technologies, Santa Clara, CA, USA) and retention times are compared with authentic standards (Sigma Aldrich, St Louis, MO, USA) for volatile identification [33]. Chemical Abstract Services (CAS) registry numbers were used to query SciFinder® substances database for associated chemical name and molecular formula presented in Table S1.

Sugars and Acids Quantification

Titratable acidity, pH, and soluble solids content [26] are averaged from four replicates of the supernatant of centrifuged thawed homogenates [1]. An appropriate dilution of the supernatant from a separate homogenate (centrifugation of 1.5 ml at 16,000 x *g* for 20 min) is analyzed using biochemical kits (per manufacturer's instructions) for quantification of citric acid, L-malic acid, D-glucose, D-fructose, and sucrose (CAT# 10-139-076-035, CAT# 10-139-068-035, and CAT# 10-716-260-035; R-Biopharm, Darmstadt, Germany) with absorbance measured at 365 nm on an Epoch Microplate Spectrophotometer (BioTek, Winooksi, VT, USA). Metabolite average concentration (mg^1 100 gFW^{-1}) is determined from two to six technical replicates per pooled sample. Derived sucrose concentrations via D-glucose and D-fructose are mathematically pooled.

Firmness Determination

Firmness of the strawberries is determined as the resistance of the fruit to penetration at its equator with a TA.XTPlus Texture Analyzer (Texture Technologies Corp., Scarsdale, NY, USA; Stable Micro Systems, Godalming, Surrey, UK). The Texture Analyzer is equipped with a 50 kg load cell and an 8 mm diameter convex tip probe. Whole fruit is penetrated on the side to 7 mm down from the epidermis at a test speed of 2 mm^1 sec^{-1}; a flap cut off the opposite provides stability. Maximum force in kg for eight fruit is averaged and reported as a measure of firmness.

Sensory Analysis

Over the course of two annual seasons, 166 recruited strawberry consumers (58 male, 108 female) evaluate strawberry cultivars. Ages of panelist ranged from 18 to 71, with a median age of 24. Panelists self-classified themselves as 98 White or Caucasian, 11 Black or African-American, 1 Native American, Alaska Native or Aleutian, 41 Asian/Pacific Islander, and 15 Other. An average of 106 (range of 98–113) panelists evaluated between three and five cultivars per session [34]. Fresh, fully-ripe strawberry fruit is removed from overnight 4°C dark storage and allowed to warm to room temperature prior to sensory analysis. Each panelist is given one to two whole strawberries for evaluation, depending on cultivar availability. Panelists bite each sample, chew, and swallow it. Ratings for overall liking and texture liking are scaled on hedonic general labeled magnitude scale (gLMS) from –100 to +100, *i.e.* least to most pleasurable experience [34]–[37]. Perceived intensity of sweetness, sourness, and strawberry flavor are scaled in context of all sensory experiences using sensory gLMS that ranges from 0 to +100, *i.e.* none to most intense sensory stimulus[34]–[37]. Scales are employed to mediate valid comparisons across subjects and sessions.

Statistical Analysis

Means and standard errors for consumer, physical, and metabolite measurements are determined from all replicates using JMP (Version 8, SAS Institute Inc., Cary, NC, USA). One-way analysis of early and late season fruit quality and consumer response measures was subjected to mean comparison using Tukey's HSD (α=0.05). Bivariate analysis among individual measurements of samples allows for linear fit, which includes summary of fit, analysis of variance, t-test, and correlation analysis for density ellipse. Two-way Ward hierarchical cluster analysis of all quantified metabolite and strawberry samples is accomplished in JMP. Amounts of individual volatile compounds are regressed using the "enter" method in SPSS (IBM Corp., Armonk, NY, USA). This is done individually for each of the three sugars: glucose, fructose or sucrose to identify which compounds have an effect on sweetness intensity [14] independent of each of the sugars. For p-values ≤ 0.05, the volatile makes a contribution to perceived sweetness that is independent of the sugar tested.

RESULTS

The inventory of 54 fully ripe (Fig. 1C) unique strawberry samples (35 cultivars, 12 harvests, two seasons) assayed for TA, pH, firmness, as well as the concentrations of malic acid, citric acid, glucose, fructose, sucrose, and quantity of 81 volatile compounds is reported. Cluster analysis of relative chemical composition of all samples and derived hierarchy of both cultivar and metabolite relatedness is displayed (Fig. 2). The vertical dendrogram (Fig. 2) demonstrates the lack of relatedness among volatile compound quantities through large distances of initial segments, as well as the high number of clusters. Slightly more structure is observed among the samples, horizontal dendrogram (Fig. 2), due to genetic or environmental effects.

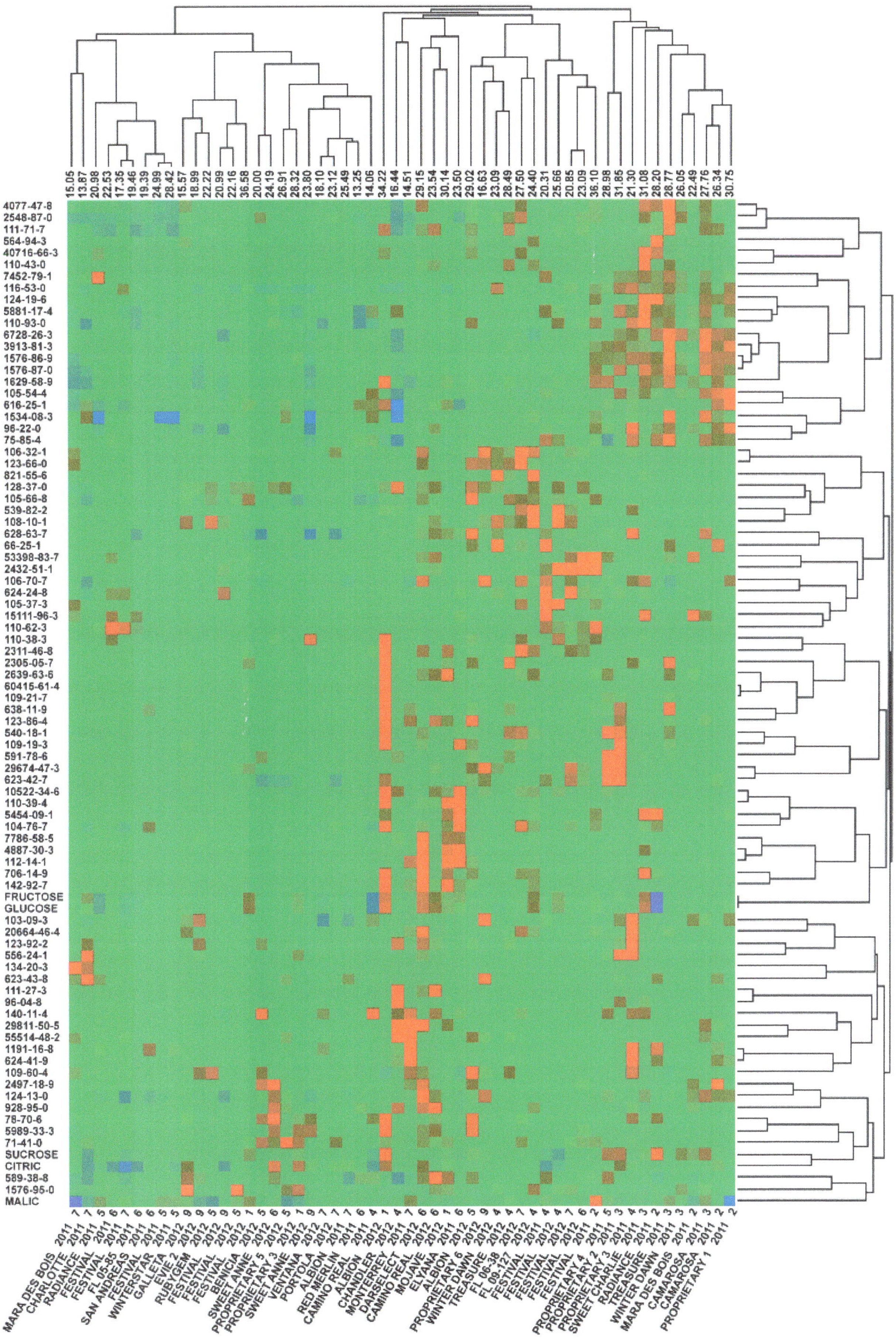

Figure 2. Cluster analysis of strawberry samples and quantified metabolites. Two-way Ward cluster analysis of strawberry samples (bottom) and quantified single metabolites (right) with overall liking score of sample (top) constructed using JMP 8. Standardization of metabolite content is by row mean and standard deviation, with high values represented as red, average as green, and low as blue. The hierarchy and distance of segments within the vertical dendrogram indicates relatedness of content across samples for single metabolites. Structure of the horizontal dendrogram indicates relatedness of all metabolite contents among individual samples.

Progression of Harvest Season Affects Perceived Quality and Metabolite Content of Strawberry

Overall liking is a measure of pleasure derived from consuming a strawberry sample. The two samples with the greatest overall liking ratings are of cultivar 'Festival'. Fruit harvested early in the season, week 2 of season 1 and week 1 of season 2, elicit overall likings of 36.1 and 36.6, respectively (Table 1). Five weeks following both early samplings of 'Festival' the overall liking of the same cultivar decreases below the sample set median of 23.5 to 17.3 in season 1 week 7 and to 23.1 in season 2 week 6 (Table 1). Therefore the earlier season samples elicit a greater hedonic response than late season samples. Overall likings are determined using the hedonic general labeled magnitude scale that ranges from –100 to +100, *i.e.* least to most pleasurable experience [34]–[37]. Conversely, sweetness, sourness, and strawberry flavor are measured using the sensory intensity general labeled magnitude scale that ranges from 0 to +100, *i.e.* none to most intense sensory stimulus[34]–[37]. Consumer perception of sweetness and strawberry flavor intensity decrease significantly between the same pairs of early and late season 'Festival' fruit (Table 1). Significant biochemical differences between early and late samples include decreased content of glucose, fructose, sucrose, and total volatiles. The early 'Festival' from the first season contains 88% more total sugar and 65% more total volatiles than the late 'Festival' of the same season (Table 1), demonstrating the disparity between early and late harvest week fruit quality and its effect on consumer sensory perception and acceptability.

Table 1. Comparison of early and late season strawberry fruit.

			Season 1				Season 2			
			Week 2		Week 7		Week 1		Week 6	
Mean week temperature										
	Daily maximum	°C	21.6	B	28.2	A	21.3	B	26.1	A
	Daily minimum	°C	7.4	B	13.3	A	6.7	B	13.1	A
	Daily average	°C	14.9	B	20.3	A	14.0	B	19.0	A
Consumer ratings										
	Overall liking	–100 to +100	36.1	A	17.3	B	36.6	A	23.1	B
	Texture liking	–100 to +100	35.7	A	23.8	B	34.8	A	24.3	B
	Sweetness intensity	0 to +100	30.3	A	15.9	B	34.0	A	22.2	B
	Sourness intensity	0 to +100	17.9	A	15.9	A	18.2	A	17.9	A
	Strawberry flavor intensity	0 to +100	34.3	A	20.4	B	37.5	A	25.2	B
Biochemical measures										
	Glucose	$(mg^1\ 100\ gFW^{-1})$	1903	A	1127	B	2187	A	1807	B
	Fructose	$(mg^1\ 100\ gFW^{-1})$	2048	A	1311	B	2327	A	1973	B
	Sucrose	$(mg^1\ 100\ gFW^{-1})$	1218	A	309	B	1902	A	450	B
	Total sugar	$(mg^1\ 100\ gFW^{-1})$	5169	-	2747	-	6417	-	4229	-
	Relative sucrose	-	0.37	B	0.41	A	0.34	B	0.43	A
	Relative fructose	-	0.40	B	0.48	A	0.36	B	0.47	A
	Relative sucrose	-	0.24	A	0.11	B	0.30	A	0.11	B
	Total volatiles	$(ng^1\ gFW^{-1}\ h^{-1})$	19097	A	11543	B	16843	A	16001	A

Comparison of means for temperature (mean of 7 days prior to harvest), consumer ratings, and biochemical measures between early and late season strawberry fruit cultivar 'Festival' from season 1 and season 2. Mean comparison accomplished in JMP 8 using Tukey's HSD. Mean marked A is significantly greater than mean marked B ($\alpha = 0.05$).

doi:10.1371/journal.pone.0088446.t001

Solar radiation, minimum temperature and maximum temperature increase gradually within the limits of similar ranges in season 1 and season 2 (Fig. S1 A-D). Relative humidity remains constant during and across seasons (Fig. S1 E, F). Slightly more rain fell in early season 1 than season 2 (Fig. S1 G-H) One manifestation of these environmental changes over a harvest season is the negative relationship between total sugar and harvest week (Table 1). The content of all individual sugars measured decreases between early and late season 'Festival' samples; however there is a significant decrease in the proportion of sucrose to total sugar (Table 1). The disproportionate decrease is observed for the collection of samples as well (Fig. S2A-C) (Table S3). Also, a significant correlation is observed across all 54 samples among total volatiles and sucrose (R^2=0.305*) (Fig. S2E) but not glucose (R^2=0.005) (data not shown) or fructose (R^2=0.001) (Fig. S2F). A harvest week associated decrease in total sugars, predominantly sucrose, results in a decrease in volatile content, which ultimately undermines late season overall liking (R^2=0.422*) (Fig. 3E) through sweetness and strawberry flavor intensity.

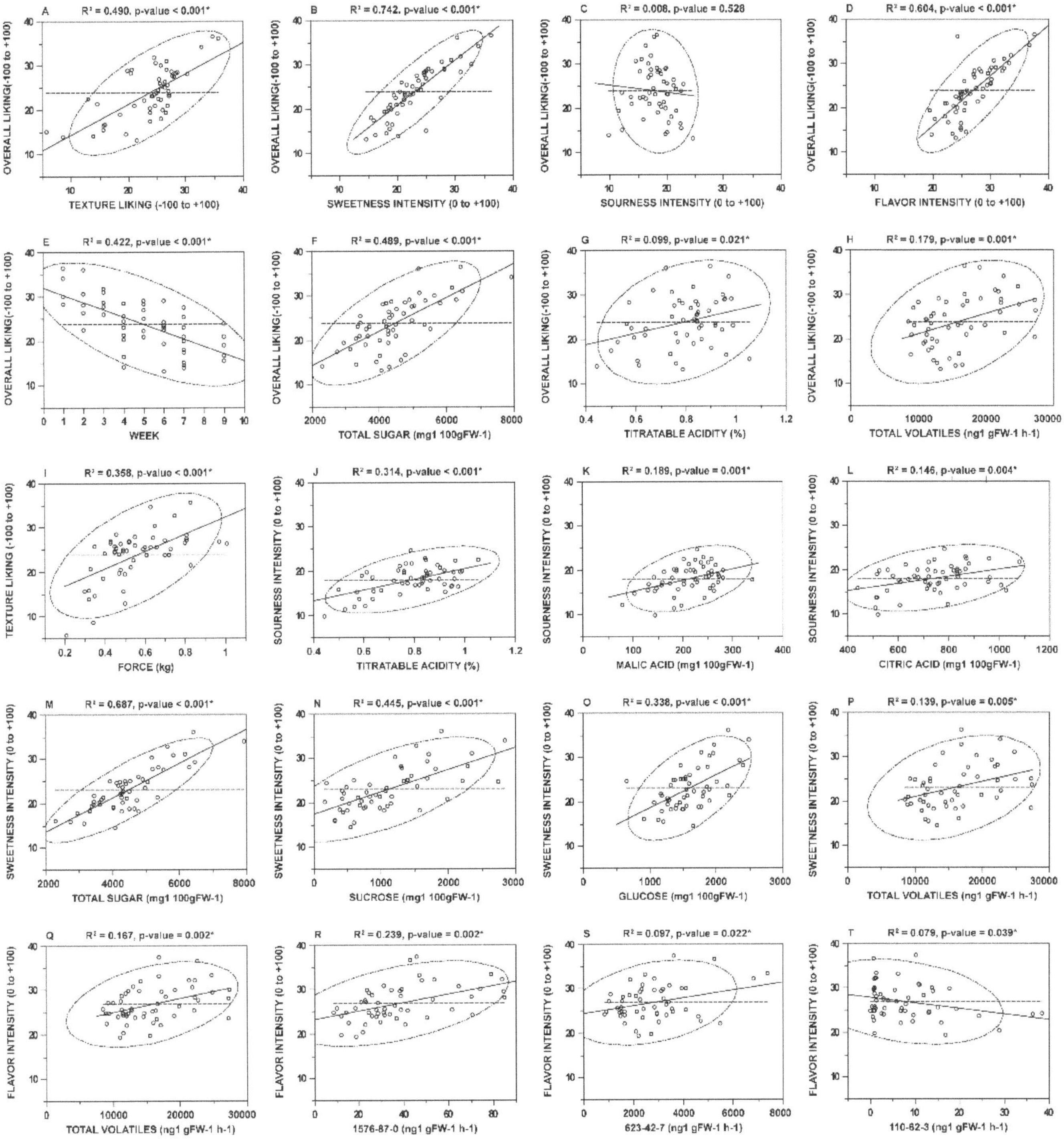

Figure 3. Regression of hedonic and sensory measures to physical and chemical fruit attributes. Hedonic overall liking is regressed against hedonic texture liking (A), sweetness intensity (B), sourness intensity (C), and strawberry flavor intensity (D). Overall liking is fitted to harvest week (E), total sugars (F), titratable acidity (G), and total volatiles (H). Texture liking is examined against puncture force (I). Intensity of sourness is fitted to titratable acidity (J), malic acid (K), and citric acid (L). Sweetness intensity is regressed against total sugars (M), sucrose (N), glucose (O), and total volatiles (P). Strawberry flavor intensity is regressed by total volatiles (Q) and select single volatile compounds 1576-87-0 (R), 623-42-7 (S), and 110-62-3 (T). Coefficient of determination (R^2) and p-value of fit is listed above individual scatterplots and is calculated using bivariate fit in JMP 8. Dashed line represents mean of independent variable, solid line represents linear fit, dashed/dotted ellipse indicates 95% confidence range of data, and asterisk denotes significant fit (α=0.05).

Overall Liking is Subject to Ratings of Sweetness, Flavor, and Texture but not Sourness

In order to elucidate factors contributing to a positive strawberry experience, overall liking of strawberry samples is fit against the hedonic measure of texture liking and the sensory intensities of sweetness, sourness, and strawberry flavor intensity (Fig. 3A-D). High correlation with significant fit exists for texture liking (R^2=0.490*) (Fig. 3A), sweetness intensity (R^2=0.742*) (Fig. 3B), and strawberry flavor intensity (R^2=0.604*) (Fig. 3D). However, sourness intensity shows no correlation to overall liking (R^2=0.008) (Fig. 3C). Increasing firmness contributes to greater texture liking (R^2=0.358*) (Fig. 3I), and texture liking has a significant influence on overall liking. Sweetness intensity is the strongest driver of overall liking measured in this study. The correlation between total sugar and overall liking (R^2=0.488*) (Fig. 3F) demonstrates the aggregate sugar metabolites effect on hedonic response to strawberry fruit. Total sugar concentration accounts for nearly half of the observed overall liking variation but is far from a complete measure. Sourness intensity appears to have no influence on the hedonic response to strawberry fruit, but fit of TA to overall liking is significant, even if minor (R^2=0.099*) (Fig. 3G). Total volatiles is the second aggregate metabolite measure having a significant enhancing effect on the overall liking of strawberry (R^2=0.179*) (Fig. 3H). This is not surprising, as strawberry flavor intensity exhibits the second highest correlation to overall liking (Fig. 3D).

Texture Liking Correlates to Fruit Firmness

The upper limit for hedonics of texture is comparable to that of overall liking and is observed in 'Festival' (sn 1, wk 2) with an average of 35.7, however, the low texture liking value of 5.8 for 'Mara Des Bois' (sn 1, wk 7) indicates a more drastic disliking of "off" textures than the overall liking of even the lowest rating fruit. Firmness of samples is assayed by measuring the force required for a set penetration of the fruit, acting as a proxy for texture. The firmness of the fresh strawberry exhibited nearly a five-fold difference in force, 0.2 kg for 'Mara des Bois' (sn 1, wk 7) and 1.0 kg for 'Festival' (sn 1, wk 5). Increasing force of penetration, i.e. increasing firmness of berries, is positively correlated with texture liking, indicating a hedonic response to firmer fruit (Fig. 3I). However, the texture liking is less than the expected rating for the two samples with greatest firmness (Fig. 3I).

Sweetness Intensity is a Result of Sugar Content

Perceived sweetness intensity is the greatest predictor of overall liking. In fact, the same samples scoring the highest and lowest for overall liking, 'Festival' (sn 2, wk 1) and 'Red Merlin' (sn 1, wk 6), elicit the greatest (36.2) and least (14.59) intense sensations of sweetness. The early and late harvest week samples support the observable decline in perceived sweetness intensity across harvest weeks, which is also observable for multiple sugar measures (Fig. S2A-C) (Table 1).

In the 54 samples assayed, the total sugar concentration ranged from 2.29 – 7.93%, a 3.5-fold difference. Glucose and fructose concentrations exhibit highly similar ranges to each other, 0.66 – 2.48% and 0.75 – 2.61%, respectively, and near-perfect correlation (R^2=0.984*) (data not shown) within a sample. However, the concentration of glucose or fructose is not predictive of sucrose concentration (R^2=0.011 and 0.004, respectively) (data not shown). Sucrose demonstrated a more dynamic state as its concentration dips as low as 0.16% and up to 2.84%, nearly a seventeen-fold difference among all samples.

Sucrose is the single metabolite with the most significant contribution to overall liking (R^2=0.442*) (Table S4). Individually, sucrose (R^2=0.445*) (Fig. 3N), glucose (R^2=0.337*) (Fig. 3O), and fructose (R^2=0.300*) (Table S4) all significantly influence the variation in sweetness intensity. However, total sugar actually only accounts for slightly more than two-thirds of sweetness intensity variation (R^2=0.687*) (Fig. 3M) likely a result of covariation of glucose and fructose. Interestingly, the total volatile content of a sample correlates positively with sweetness intensity, potentially accounting for up to 13.9%* of variation in sweetness intensity (Fig. 3P).

Sourness Intensity is Partially Explained by Titratable Acidity

Cultivar 'Red Merlin' (sn 1, wk 6) elicited the most intense sourness response at 24.6. This same sample rates as the lowest in terms of overall liking and sweetness. Acidity of strawberry fruit is assayed using measures of pH, TA, citric acid and malic acid. The pH of strawberry samples ranges from 3.35 to 4.12, while TA ranges from 0.44% to 1.05%. The range of malic acid across samples is 0.078% to 0.338% while citric acid ranged from 0.441% to 1.080%. TA has the greatest correlation to sourness intensity (R^2=0.314*) (Fig. 3J), when compared to pH (R^2=0.118*), malic acid (R^2=0.189*) (Fig. 3K), or citric acid (R^2=0.146*) (Fig. 3L) concentration. Citric acid concentration is approximately three-fold greater than malic acid and has a significant effect on TA (R^2=0.49*) (data not shown). There is no correlation of malic acid to TA (R^2=0.01) (data not shown). The lack of relationship among sourness intensity and overall liking (Fig. 3C) is shadowed by the strong correlations of sweetness intensity (Fig. 3B) and flavor intensity (Fig. 3D) to overall liking. Deficiencies in perceived sweetness and flavor intensity as observed in 'Red Merlin' can result in a fruit that is negatively perceived as intensely sour.

Flavor Intensity Is Influenced by Total and Specific Volatile Content

In this study, strawberry flavor intensity accounts for the retronasal olfaction component of chemical senses, which compliments sourness and sweetness intensities' contribution to taste. The overall highest sensory intensity is 37.5 for strawberry flavor of 'Festival' (sn 2, wk 1), which also rates highest for overall liking and sweetness intensity. Opposite this, FL- 05-85 (sn 1, wk 6) delivers the least intense strawberry flavor experience with a score of 19.4Total volatiles in 'Festival' (sn 2, wk 1) is over 50% greater than in FL 05-85 and seven more volatiles compounds are detected. Total volatiles within a sample contribute to strawberry flavor intensity (R^2=0.167*) (Fig. 3Q), but it is not simply the sum of volatile constituents that explain the effect. For instance, the maximum total volatile content detected within a sample, 27.3 μg^1 gFW^{-1} hr^{-1} from 'Camarosa' (sn 1, wk 2), does not result in the greatest flavor intensity (30.5) and the minimum, 8.5 μg^1 gFW^{-1} hr^{-1} from 'Sweet Anne' (sn 2, wk 9), does not rate as the least flavorful (25.8).

The chemical diversity of the resources analyzed allows for the identification of 81 volatile compounds from fresh strawberry fruit (Fig. S3). The majority of compounds are lipid related esters, while lipid related aldehydes account for the majority of volatile mass. Terpenes, furans, and ketones are also represented in the headspace of strawberry. Forty-three of the 81 volatile compounds are not detected (<0.06 ng^1 gFW^{-1} hr^{-1}) in at least one sample. Therefore, 38 volatiles are measured in all samples; appearing to be constant in the genetic resources analyzed. No cultivar has detectable amounts of all 81 volatiles. Samples of 'Festival', 'Camino Real', PROPRIETARY 6, and FL 06-38 are the most volatile diverse, but are lacking detectable amounts benzoic acid, 2-amino-, methyl ester (134-20-3) [5]. This methyl ester of anthranilic acid is detectable in only 'Mara des Bois' and 'Charlotte' from the final harvest (wk 7) of season 1. 'Chandler' (sn 2, wk 4) and 'Red Merlin' (sn 1, wk 6) are the least volatile diverse samples lacking detectable amounts of 19 and 17 compounds, respectively.

The most abundant ester, butanoic acid, methyl ester (623-42-7)is measured at over 7 µg[1]gFW^{-1} hr^{-1} from PROPRIETARY 2 (sn 1, wk 3) and has a significant correlation to flavor (R^2=0.097*) (Fig. 3S). A terpene alcohol, 1,6,10-Dodecatrien-3-ol, 3,7,11-trimethyl-, (6E)- (40716-66-3) (nerolidol), with maximum content of over 600 ng[1] gFW^{-1} hr^{-1} in 'Sweet Charlie' is not detected in 'Red Merlin'. The nerolidol rich 'Sweet Charlie' garners greater flavor intensity at 32.2 than deficient 'Red Merlin' at 23.95. The impact on flavor intensity by nerolidol (R^2=0.112*) (Table S4) is greater than butanoic acid, methyl ester despite having maximum contents lower by one order of magnitude. Hexanal (66-25-1) is the second most abundant individual compound, an aldehyde detected in all samples, exceeds 11 µg[1] gFW^{-1} hr^{-1}, and does not have a significant correlation to flavor intensity (R^2=0.016) (Table S4). Hexanoic acid, ethyl ester (123-66-0) exhibits over 200-fold difference across samples, and also has no bearing on sensory perception (Table S4). Conversely, two minor level aldehydes demonstrate a disparity in effect: 2-pentenal, (2E)- (1576-87-0) is enhancing toward flavor intensity (R^2=0.239*) (Fig. 3R), while pentanal (110-62-3) is the only compound that negatively correlates to flavor (R^2=0.079*) (Fig. 3T). The significant contribution of the 1,6-octadien-3-ol, 3,7-dimethyl- (78-70-6) (linalool) to flavor intensity positively correlates with increasing content (R2=0.074*) (Table S4). In 'Chandler' 3(2H)-furanone, 4-methoxy-2,5-dimethyl- (4077-47-8) is not detectable, and only has maximum content of 40 ng[1] gFW^{-1} hr^{-1} in 'Treasure' (sn 1 wk 3). The level of this characteristic strawberry furan is significantly impactful on perceived flavor intensity (R^2=0.108*) (Table S4). In total, thirty volatile compounds diverse in structure have a positive relationship to flavor intensity and their significance cannot be derived from content alone.

Specific Volatiles Enhance Sweetness Intensity Independent of Sugars

Multiple regression of individual volatile compounds against perceived intensity of sweetness is performed independent of glucose, fructose, or sucrose concentration (Table S5). Twenty four volatile compounds show significant correlations (α=0.05) to perceived sweetness intensity independent of glucose or fructose concentration, twenty-two of which are mutual between the two monosaccharides. Twenty volatiles are found to enhance sweetness intensity independent of sucrose concentration; only six of these volatiles are shared with those independent of glucose and fructose: 1-penten-3-one (1629-58-9); 2(3*H*)-furanone, dihydro-5-octyl- (2305-05-7) (γ-dodecalactone); butanoic acid, pentyl ester (540-18-1); butanoic acid, hexyl ester (2639-63-6); acetic acid, hexyl ester (142-92-7); and butanoic acid, 1-methylbutyl ester. Only three compounds are found to be negatively related to sweetness independent of at least one of the sugars: octanoic acid, ethyl ester (106-32-1) exclusively independent of glucose; 2-pentanone, 4-methyl- (108-10-1) mutually independent of glucose and fructose; and 2-buten-1-ol, 3-methyl-, 1-acetate (1191-16-8) exclusively independent of sucrose.

DISCUSSION

Exploitation of genetic diversity and environmental variation allows for a wide range of consumer hedonic and sensory responses. The cultivars in this study represent a large proportion of commercial strawberry acreage in North America, advanced breeding selections, and European cultivars. A genetic collection aimed at enhancing the diversity of physical and chemical constituents, as well as consumer experiences. Despite the perennial life cycle of strawberry much commercial production uses annual methods, which in sub-tropical Florida allows for continual harvest of ripe fruit from late November through March. A nearly three-fold difference in overall liking of strawberry is observable within all samples. The highest and lowest rating samples are 'Festival' of the first week in the second season and 'Red Merlin' of the sixth week in the first season. These two cultivars are the product of separate breeding programs, have distinct genetic backgrounds, and therefore distinct biochemical inventories. Harvested at opposite ends of the seasons the early and late season fruit are subjected to different environmental conditions, further attenuating genetic differences. The diversity of strawberries samples assayed and range of consumer liking captured (Fig. 2) indicates the chemical diversity of strawberry cultivars is not only perceivably different but certain profiles are more highly preferable.

Elevated texture liking, sweetness intensity, and strawberry flavor intensity significantly increases overall liking, while sourness intensity alone has no detectable relationship to overall liking (Fig. 3A-D). Integration and synthesis of response to sensory signals of taste, olfaction, and tactile sensation constitute an eating experience [9] and drive overall liking. The senses of taste and olfaction sample the chemicals present in food *e.g.* sugars, acids, and volatile chemical compounds. These elicitors attenuate the perception and hedonics of food [38], [39]. Ratings of strawberry fruit are correlated to specific chemical or physical attributes, especially sweetness (Fig. 3B) and flavor intensity (Fig. 3D), the two greatest drivers of overall liking.

Much work has been done to measure sugars and volatile compounds in strawberry fruit in an attempt to understand sweetness and flavor, and these aims are in line with consumer demand. A consumer survey using 36 attributes of strawberry determined "sweetness" and "complex flavor" as consistent favorable characteristics of the ideal strawberry experience[14]. Previous work in tomato [34] and this current study on strawberry surveyed participants for ideal ratings of the respective fruits. Using the same gLMS scales employed in the current study, scores for ideal strawberry and tomato overall liking, sourness intensity, and flavor intensity are highly similar. Ideal flavor evoked the highest mean response of 45 for both, exemplifying its importance to the consumer. Interestingly, a large disparity for ideal sweetness intensity is found; 42 and 33 for strawberry and tomato, respectively. Ideal sweetness intensity is much greater in strawberry, potentially due to differences in consumption. Strawberry is often consumed fresh and is a delicacy or dessert fruit, while tomato is savory and often an ingredient in complex recipes. Therefore, the desire for sweetness may be greater in strawberry. The overall liking of strawberry fruit is significantly related to texture liking (Fig. 3A), and increasing fruit firmness accounts for more than a third of increasing texture liking (Fig. 3I). The five-fold variation in firmness can be attributed to variation in fruit development or softening. Strawberry fruit development consists of division, expansion, and ripening [7]. Developmentally regulated, ripening associated fruit softening is multifaceted[19], including catalytic cell wall disassembly [20] and dissolution of cell-to-cell adhesion [3]. The relationship between texture liking and firmness does not appear entirely linear, because the two firmest samples are close to average texture liking (Fig. 3I). Excessively firm fruits may be perceived as under ripe while those with less firmness may be considered over ripe; affecting texture liking. Fruit can progress through ripening, from under to over ripe, in ten days [7], exemplifying the narrow window in which multiple facets of fruit quality must synchronize.

Despite a moderate range of intensity, perceived sourness has little to no bearing on overall liking (Fig. 3C). Just over 30% of sourness intensity variation can be accounted for by positive correlation with TA. The concentrations of citric acid and malic acid metabolites are likely additive toward the effect of TA on sourness intensity, and in fact both organic acids have significant correlations to sourness intensity (Fig. 3K-L). Despite a lack of influence by sourness intensity on overall liking, metabolites of sourness have a critical role in fruit biochemistry. Increased TA shows a significant minor correlation with overall liking (Table S4) and correlates significantly with total sugar (data not shown). This co-linearity may be due to accumulation of sugars and subsequent biosynthesis of organic acids during ripening of fruit [7], [8], [16]. Citric acid is the predominant organic acid in ripe fruit [40] and its concentration is fairly stable during ripening. Also, it is known to act as an intermediate between imported sucrose and fatty acid biosynthesis [16], which may facilitate enhancement of overall liking through volatile biosynthesis.

The consumer rating of sweetness intensity is the primary factor contributing to overall liking, and sweetness is the component of taste perception facilitating the detection of sugars. Sugars are simple carbohydrates, a readily available form of energy, and the degree of correlation among sweetness and overall liking is due to hedonic effect [38]. Variation in sweetness intensity is best explained by sugar content (Fig. 3L). Previously, soluble solid content (SSC) has been used as a valid indicator of sweetness in strawberry[1], [28]. However this is an aggregate measure, as previous quantification of individual sugars within a strawberry identifies sucrose, glucose, and fructose as the predominant soluble solids [1], [8], [15], [40]. Sucrose concentrations observed across samples is responsible for more variation in total sugar, sweetness intensity and overall liking than any other individual compound (Table S4). Metabolites contributing to perceived sweetness intensity have the greatest influence on the overall hedonics of strawberry. A significant decrease in sweetness intensity occurs between early and late season fruit, and unfortunately overall liking decreases as well (Table 1) (Fig. 3E).

Drastic fruit quality differences between early and late season fruit result in lower consumer response (Table 1) (Fig. 3E), which is likely due to environmental changes (Fig. S1) or plant maturity. A significant difference in the mean temperature one week prior to harvest is likely a causative factor (Table 1). Monitored development of 'Festival' fruit under elevated temperature decreases the fruit development period from 36 days at 15°C to 24 days at 22°C. Also, a simultaneous decrease in SSC is observed, both independent of flowering date i.e. plant maturity [26], [27]. The mean temperature of the week prior to harvest for early and late season 'Festival' fruit are 15°C and 20°C for the first season and 14°C and 19°C for the second season (Table 1). These differences in environment likely alter whole plant physiology and more specifically fruit biochemistry during development and ripening, affecting fruit quality. During strawberry fruit development sucrose is continually translocated from photosynthetic tissue, while a consistently high sucrose invertase activity in fruit hydrolyzes sucrose into glucose and fructose, maintaining sink strength of fruit [15] and in turn feed biosynthetic pathways [16]. Total and individual sugars decrease in ripe fruit during both seasons as the plant is subjected to increasing temperatures (Table 1). Increased maturation rate hastens fruit development, potentially decreasing cumulative period sucrose is imported to fruit, and inhibiting sucrose accumulation to affect other fruit quality attributes. These factors are likely causative of the observable decrease in sweetness and flavor intensity as the season progresses.

Although total sugar decreases between early and late fruit, a disproportionate amount of the decrease is attributed to sucrose (Table 1), which indicates sucrose as the waning constituent of sugar content (Fig. S2A-C). Glucose and fructose concentrations are tightly correlated to each other, show less seasonal influence than sucrose, and lack correlation to sucrose. These observations are indicative of tighter biochemical regulation of glucose and fructose than sucrose, which has the greatest variability in concentration among the three sugars. Total volatile content has an indirect dependence on sucrose concentration (Fig. S2E), and a decrease in total volatiles is observed between early and late season strawberry (Table 1). Influence of harvest date on headspace of fresh strawberry fruit is known [41], [42]. Increased volatile content is likely dependent on more free sucrose, *i.e.* a larger imported reserve, facilitating greater flux through primary and secondary metabolism. Generation of glucose and fructose initiates a complex network of primary and secondary metabolism specific to ripening strawberry fruit, in which sucrose is principal and limiting to the strawberry fruit biosynthetic pathways [16]. Upregulation of biosynthetic genes associated with volatile secondary metabolites [43] and the consumption of primary metabolite classes of fatty acids and amino acids, precursors of volatile compounds, happens in the final stages of ripening [16]. This sucrose dependent metabolic shift culminates in peak volatile content and diversity [8].

Strawberry flavor intensity is the second greatest determinant of overall liking (Fig. 3D) and accounts for perception of volatile compounds through retronasal olfaction. A significant positive relationship exists among total volatile content and the flavor intensity for a given sample, however, total volatile content is not entirely explanatory of flavor intensity. The maximum rating for strawberry flavor intensity by 'Festival' (sn 2, wk1) is the greatest consumer response evoked within this study, highlighting the significance of sensory perception of aroma. However, this sample only has slightly more than 60% of total volatile mass of the greatest sample. The extent of volatile phenotype diversity is great enough across strawberry fruit to not only be discerned but be preferred.

Within the genetic resources of *Fragaria x ananassa* analyzed in this study 81 compounds are reproducibly detected, but not one cultivar has detectable amounts of all compounds. The amount of individual volatile compounds within fruit can have a significant influence on flavor intensity, but which volatiles are determinant of flavor has a lack of consensus. Previous determination of flavor relevance relied on approaches in which importance of volatiles is at least initially based on abundance. Determination of flavor descriptors or thresholds of isolated compounds were determined using human perception via orthonasal olfaction [4], [5], [23]–[25], negating the complex system of strawberry fruit or actual flavor relevant retronasal olfaction.

Of the forty-six volatile compounds cited as relevant to strawberry flavor in five studies [4],[5], [23]–[25] only seven are mutual to at least three of the studies, exemplifying the lack of agreement in defining flavor-relevant constituents. This consensus includes butanoic acid, methyl ester; butanoic acid, ethyl ester; hexanoic acid, methyl ester (106-70-7); hexanoic acid, ethyl ester; linalool; butanoic acid, 2-methyl- (116-53-0); and 3(2*H*)-furanone, 4-methoxy-2,5-dimethyl-, all of which are quantified in this report. These compounds exhibit adequate variability in fruit samples to discern dose dependent effect on flavor intensity. However, only linalool; butanoic acid, ethyl ester; butanoic acid, methyl ester; and 3(2*H*)-furanone, 4-methoxy-2,5-dimethyl- show significant positive correlation with flavor intensity (Table S4). These compounds that are found to influence flavor intensity represent diverse classes, terpene alcohol, two esters, and a furan, respectively, while the three compounds not fitting to flavor are all esters. With esters accounting for the majority of chemical compounds detected in strawberry it is possible that too much emphasis is placed on the chemical class for flavor, or that in a complex mixture less are perceivable than when smelled individually.

Over one third of volatiles in this study significantly correlate with strawberry flavor intensity, potentially enhancing perception of a complex and highly variable volatile mixture (Table S4), seventeen of which are not of previous strawberry flavor focus. Two of these unrecognized compounds, 1-hexanol (111-71-7) and butanoic acid, 3-methyl-, butyl ester (109-19-3), are present in the most flavorful strawberry sample but undetected in the least flavorful. This pair of compounds as well as pentanoic acid, ethyl ester (539-82-2) and butanoic acid, 3-methyl-, octyl ester (7786-58-5), also present/absent in the most/least flavorful, have relatively minor amounts but show evidence of enhancing perceived sweetness intensity independent of individual sugars. Relatively low abundance volatiles are indicated as new impactful components of strawberry flavor.

Thirty-eight volatile compounds are observed to significantly enhance the perceived intensity of sweetness; twenty-two mutually independent of glucose and fructose, fourteen uniquely independent of sucrose, and six compounds mutually independent of all three sugar: 1-penten-3-one; 2(3*H*)-furanone, dihydro-5-octyl- (γ-dodecalactone); butanoic acid, pentyl ester; butanoic acid, hexyl ester; acetic acid, hexyl ester; and butanoic acid, 1-methylbutyl ester (Table S5). In tomato, similar analysis of a volatile subset identifies three compounds enhancing sweetness intensity independent of fructose: geranial; 1-butanol, 3-methyl- (123-51-3); and butanal, 2-methyl- (96-17-3) [34]. These compounds are not identified in the current study; therefore the effect cannot be confirmed in a second system. Botanically, tomato is considered a true fruit and demonstrates climacteric ripening, while strawberry fruit is non-climacteric and considered an aggregate accessory fruit. The developmental origin of the flesh which is consumed is divergent, exhibiting unique biochemistries, but the observance of volatile compounds potentially enhancing perceived sweetness appears to be widespread in fruit.

Orthonasal olfaction is the result of smelling *i.e.* bringing odor in through the nose, while retronasal olfaction is elicited by odorants traveling from oral cavity or esophagus up to nasal cavity [44]. Orthonasal olfaction introduces volatile compounds to the nasal epithelium via inhalation, while retronasal olfaction is achieved during exhalation [45]. Specifically, the path of odorants distinguishes the manner of interaction between consumer and potential food, with orthonasal contributing to aroma and retronasal to flavor. Integration of sensory stimuli relies on projection of signals to various structures of the brain. Interestingly, portions of orthonasal (smell) and retronasal (flavor) olfaction project to different brain areas for processing [46], while taste activation partly overlaps that of retronasal olfaction for integration to produce flavor [47]. Co-activation of taste and retronasal olfaction, but not orthonasal, is shown to elicit responses at otherwise independently sub-threshold levels, exemplifying the ability of multiple sensory integration to intensify one another [48]. Mechanical blockage of retronasal olfaction during tasting of solutions significantly reduces the ability to correctly identify solute, including sucrose [45]. Combination of taste and retronasal olfaction results in a sensory system more adapt at analyzing the chemical content of food, but cross communication also facilitates manipulation of the system.

The food industry knows of the intensification of volatile sensations by the addition of small amounts of sweeteners to solutions containing volatiles [49]. The ability of volatiles to enhance taste is also a known phenomenon [38]. Enhancement of perceived sweetness is demonstrated by addition of volatiles amyl acetate (banana) [50] and citral [51]. Multiple studies show the ability of strawberry aroma to intensify the sweetness of a sugar solution[52], [53], as well as pineapple, raspberry, passion fruit, lychee, and peach [53], [54]. Also, sweetness enhancement has been achieved with vanilla [55], caramel [53], [56], and chocolate [45] indicating this phenomenon is not only associated with fruit volatiles. Studies to determine perceptional differences when tomato is spiked with sugars, acids, and volatiles indicates cross talk between taste and olfaction, in which volatiles impact perception of sweetness and vice versa [57]. Individual volatile compounds have been implicated in tomato to intensify perceived sweetness independent of sugar content [34],[58]. The results here narrow the previous effect of enhanced sweetness by strawberry aroma, a variable mixture, to individual compounds in the fruit. These volatiles are not present at the highest amounts in fruits and most have not been targets of flavor analysis. Also, most appear to be associated with lipid metabolism, like many other volatiles quantified in this work, yet their presence or increased content has an enhancing effect on perceived sweetness independent of sugars. Technically, sweetness is a facet of taste [38]. Therefore a means to convey sweetness via aroma can serve as an attractant to seed dispersers of wild strawberry, or perhaps it is a result of artificial selection [59] to enhance a limited sugar capacity in commercial fruit.

CONCLUSIONS

Strawberry fruit ripening culminates as the flesh softens, volatile emission peaks, and sugars accumulate. This highly coordinated process results in fruit with strong liking due primarily to texture, flavor, and sweetness. However, cultivar, environmental conditions, and their interactions influence fruit attributes, altering the composition of strawberry. This diversity allows for a spectrum of experiences such that the hedonics and intensities of these sensations can vary greatly. The importance of sucrose to sweetness intensity is evident, and the correlation of total volatiles to sucrose highlights the dependence of secondary metabolism to primary metabolism. Individual volatiles correlate to strawberry flavor intensity, helping to better define distinct, perceptually impactful compounds from the larger mixture of the fruit. The dependence of liking on sweetness and strawberry flavor is undermined by environmental pressures that reduce sucrose and total volatile content. A cultivar that exhibits minimal seasonal environmental influence presents itself as a breeding ideotype, as maintenance of sucrose concentration may alleviate loss of overall liking. Selection for increased amounts of volatile compounds that act independently of sugars to enhance sweetness can serve as an alternate approach. The volatiles described herein are sampled mainly from current commercial cultivars and are therefore feasible targets for varietal improvement in the short-term, whereas future studies will be necessary to identify sweet-enhancing volatiles not already present in elite germplasm.

SUPPORTING INFORMATION

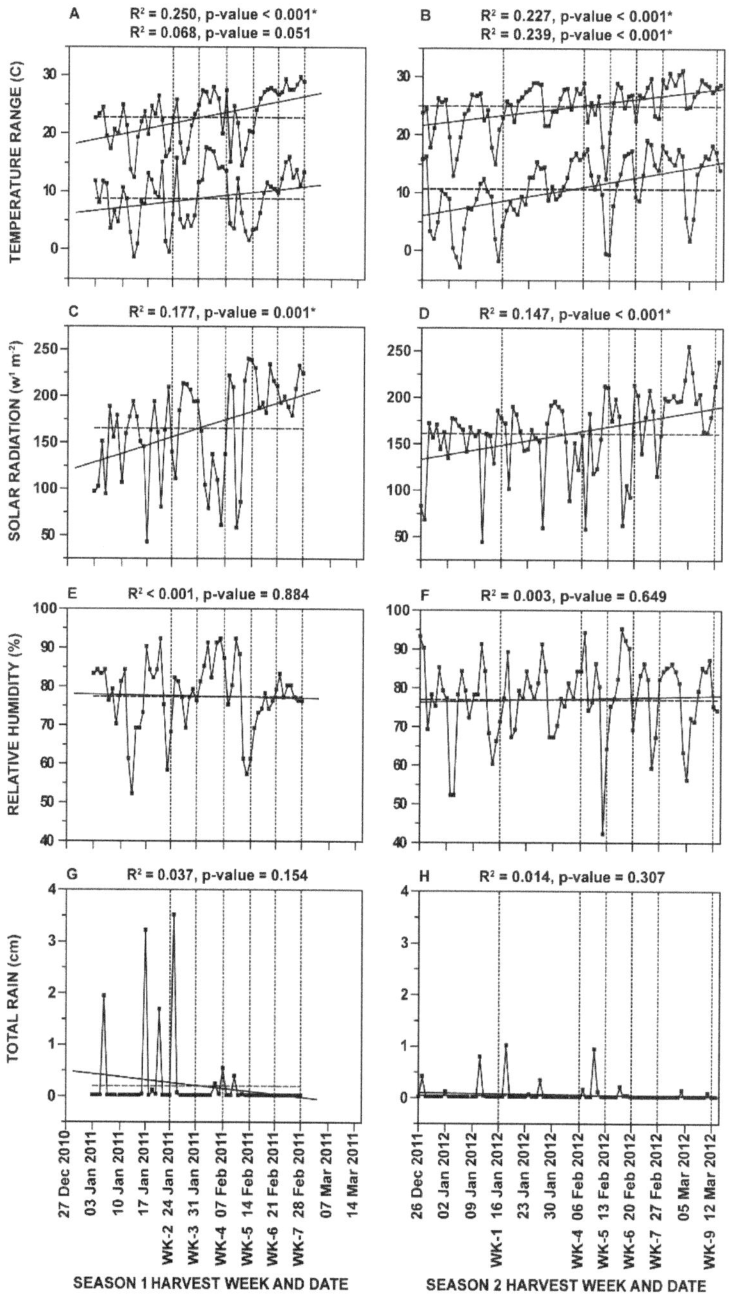

Figure S1. Season Environmental Conditions. Daily maximum and minimum temperatures (A and B), daily average solar radiation (C and D), daily average relative humidity (E and F), and daily total rain fall (G and H) during the 2011 (A, C, E, and G) and 2012 (B, D, F, and H)

seasons. Data for Balm, FL obtained from Florida Automated Weather Network (http://fawn.ifas.ufl.edu/data/reports). Data spans three weeks prior to first harvest through last harvest of each season with individual harvests indicated by dotted vertical line and harvest week number. Dashed horizontal lines represent means of environmental measures. Solid lines are the bivariate fit of environmental measure across season. Coefficients of determination (R^2) and p-value of fit is listed above individual scatterplots and are calculated using bivariate fit in JMP 8.

doi:10.1371/journal.pone.0088446.s001

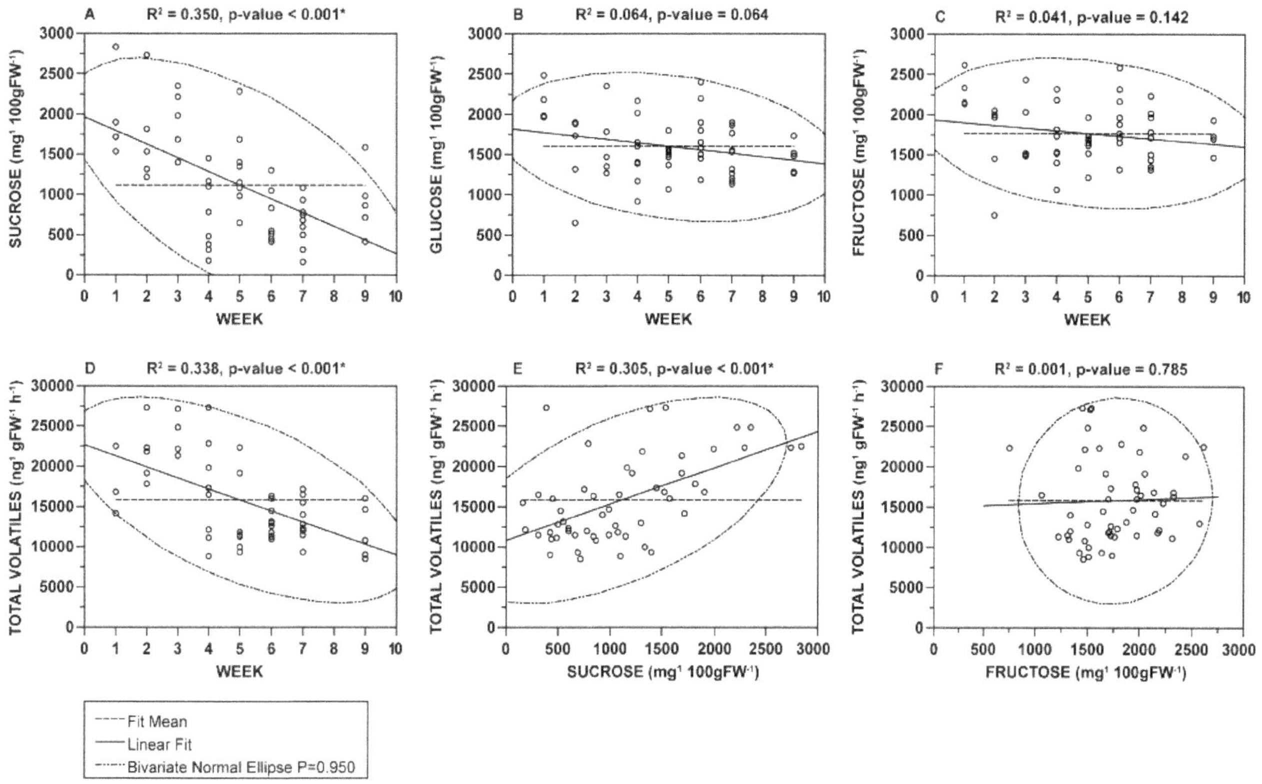

Figure S2. Individual sugars and total volatiles regressed against season progression. Regression of sucrose (A), glucose (B), fructose (C), and total volatiles (D) by harvest week during the seasons. Total volatile content is regressed against sucrose (E) and fructose (F). Sucrose (A) and total volatiles (D) demonstrate a significant negative fit to harvest week, unlike glucose (B) and fructose (C). A strong relationship between total volatile emmission and sucrose concentration is found (E) that is not observed between total volatiles and glucose (data not shown) and fructose (F). Coefficient of determination (R^2) and p-value of fit is listed above individual scatterplots and is calculated using bivariate fit in JMP 8. Dashed line represents mean of independent variable, solid line represents linear fit, dashed/dotted ellipse indicates 95% confidence range of data, and asterisk denotes significant fit (α=0.05).

doi:10.1371/journal.pone.0088446.s002

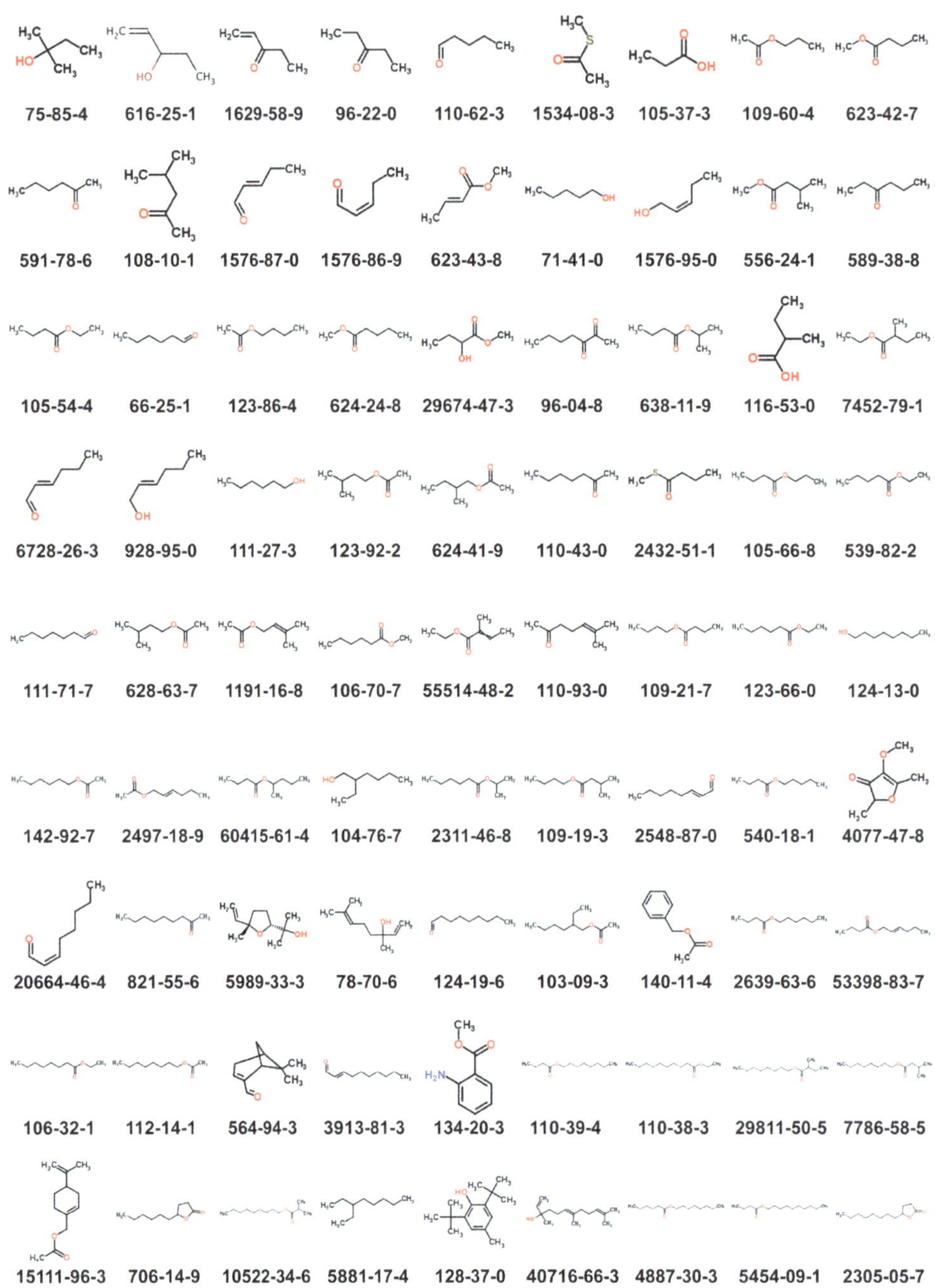

Figure S3. Chemical structure of volatile compounds. Chemical structure of volatile compounds quantified in strawberry. Sorted by increasing retention time (left to right, top row to bottom row), identified by CAS Registry Number.

doi:10.1371/journal.pone.0088446.s003

Table S1. CAS registry number, chemical name, and formula index. Chemical Abstract Services (CAS) registry numbers were used to query SciFinder® substances database for associated chemical name and molecular formula.

doi:10.1371/journal.pone.0088446.s004

CAS Registry Number	Chemical Name	Formula
75-85-4	2-Butanol, 2-methyl-	$C_5 H_{12} O$
616-25-1	1-Penten-3-ol	$C_5 H_{10} O$
1629-58-9	1-Penten-3-one	$C_5 H_8 O$
96-22-0	3-Pentanone	$C_5 H_{10} O$
110-62-3	Pentanal	$C_5 H_{10} O$
1534-08-3	Ethanethioic acid, *S*-methyl ester (9CI)	$C_3 H_6 O S$
105-37-3	Propanoic acid, ethyl ester	$C_5 H_{10} O_2$

109-60-4	Acetic acid, propyl ester	$C_5H_{10}O_2$
623-42-7	Butanoic acid, methyl ester	$C_5H_{10}O_2$
591-78-6	2-Hexanone	$C_6H_{12}O$
108-10-1	2-Pentanone, 4-methyl-	$C_6H_{12}O$
1576-87-0	2-Pentenal, (2*E*)-	C_5H_8O
1576-86-9	2-Pentenal, (2*Z*)-	C_5H_8O
623-43-8	2-Butenoic acid, methyl ester, (2*E*)-	$C_5H_8O_2$
71-41-0	1-Pentanol	$C_5H_{12}O$
1576-95-0	2-Penten-1-ol, (2*Z*)-	$C_5H_{10}O$
556-24-1	Butanoic acid, 3-methyl-, methyl ester	$C_6H_{12}O_2$
589-38-8	3-Hexanone	$C_6H_{12}O$
105-54-4	Butanoic acid, ethyl ester	$C_6H_{12}O_2$
66-25-1	Hexanal	$C_6H_{12}O$
123-86-4	Acetic acid, butyl ester	$C_6H_{12}O_2$
624-24-8	Pentanoic acid, methyl ester	$C_6H_{12}O_2$
29674-47-3	Butanoic acid, 2-hydroxy-, methyl ester	$C_5H_{10}O_3$
96-04-8	2,3-Heptanedione	$C_7H_{12}O_2$
638-11-9	Butanoic acid, 1-methylethyl ester	$C_7H_{14}O_2$
116-53-0	Butanoic acid, 2-methyl-	$C_5H_{10}O_2$
7452-79-1	Butanoic acid, 2-methyl-, ethyl ester	$C_7H_{14}O_2$
6728-26-3	2-Hexenal, (2*E*)-	$C_6H_{10}O$
928-95-0	2-Hexen-1-ol, (2*E*)-	$C_6H_{12}O$
111-27-3	Heptanal	$C_7H_{14}O$
123-92-2	1-Butanol, 3-methyl-, 1-acetate	$C_7H_{14}O_2$
624-41-9	1-Butanol, 2-methyl-, 1-acetate	$C_7H_{14}O_2$
110-43-0	2-Heptanone	$C_7H_{14}O$
2432-51-1	Butanethioic acid, *S*-methyl ester	$C_5H_{10}OS$
105-66-8	Butanoic acid, propyl ester	$C_7H_{14}O_2$
539-82-2	Pentanoic acid, ethyl ester	$C_7H_{14}O_2$
111-71-7	1-Hexanol	$C_6H_{14}O$
628-63-7	Acetic acid, pentyl ester	$C_7H_{14}O_2$
1191-16-8	2-Buten-1-ol, 3-methyl-, 1-acetate	$C_7H_{12}O_2$
106-70-7	Hexanoic acid, methyl ester	$C_7H_{14}O_2$
55514-48-2	2-Butenoic acid, 2-methyl-, ethyl ester	$C_7H_{12}O_2$
110-93-0	5-Hepten-2-one, 6-methyl-	$C_8H_{14}O$
109-21-7	Butanoic acid, butyl ester	$C_8H_{16}O_2$
123-66-0	Hexanoic acid, ethyl ester	$C_8H_{16}O_2$
124-13-0	Octanal	$C_8H_{16}O$
142-92-7	Acetic acid, hexyl ester	$C_8H_{16}O_2$
2497-18-9	2-Hexen-1-ol, 1-acetate, (2*E*)-	$C_8H_{14}O_2$
60415-61-4	Butanoic acid, 1-methylbutyl ester	$C_9H_{18}O_2$
104-76-7	1-Hexanol, 2-ethyl-	$C_8H_{18}O$
2311-46-8	Hexanoic acid, 1-methylethyl ester	$C_9H_{18}O_2$
109-19-3	Butanoic acid, 3-methyl-, butyl ester	$C_9H_{18}O_2$
2548-87-0	2-Octenal, (2*E*)-	$C_8H_{14}O$
540-18-1	Butanoic acid, pentyl ester	$C_9H_{18}O_2$
4077-47-8	3(2*H*)-Furanone, 4-methoxy-2,5-dimethyl-	$C_7H_{10}O_3$
20664-46-4	2-Octenal, (2*Z*)-	$C_8H_{14}O$
821-55-6	2-Nonanone	$C_9H_{18}O$

5989-33-3	2-Furanmethanol, 5-ethenyltetrahydro-α,α,5-trimethyl-, (2*R*,5*S*)-*rel*-	$C_{10}H_{18}O_2$
78-70-6	1,6-Octadien-3-ol, 3,7-dimethyl-	$C_{10}H_{18}O$
124-19-6	Nonanal	$C_9H_{18}O$
103-09-3	Acetic acid, 2-ethylhexyl ester	$C_{10}H_{20}O_2$
140-11-4	Acetic acid, phenylmethyl ester	$C_9H_{10}O_2$
2639-63-6	Butanoic acid, hexyl ester	$C_{10}H_{20}O_2$
53398-83-7	Butanoic acid, (2*E*)-2-hexen-1-yl ester	$C_{10}H_{18}O_2$
106-32-1	Octanoic acid, ethyl ester	$C_{10}H_{20}O_2$
112-14-1	Acetic acid, octyl ester	$C_{10}H_{20}O_2$
564-94-3	Bicyclo[3.1.1]hept-2-ene-2-carboxaldehyde, 6,6-dimethyl-	$C_{10}H_{14}O$
3913-81-3	2-Decenal, (2*E*)-	$C_{10}H_{18}O$
134-20-3	Benzoic acid, 2-amino-, methyl ester	$C_8H_9NO_2$
110-39-4	Butanoic acid, octyl ester	$C_{12}H_{24}O_2$
110-38-3	Decanoic acid, ethyl ester	$C_{12}H_{24}O_2$
29811-50-5	Butanoic acid, 2-methyl-, octyl ester	$C_{13}H_{26}O_2$
7786-58-5	Butanoic acid, 3-methyl-, octyl ester	$C_{13}H_{26}O_2$
15111-96-3	1-Cyclohexene-1-methanol, 4-(1-methylethenyl)-, 1-acetate	$C_{12}H_{18}O_2$
706-14-9	2(3*H*)-Furanone, 5-hexyldihydro-	$C_{10}H_{18}O_2$
10522-34-6	Propanoic acid, 2-methyl-, nonyl ester	$C_{13}H_{26}O_2$
5881-17-4	Octane, 3-ethyl-	$C_{10}H_{22}$
128-37-0	Phenol, 2,6-bis(1,1-dimethylethyl)-4-methyl-	$C_{15}H_{24}O$
40716-66-3	1,6,10-Dodecatrien-3-ol, 3,7,11-trimethyl-, (6*E*)-	$C_{15}H_{26}O$
4887-30-3	Hexanoic acid, octyl ester	$C_{14}H_{28}O_2$
5454-09-1	Butanoic acid, decyl ester	$C_{14}H_{28}O_2$
2305-05-7	2(3*H*)-Furanone, dihydro-5-octyl-	$C_{12}H_{22}O_2$

Table S3. Fruit attributes bivariate fit during season. Regression of harvest week during season (X) on panel responses and metabolite concentration (Y). Coefficient of determination (R^2), correlation coefficient, p-value, sample size (n), mean and standard deviation of X and Y derived from bivariate fit in JMP 8.

doi:10.1371/journal.pone.0088446.s006

X	Y	R^2	CORR COEFF	p-VALUE	n	MEAN X	STD DEV X	MEAN Y	STD DEV Y
WEEK	1629-58-9	0.489	-0.699	0.000	54	5.0	2.2	117.9	65.9
WEEK	1576-87-0	0.485	-0.697	0.000	54	5.0	2.2	37.5	20.1
WEEK	SWEETNESS INTENSITY	0.471	-0.686	0.000	54	5.0	2.2	23.0	4.8
WEEK	SSC	0.444	-0.666	0.000	54	5.0	2.2	7.4	1.4
WEEK	STRAWBERRY FLAVOR INTENSITY	0.430	-0.656	0.000	54	5.0	2.2	26.9	3.9
WEEK	OVERALL LIKING	0.422	-0.650	0.000	54	5.0	2.2	23.8	5.7
WEEK	1576-86-9	0.402	-0.634	0.000	54	5.0	2.2	37.8	21.7
WEEK	SUCROSE	0.350	-0.592	0.000	54	5.0	2.2	1112.6	646.5
WEEK	6728-26-3	0.348	-0.590	0.000	54	5.0	2.2	8666.5	3359.7
WEEK	TOTAL VOLATILES	0.338	-0.581	0.000	54	5.0	2.2	15814.0	5238.5
WEEK	5881-17-4	0.315	-0.561	0.000	54	5.0	2.2	6.2	2.5
WEEK	105-54-4	0.294	-0.542	0.000	54	5.0	2.2	42.0	17.1
WEEK	TOTAL SUGAR	0.287	-0.536	0.000	54	5.0	2.2	4473.9	1037.2

WEEK	124-19-6	0.264	-0.513	0.000	54	5.0	2.2	8.5	7.5
WEEK	3913-81-3	0.252	-0.502	0.000	54	5.0	2.2	1.9	1.4
WEEK	96-22-0	0.242	-0.492	0.000	54	5.0	2.2	51.2	18.6
WEEK	2305-05-7	0.240	-0.490	0.000	54	5.0	2.2	6.7	6.7
WEEK	110-93-0	0.179	-0.423	0.001	54	5.0	2.2	2.7	1.5
WEEK	616-25-1	0.158	-0.398	0.003	54	5.0	2.2	15.9	7.3
WEEK	75-85-4	0.148	-0.385	0.004	54	5.0	2.2	3.9	2.2
WEEK	2639-63-6	0.141	-0.375	0.005	54	5.0	2.2	10.8	13.0
WEEK	116-53-0	0.139	-0.372	0.006	54	5.0	2.2	19.7	16.0
WEEK	142-92-7	0.137	-0.370	0.006	54	5.0	2.2	53.5	49.7
WEEK	111-71-7	0.134	-0.366	0.006	54	5.0	2.2	3.2	2.3
WEEK	TEXTURE LIKING	0.132	-0.364	0.007	54	5.0	2.2	23.8	5.7
WEEK	L* int	0.127	-0.357	0.008	54	5.0	2.2	54.9	5.8
WEEK	60415-61-4	0.118	-0.343	0.011	54	5.0	2.2	0.7	1.6
WEEK	5454-09-1	0.112	-0.335	0.013	54	5.0	2.2	3.7	6.2
WEEK	66-25-1	0.106	-0.325	0.017	54	5.0	2.2	2545.9	1722.0
WEEK	106-32-1	0.101	0.317	0.019	54	5.0	2.2	2.2	3.0
WEEK	7452-79-1	0.098	-0.312	0.021	54	5.0	2.2	50.0	31.0
WEEK	109-21-7	0.097	-0.311	0.022	54	5.0	2.2	72.1	157.2
WEEK	4077-47-8	0.089	-0.298	0.028	54	5.0	2.2	11.7	8.3
WEEK	123-86-4	0.087	-0.296	0.030	54	5.0	2.2	73.5	85.0
WEEK	1534-08-3	0.082	-0.286	0.036	54	5.0	2.2	0.4	0.2
WEEK	638-11-9	0.078	-0.280	0.041	54	5.0	2.2	72.7	67.2
WEEK	628-63-7	0.071	-0.266	0.052	54	5.0	2.2	4.6	1.7
WEEK	A* ext	0.070	-0.265	0.053	54	5.0	2.2	36.4	3.1
WEEK	40716-66-3	0.069	-0.262	0.055	54	5.0	2.2	84.3	107.5
WEEK	GLUCOSE	0.064	-0.254	0.064	54	5.0	2.2	1594.6	378.2
WEEK	110-43-0	0.060	-0.244	0.075	54	5.0	2.2	14.8	19.6
WEEK	53398-83-7	0.050	-0.223	0.105	54	5.0	2.2	5.0	4.6
WEEK	110-39-4	0.049	-0.222	0.107	54	5.0	2.2	40.8	70.0
WEEK	109-60-4	0.044	0.210	0.128	54	5.0	2.2	3.8	2.7
WEEK	564-94-3	0.043	-0.208	0.132	54	5.0	2.2	6.5	7.4
WEEK	591-78-6	0.043	-0.207	0.134	54	5.0	2.2	10.3	13.9
WEEK	FRUCTOSE	0.041	-0.203	0.141	54	5.0	2.2	1766.7	381.5
WEEK	TA	0.041	-0.201	0.144	54	5.0	2.2	0.8	0.1
WEEK	706-14-9	0.040	-0.200	0.146	54	5.0	2.2	44.5	81.1
WEEK	5989-33-3	0.040	-0.199	0.149	54	5.0	2.2	2.8	2.4
WEEK	104-76-7	0.038	-0.195	0.158	54	5.0	2.2	6.0	4.9
WEEK	2548-87-0	0.036	-0.191	0.167	54	5.0	2.2	2.4	1.5
WEEK	15111-96-3	0.030	-0.173	0.212	54	5.0	2.2	1.2	1.2
WEEK	109-19-3	0.029	-0.171	0.217	54	5.0	2.2	2.7	3.8
WEEK	pH	0.029	-0.169	0.221	54	5.0	2.2	3.7	0.2
WEEK	108-10-1	0.028	0.167	0.228	54	5.0	2.2	1.5	2.6
WEEK	10522-34-6	0.025	-0.159	0.252	54	5.0	2.2	1.1	1.1
WEEK	L* ext	0.025	-0.158	0.255	54	5.0	2.2	33.6	2.6
WEEK	A* int	0.024	0.156	0.261	54	5.0	2.2	28.8	7.6
WEEK	134-20-3	0.024	0.156	0.261	54	5.0	2.2	0.1	0.7
WEEK	123-66-0	0.021	0.146	0.292	54	5.0	2.2	108.2	128.0

WEEK	623-42-7	0.019	-0.139	0.316	54	5.0	2.2	2780.2	1376.8
WEEK	FORCE	0.019	-0.139	0.316	54	5.0	2.2	0.6	0.2
WEEK	4887-30-3	0.019	-0.138	0.319	54	5.0	2.2	16.9	32.5
WEEK	123-92-2	0.018	0.132	0.340	54	5.0	2.2	23.0	21.5
WEEK	96-04-8	0.016	-0.127	0.359	54	5.0	2.2	3.1	8.0
WEEK	821-55-6	0.015	-0.123	0.374	54	5.0	2.2	3.5	8.2
WEEK	539-82-2	0.012	-0.111	0.425	54	5.0	2.2	3.3	3.9
WEEK	540-18-1	0.012	-0.108	0.435	54	5.0	2.2	3.5	3.2
WEEK	128-37-0	0.010	0.099	0.475	54	5.0	2.2	4.1	3.7
WEEK	55514-48-2	0.009	0.097	0.485	54	5.0	2.2	0.5	0.5
WEEK	124-13-0	0.009	-0.096	0.491	54	5.0	2.2	5.9	3.0
WEEK	112-14-1	0.009	-0.095	0.495	54	5.0	2.2	18.3	24.0
WEEK	CITRIC ACID	0.009	-0.094	0.499	54	5.0	2.2	741.0	147.2
WEEK	SOURNESS INTENSITY	0.009	0.093	0.505	54	5.0	2.2	18.1	3.1
WEEK	623-43-8	0.008	0.090	0.518	54	5.0	2.2	3.4	3.4
WEEK	105-66-8	0.008	-0.088	0.528	54	5.0	2.2	5.0	3.7
WEEK	B* int	0.007	0.084	0.544	54	5.0	2.2	25.8	4.5
WEEK	110-62-3	0.005	-0.074	0.597	54	5.0	2.2	7.9	8.9
WEEK	1576-95-0	0.005	-0.072	0.604	54	5.0	2.2	2.1	2.0
WEEK	B* ext	0.005	-0.070	0.613	54	5.0	2.2	19.0	3.3
WEEK	2311-46-8	0.004	-0.066	0.637	54	5.0	2.2	3.9	4.4
WEEK	MALIC ACID	0.004	-0.065	0.641	54	5.0	2.2	212.4	51.6
WEEK	29811-50-5	0.004	-0.064	0.643	54	5.0	2.2	3.2	5.3
WEEK	103-09-3	0.003	0.057	0.680	54	5.0	2.2	3.0	1.1
WEEK	140-11-4	0.003	-0.054	0.700	54	5.0	2.2	11.1	8.6
WEEK	2432-51-1	0.002	0.048	0.729	54	5.0	2.2	4.4	5.8
WEEK	556-24-1	0.002	-0.044	0.752	54	5.0	2.2	46.6	57.0
WEEK	78-70-6	0.001	-0.038	0.787	54	5.0	2.2	128.8	113.0
WEEK	589-38-8	0.001	0.036	0.798	54	5.0	2.2	1.9	1.2
WEEK	2497-18-9	0.001	0.033	0.815	54	5.0	2.2	24.9	21.5
WEEK	1191-16-8	0.001	0.029	0.837	54	5.0	2.2	5.5	7.0
WEEK	928-95-0	0.001	0.028	0.838	54	5.0	2.2	66.8	61.7
WEEK	106-70-7	0.001	-0.027	0.845	54	5.0	2.2	252.7	164.0
WEEK	71-41-0	0.001	-0.026	0.854	54	5.0	2.2	1.0	1.3
WEEK	20664-46-4	0.000	-0.020	0.887	54	5.0	2.2	20.6	19.0
WEEK	7786-58-5	0.000	-0.019	0.890	54	5.0	2.2	12.2	28.4
WEEK	29674-47-3	0.000	-0.019	0.892	54	5.0	2.2	5.0	5.0
WEEK	110-38-3	0.000	-0.019	0.893	54	5.0	2.2	2.0	2.6
WEEK	105-37-3	0.000	0.017	0.903	54	5.0	2.2	10.1	14.1
WEEK	624-41-9	0.000	-0.010	0.945	54	5.0	2.2	18.9	18.1
WEEK	624-24-8	0.000	-0.007	0.960	54	5.0	2.2	5.6	3.7
WEEK	111-27-3	0.000	0.004	0.976	54	5.0	2.2	45.5	94.6

Table S4. Fruit quality bivariate fit. Regression of chemical and physical measures of fruit (X) to panel responses (Y). Coefficient of determination (R^2), correlation coefficient, p-value, sample size (n), mean and standard deviation of X and Y derived from bivariate fit in JMP 8.

doi:10.1371/journal.pone.0088446.s007

X	Y	R^2	CORR COEFF	p-VALUE	n	MEAN X	STD DEV X	MEAN Y	STD DEV Y
TOTAL SUGAR	OVERALL LIKING	0.489	0.699	0.000	54	4473.9	1037.2	23.8	5.7
SSC	OVERALL LIKING	0.457	0.676	0.000	54	7.4	1.4	23.8	5.7
SUCROSE	OVERALL LIKING	0.442	0.665	0.000	54	1112.6	646.5	23.8	5.7
1629-58-9	OVERALL LIKING	0.437	0.661	0.000	54	117.9	65.9	23.8	5.7
1576-87-0	OVERALL LIKING	0.371	0.609	0.000	54	37.5	20.1	23.8	5.7
2305-05-7	OVERALL LIKING	0.310	0.557	0.000	54	6.7	6.7	23.8	5.7
1576-86-9	OVERALL LIKING	0.301	0.549	0.000	54	37.8	21.7	23.8	5.7
111-71-7	OVERALL LIKING	0.288	0.537	0.000	54	3.2	2.3	23.8	5.7
540-18-1	OVERALL LIKING	0.244	0.494	0.000	54	3.5	3.2	23.8	5.7
3913-81-3	OVERALL LIKING	0.241	0.491	0.000	54	1.9	1.4	23.8	5.7
110-93-0	OVERALL LIKING	0.228	0.477	0.000	54	2.7	1.5	23.8	5.7
2639-63-6	OVERALL LIKING	0.200	0.447	0.001	54	10.8	13.0	23.8	5.7
124-19-6	OVERALL LIKING	0.196	0.443	0.001	54	8.5	7.5	23.8	5.7
2548-87-0	OVERALL LIKING	0.191	0.437	0.001	54	2.4	1.5	23.8	5.7
4077-47-8	OVERALL LIKING	0.189	0.434	0.001	54	11.7	8.3	23.8	5.7
104-76-7	OVERALL LIKING	0.187	0.432	0.001	54	6.0	4.9	23.8	5.7
TOTAL VOLATILES	OVERALL LIKING	0.179	0.424	0.001	54	15814.0	5238.5	23.8	5.7
638-11-9	OVERALL LIKING	0.179	0.423	0.001	54	72.7	67.2	23.8	5.7
60415-61-4	OVERALL LIKING	0.177	0.421	0.002	54	0.7	1.6	23.8	5.7
GLUCOSE	OVERALL LIKING	0.175	0.419	0.002	54	1594.6	378.2	23.8	5.7
5989-33-3	OVERALL LIKING	0.168	0.410	0.002	54	2.8	2.4	23.8	5.7
40716-66-3	OVERALL LIKING	0.164	0.405	0.002	54	84.3	107.5	23.8	5.7
5881-17-4	OVERALL LIKING	0.157	0.397	0.003	54	6.2	2.5	23.8	5.7

109-19-3	OVERALL LIKING	0.154	0.392	0.003	54	2.7	3.8	23.8	5.7
L* int	OVERALL LIKING	0.147	0.384	0.004	54	54.9	5.8	23.8	5.7
109-21-7	OVERALL LIKING	0.143	0.378	0.005	54	72.1	157.2	23.8	5.7
142-92-7	OVERALL LIKING	0.138	0.371	0.006	54	53.5	49.7	23.8	5.7
110-43-0	OVERALL LIKING	0.137	0.370	0.006	54	14.8	19.6	23.8	5.7
FRUCTOSE	OVERALL LIKING	0.129	0.359	0.008	54	1766.7	381.5	23.8	5.7
5454-09-1	OVERALL LIKING	0.125	0.353	0.009	54	3.7	6.2	23.8	5.7
706-14-9	OVERALL LIKING	0.122	0.349	0.010	54	44.5	81.1	23.8	5.7
591-78-6	OVERALL LIKING	0.118	0.343	0.011	54	10.3	13.9	23.8	5.7
123-86-4	OVERALL LIKING	0.108	0.329	0.015	54	73.5	85.0	23.8	5.7
6728-26-3	OVERALL LIKING	0.108	0.329	0.015	54	8666.5	3359.7	23.8	5.7
53398-83-7	OVERALL LIKING	0.106	0.325	0.016	54	5.0	4.6	23.8	5.7
A* int	OVERALL LIKING	0.105	-0.324	0.017	54	28.8	7.6	23.8	5.7
TA	OVERALL LIKING	0.099	0.314	0.021	54	0.8	0.1	23.8	5.7
110-38-3	OVERALL LIKING	0.094	0.307	0.024	54	2.0	2.6	23.8	5.7
105-66-8	OVERALL LIKING	0.094	0.306	0.024	54	5.0	3.7	23.8	5.7
105-54-4	OVERALL LIKING	0.088	0.297	0.029	54	42.0	17.1	23.8	5.7
616-25-1	OVERALL LIKING	0.082	0.286	0.036	54	15.9	7.3	23.8	5.7
96-22-0	OVERALL LIKING	0.079	0.280	0.040	54	51.2	18.6	23.8	5.7
134-20-3	OVERALL LIKING	0.076	-0.276	0.043	54	0.1	0.7	23.8	5.7
10522-34-6	OVERALL LIKING	0.076	0.276	0.044	54	1.1	1.1	23.8	5.7
623-42-7	OVERALL LIKING	0.075	0.275	0.045	54	2780.2	1376.8	23.8	5.7
1191-16-8	OVERALL LIKING	0.072	-0.268	0.050	54	5.5	7.0	23.8	5.7
110-39-4	OVERALL LIKING	0.067	0.258	0.059	54	40.8	70.0	23.8	5.7
55514-48-2	OVERALL LIKING	0.065	-0.254	0.064	54	0.5	0.5	23.8	5.7
2311-46-8	OVERALL LIKING	0.061	0.247	0.072	54	3.9	4.4	23.8	5.7
B* int	OVERALL LIKING	0.059	-0.242	0.078	54	25.8	4.5	23.8	5.7

116-53-0	OVERALL LIKING	0.058	0.240	0.080	54	19.7	16.0	23.8	5.7
CITRIC ACID	OVERALL LIKING	0.056	0.237	0.084	54	741.0	147.2	23.8	5.7
pH	OVERALL LIKING	0.053	0.231	0.094	54	3.7	0.2	23.8	5.7
29674-47-3	OVERALL LIKING	0.051	0.226	0.100	54	5.0	5.0	23.8	5.7
4887-30-3	OVERALL LIKING	0.051	0.226	0.101	54	16.9	32.5	23.8	5.7
564-94-3	OVERALL LIKING	0.047	0.218	0.114	54	6.5	7.4	23.8	5.7
623-43-8	OVERALL LIKING	0.041	-0.203	0.142	54	3.4	3.4	23.8	5.7
539-82-2	OVERALL LIKING	0.039	0.196	0.155	54	3.3	3.9	23.8	5.7
66-25-1	OVERALL LIKING	0.035	0.187	0.175	54	2545.9	1722.0	23.8	5.7
628-63-7	OVERALL LIKING	0.035	0.187	0.177	54	4.6	1.7	23.8	5.7
FORCE	OVERALL LIKING	0.034	0.185	0.181	54	0.6	0.2	23.8	5.7
7786-58-5	OVERALL LIKING	0.033	0.182	0.188	54	12.2	28.4	23.8	5.7
A* ext	OVERALL LIKING	0.033	0.182	0.189	54	36.4	3.1	23.8	5.7
71-41-0	OVERALL LIKING	0.029	0.172	0.215	54	1.0	1.3	23.8	5.7
624-41-9	OVERALL LIKING	0.028	-0.168	0.225	54	18.9	18.1	23.8	5.7
109-60-4	OVERALL LIKING	0.027	-0.166	0.231	54	3.8	2.7	23.8	5.7
MALIC ACID	OVERALL LIKING	0.027	0.165	0.234	54	212.4	51.6	23.8	5.7
111-27-3	OVERALL LIKING	0.027	-0.164	0.236	54	45.5	94.6	23.8	5.7
124-13-0	OVERALL LIKING	0.024	0.154	0.267	54	5.9	3.0	23.8	5.7
78-70-6	OVERALL LIKING	0.023	0.153	0.270	54	128.8	113.0	23.8	5.7
821-55-6	OVERALL LIKING	0.022	0.148	0.286	54	3.5	8.2	23.8	5.7
624-24-8	OVERALL LIKING	0.022	0.147	0.289	54	5.6	3.7	23.8	5.7
7452-79-1	OVERALL LIKING	0.022	0.147	0.289	54	50.0	31.0	23.8	5.7
106-70-7	OVERALL LIKING	0.021	0.145	0.296	54	252.7	164.0	23.8	5.7
112-14-1	OVERALL LIKING	0.021	0.144	0.298	54	18.3	24.0	23.8	5.7
1534-08-3	OVERALL LIKING	0.018	0.135	0.332	54	0.4	0.2	23.8	5.7
20664-46-4	OVERALL LIKING	0.018	0.133	0.339	54	20.6	19.0	23.8	5.7

589-38-8	OVERALL LIKING	0.018	0.133	0.340	54	1.9	1.2	23.8	5.7
123-66-0	OVERALL LIKING	0.017	0.132	0.341	54	108.2	128.0	23.8	5.7
2432-51-1	OVERALL LIKING	0.016	0.127	0.360	54	4.4	5.8	23.8	5.7
75-85-4	OVERALL LIKING	0.014	0.119	0.390	54	3.9	2.2	23.8	5.7
928-95-0	OVERALL LIKING	0.012	-0.111	0.422	54	66.8	61.7	23.8	5.7
140-11-4	OVERALL LIKING	0.012	-0.111	0.423	54	11.1	8.6	23.8	5.7
103-09-3	OVERALL LIKING	0.011	-0.106	0.446	54	3.0	1.1	23.8	5.7
123-92-2	OVERALL LIKING	0.010	-0.099	0.475	54	23.0	21.5	23.8	5.7
B* ext	OVERALL LIKING	0.009	-0.095	0.492	54	19.0	3.3	23.8	5.7
128-37-0	OVERALL LIKING	0.007	0.086	0.536	54	4.1	3.7	23.8	5.7
2497-18-9	OVERALL LIKING	0.006	-0.075	0.590	54	24.9	21.5	23.8	5.7
556-24-1	OVERALL LIKING	0.004	0.065	0.638	54	46.6	57.0	23.8	5.7
29811-50-5	OVERALL LIKING	0.004	-0.062	0.658	54	3.2	5.3	23.8	5.7
110-62-3	OVERALL LIKING	0.003	0.057	0.680	54	7.9	8.9	23.8	5.7
96-04-8	OVERALL LIKING	0.003	-0.052	0.711	54	3.1	8.0	23.8	5.7
1576-95-0	OVERALL LIKING	0.002	0.040	0.773	54	2.1	2.0	23.8	5.7
106-32-1	OVERALL LIKING	0.001	-0.032	0.817	54	2.2	3.0	23.8	5.7
108-10-1	OVERALL LIKING	0.001	-0.023	0.870	54	1.5	2.6	23.8	5.7
15111-96-3	OVERALL LIKING	0.000	0.018	0.897	54	1.2	1.2	23.8	5.7
105-37-3	OVERALL LIKING	0.000	-0.009	0.946	54	10.1	14.1	23.8	5.7
L* ext	OVERALL LIKING	0.000	0.004	0.976	54	33.6	2.6	23.8	5.7
FORCE	TEXTURE LIKING	0.358	0.598	0.000	54	0.6	0.2	23.8	5.7
MALIC ACID	TEXTURE LIKING	0.282	0.531	0.000	54	212.4	51.6	23.8	5.7
134-20-3	TEXTURE LIKING	0.277	-0.526	0.000	54	0.1	0.7	23.8	5.7
104-76-7	TEXTURE LIKING	0.231	0.480	0.000	54	6.0	4.9	23.8	5.7
623-43-8	TEXTURE LIKING	0.179	-0.423	0.001	54	3.4	3.4	23.8	5.7
1629-58-9	TEXTURE LIKING	0.151	0.388	0.004	54	117.9	65.9	23.8	5.7

103-09-3	TEXTURE LIKING	0.145	-0.381	0.004	54	3.0	1.1	23.8	5.7
1576-87-0	TEXTURE LIKING	0.145	0.381	0.005	54	37.5	20.1	23.8	5.7
638-11-9	TEXTURE LIKING	0.141	0.375	0.005	54	72.7	67.2	23.8	5.7
40716-66-3	TEXTURE LIKING	0.132	0.364	0.007	54	84.3	107.5	23.8	5.7
2548-87-0	TEXTURE LIKING	0.129	0.360	0.008	54	2.4	1.5	23.8	5.7
111-71-7	TEXTURE LIKING	0.122	0.349	0.010	54	3.2	2.3	23.8	5.7
110-38-3	TEXTURE LIKING	0.117	0.342	0.011	54	2.0	2.6	23.8	5.7
2305-05-7	TEXTURE LIKING	0.116	0.341	0.012	54	6.7	6.7	23.8	5.7
60415-61-4	TEXTURE LIKING	0.104	0.322	0.018	54	0.7	1.6	23.8	5.7
SUCROSE	TEXTURE LIKING	0.102	0.320	0.018	54	1112.6	646.5	23.8	5.7
556-24-1	TEXTURE LIKING	0.097	-0.312	0.022	54	46.6	57.0	23.8	5.7
123-92-2	TEXTURE LIKING	0.090	-0.300	0.027	54	23.0	21.5	23.8	5.7
5454-09-1	TEXTURE LIKING	0.080	0.282	0.039	54	3.7	6.2	23.8	5.7
5989-33-3	TEXTURE LIKING	0.078	0.280	0.040	54	2.8	2.4	23.8	5.7
110-62-3	TEXTURE LIKING	0.076	0.276	0.043	54	7.9	8.9	23.8	5.7
55514-48-2	TEXTURE LIKING	0.076	-0.275	0.044	54	0.5	0.5	23.8	5.7
109-21-7	TEXTURE LIKING	0.067	0.260	0.058	54	72.1	157.2	23.8	5.7
TOTAL SUGAR	TEXTURE LIKING	0.067	0.260	0.058	54	4473.9	1037.2	23.8	5.7
124-19-6	TEXTURE LIKING	0.065	0.256	0.062	54	8.5	7.5	23.8	5.7
110-93-0	TEXTURE LIKING	0.065	0.256	0.062	54	2.7	1.5	23.8	5.7
540-18-1	TEXTURE LIKING	0.063	0.251	0.068	54	3.5	3.2	23.8	5.7
105-66-8	TEXTURE LIKING	0.062	0.250	0.068	54	5.0	3.7	23.8	5.7
110-43-0	TEXTURE LIKING	0.062	0.249	0.070	54	14.8	19.6	23.8	5.7
109-60-4	TEXTURE LIKING	0.059	-0.244	0.076	54	3.8	2.7	23.8	5.7
1576-86-9	TEXTURE LIKING	0.058	0.241	0.079	54	37.8	21.7	23.8	5.7
4077-47-8	TEXTURE LIKING	0.058	0.241	0.080	54	11.7	8.3	23.8	5.7
pH	TEXTURE LIKING	0.056	0.237	0.084	54	3.7	0.2	23.8	5.7

SSC	TEXTURE LIKING	0.055	0.234	0.089	54	7.4	1.4	23.8	5.7
10522-34-6	TEXTURE LIKING	0.048	0.218	0.113	54	1.1	1.1	23.8	5.7
109-19-3	TEXTURE LIKING	0.045	0.213	0.123	54	2.7	3.8	23.8	5.7
706-14-9	TEXTURE LIKING	0.044	0.210	0.128	54	44.5	81.1	23.8	5.7
2639-63-6	TEXTURE LIKING	0.040	0.200	0.147	54	10.8	13.0	23.8	5.7
111-27-3	TEXTURE LIKING	0.039	-0.199	0.150	54	45.5	94.6	23.8	5.7
2432-51-1	TEXTURE LIKING	0.035	0.187	0.176	54	4.4	5.8	23.8	5.7
564-94-3	TEXTURE LIKING	0.035	0.186	0.177	54	6.5	7.4	23.8	5.7
110-39-4	TEXTURE LIKING	0.034	0.184	0.183	54	40.8	70.0	23.8	5.7
3913-81-3	TEXTURE LIKING	0.034	0.184	0.183	54	1.9	1.4	23.8	5.7
539-82-2	TEXTURE LIKING	0.030	0.173	0.212	54	3.3	3.9	23.8	5.7
624-24-8	TEXTURE LIKING	0.028	0.168	0.226	54	5.6	3.7	23.8	5.7
116-53-0	TEXTURE LIKING	0.026	0.163	0.240	54	19.7	16.0	23.8	5.7
TA	TEXTURE LIKING	0.026	0.161	0.244	54	0.8	0.1	23.8	5.7
123-86-4	TEXTURE LIKING	0.026	0.160	0.247	54	73.5	85.0	23.8	5.7
TOTAL VOLATILES	TEXTURE LIKING	0.025	0.158	0.254	54	15814.0	5238.5	23.8	5.7
5881-17-4	TEXTURE LIKING	0.025	0.157	0.256	54	6.2	2.5	23.8	5.7
66-25-1	TEXTURE LIKING	0.023	0.153	0.269	54	2545.9	1722.0	23.8	5.7
78-70-6	TEXTURE LIKING	0.023	0.152	0.272	54	128.8	113.0	23.8	5.7
124-13-0	TEXTURE LIKING	0.023	-0.152	0.273	54	5.9	3.0	23.8	5.7
142-92-7	TEXTURE LIKING	0.023	0.151	0.274	54	53.5	49.7	23.8	5.7
2311-46-8	TEXTURE LIKING	0.023	0.151	0.276	54	3.9	4.4	23.8	5.7
928-95-0	TEXTURE LIKING	0.020	-0.143	0.302	54	66.8	61.7	23.8	5.7
96-04-8	TEXTURE LIKING	0.019	-0.138	0.319	54	3.1	8.0	23.8	5.7
20664-46-4	TEXTURE LIKING	0.019	-0.137	0.324	54	20.6	19.0	23.8	5.7
821-55-6	TEXTURE LIKING	0.018	0.135	0.329	54	3.5	8.2	23.8	5.7
53398-83-7	TEXTURE LIKING	0.018	0.133	0.338	54	5.0	4.6	23.8	5.7

29811-50-5	TEXTURE LIKING	0.017	-0.131	0.346	54	3.2	5.3	23.8	5.7
128-37-0	TEXTURE LIKING	0.017	0.129	0.352	54	4.1	3.7	23.8	5.7
140-11-4	TEXTURE LIKING	0.016	-0.127	0.359	54	11.1	8.6	23.8	5.7
4887-30-3	TEXTURE LIKING	0.016	0.126	0.363	54	16.9	32.5	23.8	5.7
75-85-4	TEXTURE LIKING	0.015	0.124	0.371	54	3.9	2.2	23.8	5.7
L* ext	TEXTURE LIKING	0.015	-0.123	0.377	54	33.6	2.6	23.8	5.7
1191-16-8	TEXTURE LIKING	0.015	-0.121	0.385	54	5.5	7.0	23.8	5.7
GLUCOSE	TEXTURE LIKING	0.013	0.115	0.408	54	1594.6	378.2	23.8	5.7
B* ext	TEXTURE LIKING	0.013	-0.113	0.417	54	19.0	3.3	23.8	5.7
624-41-9	TEXTURE LIKING	0.012	-0.111	0.424	54	18.9	18.1	23.8	5.7
6728-26-3	TEXTURE LIKING	0.012	0.108	0.435	54	8666.5	3359.7	23.8	5.7
A* ext	TEXTURE LIKING	0.010	0.099	0.475	54	36.4	3.1	23.8	5.7
2497-18-9	TEXTURE LIKING	0.008	-0.092	0.508	54	24.9	21.5	23.8	5.7
71-41-0	TEXTURE LIKING	0.008	0.087	0.530	54	1.0	1.3	23.8	5.7
589-38-8	TEXTURE LIKING	0.007	0.087	0.534	54	1.9	1.2	23.8	5.7
106-32-1	TEXTURE LIKING	0.007	-0.085	0.542	54	2.2	3.0	23.8	5.7
616-25-1	TEXTURE LIKING	0.007	0.082	0.555	54	15.9	7.3	23.8	5.7
106-70-7	TEXTURE LIKING	0.006	-0.076	0.587	54	252.7	164.0	23.8	5.7
B* int	TEXTURE LIKING	0.006	-0.075	0.588	54	25.8	4.5	23.8	5.7
7452-79-1	TEXTURE LIKING	0.005	0.072	0.607	54	50.0	31.0	23.8	5.7
105-54-4	TEXTURE LIKING	0.005	0.068	0.626	54	42.0	17.1	23.8	5.7
A* int	TEXTURE LIKING	0.004	-0.060	0.666	54	28.8	7.6	23.8	5.7
628-63-7	TEXTURE LIKING	0.003	-0.055	0.693	54	4.6	1.7	23.8	5.7
591-78-6	TEXTURE LIKING	0.003	0.054	0.700	54	10.3	13.9	23.8	5.7
7786-58-5	TEXTURE LIKING	0.003	-0.052	0.707	54	12.2	28.4	23.8	5.7
FRUCTOSE	TEXTURE LIKING	0.002	0.050	0.720	54	1766.7	381.5	23.8	5.7
1576-95-0	TEXTURE LIKING	0.002	0.047	0.736	54	2.1	2.0	23.8	5.7

CITRIC ACID	TEXTURE LIKING	0.002	-0.047	0.736	54	741.0	147.2	23.8	5.7
112-14-1	TEXTURE LIKING	0.001	-0.028	0.843	54	18.3	24.0	23.8	5.7
105-37-3	TEXTURE LIKING	0.001	-0.026	0.851	54	10.1	14.1	23.8	5.7
29674-47-3	TEXTURE LIKING	0.001	0.023	0.867	54	5.0	5.0	23.8	5.7
108-10-1	TEXTURE LIKING	0.001	0.023	0.869	54	1.5	2.6	23.8	5.7
15111-96-3	TEXTURE LIKING	0.000	-0.022	0.875	54	1.2	1.2	23.8	5.7
623-42-7	TEXTURE LIKING	0.000	0.017	0.905	54	2780.2	1376.8	23.8	5.7
96-22-0	TEXTURE LIKING	0.000	0.015	0.915	54	51.2	18.6	23.8	5.7
L* int	TEXTURE LIKING	0.000	-0.014	0.919	54	54.9	5.8	23.8	5.7
123-66-0	TEXTURE LIKING	0.000	-0.007	0.957	54	108.2	128.0	23.8	5.7
1534-08-3	TEXTURE LIKING	0.000	0.002	0.988	54	0.4	0.2	23.8	5.7
SSC	SWEETNESS INTENSITY	0.690	0.831	0.000	54	7.4	1.4	23.0	4.8
TOTAL SUGAR	SWEETNESS INTENSITY	0.687	0.829	0.000	54	4473.9	1037.2	23.0	4.8
SUCROSE	SWEETNESS INTENSITY	0.445	0.667	0.000	54	1112.6	646.5	23.0	4.8
1629-58-9	SWEETNESS INTENSITY	0.377	0.614	0.000	54	117.9	65.9	23.0	4.8
GLUCOSE	SWEETNESS INTENSITY	0.338	0.581	0.000	54	1594.6	378.2	23.0	4.8
FRUCTOSE	SWEETNESS INTENSITY	0.300	0.548	0.000	54	1766.7	381.5	23.0	4.8
2639-63-6	SWEETNESS INTENSITY	0.296	0.544	0.000	54	10.8	13.0	23.0	4.8
2305-05-7	SWEETNESS INTENSITY	0.295	0.543	0.000	54	6.7	6.7	23.0	4.8
540-18-1	SWEETNESS INTENSITY	0.254	0.504	0.000	54	3.5	3.2	23.0	4.8
1576-87-0	SWEETNESS INTENSITY	0.242	0.492	0.000	54	37.5	20.1	23.0	4.8
142-92-7	SWEETNESS INTENSITY	0.239	0.489	0.000	54	53.5	49.7	23.0	4.8
60415-61-4	SWEETNESS INTENSITY	0.233	0.482	0.000	54	0.7	1.6	23.0	4.8
1576-86-9	SWEETNESS INTENSITY	0.217	0.466	0.000	54	37.8	21.7	23.0	4.8
109-21-7	SWEETNESS INTENSITY	0.198	0.445	0.001	54	72.1	157.2	23.0	4.8
111-71-7	SWEETNESS INTENSITY	0.195	0.441	0.001	54	3.2	2.3	23.0	4.8
3913-81-3	SWEETNESS INTENSITY	0.192	0.438	0.001	54	1.9	1.4	23.0	4.8

L* int	SWEETNESS INTENSITY	0.187	0.432	0.001	54	54.9	5.8	23.0	4.8
109-19-3	SWEETNESS INTENSITY	0.184	0.429	0.001	54	2.7	3.8	23.0	4.8
5989-33-3	SWEETNESS INTENSITY	0.171	0.414	0.002	54	2.8	2.4	23.0	4.8
123-86-4	SWEETNESS INTENSITY	0.158	0.397	0.003	54	73.5	85.0	23.0	4.8
706-14-9	SWEETNESS INTENSITY	0.151	0.388	0.004	54	44.5	81.1	23.0	4.8
638-11-9	SWEETNESS INTENSITY	0.151	0.388	0.004	54	72.7	67.2	23.0	4.8
110-93-0	SWEETNESS INTENSITY	0.151	0.388	0.004	54	2.7	1.5	23.0	4.8
591-78-6	SWEETNESS INTENSITY	0.144	0.379	0.005	54	10.3	13.9	23.0	4.8
A* int	SWEETNESS INTENSITY	0.139	-0.373	0.005	54	28.8	7.6	23.0	4.8
TOTAL VOLATILES	SWEETNESS INTENSITY	0.139	0.373	0.005	54	15814.0	5238.5	23.0	4.8
124-19-6	SWEETNESS INTENSITY	0.139	0.372	0.006	54	8.5	7.5	23.0	4.8
5454-09-1	SWEETNESS INTENSITY	0.132	0.363	0.007	54	3.7	6.2	23.0	4.8
CITRIC ACID	SWEETNESS INTENSITY	0.124	0.353	0.009	54	741.0	147.2	23.0	4.8
110-39-4	SWEETNESS INTENSITY	0.123	0.351	0.009	54	40.8	70.0	23.0	4.8
53398-83-7	SWEETNESS INTENSITY	0.123	0.351	0.009	54	5.0	4.6	23.0	4.8
104-76-7	SWEETNESS INTENSITY	0.118	0.343	0.011	54	6.0	4.9	23.0	4.8
5881-17-4	SWEETNESS INTENSITY	0.112	0.334	0.014	54	6.2	2.5	23.0	4.8
4077-47-8	SWEETNESS INTENSITY	0.109	0.330	0.015	54	11.7	8.3	23.0	4.8
110-43-0	SWEETNESS INTENSITY	0.106	0.326	0.016	54	14.8	19.6	23.0	4.8
TA	SWEETNESS INTENSITY	0.094	0.307	0.024	54	0.8	0.1	23.0	4.8
616-25-1	SWEETNESS INTENSITY	0.094	0.307	0.024	54	15.9	7.3	23.0	4.8
10522-34-6	SWEETNESS INTENSITY	0.092	0.304	0.026	54	1.1	1.1	23.0	4.8
40716-66-3	SWEETNESS INTENSITY	0.092	0.303	0.026	54	84.3	107.5	23.0	4.8
623-42-7	SWEETNESS INTENSITY	0.085	0.292	0.032	54	2780.2	1376.8	23.0	4.8
2311-46-8	SWEETNESS INTENSITY	0.085	0.291	0.033	54	3.9	4.4	23.0	4.8
6728-26-3	SWEETNESS INTENSITY	0.078	0.278	0.041	54	8666.5	3359.7	23.0	4.8
105-66-8	SWEETNESS INTENSITY	0.077	0.278	0.042	54	5.0	3.7	23.0	4.8

B* int	SWEETNESS INTENSITY	0.075	-0.275	0.044	54	25.8	4.5	23.0	4.8
4887-30-3	SWEETNESS INTENSITY	0.075	0.274	0.045	54	16.9	32.5	23.0	4.8
110-38-3	SWEETNESS INTENSITY	0.070	0.264	0.054	54	2.0	2.6	23.0	4.8
7786-58-5	SWEETNESS INTENSITY	0.069	0.262	0.056	54	12.2	28.4	23.0	4.8
112-14-1	SWEETNESS INTENSITY	0.066	0.257	0.061	54	18.3	24.0	23.0	4.8
2548-87-0	SWEETNESS INTENSITY	0.064	0.253	0.064	54	2.4	1.5	23.0	4.8
96-22-0	SWEETNESS INTENSITY	0.063	0.250	0.068	54	51.2	18.6	23.0	4.8
1191-16-8	SWEETNESS INTENSITY	0.063	-0.250	0.068	54	5.5	7.0	23.0	4.8
105-54-4	SWEETNESS INTENSITY	0.062	0.249	0.069	54	42.0	17.1	23.0	4.8
124-13-0	SWEETNESS INTENSITY	0.053	0.231	0.093	54	5.9	3.0	23.0	4.8
29674-47-3	SWEETNESS INTENSITY	0.051	0.226	0.101	54	5.0	5.0	23.0	4.8
pH	SWEETNESS INTENSITY	0.048	0.220	0.110	54	3.7	0.2	23.0	4.8
71-41-0	SWEETNESS INTENSITY	0.045	0.213	0.122	54	1.0	1.3	23.0	4.8
628-63-7	SWEETNESS INTENSITY	0.040	0.199	0.149	54	4.6	1.7	23.0	4.8
556-24-1	SWEETNESS INTENSITY	0.039	0.198	0.151	54	46.6	57.0	23.0	4.8
564-94-3	SWEETNESS INTENSITY	0.030	0.173	0.211	54	6.5	7.4	23.0	4.8
20664-46-4	SWEETNESS INTENSITY	0.029	0.171	0.215	54	20.6	19.0	23.0	4.8
624-41-9	SWEETNESS INTENSITY	0.026	-0.162	0.243	54	18.9	18.1	23.0	4.8
539-82-2	SWEETNESS INTENSITY	0.025	0.158	0.255	54	3.3	3.9	23.0	4.8
1534-08-3	SWEETNESS INTENSITY	0.021	0.146	0.292	54	0.4	0.2	23.0	4.8
106-70-7	SWEETNESS INTENSITY	0.020	0.142	0.307	54	252.7	164.0	23.0	4.8
B* ext	SWEETNESS INTENSITY	0.019	-0.137	0.324	54	19.0	3.3	23.0	4.8
116-53-0	SWEETNESS INTENSITY	0.018	0.133	0.339	54	19.7	16.0	23.0	4.8
A* ext	SWEETNESS INTENSITY	0.017	0.130	0.348	54	36.4	3.1	23.0	4.8
78-70-6	SWEETNESS INTENSITY	0.017	0.130	0.350	54	128.8	113.0	23.0	4.8
123-66-0	SWEETNESS INTENSITY	0.017	0.129	0.352	54	108.2	128.0	23.0	4.8
55514-48-2	SWEETNESS INTENSITY	0.014	-0.120	0.387	54	0.5	0.5	23.0	4.8

7452-79-1	SWEETNESS INTENSITY	0.013	0.116	0.405	54	50.0	31.0	23.0	4.8
821-55-6	SWEETNESS INTENSITY	0.013	0.113	0.416	54	3.5	8.2	23.0	4.8
66-25-1	SWEETNESS INTENSITY	0.011	0.104	0.455	54	2545.9	1722.0	23.0	4.8
928-95-0	SWEETNESS INTENSITY	0.010	-0.101	0.468	54	66.8	61.7	23.0	4.8
589-38-8	SWEETNESS INTENSITY	0.009	0.097	0.484	54	1.9	1.2	23.0	4.8
110-62-3	SWEETNESS INTENSITY	0.009	-0.095	0.496	54	7.9	8.9	23.0	4.8
109-60-4	SWEETNESS INTENSITY	0.009	-0.095	0.496	54	3.8	2.7	23.0	4.8
111-27-3	SWEETNESS INTENSITY	0.009	-0.094	0.501	54	45.5	94.6	23.0	4.8
140-11-4	SWEETNESS INTENSITY	0.009	-0.093	0.501	54	11.1	8.6	23.0	4.8
MALIC ACID	SWEETNESS INTENSITY	0.009	-0.093	0.506	54	212.4	51.6	23.0	4.8
FORCE	SWEETNESS INTENSITY	0.005	-0.074	0.594	54	0.6	0.2	23.0	4.8
128-37-0	SWEETNESS INTENSITY	0.004	0.061	0.664	54	4.1	3.7	23.0	4.8
106-32-1	SWEETNESS INTENSITY	0.003	-0.055	0.694	54	2.2	3.0	23.0	4.8
623-43-8	SWEETNESS INTENSITY	0.003	-0.050	0.718	54	3.4	3.4	23.0	4.8
108-10-1	SWEETNESS INTENSITY	0.002	-0.050	0.721	54	1.5	2.6	23.0	4.8
29811-50-5	SWEETNESS INTENSITY	0.002	0.048	0.728	54	3.2	5.3	23.0	4.8
15111-96-3	SWEETNESS INTENSITY	0.002	0.048	0.731	54	1.2	1.2	23.0	4.8
2432-51-1	SWEETNESS INTENSITY	0.002	0.041	0.770	54	4.4	5.8	23.0	4.8
624-24-8	SWEETNESS INTENSITY	0.002	0.039	0.779	54	5.6	3.7	23.0	4.8
134-20-3	SWEETNESS INTENSITY	0.001	0.030	0.827	54	0.1	0.7	23.0	4.8
1576-95-0	SWEETNESS INTENSITY	0.000	0.018	0.899	54	2.1	2.0	23.0	4.8
123-92-2	SWEETNESS INTENSITY	0.000	-0.017	0.905	54	23.0	21.5	23.0	4.8
2497-18-9	SWEETNESS INTENSITY	0.000	-0.012	0.934	54	24.9	21.5	23.0	4.8
L* ext	SWEETNESS INTENSITY	0.000	0.010	0.943	54	33.6	2.6	23.0	4.8
103-09-3	SWEETNESS INTENSITY	0.000	-0.007	0.960	54	3.0	1.1	23.0	4.8
105-37-3	SWEETNESS INTENSITY	0.000	0.005	0.973	54	10.1	14.1	23.0	4.8
75-85-4	SWEETNESS INTENSITY	0.000	-0.001	0.995	54	3.9	2.2	23.0	4.8

96-04-8	SWEETNESS INTENSITY	0.000	-0.001	0.996	54	3.1	8.0	23.0	4.8
TA	SOURNESS INTENSITY	0.314	0.561	0.000	54	0.8	0.1	18.1	3.1
MALIC ACID	SOURNESS INTENSITY	0.189	0.435	0.001	54	212.4	51.6	18.1	3.1
CITRIC ACID	SOURNESS INTENSITY	0.146	0.382	0.004	54	741.0	147.2	18.1	3.1
134-20-3	SOURNESS INTENSITY	0.137	-0.370	0.006	54	0.1	0.7	18.1	3.1
pH	SOURNESS INTENSITY	0.118	-0.344	0.011	54	3.7	0.2	18.1	3.1
15111-96-3	SOURNESS INTENSITY	0.106	-0.325	0.016	54	1.2	1.2	18.1	3.1
624-41-9	SOURNESS INTENSITY	0.097	0.311	0.022	54	18.9	18.1	18.1	3.1
FRUCTOSE	SOURNESS INTENSITY	0.089	-0.298	0.029	54	1766.7	381.5	18.1	3.1
GLUCOSE	SOURNESS INTENSITY	0.075	-0.274	0.045	54	1594.6	378.2	18.1	3.1
4887-30-3	SOURNESS INTENSITY	0.073	-0.270	0.048	54	16.9	32.5	18.1	3.1
1191-16-8	SOURNESS INTENSITY	0.068	0.261	0.056	54	5.5	7.0	18.1	3.1
78-70-6	SOURNESS INTENSITY	0.068	0.260	0.058	54	128.8	113.0	18.1	3.1
5454-09-1	SOURNESS INTENSITY	0.067	-0.259	0.058	54	3.7	6.2	18.1	3.1
110-39-4	SOURNESS INTENSITY	0.066	-0.257	0.061	54	40.8	70.0	18.1	3.1
589-38-8	SOURNESS INTENSITY	0.062	0.248	0.070	54	1.9	1.2	18.1	3.1
624-24-8	SOURNESS INTENSITY	0.054	-0.233	0.090	54	5.6	3.7	18.1	3.1
928-95-0	SOURNESS INTENSITY	0.050	0.223	0.106	54	66.8	61.7	18.1	3.1
128-37-0	SOURNESS INTENSITY	0.049	0.221	0.108	54	4.1	3.7	18.1	3.1
110-62-3	SOURNESS INTENSITY	0.047	-0.218	0.113	54	7.9	8.9	18.1	3.1
2497-18-9	SOURNESS INTENSITY	0.036	0.189	0.171	54	24.9	21.5	18.1	3.1
111-27-3	SOURNESS INTENSITY	0.036	0.189	0.172	54	45.5	94.6	18.1	3.1
96-04-8	SOURNESS INTENSITY	0.033	0.182	0.188	54	3.1	8.0	18.1	3.1
104-76-7	SOURNESS INTENSITY	0.032	-0.178	0.198	54	6.0	4.9	18.1	3.1
10522-34-6	SOURNESS INTENSITY	0.031	-0.177	0.201	54	1.1	1.1	18.1	3.1
140-11-4	SOURNESS INTENSITY	0.031	0.176	0.203	54	11.1	8.6	18.1	3.1
L* ext	SOURNESS INTENSITY	0.030	0.172	0.212	54	33.6	2.6	18.1	3.1

556-24-1	SOURNESS INTENSITY	0.028	-0.167	0.226	54	46.6	57.0	18.1	3.1
55514-48-2	SOURNESS INTENSITY	0.027	0.165	0.232	54	0.5	0.5	18.1	3.1
2548-87-0	SOURNESS INTENSITY	0.027	0.165	0.233	54	2.4	1.5	18.1	3.1
623-43-8	SOURNESS INTENSITY	0.024	-0.154	0.267	54	3.4	3.4	18.1	3.1
TOTAL SUGAR	SOURNESS INTENSITY	0.022	-0.148	0.287	54	4473.9	1037.2	18.1	3.1
L* int	SOURNESS INTENSITY	0.021	-0.145	0.296	54	54.9	5.8	18.1	3.1
SSC	SOURNESS INTENSITY	0.020	-0.142	0.305	54	7.4	1.4	18.1	3.1
112-14-1	SOURNESS INTENSITY	0.020	-0.141	0.311	54	18.3	24.0	18.1	3.1
B* ext	SOURNESS INTENSITY	0.020	0.140	0.312	54	19.0	3.3	18.1	3.1
B* int	SOURNESS INTENSITY	0.020	0.140	0.314	54	25.8	4.5	18.1	3.1
103-09-3	SOURNESS INTENSITY	0.017	0.129	0.354	54	3.0	1.1	18.1	3.1
1534-08-3	SOURNESS INTENSITY	0.016	-0.127	0.359	54	0.4	0.2	18.1	3.1
106-70-7	SOURNESS INTENSITY	0.016	-0.125	0.366	54	252.7	164.0	18.1	3.1
110-43-0	SOURNESS INTENSITY	0.015	-0.121	0.381	54	14.8	19.6	18.1	3.1
53398-83-7	SOURNESS INTENSITY	0.014	-0.120	0.389	54	5.0	4.6	18.1	3.1
5989-33-3	SOURNESS INTENSITY	0.014	0.117	0.398	54	2.8	2.4	18.1	3.1
5881-17-4	SOURNESS INTENSITY	0.014	0.117	0.401	54	6.2	2.5	18.1	3.1
7452-79-1	SOURNESS INTENSITY	0.013	-0.114	0.411	54	50.0	31.0	18.1	3.1
1576-95-0	SOURNESS INTENSITY	0.013	0.114	0.413	54	2.1	2.0	18.1	3.1
20664-46-4	SOURNESS INTENSITY	0.012	0.111	0.423	54	20.6	19.0	18.1	3.1
29811-50-5	SOURNESS INTENSITY	0.012	0.108	0.436	54	3.2	5.3	18.1	3.1
110-38-3	SOURNESS INTENSITY	0.012	-0.108	0.437	54	2.0	2.6	18.1	3.1
2639-63-6	SOURNESS INTENSITY	0.011	-0.107	0.443	54	10.8	13.0	18.1	3.1
123-92-2	SOURNESS INTENSITY	0.011	0.105	0.450	54	23.0	21.5	18.1	3.1
109-21-7	SOURNESS INTENSITY	0.011	-0.105	0.451	54	72.1	157.2	18.1	3.1
110-93-0	SOURNESS INTENSITY	0.011	0.105	0.452	54	2.7	1.5	18.1	3.1
29674-47-3	SOURNESS INTENSITY	0.011	0.104	0.454	54	5.0	5.0	18.1	3.1

SUCROSE	SOURNESS INTENSITY	0.010	0.099	0.476	54	1112.6	646.5	18.1	3.1
4077-47-8	SOURNESS INTENSITY	0.009	0.092	0.507	54	11.7	8.3	18.1	3.1
539-82-2	SOURNESS INTENSITY	0.008	0.092	0.510	54	3.3	3.9	18.1	3.1
591-78-6	SOURNESS INTENSITY	0.008	0.091	0.514	54	10.3	13.9	18.1	3.1
A* int	SOURNESS INTENSITY	0.008	0.088	0.525	54	28.8	7.6	18.1	3.1
109-19-3	SOURNESS INTENSITY	0.007	-0.086	0.537	54	2.7	3.8	18.1	3.1
60415-61-4	SOURNESS INTENSITY	0.007	-0.084	0.545	54	0.7	1.6	18.1	3.1
142-92-7	SOURNESS INTENSITY	0.007	-0.081	0.559	54	53.5	49.7	18.1	3.1
111-71-7	SOURNESS INTENSITY	0.006	0.075	0.589	54	3.2	2.3	18.1	3.1
106-32-1	SOURNESS INTENSITY	0.006	0.075	0.591	54	2.2	3.0	18.1	3.1
66-25-1	SOURNESS INTENSITY	0.005	0.072	0.605	54	2545.9	1722.0	18.1	3.1
564-94-3	SOURNESS INTENSITY	0.005	0.072	0.606	54	6.5	7.4	18.1	3.1
116-53-0	SOURNESS INTENSITY	0.005	-0.068	0.625	54	19.7	16.0	18.1	3.1
109-60-4	SOURNESS INTENSITY	0.005	0.068	0.626	54	3.8	2.7	18.1	3.1
2432-51-1	SOURNESS INTENSITY	0.004	-0.064	0.644	54	4.4	5.8	18.1	3.1
1576-86-9	SOURNESS INTENSITY	0.004	0.064	0.645	54	37.8	21.7	18.1	3.1
124-13-0	SOURNESS INTENSITY	0.004	0.061	0.660	54	5.9	3.0	18.1	3.1
3913-81-3	SOURNESS INTENSITY	0.004	-0.060	0.669	54	1.9	1.4	18.1	3.1
1629-58-9	SOURNESS INTENSITY	0.003	0.058	0.678	54	117.9	65.9	18.1	3.1
628-63-7	SOURNESS INTENSITY	0.003	-0.052	0.709	54	4.6	1.7	18.1	3.1
7786-58-5	SOURNESS INTENSITY	0.003	-0.051	0.713	54	12.2	28.4	18.1	3.1
96-22-0	SOURNESS INTENSITY	0.002	0.050	0.721	54	51.2	18.6	18.1	3.1
616-25-1	SOURNESS INTENSITY	0.002	0.048	0.731	54	15.9	7.3	18.1	3.1
75-85-4	SOURNESS INTENSITY	0.002	0.047	0.737	54	3.9	2.2	18.1	3.1
706-14-9	SOURNESS INTENSITY	0.002	-0.040	0.772	54	44.5	81.1	18.1	3.1
A* ext	SOURNESS INTENSITY	0.002	-0.040	0.774	54	36.4	3.1	18.1	3.1
2311-46-8	SOURNESS INTENSITY	0.002	0.039	0.779	54	3.9	4.4	18.1	3.1

TOTAL VOLATILES	SOURNESS INTENSITY	0.001	0.029	0.834	54	15814.0	5238.5	18.1	3.1
40716-66-3	SOURNESS INTENSITY	0.001	-0.029	0.838	54	84.3	107.5	18.1	3.1
123-66-0	SOURNESS INTENSITY	0.001	-0.027	0.849	54	108.2	128.0	18.1	3.1
1576-87-0	SOURNESS INTENSITY	0.001	0.026	0.851	54	37.5	20.1	18.1	3.1
71-41-0	SOURNESS INTENSITY	0.001	-0.025	0.855	54	1.0	1.3	18.1	3.1
124-19-6	SOURNESS INTENSITY	0.000	0.021	0.883	54	8.5	7.5	18.1	3.1
108-10-1	SOURNESS INTENSITY	0.000	0.020	0.886	54	1.5	2.6	18.1	3.1
638-11-9	SOURNESS INTENSITY	0.000	-0.018	0.899	54	72.7	67.2	18.1	3.1
6728-26-3	SOURNESS INTENSITY	0.000	0.016	0.909	54	8666.5	3359.7	18.1	3.1
123-86-4	SOURNESS INTENSITY	0.000	-0.016	0.910	54	73.5	85.0	18.1	3.1
105-66-8	SOURNESS INTENSITY	0.000	0.011	0.938	54	5.0	3.7	18.1	3.1
105-54-4	SOURNESS INTENSITY	0.000	-0.011	0.939	54	42.0	17.1	18.1	3.1
FORCE	SOURNESS INTENSITY	0.000	0.008	0.953	54	0.6	0.2	18.1	3.1
540-18-1	SOURNESS INTENSITY	0.000	-0.008	0.957	54	3.5	3.2	18.1	3.1
2305-05-7	SOURNESS INTENSITY	0.000	-0.007	0.962	54	6.7	6.7	18.1	3.1
105-37-3	SOURNESS INTENSITY	0.000	-0.006	0.964	54	10.1	14.1	18.1	3.1
623-42-7	SOURNESS INTENSITY	0.000	-0.006	0.966	54	2780.2	1376.8	18.1	3.1
821-55-6	SOURNESS INTENSITY	0.000	0.001	0.997	54	3.5	8.2	18.1	3.1
SSC	STRAWBERRY FLAVOR INTENSITY	0.584	0.764	0.000	54	7.4	1.4	26.9	3.9
TOTAL SUGAR	STRAWBERRY FLAVOR INTENSITY	0.569	0.755	0.000	54	4473.9	1037.2	26.9	3.9
SUCROSE	STRAWBERRY FLAVOR INTENSITY	0.498	0.705	0.000	54	1112.6	646.5	26.9	3.9
1629-58-9	STRAWBERRY FLAVOR INTENSITY	0.356	0.597	0.000	54	117.9	65.9	26.9	3.9
2305-05-7	STRAWBERRY FLAVOR INTENSITY	0.283	0.532	0.000	54	6.7	6.7	26.9	3.9
540-18-1	STRAWBERRY FLAVOR INTENSITY	0.260	0.509	0.000	54	3.5	3.2	26.9	3.9

TA	STRAWBERRY FLAVOR INTENSITY	0.256	0.506	0.000	54	0.8	0.1	26.9	3.9
1576-87-0	STRAWBERRY FLAVOR INTENSITY	0.239	0.488	0.000	54	37.5	20.1	26.9	3.9
2639-63-6	STRAWBERRY FLAVOR INTENSITY	0.235	0.485	0.000	54	10.8	13.0	26.9	3.9
CITRIC ACID	STRAWBERRY FLAVOR INTENSITY	0.235	0.485	0.000	54	741.0	147.2	26.9	3.9
60415-61-4	STRAWBERRY FLAVOR INTENSITY	0.233	0.482	0.000	54	0.7	1.6	26.9	3.9
142-92-7	STRAWBERRY FLAVOR INTENSITY	0.227	0.477	0.000	54	53.5	49.7	26.9	3.9
1576-86-9	STRAWBERRY FLAVOR INTENSITY	0.222	0.472	0.000	54	37.8	21.7	26.9	3.9
5989-33-3	STRAWBERRY FLAVOR INTENSITY	0.208	0.456	0.001	54	2.8	2.4	26.9	3.9
GLUCOSE	STRAWBERRY FLAVOR INTENSITY	0.205	0.453	0.001	54	1594.6	378.2	26.9	3.9
123-86-4	STRAWBERRY FLAVOR INTENSITY	0.205	0.453	0.001	54	73.5	85.0	26.9	3.9
111-71-7	STRAWBERRY FLAVOR INTENSITY	0.201	0.448	0.001	54	3.2	2.3	26.9	3.9
109-21-7	STRAWBERRY FLAVOR INTENSITY	0.194	0.440	0.001	54	72.1	157.2	26.9	3.9
591-78-6	STRAWBERRY FLAVOR INTENSITY	0.180	0.425	0.001	54	10.3	13.9	26.9	3.9
109-19-3	STRAWBERRY FLAVOR INTENSITY	0.179	0.424	0.001	54	2.7	3.8	26.9	3.9
706-14-9	STRAWBERRY FLAVOR INTENSITY	0.179	0.423	0.001	54	44.5	81.1	26.9	3.9
TOTAL VOLATILES	STRAWBERRY FLAVOR INTENSITY	0.167	0.409	0.002	54	15814.0	5238.5	26.9	3.9
FRUCTOSE	STRAWBERRY FLAVOR INTENSITY	0.166	0.407	0.002	54	1766.7	381.5	26.9	3.9
638-11-9	STRAWBERRY FLAVOR INTENSITY	0.164	0.405	0.002	54	72.7	67.2	26.9	3.9

3913-81-3	STRAWBERRY FLAVOR INTENSITY	0.158	0.398	0.003	54	1.9	1.4	26.9	3.9
616-25-1	STRAWBERRY FLAVOR INTENSITY	0.154	0.392	0.003	54	15.9	7.3	26.9	3.9
5881-17-4	STRAWBERRY FLAVOR INTENSITY	0.153	0.391	0.003	54	6.2	2.5	26.9	3.9
110-93-0	STRAWBERRY FLAVOR INTENSITY	0.137	0.370	0.006	54	2.7	1.5	26.9	3.9
124-19-6	STRAWBERRY FLAVOR INTENSITY	0.129	0.359	0.008	54	8.5	7.5	26.9	3.9
40716-66-3	STRAWBERRY FLAVOR INTENSITY	0.112	0.335	0.013	54	84.3	107.5	26.9	3.9
L* int	STRAWBERRY FLAVOR INTENSITY	0.109	0.330	0.015	54	54.9	5.8	26.9	3.9
4077-47-8	STRAWBERRY FLAVOR INTENSITY	0.108	0.328	0.015	54	11.7	8.3	26.9	3.9
2311-46-8	STRAWBERRY FLAVOR INTENSITY	0.103	0.322	0.018	54	3.9	4.4	26.9	3.9
110-43-0	STRAWBERRY FLAVOR INTENSITY	0.101	0.318	0.019	54	14.8	19.6	26.9	3.9
623-42-7	STRAWBERRY FLAVOR INTENSITY	0.097	0.312	0.022	54	2780.2	1376.8	26.9	3.9
6728-26-3	STRAWBERRY FLAVOR INTENSITY	0.096	0.310	0.022	54	8666.5	3359.7	26.9	3.9
105-54-4	STRAWBERRY FLAVOR INTENSITY	0.089	0.299	0.028	54	42.0	17.1	26.9	3.9
A* int	STRAWBERRY FLAVOR INTENSITY	0.087	-0.295	0.030	54	28.8	7.6	26.9	3.9
110-62-3	STRAWBERRY FLAVOR INTENSITY	0.079	-0.281	0.039	54	7.9	8.9	26.9	3.9
110-39-4	STRAWBERRY FLAVOR INTENSITY	0.079	0.281	0.040	54	40.8	70.0	26.9	3.9
78-70-6	STRAWBERRY FLAVOR INTENSITY	0.074	0.272	0.046	54	128.8	113.0	26.9	3.9
29674-47-3	STRAWBERRY FLAVOR INTENSITY	0.069	0.263	0.055	54	5.0	5.0	26.9	3.9

5454-09-1	STRAWBERRY FLAVOR INTENSITY	0.067	0.259	0.059	54	3.7	6.2	26.9	3.9
96-22-0	STRAWBERRY FLAVOR INTENSITY	0.066	0.258	0.060	54	51.2	18.6	26.9	3.9
104-76-7	STRAWBERRY FLAVOR INTENSITY	0.064	0.252	0.066	54	6.0	4.9	26.9	3.9
10522-34-6	STRAWBERRY FLAVOR INTENSITY	0.060	0.246	0.073	54	1.1	1.1	26.9	3.9
7786-58-5	STRAWBERRY FLAVOR INTENSITY	0.059	0.243	0.077	54	12.2	28.4	26.9	3.9
124-13-0	STRAWBERRY FLAVOR INTENSITY	0.057	0.239	0.081	54	5.9	3.0	26.9	3.9
112-14-1	STRAWBERRY FLAVOR INTENSITY	0.056	0.237	0.084	54	18.3	24.0	26.9	3.9
2548-87-0	STRAWBERRY FLAVOR INTENSITY	0.056	0.236	0.086	54	2.4	1.5	26.9	3.9
628-63-7	STRAWBERRY FLAVOR INTENSITY	0.042	0.206	0.135	54	4.6	1.7	26.9	3.9
B* int	STRAWBERRY FLAVOR INTENSITY	0.039	-0.198	0.152	54	25.8	4.5	26.9	3.9
105-66-8	STRAWBERRY FLAVOR INTENSITY	0.037	0.193	0.162	54	5.0	3.7	26.9	3.9
20664-46-4	STRAWBERRY FLAVOR INTENSITY	0.037	0.193	0.162	54	20.6	19.0	26.9	3.9
564-94-3	STRAWBERRY FLAVOR INTENSITY	0.037	0.192	0.165	54	6.5	7.4	26.9	3.9
4887-30-3	STRAWBERRY FLAVOR INTENSITY	0.034	0.185	0.180	54	16.9	32.5	26.9	3.9
110-38-3	STRAWBERRY FLAVOR INTENSITY	0.031	0.176	0.204	54	2.0	2.6	26.9	3.9
53398-83-7	STRAWBERRY FLAVOR INTENSITY	0.029	0.172	0.215	54	5.0	4.6	26.9	3.9
539-82-2	STRAWBERRY FLAVOR INTENSITY	0.026	0.160	0.247	54	3.3	3.9	26.9	3.9
134-20-3	STRAWBERRY FLAVOR INTENSITY	0.021	-0.145	0.295	54	0.1	0.7	26.9	3.9

29811-50-5	STRAWBERRY FLAVOR INTENSITY	0.020	0.143	0.303	54	3.2	5.3	26.9	3.9
556-24-1	STRAWBERRY FLAVOR INTENSITY	0.019	0.137	0.324	54	46.6	57.0	26.9	3.9
589-38-8	STRAWBERRY FLAVOR INTENSITY	0.018	0.136	0.328	54	1.9	1.2	26.9	3.9
821-55-6	STRAWBERRY FLAVOR INTENSITY	0.017	0.130	0.348	54	3.5	8.2	26.9	3.9
2432-51-1	STRAWBERRY FLAVOR INTENSITY	0.017	-0.129	0.351	54	4.4	5.8	26.9	3.9
7452-79-1	STRAWBERRY FLAVOR INTENSITY	0.016	0.125	0.369	54	50.0	31.0	26.9	3.9
66-25-1	STRAWBERRY FLAVOR INTENSITY	0.016	0.125	0.369	54	2545.9	1722.0	26.9	3.9
116-53-0	STRAWBERRY FLAVOR INTENSITY	0.015	0.122	0.379	54	19.7	16.0	26.9	3.9
128-37-0	STRAWBERRY FLAVOR INTENSITY	0.013	0.115	0.407	54	4.1	3.7	26.9	3.9
pH	STRAWBERRY FLAVOR INTENSITY	0.010	0.102	0.464	54	3.7	0.2	26.9	3.9
B* ext	STRAWBERRY FLAVOR INTENSITY	0.009	-0.097	0.488	54	19.0	3.3	26.9	3.9
106-70-7	STRAWBERRY FLAVOR INTENSITY	0.009	0.096	0.491	54	252.7	164.0	26.9	3.9
624-24-8	STRAWBERRY FLAVOR INTENSITY	0.009	-0.094	0.498	54	5.6	3.7	26.9	3.9
623-43-8	STRAWBERRY FLAVOR INTENSITY	0.008	-0.088	0.529	54	3.4	3.4	26.9	3.9
1534-08-3	STRAWBERRY FLAVOR INTENSITY	0.007	0.086	0.535	54	0.4	0.2	26.9	3.9
71-41-0	STRAWBERRY FLAVOR INTENSITY	0.007	0.084	0.548	54	1.0	1.3	26.9	3.9
123-66-0	STRAWBERRY FLAVOR INTENSITY	0.006	0.078	0.576	54	108.2	128.0	26.9	3.9
A* ext	STRAWBERRY FLAVOR INTENSITY	0.005	0.073	0.599	54	36.4	3.1	26.9	3.9

15111-96-3	STRAWBERRY FLAVOR INTENSITY	0.005	-0.071	0.608	54	1.2	1.2	26.9	3.9
106-32-1	STRAWBERRY FLAVOR INTENSITY	0.005	-0.070	0.616	54	2.2	3.0	26.9	3.9
1191-16-8	STRAWBERRY FLAVOR INTENSITY	0.004	-0.064	0.646	54	5.5	7.0	26.9	3.9
75-85-4	STRAWBERRY FLAVOR INTENSITY	0.004	0.059	0.670	54	3.9	2.2	26.9	3.9
105-37-3	STRAWBERRY FLAVOR INTENSITY	0.002	-0.049	0.723	54	10.1	14.1	26.9	3.9
108-10-1	STRAWBERRY FLAVOR INTENSITY	0.002	-0.048	0.729	54	1.5	2.6	26.9	3.9
140-11-4	STRAWBERRY FLAVOR INTENSITY	0.002	0.048	0.729	54	11.1	8.6	26.9	3.9
FORCE	STRAWBERRY FLAVOR INTENSITY	0.002	-0.048	0.730	54	0.6	0.2	26.9	3.9
111-27-3	STRAWBERRY FLAVOR INTENSITY	0.002	-0.045	0.744	54	45.5	94.6	26.9	3.9
2497-18-9	STRAWBERRY FLAVOR INTENSITY	0.002	0.045	0.746	54	24.9	21.5	26.9	3.9
624-41-9	STRAWBERRY FLAVOR INTENSITY	0.002	0.045	0.748	54	18.9	18.1	26.9	3.9
96-04-8	STRAWBERRY FLAVOR INTENSITY	0.002	0.040	0.776	54	3.1	8.0	26.9	3.9
109-60-4	STRAWBERRY FLAVOR INTENSITY	0.001	-0.037	0.792	54	3.8	2.7	26.9	3.9
L* ext	STRAWBERRY FLAVOR INTENSITY	0.001	0.034	0.809	54	33.6	2.6	26.9	3.9
123-92-2	STRAWBERRY FLAVOR INTENSITY	0.001	0.032	0.819	54	23.0	21.5	26.9	3.9
55514-48-2	STRAWBERRY FLAVOR INTENSITY	0.000	-0.020	0.884	54	0.5	0.5	26.9	3.9
MALIC ACID	STRAWBERRY FLAVOR INTENSITY	0.000	0.015	0.914	54	212.4	51.6	26.9	3.9
928-95-0	STRAWBERRY FLAVOR INTENSITY	0.000	-0.010	0.943	54	66.8	61.7	26.9	3.9

103-09-3	STRAWBERRY FLAVOR INTENSITY	0.000	0.006	0.968	54	3.0	1.1	26.9	3.9
1576-95-0	STRAWBERRY FLAVOR INTENSITY	0.000	-0.003	0.983	54	2.1	2.0	26.9	3.9

Table S5. Multiple regression for identification of sweetness enhancing volatiles. Individual volatile compound concentrations are regressed against perceived sweetness intensity independent of effect from glucose, fructose, or sucrose, separately. Thirty compounds (*) (α=0.05) were found to enhance intensity of sweetness independent of at least one of the three sugars. Six compounds (bold) were found to significantly enhance intensity of sweetness independent of all three sugars.

doi:10.1371/journal.pone.0088446.s008

CAS #	FRUCTOSE t RATIO	FRUCTOSE *p*-VALUE		SUCROSE t RATIO	SUCROSE *p*-VALUE		GLUCOSE t RATIO	GLUCOSE *p*-VALUE	
1629-58-9	**5.097**	**0**	*	**2.41**	**0.02**	*	**4.696**	**0**	*
1576-87-0	4.566	0	*	1.024	0.311		4.301	0	*
1576-86-9	4.16	0	*	0.935	0.354		3.915	0	*
2305-05-7	**3.933**	**0**	*	**2.784**	**0.008**	*	**3.549**	**0.001**	*
3913-81-3	3.694	0.001	*	1.411	0.164		3.494	0.001	*
124-19-6	3.696	0.001	*	0.226	0.822		3.402	0.001	*
6728-26-3	3.349	0.002	*	-0.816	0.418		3.314	0.002	*
591-78-6	2.807	0.007	*	0.767	0.447		2.788	0.007	*
5881-17-4	2.894	0.006	*	0.608	0.546		2.662	0.01	*
540-18-1	**2.71**	**0.009**	*	**2.292**	**0.026**	*	**2.515**	**0.015**	*
2639-63-6	**2.865**	**0.006**	*	**2.892**	**0.006**	*	**2.512**	**0.015**	*
105-54-4	2.533	0.014	*	0.034	0.973		2.493	0.016	*
564-94-3	2.588	0.013	*	-1.322	0.192		2.455	0.018	*
111-71-7	2.599	0.012	*	1.342	0.186		2.283	0.027	*
4077-47-8	2.414	0.019	*	0.299	0.766		2.185	0.034	*
110-93-0	2.527	0.015	*	1.43	0.159		2.165	0.035	*
638-11-9	2.311	0.025	*	1.256	0.215		2.14	0.037	*
142-92-7	**2.346**	**0.023**	*	**2.943**	**0.005**	*	**2.096**	**0.041**	*
60415-61-4	**2.309**	**0.025**	*	**2.119**	**0.039**	*	**2.062**	**0.044**	*
116-53-0	2.01	0.05	*	-0.286	0.776		2.035	0.047	*
123-86-4	2.179	0.034	*	1.147	0.257		2.008	0.05	*
7452-79-1	1.959	0.056		-0.785	0.436		1.993	0.052	*
109-21-7	2.181	0.034	*	1.65	0.105		1.961	0.055	
109-19-3	2.005	0.05	*	1.662	0.103		1.954	0.056	
616-25-1	1.773	0.082		0.795	0.43		1.628	0.11	
5454-09-1	1.804	0.077		2.085	0.042	*	1.579	0.12	
96-22-0	1.82	0.075		0.912	0.366		1.576	0.121	
5989-33-3	1.869	0.067		1.953	0.056		1.561	0.125	
2548-87-0	1.76	0.084		0.067	0.947		1.509	0.138	
623-42-7	1.455	0.152		0.419	0.677		1.452	0.153	
29674-47-3	1.339	0.187		0.035	0.972		1.362	0.179	

53398-83-7	1.482	0.144		2.556	0.014	*	1.283	0.205	
40716-66-3	1.521	0.134		1	0.322		1.202	0.235	
66-25-1	1.229	0.225		0.21	0.835		1.178	0.244	
104-76-7	1.189	0.24		2.046	0.046	*	0.982	0.331	
556-24-1	0.732	0.468		0.972	0.336		0.88	0.383	
706-14-9	1.247	0.218		1.65	0.105		0.879	0.384	
110-39-4	0.96	0.341		2.645	0.011	*	0.814	0.419	
628-63-7	0.882	0.382		0.427	0.671		0.749	0.457	
78-70-6	0.872	0.387		-0.1	0.921		0.7	0.487	
124-13-0	0.871	0.388		0.169	0.866		0.685	0.497	
75-85-4	0.736	0.465		-0.306	0.761		0.667	0.508	
110-43-0	1.054	0.297		1.899	0.063		0.657	0.514	
105-66-8	0.88	0.383		2.421	0.019	*	0.638	0.526	
623-43-8	0.248	0.805		-1.396	0.169		0.54	0.592	
1534-08-3	0.488	0.628		1.035	0.305		0.429	0.67	
71-41-0	0.448	0.656		1.938	0.058		0.296	0.769	
10522-34-6	0.536	0.594		2.049	0.046	*	0.288	0.775	
112-14-1	0.413	0.681		2.292	0.026	*	0.284	0.777	
4887-30-3	0.392	0.697		2.71	0.009	*	0.254	0.801	
7786-58-5	0.362	0.718		2.027	0.048	*	0.227	0.821	
103-09-3	0.172	0.864		-1.053	0.297		0.21	0.835	
134-20-3	-0.009	0.993		1.386	0.172		0.2	0.842	
15111-96-3	0.23	0.819		0.927	0.358		0.192	0.849	
110-38-3	0.281	0.78		2.621	0.012	*	0.055	0.956	
96-04-8	0.142	0.887		-0.646	0.521		0.002	0.998	
123-66-0	0.066	0.948		1.064	0.292		-0.001	0.999	
2311-46-8	0.238	0.813		2.211	0.032	*	-0.058	0.954	
29811-50-5	0.016	0.987		0.358	0.722		-0.098	0.922	
1576-95-0	-0.186	0.853		-0.304	0.762		-0.231	0.819	
2497-18-9	-0.313	0.756		-0.056	0.955		-0.331	0.742	
20664-46-4	-0.17	0.865		1.383	0.173		-0.395	0.694	
624-24-8	-0.406	0.686		0.662	0.511		-0.416	0.679	
589-38-8	-0.217	0.829		0.592	0.557		-0.427	0.671	
109-60-4	-0.491	0.626		-0.224	0.823		-0.454	0.652	
821-55-6	-0.267	0.791		1.192	0.239		-0.467	0.642	
624-41-9	-0.433	0.667		-1.323	0.192		-0.494	0.624	
140-11-4	-0.361	0.72		-1.357	0.181		-0.513	0.61	
1191-16-8	-0.581	0.564		-2.268	0.028	*	-0.529	0.599	
106-70-7	-0.321	0.75		0.802	0.426		-0.531	0.598	
110-62-3	-0.572	0.57		0.997	0.323		-0.568	0.573	

105-37-3	-0.623	0.536		1.964	0.055		-0.674	0.503	
55514-48-2	-0.644	0.523		-0.889	0.378		-0.675	0.502	
123-92-2	-0.725	0.472		0.108	0.914		-0.771	0.444	
539-82-2	-0.48	0.633		2.273	0.027	*	-0.802	0.426	
2432-51-1	-0.956	0.344		2.033	0.047	*	-1.092	0.28	
128-37-0	-0.856	0.396		0.482	0.632		-1.17	0.247	
111-27-3	-1.253	0.216		-0.542	0.59		-1.481	0.145	
928-95-0	-1.842	0.071		-0.516	0.608		-1.521	0.134	
108-10-1	-2.185	0.034	*	0.898	0.374		-2.138	0.037	*
106-32-1	-1.436	0.157		0.193	0.848		-2.36	0.022	*

ACKNOWLEDGMENTS

The authors wish to acknowledge David Moore for fruit transport, Yanina Perez for assistance with fruit chemical data collection, and Timothy Johnson for assistance with volatile collection. Portions of the results are protected by US Patent 20130280400 and International Patent WO 2013/163272 A1.

Author Contributions

Conceived and designed the experiments: MLS TAC LMB DMT KMF HJK CAS VMW DGC. Performed the experiments: MLS JLG AZO. Analyzed the data: MLS EAJ LMB HRM. Contributed reagents/materials/analysis tools: DMT HJK VMW. Wrote the paper: MLS.

REFERENCES

1. Whitaker VM, Hasing T, Chandler CK, Plotto A, Baldwin E (2011) Historical Trends in Strawberry Fruit Quality Revealed by a Trial of University of Florida Cultivars and Advanced Selections. Hortscience 46: 553–557. doi: 10.1080/15538362.2012.698183
2. Hong V, Wrolstad RE (1990) Use of HPLC separation photodiode array detection for characterization of anthocyanins. Journal of Agricultural and Food Chemistry 38.
3. Brummell DA, Harpster MH (2001) Cell wall metabolism in fruit softening and quality and its manipulation in transgenic plants. Plant Molecular Biology 47.
4. Ulrich D, Hoberg E, Rapp A, Kecke S (1997) Analysis of strawberry flavour - discrimination of aroma types by quantification of volatile compounds. Zeitschrift Fur Lebensmittel-Untersuchung Und-Forschung a-Food Research and Technology 205.
5. Schieberle P, Hofmann T (1997) Evaluation of the character impact odorants in fresh strawberry juice by quantitative measurements and sensory studies on model mixtures. Journal of Agricultural and Food Chemistry 45.
6. Hancock JF (1999) Strawberries.
7. Zhang J, Wang X, Yu O, Tang J, Gu X, et al.. (2011) Metabolic profiling of strawberry (Fragariaxananassa Duch.) during fruit development and maturation. Journal of Experimental Botany 62.
8. Menager I, Jost M, Aubert C (2004) Changes in physicochemical characteristics and volatile constituents of strawberry (Cv. cigaline) during maturation. Journal of Agricultural and Food Chemistry 52.
9. Prescott JJ (2004) Psychological processes in flavour perception. Flavor perception.
10. Causse M, Saliba-Colombani V, Lesschaeve I, Buret M (2001) Genetic analysis of organoleptic quality in fresh market tomato. 2. Mapping QTLs for sensory attributes. Theoretical and Applied Genetics 102: 273–283. doi: 10.1007/s001220051644
11. Christensen CM (1983) Effects of color on aroma, flavor and texture judgements of foods. Journal of Food Science 48: 787–790. doi: 10.1111/j.1365-2621.1983.tb14899.x
12. Hall RL (1968) Flavor and flavoring seeking a concensus of definition. Food Technology 22: 1496–&.
13. Stommel J, Abbott JA, Saftner RA, Camp MJ (2005) Sensory and objective quality attributes of beta-carotene and lycopene-rich tomato fruit. Journal of the American Society for Horticultural Science 130: 244–251.
14. Colquhoun TA, Levin LA, Moskowitz HR, Whitaker VM, Clark DG, et al. (2012) Framing the perfect strawberry: an exercise in consumer-assisted selection of fruit crops. Journal of Berry Research 2: 45–61.
15. Basson CE, Groenewald JH, Kossmann J, Cronje C, Bauer R (2010) Sugar and acid-related quality attributes and enzyme activities in strawberry fruits: Invertase is the main sucrose hydrolysing enzyme. Food Chemistry 121.
16. Fait A, Hanhineva K, Beleggia R, Dai N, Rogachev I, et al.. (2008) Reconfiguration of the achene and receptacle metabolic networks during strawberry fruit development. Plant Physiology 148.
17. Hoffmann T, Kalinowski G, Schwab W (2006) RNAi-induced silencing of gene expression in strawberry fruit (Fragaria x ananassa) by agroinfiltration: a rapid assay for gene function analysis. Plant Journal 48.
18. Aharoni A, Keizer LCP, Bouwmeester HJ, Sun ZK, Alvarez-Huerta M, et al.. (2000) Identification of the SAAT gene involved in strawberry flavor biogenesis by use of DNA microarrays. Plant Cell 12.
19. Quesada MA, Blanco-Portales R, Pose S, Garcia-Gago

JA, Jimenez-Bermudez S, et al.. (2009) Antisense Down-Regulation of the FaPG1 Gene Reveals an Unexpected Central Role for Polygalacturonase in Strawberry Fruit Softening. Plant Physiology 150.

20. Trainotti L, Ferrarese L, Dalla Vecchia F, Rascio N, Casadoro G (1999) Two different endo-beta-1,4-glucanases contribute to the softening of the strawberry fruits. Journal of Plant Physiology 154.

21. Perez AG, Olias R, Luaces P, Sanz C (2002) Biosynthesis of strawberry aroma compounds through amino acid metabolism. Journal of Agricultural and Food Chemistry 50.

22. Maarse H (1991) Volatile Compounds in Foods and Beverages.

23. Hakala MA, Lapvetelainen AT, Kallio HP (2002) Volatile compounds of selected strawberry varieties analyzed by purge-and-trap headspace GC-MS. Journal of Agricultural and Food Chemistry 50.

24. Jetti RR, Yang E, Kurnianta A, Finn C, Qian MC (2007) Quantification of selected aroma-active compounds in strawberries by headspace solid-phase microextraction gas chromatography and correlation with sensory descriptive analysis. Journal of Food Science 72.

25. Olbricht K, Grafe C, Weiss K, Ulrich D (2008) Inheritance of aroma compounds in a model population of Fragaria x ananassa Duch. Plant Breeding 127.

26. MacKenzie SJ, Chandler CK (2009) The late season decline in strawberry fruit soluble solid content observed in Florida is caused by rising temperatures. Acta Horticulturae.

27. MacKenzie SJ, Chandler CK, Hasing T, Whitaker VM (2011) The Role of Temperature in the Late-season Decline in Soluble Solids Content of Strawberry Fruit in a Subtropical Production System. Hortscience 46.

28. Jouquand C, Chandler C, Plotto A, Goodner K (2008) A Sensory and Chemical Analysis of Fresh Strawberries Over Harvest Dates and Seasons Reveals Factors That Affect Eating Quality. Journal of the American Society for Horticultural Science 133: 859–867.

29. Santos BM, Peres NA, Price JF, Whitaker VM, Dittmar PJ, et al.. (2012) Strawberry Production in Florida. Vegetable Production Handbook. Gainesville, FL: University of Florida Institute of Food and Agricultural Sciences.

30. Strand L (2008) Integrated Pest Management for Strawberries: ANR Publications.

31. Underwood BA, Tieman DM, Shibuya K, Dexter RJ, Loucas HM, et al. (2005) Ethylene-regulated floral volatile synthesis in petunia corollas. Plant Physiology 138: 255–266. doi: 10.1104/pp.104.051144

32. Schmelz EA, Alborn HT, Banchio E, Tumlinson JH (2003) Quantitative relationships between induced jasmonic acid levels and volatile emission in Zea mays during Spodoptera exigua herbivory. Planta 216: 665–673.

33. Schmelz EA, Alborn HT, Tumlinson JH (2001) The influence of intact-plant and excised-leaf bioassay designs on volicitin- and jasmonic acid-induced sesquiterpene volatile release in Zea mays. Planta 214: 171–179. doi: 10.1007/s004250100603

34. Tieman D, Bliss P, McIntyre LM, Blandon-Ubeda A, Bies D, et al. (2012) The Chemical Interactions Underlying Tomato Flavor Preferences. Current Biology 22: 1035–1039. doi: 10.1016/j.cub.2012.04.016

35. Bartoshuk LM, Duffy V, Green BG, Hoffman HJ, Ko CW, et al. (2004) Valid across-group comparisons with labeled scales: the gLMS versus magnitude matching. Physiology & Behavior 82: 109–114. doi: 10.1016/j.physbeh.2004.02.033

36. Bartoshuk LM, Duffy VB, Fast K, Green BG, Prutkin J, et al. (2003) Labeled scales (eg, category, Likert, VAS) and invalid across-group comparisons: what we have learned from genetic variation in taste. Food Quality and Preference 14: 125–138. doi: 10.1016/s0950-3293(02)00077-0

37. Bartoshuk LM, Fast K, Snyder DJ (2005) Differences in our sensory - Invalid comparisons with labeled scales. Current Directions in Psychological Science 14: 122–125. doi: 10.1111/j.0963-7214.2005.00346.x

38. Lindemann B (2001) Receptors and transduction in taste. Nature 413: 219–225. doi: 10.1038/35093032

39. Fujimaru T, Lim J (2013) Effects of Stimulus Intensity on Odor Enhancement by Taste. Chemosensory Perception 6: 1–7. doi: 10.1007/s12078-013-9140-5

40. Mikulic-Petkovsek M, Schmitzer V, Slatnar A, Stampar F, Veberic R (2012) Composition of Sugars, Organic Acids, and Total Phenolics in 25 Wild or Cultivated Berry Species. Journal of Food Science 77: C1064–C1070. doi: 10.1111/j.1750-3841.2012.02896.x

41. Watson R, Wright CJ, McBurney T, Taylor AJ, Linforth RST (2002) Influence of harvest date and light integral on the development of strawberry flavour compounds. Journal of Experimental Botany 53.

42. Pelayo-Zaldivar C, Ebeler SE, Kader AA (2005) Cultivar and harvest date effects on flavor and other quality attributes of California strawberries. Journal of Food Quality 28.

43. Cumplido-Laso G, Medina-Puche L, Moyano E, Hoffmann T, Sinz Q, et al.. (2012) The fruit ripening-related gene FaAAT2 encodes an acyl transferase involved in strawberry aroma biogenesis. Journal of Experimental Botany 63.

44. Pierce J, Halpern BP (1996) Orthonasal and retronasal odorant identification based upon vapor phase input from common substances. Chemical Senses 21: 529–543. doi: 10.1093/chemse/21.5.529

45. Masaoka Y, Satoh H, Akai L, Homma I (2010) Expiration: The moment we experience retronasal olfaction in flavor. Neuroscience Letters 473: 92–96. doi: 10.1016/j.neulet.2010.02.024

46. Small DM, Jones-Gotman M (2001) Neural substrates of taste/smell interactions and flavour in the human brain. Chemical Senses. pp. 1034.

47. Small DM, Voss J, Mak YE, Simmons KB, Parrish T, et al.. (2004) Experience-dependent neural integration of taste and smell in the human brain. Journal of

Neurophysiology 92.

48. Veldhuizen MG, Shepard TG, Wang MF, Marks LE (2010) Coactivation of Gustatory and Olfactory Signals in Flavor Perception. Chemical Senses 35: 121–133. doi: 10.1093/chemse/bjp089

49. SjÖStrÖM Loren B, Cairncross Stanley E (1955) Role of Sweeteners in Food Flavor. Use of Sugars and Other Carbohydrates in the Food Industry: American Chemical Society. pp. 108–113.

50. Burdach KJ, Kroeze JHA, Koster EP (1984) Nasal, Retronasal, and Gustatory Perception - An Experimental Comparison. Perception & Psychophysics 36.

51. Murphy C, Cain WS (1980) Taste and Olfaction - Independence vs Interaction. Physiology & Behavior 24.

52. Frank RA, Byram J (1988) Taste smell interactions are tastant and odorant dependent. Chemical Senses 13.

53. Stevenson RJ, Prescott J, Boakes RA (1999) Confusing tastes and smells: How odours can influence the perception of sweet and sour tastes. Chemical Senses 24.

54. Cliff M, Noble AC (1990) Time-intensity evaluation of sweetness and fruitiness and their interaction in a model solution. Journal of Food Science 55.

55. Lavin JG, Lawless HT (1998) Effects of color and odor on judgments of sweetness among children and adults. Food Quality and Preference 9.

56. Prescott J (1999) Flavour as a psychological construct: implications for perceiving and measuring the sensory qualities of foods. Food Quality and Preference 10.

57. Baldwin EA, Goodner K, Plotto A (2008) Interaction of volatiles, sugars, and acids on perception of tomato aroma and flavor descriptors. Journal of Food Science 73.

58. Bartoshuk LM, Klee HJ (2013) Better Fruits and Vegetables through Sensory Analysis. Current Biology 23: R374–R378. doi: 10.1016/j.cub.2013.03.038

59. Aharoni A, Giri AP, Verstappen FWA, Bertea CM, Sevenier R, et al.. (2004) Gain and loss of fruit flavor compounds produced by wild and cultivated strawberry species. Plant Cell 16.

CHAPTER 11

STUDIES ON EXTRACTION OF SAFFLOWER PIGMENTS AND ITS UTILIZATION IN ICE CREAM

Machewad GM[1], Ghatge P[2], Chappalwar V[2], Jadhav B[2] and Chappalwar A[3]

[1] Department of Food and Industrial Microbiology, College of Food Technology, Marathwada Krishi Vidyapeeth, Parbhani, India

[2] Department of Food Chemistry and Nutrition, College of Food Technology, Marathwada Krishi Vidyapeeth, Parbhani, Maharashtra

[3] Department of Livestock Product Technology, College of Veterinary and Animal Sciences, Maharashtra Animal and Fishery Sciences University, Parbhani, Maharashtra

ABSTRACT

The extractions of pigments from safflower petals were studied. The pigments which yield from yellow safflower (carthamidin) were 29.59% and safflower red (carthamin) was 0.77%. The pigment extracted was further utilized as a natural color in ice cream to replace the synthetic color. The chemical characteristics of ice cream were significantly increased (P< 0.05) with the addition of carthamidin extract in ice cream. The sensory evaluation of the ice cream fortified with carthamidin was carried out by the panel of 10 trained judges using 9-point Hedonic scale. The addition of carthamidin (0.06 mL) in the ice cream was found to score higher overall acceptability. Further addition of the safflower yellow 0.09 mL scored less for color of the ice cream.

INTRODUCTION

Safflower (Carthamus tincttiorius L.) is one of the world's oldest crops, highly branched, herbaceous, thistle like annual herb with yellow to red petals. It belongs to the family Asteraceae of the broad group Composites. The safflower is a warm temperature crop, cultivated over the greater parts of tropical Asia, Africa, Russia and China. It is commonly known as "Kardi" in Marathi and "Kussum" in Hindi. The safflower seeds contain oil (35-45%). The colour of flower varies from whitish yellow to red orange, the most common being deep yellow. Safflower flowers contain two pigments viz. red (carthamin) which is insoluble in water and yellow (carthamidin) which is soluble in water and mainly used as a material for dye and is currently being used as a natural food colorant. The alkaline extracts were used for dyeing silk, wool, cotton and paper, to make the pigment in the state of the precipitate and ancient Chinese manufactured and produced it as red paint for cosmetics [1]. A growing interest in the use of natural dyes in textile coloration has gained. The stringent environmental standards are imposed by many countries. Eco-friendly and biodegradable dyes derived from natural resources have emerged as important alternative to synthetic dyes. Safflower petals contains about yellow (30%) and red pigment (0.83%) [2,3]. Safflower pigments are widely used as stain, additive in beverages and cosmetics, printing and dyeing [4-6].

Citation: Machewad GM, Ghatge P, Chappalwar V, Jadhav B, Chappalwar A (2012) Studies on Extraction of Safflower Pigments and its Utilization in Ice Cream. J Food Process Technol 3:172. doi:10.4172/2157-7110.1000172.

A wide range of colours, expensive dyes were used to dye the analyzed silk yarns, coccid dyestuffs, madder, weld, young fustic, tannins and an indigoid dye were identified in a 16th century tapestry. Moreover, the use of safflower has been assessed for the first time in a European fabric [7]. Many Chinese medicines are prepared by using dried flowers and extract of flowers. Now-a-days the medicinal uses of flowers in China have become known to the rest of the world. The extract of florets containing pigments are used in treatment of many illnesses such as menstrual problem, cardio vascular diseases pain and swelling associated with trauma. These medicinal preparations have been widely accepted which helps in increasing the demand for the safflower petals. The administration of 1% concentration of safflower petals exerted a positive influence on blood pressure and serum lipid profile of hypertensive and hyperlipdemic subjects [8]. Safflower pigments are safe for food and have curative effects on diseases such as lack of oxygen coronary heart diseases, myocardial infarction, cerebral thrombosis and renal thrombosis etc [9]. India is the largest producer of safflower (2 lakh tones) in the world with highest acreage (4.3 lakh hectares), but it is cultivated as oil seed crop. The valuable safflower petals are being wasted although they are known to be used as coloring agent. It is clear that safflower petals in India are wasted in large quantity and its therapeutic value is also ignored. The carthamidin (yellow pigment) extracted from petals is utilized in food products for preventing petal waste, use of artificial colour and improving therapeutic value of food product. Besides these the safflower petal extracted color can be used up to any extent. Therefore, it is considered worthwhile to prepare value added food product like ice cream by incorporating extracted safflower yellow pigment.

MATERIAL AND METHODS

The safflower petals (variety PBNS-12) were collected from All India Co-ordinated Research Project on Safflower, Marathwada Krishi Vidyapeeth, Parbhani. The buffalo milk required for the preparation of ice cream was obtained from the Dairy Farm, College of Agriculture, Marathwada Krishi Vidyapeeth, Parbhani. The cream was prepared by using centrifugal cream separator.

Extraction of safflower yellow

The extraction of safflower yellow (water soluble pigment) from safflower florets was done [10] (Figure 1-3).

Determination of Chemicals Characteristics of Safflower Petals

The chemicals characteristics such as moisture content, protein, fat, ash and fiber content were determined as per standard procedure [11].

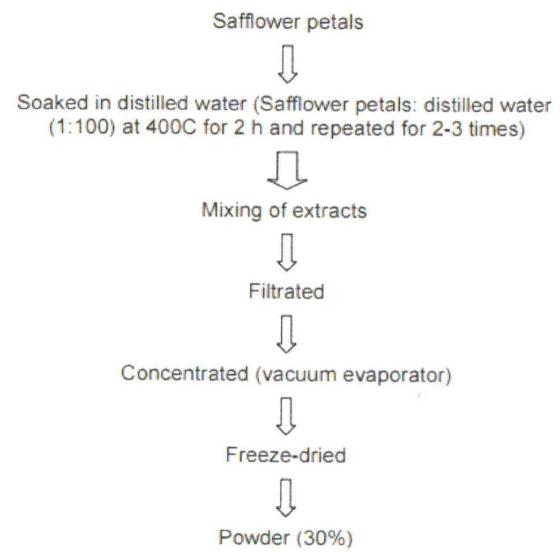

Figure 1: Flow sheet for the extraction of carthamidin from safflower petals.

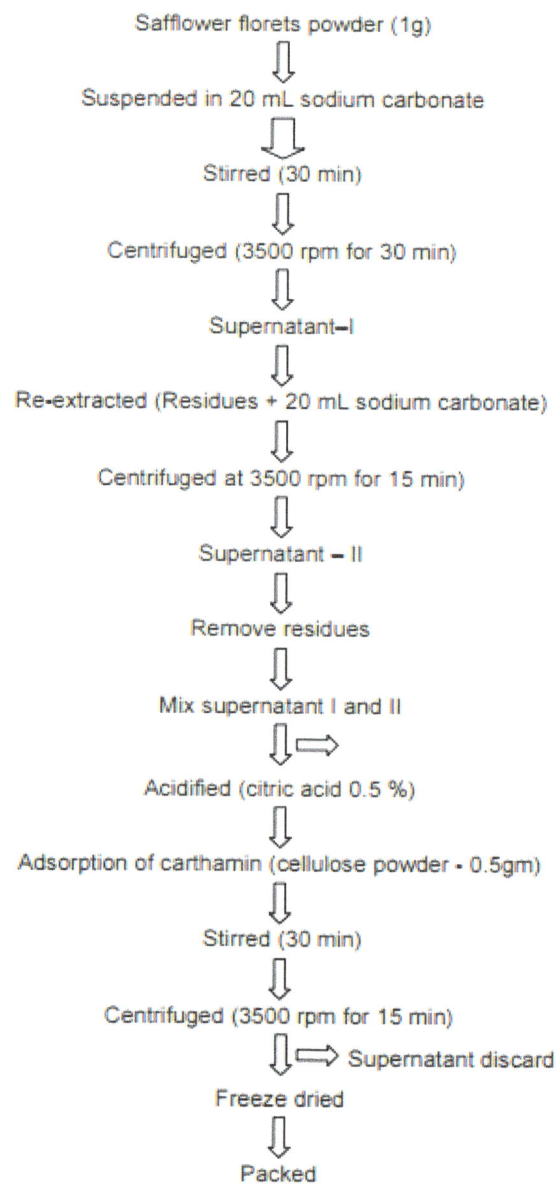

Figure 2: Flow sheet for the extraction of carthamin from safflower petals.

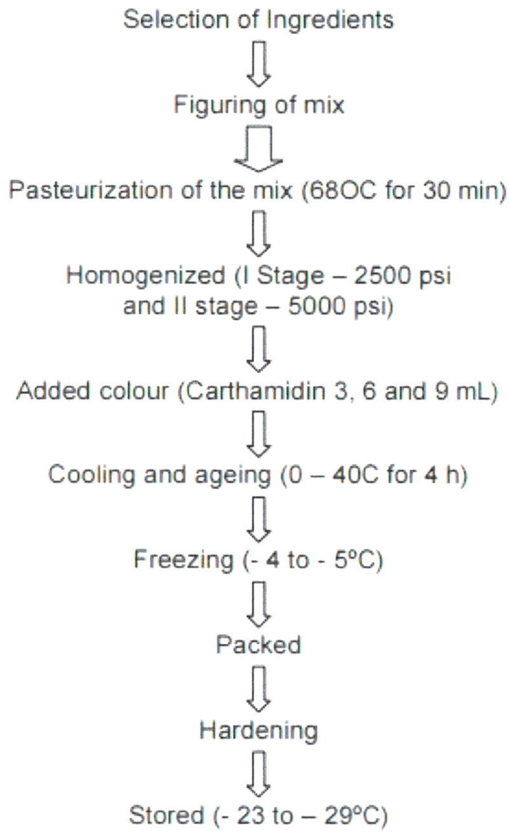

Figure 3: Flow sheet for the preparation of carthamidin extract fortified ice cream.

Table 1: Chemical composition of safflower petals.

Constituents	Value (%)
Moisture	4.70
Crude Protein	1.82
Fat	4.80
Ash	10.80
Crude Fibre	11.60
Carthamin (Red)	0.77
Carthamidin (Yellow)	29.59

Each value is average of three determinations

The ice cream was prepared by incorporation of carthamidin extract (0.03, 0.06 and 0.09 mL).

Organoleptic Evaluation

The sensory evaluation of ice cream fortified with varying levels of carthamidin extract was carried out by trained panel of ten judges on a 9.0 point Hedonic scale [12].

Statistical Analysis

The data generated in the experiments were recorded and subjected to statistical analysis using standard procedure [13]. The standard errors (SE) and critical differences (CD) at 5% level of significance were worked out for comparison of treatments and presented in the respective tables.

RESULTS AND DISCUSSION

Chemical Composition of Safflower Petals

The chemical composition of safflower petals revealed that higher quantity of constituents (Table 1). The safflower petals are nutritive and containing natural colouring pigment. Further the results are in good agreement with the findings of others [2,3,14].

Table 2: Chemical composition of the ice cream added with carthamidin extract.

Ice cream	Moisture (%)	Protein (%)	Fat (%)	Total carbohydrates (%)	Ash (%)
Control	62.05	4.85	10.00	21.38	1.52
A	62.40	5.05	10.05	21.78	1.60
B	62.65	5.03	10.10	21.75	1.64
C	63.05	5.00	10.12	21.77	1.69
SE±	0.122	0.076	0.103	0.112	0.058
CD at 5%	0.399	0.249	0.335	0.364	0.188

SE, standard error of the mean; CD, critical difference
A-Ice cream containing 0.03 mL carthamidin; B-Ice cream containing 0.06 mL carthamidin; C-Ice cream containing 0.09 mL carthamidin
Each value is average of three determinations

Table 3: Organoleptic evaluation of ice cream added with carthamidin extract.

Ice cream	Flavour	Body and texture	Colour	Taste	Overall acceptability
Control	8.5	8.8	8.0	8.8	8.3
A	8.3	8.3	8.3	8.1	8.6
B	8.9	8.4	8.6	8.3	8.8
C	8.1	8.0	8.3	8.3	8.3
SE±	0.07	0.06	0.09	0.10	0.09
CD at 5%	0.25	0.19	0.30	0.33	0.30

SE, standard error of the mean; CD, critical difference
A-Ice cream containing 0.03 mL carthamidin; B-Ice cream containing 0.06 mL carthamidin; C-Ice cream containing 0.09 mL carthamidin
Each value is average of ten determinations.

Chemical Composition of Ice Cream Added With Carthamidin

The chemical characteristics of ice cream were significantly increased ($P<0.05$) with the addition of carthamidin extract (Table 2). The moisture content was found to be higher in the ice cream incorporated with carthamidin extract than control. Similar trend was observed for fat and ash content of ice cream. The carbohydrates and protein content were found to increase with increase in concentration of carthamidin extract.

Organoleptic Evaluation of Ice Cream Added With Carthamidin Extract

The sensory evaluation of the ice-cream was done on the basis of 9-point Hedonic scale by the panel of 10 trained judges. The color added was 0.03, 0.06, and 0.09 mL carthamidin extract (Table 3). The addition of carthamidin (0.06 mL) scored significantly higher ($P< 0.05$) for flavour and overall acceptability. Further, it was found that the texture of all samples was soft with even mouthfeel. The texture among the samples was much uniform with slight variations due to addition of the color. The addition of the more carthamidin in the ice cream has some safflower petals like flavor. Finally it can be concluded that the addition of the carthamidin extract to the value added product i.e. icecream could give the natural way of color pigment in the food products as well as give the health benefits of the safflower.

REFERENCES

1. Wouters J, Grzywacz CM, Claro A (2010) Markers for Identification of Faded Safflower (Carthamus tinctorius L.) Colorants by HPLC-PDA-MS: ANCIENT FIBERS, PIGMENTS, PAINTS AND COSMETICS DERIVED FROM ANTIQUE RECIPES. Studies in Conservation 55: 186-203.
2. Nagaraj G, Devi GN, Srinivas CVS (2001) Safflower petals and their chemical composition. Proceedings of the 5th International Conference on Safflower, Williston, North Dakota and Sidney, Montana 301-302.
3. Kulkarni DN, Kulkarni KD, Tathe S (2001) Studies on the extraction of safflower yellow B and carthamin red pigment from safflower florets as food colorant. Proceedings of the 5th International Conference on Safflower, Williston, North Dakota and Sidney, Montana 321-324.
4. Wang Z, Lijie D (2001) Current situation and prospects of safflower products development in China. Proceedings of the 5th International Conference on Safflower, Williston, North Dakota and Sidney, Montana 315-319.
5. Obara H and Onodera J (1979) Structure of carthamin. Chem Lett 201-204.
6. Takahashi Y, Miyasaka N, Tasaka S, Miura I, Urano S, et al., (1982) Constitution of two colouring matters in the flower petals of Carthamus tinctorius L. Tetrahedron Lett 23: 5163-5166.
7. Degano I, Lucejko JJ, Colombini MP (2011) The unprecedented identification of Safflower dyestuff in a 16th century tapestry through the application of a new reliable diagnostic procedure. J Cul Heri 12: 295-299.

8. Deodhar SK (2001) Evaluation of medicinal value of safflower petals in hypertensive and hyperlipdemic subjects. M.Sc Thesis, College of Agriculture, Marathwada Krishi Vidyapeeth, Parbhani, Maharashtra, India.

9. Shouchun W, Jianxiang F and Rui Z (1993) The research and production of yellow pigments from safflower. Proceedings of the 3rd International Conference on Safflower, Beijing 869-879.

10. Fatahi N, Carapetian J, Heidari R (2008) Sphectrophotometric Measurement of Valuable Pigments from Petals of Safflower (Carthamus tinctorius L.) and their Identification by TLC Method. Res J Biological Sci 3: 761-763.

11. Association of Analytical Chemists (2000) Official Methods of Analysis of AOAC International,(17th Edn. Gaithersburg, USA.

12. Amerine MA, Pangborn RM, Roessler EB (1965) Principles of sensory evaluation of food. Academic Press, New York, USA.

13. Snedecor GW, Cochran WG (1989) Statistical Methods, 8th Edn. Iowa State University Press, Ames, IA, USA.

14. Zehra E (2005) Resurgence of safflower (Carthamus tinctorius L.) Utilization: A Global View. J Agronomy 4: 83-87.

CHAPTER 12 ISOLATION AND CHARACTERIZATION OF FLAVONOIDS IN URENA LOBATA LEAVES

Dixa Singh[1] and V. S. Singh[1]

[1]Department of Chemistry, H.N.B. Garhwal University Campus-Pauri (Uttarakhand), India

ABSTRACTS

Aim: *Urena lobata* **L. a member of Malvaceae family, is widely used as famine food in Africa. It is also used in traditional medicine system to cure gonorrhea, fever, wounds, toothache & rheumatism. The plant has already been tested for its antioxidant activity. The work was extended to investigate and identify the flavonoid glycosides present in the plant.**

Methodology: *Urena lobata* **leaves were dried, powdered and extracted with petroleum ether followed by methanol. The methanolic extract, after processing through different solvents, was used to obtain the Chemically active constituents .Column Chromatography and Thin Layer Chromatography (TLC) were used to isolate four compounds which were, in turn, subjected to spectroscopic analysis.**

Results: The spectroscopic studies indicated that the four isolated compounds are (1) Quercetin, (2) Kaempferol, (3) Quercetin 3-0-rutinoside and (4) Kaempferol 3-0- β glucopyranoside.

Conclusion: The presence of flavonoids in the plant makes it an important ingredient of the traditional medicinal system. Since flavonoids are associated with antioxidant activity, their presence in the plant makes it an important food material.

INTRODUCTION

Flavonoids are a group of naturally occurring compounds having phenolic groups. These are also known as plant pigments and are found in all parts of a plant namely roots, stem .bark, leaves, flowers and fruits [1]. Flavonoids together with carotenes provide different colours to fruits, vegetables & herbs. The chemical structure of flavonoids is based on C15 skeleton with two benzene rings joined by three linear carbon atoms which may or may not be a part of third ring [2]. The skeleton of flavonoids consists of six major group namely flavone, chalcone, flavonol, flavonone, anthocyanin & iso-flavonone [3]. Flavonoids occur in the plants as aglycone (sugar free), glycosides (one or more sugar molecules attached to it) or methylated derivatives [4]. All flavonoids are water soluble. Flavonoids may acquire phenolic groups on 3, 5, 7, 2', 3', 4' & 5' positions in the skeleton. Flavonoids are

Citation: Singh, D and Singh VS. 2016. Isolation and Characterization of Flavonoids in Urena Iobata Leaves. European Journal of Medicinal Plants, ISSN: 2231-0894,Vol.: 11, Issue.: 1. 1-6. DOI : 10.9734/ EJMP/2016/21502.

secondary metabolites in plants. These are synthesized by phenyl -propenoid pathway. They are responsible for various biological and pharmacological activities in mammals [5], the most important being the antioxidant activity [5,6]. Beside, antioxidant activity, flavonoid show some other activities like microbial [7,8], antifungal, antiviral [8], antiinflammatory [9], anticancer [10,11] and antidiabetic [12].

Urena lobata L. plant is a member of the Malvaceae family and is commonly known as Bachita, Congo jute or Caesarweed [13]. It is a woody, erect perennial shrub reaching upto 2-3 m height. It is widely distributed in temperate and tropical zone of India. It is more or less hairy, the leaves varying in shape and size. The Urena lobata flowers are axillary, pink or purple of about 2.5 cm diameter. Various extracts of leaves and roots of the plant are used in ancient medicine to treat diseases like diabetes, rheumatism, wounds, gonorrhea, tooth ache etc [14]. The leaves and flowers are eaten as famine food in Africa [15].

Earlier workers [16] have shown the presence of Mangeferin and Quercetin in aerial part of Urena lobata. Triglycerides were also isolated from the plant [17], imperatorin and furocoramin were isolated from the roots of the plant [18]. The authors have also determined presence of Quercetin and Kaempferol and examined anti - oxidant activity of the plant extract in their preceding work [6]. The work has been extended to ascertain the nature of bound flavonoids in the leaves of the plant extract.

MATERIALS AND METHODS

Plant Material

The plant leaves were collected from Kotdwar region in Pauri District. The plant material was identified taxonomically by Dr. A. K. Agarwal, Associate Professor in Botany Department, Govt. P. G. College, Uttarkashi.

Extraction and Isolation

The plant leaves powder (2 Kg) was firstly extracted with petroleum ether (60°C- 80°C), then with methyl alcohol at 60°C in a soxhlet apparatus for 36 hours. The methanolic extract was concentrated by distilling off the solvent. The methanolic residue (about 50 gm) was dissolved in water and then partitioned withn-hexane, chloroform and ethyl acetate respectively. The ethyl acetate fraction gave positive test for flavonoids [19]. Since the authors were interested in chemically active constituents, the fractions were subjected to chemical analysis. The ethyl acetate extract was subjected to column chromatography over silica gel. The column was eluted first with chloroform, then a chloroform – methanol mixture with increasing amount of methanol upto 20%. The fractions obtained from the column chromatography were subjected to TLC using $CHCl_3$ – CH_3OH (10:2) as the solvent system and examined under UV light alone and with ammonia vapours followed by locating reagent $FeCl_3$. The fractions eluted with 2 - 5% methyl alcohol were found to be the same after TLC analysis and therefore were combined. This combined fraction containing a mixture of three compounds, was again chromatographed on preparative silica gel TLC plates and eluted with $CHCl_3$: CH_3OH (10:2). Repeated purification of each compound by silica gel chromatography with methyl alcohol produced compound 1 (12.55 mg) and compound 2(14.20 mg). The fractions eluted with silica gel column chromatography with 6 - 8% methyl alcohol, were found to be the same combined after TLC examination and so combined together. The major component found was purified by further column chromatography eluted with CH_3OH. This was named compound 3(8.20 mg). The fractions eluted with 10 -20% were combined as before after TLC analysis. The fraction was subjected to PTLC eluted with $CHCl_3$: CH_3OH (10:2) to give one major spot. Further purified on silica gel column chromatography eluted with methyl alcohol to give pure compound 4 (9.82 mg).

RESULTS AND DISCUSSION

The four compounds obtained, as mentioned in the experimental section, were subjected to spectroscopic analysis for identification. The Techniques used were UV, IR, 1HNMR & 13CNMR spectroscopy. The details of the spectral peaks were noted as follows.

Compound (1):

Amorphous yellow powder UV λmax nm (MeOH): 254 sh, 268, 322 sh,365;+NaOMe : 275, 320, 416; +$AlCl_3$: 262 sh, 270, 352, 426; +$AlCl_3$ / HCl: 260, 271, 350, 426; +NaOAc: 275, 300, 385; +NaOAc /H_3BO_3:269, 295sh,320sh,370. 1HNMR (200 MHz, DMSO-d_6) δ: 8.00 (2H, d, J =8.1 Hz,H-3', H-5'),6.90 (2H,d,J=8.1Hz, H-2', H- 6'),6.40 (1H,d J=2.5 Hz, H-8), 6.21 (1H,d, J=2.5Hz, H-6).13CNMR (CD3OD): δ:94.2(C-8), 99.0 (C-6), 104.1(C-10),115.9 (C-3', C-5'), 123.4 (C-1'),130.4 (C-2',C-6'),136.9 (C-3), 147.8 (C- 9),158 (C-2),160.4(C-5),161.8 (C-7),165.4(C-4')& 177.2(C-4).

Compound (2):

Yellow crystal, UV λmax nm (MeOH): 258, 266sh, 299 sh, 360; + NaOMe: 272,327,416; +AlCl3: 275, 303 sh,430; +$AlCl_3$ / HCl: 271,300, 364sh, 402; +NaOAc: 270, 325 ,393; + NaOAc / H_3BO_3: 262,298,387. 1HNMR (200 MHz, DMSOd6) δ:7.28 (dd,J= 1.4 H-6') &.7.51 (d J=1.4, H-2'), 6.88(1H,d,J=8.0Hz, H-5'),6.40 (1H,d,J=2.5Hz, H- 8),620 (1H,d,J=2.5Hz, H-6).

Compound (3):

Yellow needle. UV λmax nm (MeOH): 254, 350; +NaOAc: 275, 306, 366; +NaOAc / H3BO3: 268, 351; +NaOMe: 276,327,404; +AlCl3: 265, 305,352; +AlCl3 /HCl: 272,(300), 400nm.IR (KBr): 3240,1710, 1660, 1182 - 1081 & 832 cm-1 . 1H NMR (300 MHz, CD3 OD) δ: 11.92 (OH -5), 10.86 (7- OH), 6.68 (H-8,d,j=2.1hz), 6.50 (H- 6,d,j=2.8hz), 7.85 (H-2',d,J=8.5hz), 6.94 (H- 3',d,j=8.9), 9.54 (4'-OH), 6.94 (H -5',d,J=8.9), 7.85(H-6',d,j=8.5), 4.64(H-1",d,j=7.8),4.07 (H-2",dd,j=9.6),3.77(H-3",d, j =9.6),3.48 (H- 4",t,j=8.9),3.68(H-5",d,j=9.6,2.6hz) & 4.55(H-6", dd, J=10,2.6hz). C^{13}NMR (75MHz,CD_3OD),δ:156.91(C- 2),135.51(C-3),175.90(C-4),165.53(C-5), 99.52(C-6), 160.0(C-7), 91.80 (C-8),105.13(C-9),152.52(C-10),121.82(C-1'),130.10(C-2'), 115.41(C-3'),157.53(C-4'),115.42 (C-5'), 130.10 (C-6'),SUGAR MOIETY- δ:101.42(C-1"),74(C-2"),78.51(C-3"),71.31(C-4"),78.40(C-5") & 62.32(C-6").

Compound (4):

Amorphous yellow powder, UV λmax nm (MeOH): 255, 352 IR (KBr): 3260, 1660.85, 1606.85, 1504.61,1454,13 59,1304,1197,1060,1010,933, 798, 709, 644,592 cm^{-1}. ^1H NMR (300 MHz, DMSO-d_6) δ: 6.16 (1H,d,J=2.2,H-6), 6.35 (1H,d,J=2.2,H-8),7.35 (1H,d,J=1.8,H-2'),6.81 (1H,d,J=8,H-5'),7.50(1H,dd,J=8.0,H-6'), 5.32 (1H,d,J=7.4,H-1''), 3.01-3.17 (4H,m,H-2'',H-3'',H- 4'',H-5''), 3.30 (1H,m,Ha -6''), 3.68 (1H, d, J=10.3 Hz, Hb -6''), 4.36 (1H, d,J=1.8,H-1'''), 3.01 - 3.07 (4H, m, H-2''', H-3''', H -4''' & H -5'''), 0.97 (3H,d,J=6.2Hz, CH3- 6'''),12.59 (1H,S,5-OH), 9.39 (1H,s,OH-5'),9.45 (1H,s,OH-4')&10.73 (1H,s, OH-7). C13NMR: (125 MHz, DMSO - d 6) δ: 177.40 (C-4), 164.25 (C-7),161.25 (C- 5),156.52 (C-2) 156.48 (C-9), 133.34 (C-3), 104.12(C-10),121.3 (C-1'),148.53 (C-4'), 144.8(C-3'), 121.33(C-6'), 115.36 (C-2'),116.38 (C-5'),104.12 (C-10), 98.84 (C-6),Glucose moiety δ:101.79 (C-1''),74.42 (C-2''), 77.12 (C- 3''), 70.48(C-4''), 77.05(C-5''),60.85(C-6''), Rhamnose moiety δ:100.72 (C-1'''),70.48(C-2'''), 70.48 (C-3'''),72.17(C-4'''),68.30(C-5''') & 17.52 (C-6''').

The interpretation of the spectral peaks may be as follows.

Compound (1):

The proton NMR spectrum of the compound produced four proton doublet peaks at the aromatic region. The aromatic proton doublets at δ 6.21 and δ 6.40 were due to the meta coupled protons of a 5,7 substituted ring A. This accounted for protons H-6 & H-8 respectively. The two proton doublets at δ 6.90 & δ8.00 accounted for the protons H-2',H-6' and H-3',H-5' of a 4'–substituted ring B respectively. The spectral data were in agreement with Kaempferol reported in the literature [20]. Thus compound (1) is identified as Kaempferol (Fig. 1).

Compound (2):

The proton NMR spectrum showed five aromatic signals. The aromatic proton doublet at δ6 .20 & δ6.40 were due to the meta coupled protons of a 5,7 substituted ring A, for protons H-6 & H-8 respectively. The peaks at δ6.88, δ7.28 & 7.51 were assigned for protons H-5', H-6' & H-2' showing an ABX coupling system of a 3', 4' substituted ring B of a flavonol. The structure of this compound was confirmed (Fig. 2) by comparing 1HNMR of Quercetin from the literature [21].

Compound (3):

The IR (KBr) spectrum showed that the absorption band at 3240cm-1 is the stretching vibration of hydroxyl group (OH), 1710cm-1 is due to C=O stretching vibration ,1660cm-1 is the stretching vibration of C=C bond of aromatic ring,1450-1356 cm-1 is bending vibration of C-H bond, 1182-1081 cm^{-1}-is due to the presence of C-O-C bond of ether ,832cm-1is the C-H bond outside the field. 1HNMR (300MHz,CD3OD) indicates that the compound is a group of flavonoid glycosides. 3 sharp peaks as singlet δ 11.92, 10.86 & 9.54 shows that there are three hydroxyl group present at 5-OH, 7-OH & 4'OH respectively. The presence of aromatic proton at δ 7.85(2H,d, ,J=8.5Hz, H-2' & H-6') & δ 6.94 (2H,d, J=8.9Hz for H-3' & H-5'). Besides, the two aromatic proton as singlet at δ 6.68 (1H,s) for H- 8 and δ 6.50 (1H,s) for H-6. C13NMR (75MHz,CD 3OD) showed that compound has 21 carbon atoms in which, there were nine quaternary carbon (C-2,C-3,C-4,C-5,C-7,C-9,C-10,C-1' & C- 4') and one CH2 group of sugar at δ 62.32 7 & five methyne group for sugar at C-1'', C-2'', C-3'', C-4'' & C-5''. Remaining four would be C-6 C-8, C-2' C-3',C-5' C-6'. The HMBC and HMQC spectra showed an aromatic proton at δ 6.68 (at C-8) correlates to carbon signal at 95.3 (atC-6), correlates to carbon (at C-10,C-9,C-6 & C-7). The correlation between proton at δ 6.50 (at C-6) to carbon signal at 103.9 (at C-9), correlated to the C at C-5, C-10, C-7 & C-8. The aromatic proton at δ 6.94 correlates to carbon signal at C- 3 & C-5') and at C-4' & C-2''.Glucoside proton present at C-1'''at δ 4.64 correlates to carbon signal at C-1'', also correlates to the C at C-2'',C- 3'',C-1,C-2 & C-3.Similarly, proton of C-2''at δ 4.07 correlates to the carbon signal at C- 2'',correlates to the C at C-3''. The correlation between C-3' at δ 3.77 to carbon signal at C-3'', correlate to the C at C-4'' & H of the C-4''at δ 3.48 correlates to the Carbon signal at C-4'', also correlates to C at C-3'' & H of C-5 at δ3.68 correlates with C-5'',C- 4'' & C-3''. On the basis of HMBC spectra identified that a correlation between the anomeric proton present at C-1''' with C-1,C-2& C-3 indicate that the group is located at the C-6 position. On the basis of UV, IR, 1HNMR, C13NMR and HMBC data analysis the compound 3 is identified as Kaempferol 3- O–beta glucopyranoside (Fig. 3).

Compound (4):

The UV spectrum showed absorption band at 255 & 352 nm for 3-substituted flavanol. IR spectrum showed a strong absorption peak at 1660cm-1 for a carbonyl group & intense broad band at 3260 cm-1 due to OH group. A peak at 1606 cm-1 due to stretching vibration of C =C bond of aromatic ring at 1450-1356 cm-1 due to bending vibration of CH bond. The aromatic region of 1HNMR spectrum of the compound has the characteristic of Quercetin moiety. 1HNMR signals at δ 6.16 (1H,s),6.35 (1H,s) are due to meta coupled protons of ring A (H-6 & H-8) of a flavonoid nucleus. It clearly indicates that the compound must be 5,7 dihydroxy flavonol. A signal at δ 7.53 (d,J=1.8),6.81 (d,J=8), 7.50 (d, J=8) are assigned to H-2', H-5' & H-6' of ring B respectively of Quercetin nucleus. The 13C NMR spectrum showed signals which indicated the presence of flavonoids moiety along with two different sugar moieties & these two sugar moieties were confirmed as β glucose and α rhaminose by the spectroscopic analysis of 1HNMR & 13C NMR data. The 1HNMR spectra showed a singlet peak at δ 5.08 ppm attributed to the anomeric α rhamnopyranosyl proton & a doublet at δ 5.50 (J=7.4 hz) with diaxial coupling assignable to the anomeric β glucopyanosyl proton with chemical shift characteristics of glycosidation in a close position i.e.in C-2 position of glycosyl residue (Δδ -0.5 ppm) downfield shift in comparing with chemical shift of anomeric proton signal in quercetin 3-glucoside (2.22). In 13C NMR of compound the C-1 & C-2 carbon signal of the glucosyl moiety were observed at 98.97 and 77.42 ppm respectively which proved that the suggested interglycosidic bond (1→2) by recongnisable downfield shift of C-2 carbon signal (Δδ -3.3 ppm) upfield shift of the C-1 signal (Δδ -2.1 ppm) in comparison with quercetin 3 glucoside (2.23).

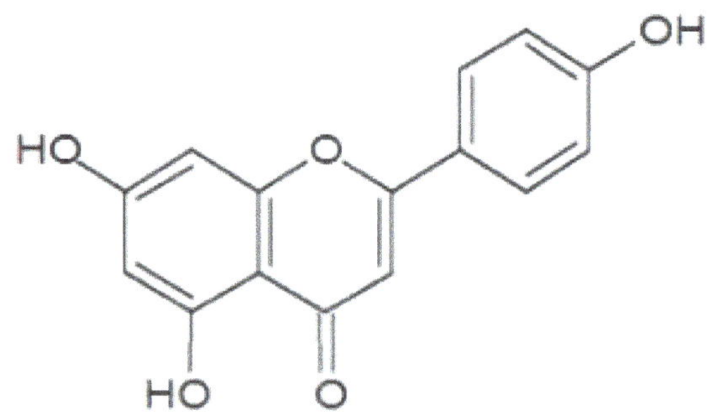

Fig. 1. Kaempferol

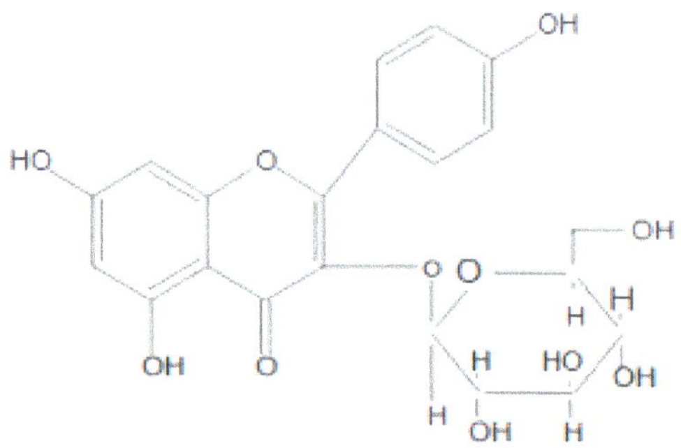

Figure 3. Kaempferol 3 –O-beta glucopyranoside

^{13}C NMR showed that the compound containing 28 carbon atoms consisting one CH2 group at δ 60.85 and one CH3 group at δ 17.52. Downfield shift of C-6" in 13C NMR spectra & HMBC correlation between H-1"' (δ 4.36) & C-6" (δ 60.85) indicated that rhamnose was linked to the C-6" of glucose, confirming the rutinose unit as a sugar moiety. The attachment of rutinose moiety to the aglycone was determined from the correlation between H-1"(δ 5.32,d) & C-3 (δ 133) of aglycone. On the basis of UV, IR, ^{1}HNMR, ^{13}C NMR, HMBC data analysis the compound 4 is identified as Quercertin 3-O–rutinoside (see above Fig. 4).

CONCLUSION

It has been established that Urena lobata leaves contain four important flavonoids, two in aglycone and two in glycoside form. The leaves can serve as good food supplement due to presence of the flavonoids. Since there are four types of flavonoids present in the leaves, these can work as a good antioxidant.

CONSENT

It is not applicable.

ETHICAL APPROVAL

It is not applicable.

COMPETING INTERESTS

Authors have declared that no competing interests exist.

REFERENCES

1. Tsukasa I, The structure & distribution of flavonoids in plants. J of Plant Research. 2000;113:287-299.
2. Middleton E, The flavonoids. Trends in Pharmacological Sciences. 1984;5:335- 338.
3. Narayana KR, Reddy MS, Chauvedi MR, Krishna DR. Biflavonoids, classification, pharmacological & biochemical effects & therapeutic potential. Indian J of Pharmacology. 2001;33:2-16.
4. Harbone JB. The flavonoids. Advances in Research Sciences, 1980.ed.1.London: Chapman & Hall; 1998.
5. Robak J, Gryglewski RJ, Biocativity of flavonoids. Pol J Pharmacol. 1996;48:555- 564.
6. Singh D, Singh VS. Urena lobata: A green source of antioxidant. J of Plant Sciences. 2014;2(6):299-303.
7. Cushine TP, Lamb AJ, Antimicrobial activity of flavonoids. International J of Antimicrobial Agents. 2005;26(5):343-356.
8. Friedman M, Overview of antimicrobial, antitoxin, antiviral & antifungal activities of the flavonoids & tea. Molecular Nutrition & Food Research. 2007;51(1):116-134.
9. Murlidhar A, Babu KS, Sankar TR, Redenna P, Reddy GV, Latha J. Antiflammatory activity of flavonoid fraction isolated from stem bark of Butea monosperma (Lam): A mechanism based study. International J of Phytopharmacology. 2010;1:124-132.
10. Davis WL, Mathew SB. Antioxiodants and Cancer III: Quercetin. Alternative Medicine Review. 2000;5(3):196-208.
11. Romagnolo DF, Selmini OI. Flavonoid & Cancer prevention: A review of the evidences. J Nutr Gerontol Geriatr. 2012;31(3):206-238.
12. Josline YS, Halana NM. Antidiabetic properties of flavonoid compound isolated from Hyphaena thebaica epicarp on alloxan induced diabetic rats, Pharmacognosy Research. 2013;5(1):22- 29.
13. Carl Linnaeus, Species Plantarum, Laurentius Salvius. 1753;2:692.
14. Francis JK. Atlas of florida vascular plants, Institute of systematic Botany, University of South Florida, Tampa, Fl; 2003.
15. Freedman R. Famine foods: Malvaceae, Predue University, Perdue, IN; 1998.
16. Srinivasan KK, Subramanian SS. Isolation of Mangiferin from Urena lobata. Arog J Health Sci. 1981;7(1):140-141.
17. Morelli CF, Cairoli P, Speranza GM, Rajia S. Triglycerides from U.lobata. Fitoterapia. 2006;77(4):296-299.
18. Keshab G. A Furocouramin, Imperatoin isolated from Urena lobata. Molebank 2004;M382.
19. Chatwal GR. Organic chemistry of natural product. 4ed. Mumbai: Himalaya Publication House; 2007.
20. Markham KR, Ternal B, Stanley R, Geiger H, Mabry TJ. C13 NMR studies of flavonoids-III: Naturally ouccuring flavonoids glycosides and their acylated derivatives. Tetrahedron. 1978;34:1389- 1397.
21. Wollenweber E, Dietz VH, Occurrence and distribution of free flavonoid aglycones in plants. Phytochem. 1981;20(5):859–932.

CHAPTER 13

PHYTOCHEMICAL INVESTIGATIONS OF THE MEDICINAL PLANT SWERTIA CHIRATA HAM

Mst. Jesmin Sultana[1] and Fazle Rabbi Shakil Ahmed[2]

[1] Department of Materials Science and Engineering, Rajshahi University, Rajshahi, Bangladesh

[2] Department of Pharmacy, Rajshahi Science and Technology University, Natore, Bangladesh

ABSTRACT

The fresh stem of the plant Swertia chirata Ham was extracted by rectified spirit. The crude rectified spirit extract was fractionated by using standard chromatographic techniques, on alumina gave several fractions (A, B, C, D, E and F). Fraction D, when subjected to column chromatographic analysis on neutral alumina, yielded a compound, tentatively known as compound AJ-1 m.p. 178°C. In this research work only one compound is isolated. From the spectral evidences, the compound is an alkaloid containing 37 carbon atoms with 52 hydrogen along with secondary or tertiary nitrogen and several OH groups in the molecule. The pure compound AJ-1 has found antidiabetic, analgesic, anti-inflammatory and other different bioactivity tests can be performed for the pure compounds AJ-1.The plant is being used as flok medicine in the treatment of diabetes, jaundice and skin diseases.

INTRODUCTION

Natural product chemistry is an ancient science. Different medicinal compounds such as antitumor, anticancer, anti AIDS etc were isolated from natural plants. Isolation of any compound from any plants is uncertain. None can assure it before investigation. But generally secondary metabolites are obtained by proper investigation. Plants are unique in their ability to synthesize fats, carbohydrates, proteins that constitute three major food classes for human race [1]. The array of medicines derived from them is impressive and includes hypotensive drugs. analgesics, anaesthetics, anticancer and antiparasitic compounds, antiinflammatory drugs, steroids, laxatives, diuretics and many others [2]. Theophrastus [3] a student of plato, in his historia plantarum described the use of nearly 500 medicinal plants. A vast compilation of ancient knowledge of medicinal plant, can be seen in the natural history of pliny [4], the famous Roman physician. The Chinese Emperor Shen Nung (3000 BC) [5] compiled a treatise on medicinal plants. The medicinal plants that have been used since ancient time, many have yielded most useful drugs that are very much in use in current medicine. Such as- The most important natural analgesic, morphine [6] was isolated from

Citation: Sultana MJ, Ahmed FRS (2013) Phytochemical Investigations of the Medicinal Plant Swertia Chirata Ham. Biochem Anal Biochem 2: 145. doi: 10.4172/2161-1009.1000145.

opium poppy (latex of Papaver somniferum fruit). Quinine [6], an antimalarial alkaloid, was isolated from cinchona bark (Cinchona succirubra) which was used by the South Americans and the Indians. Emetine [7], which is considered as an important medicine for amoebiasis, is the main alkaloid obtained from the root of ipecacuanha (Cephaelis ipecacuanha) was used in Brazil and Far east for dysentery and diarrhoea. Reserpine [8], used as a hypotensive drug and tranquilizer was isolated from the plant Rauwolfia serpentia, considered to be a common remedy for mental illness and snake bite in the Indian sub continent. Bronchodilating effect of ephedrine from Ephedra vulgaris [6].

Curare used as arrow poison by the South American natives has now given as tubocurarine [9], a quaternary alkaloid now being used as an adjunct to anesthesia for surgery. Anti-spasmodic effect of atropine [6], from Atropa belladona. Analeptic effect of strychnine [6] from Nux vomica and pelletierine, from pomegranate. Antileprotic effect of chaulmoogra fruit [6] was well known to the ancient Indians. At present thousands of plant metabolites are also now being successfully used in the treatment of a variety of diseases. A few striking examples of plant metabolites are as follow: Taxol [10], from Taxus brevifolia, Vincristine [11] and Vinblastine [11] from Vinca rosea Linn. (Periwink Plant), all of which are important anticancer drugs. Arrow poison of foxglove [12] consisting of digitalis glycosides is cardiotonic for man. The root bark of Mexican yam [13] is used in making cortisone and other steroidal drugs. Salicylic acid [14] isolated from willow bark has a variety of pharmacological effects on platelet aggregation, pain and immune system. Artemisinin, a new antimalarial from Chinese herbal medicine [15].

The plant under investigation is Swertia chirata Ham which belongs to the family Gentianaceae. The Gentianaceae is a tropical family of small trees, herb and bitter tonic. It consists of 180 species. About 8-10 species exist in India [16]. This plant is indigenous to temperature Himalayas at altitudes above 4000 feet from Kashmir, Nepal and Bhutan [17]. In this family all plants are use as medicine. In the present investigation the plant Swertia chirata Ham was selected for phytochemical and biological studies.

The plant Swertia chirata (Family: Gentianaceae) Ham was chosen for investigation since it has a folkloric reputation. So far literature surveyed, Swertia chirata Ham has tremendous uses in traditional medicines. It has anti-microbial activity against Gram positive and Gram negative bacteria. All the plant are used as astringent, unani-tonic to heart, liver, eyes, cough, scanty-urine, melancholia, dropsy, sciatia, skin diseases, the plant is used as a bitter tonic in gastrointestinal disorders, like dyspepsia/anorexia, it is used as digestive, febrifuge and laxative. It is used to prevent malaria, particularly useful in fever. The plant is also effective against intestinal worms burning of the body, bronchial asthma, regulating the bowels [18] (Figure 1).

MATERIALS AND METHODS

Dried stem of Swertia chirata Ham (Locally known as chirata) were collected from Bhangura Kabiraji Shop of Rajshahi Shaheb Bazar. The stems of the plant were cut into small pieces by a sharp knife. About 650 gm of the plant pieces were weighed by the electric balance and made paste by pestle-moter. The paste materials (stem) were taken in a clean flat bottomed glass container (2.5 litre) and macerated with sufficient amount of rectified spirit and with occasional shaking. After 15 days the solvents was decanted and filtered by Tincture filter press (Karl Kolb, Scientific- Technical Supplies, Frankfurt/M Germany) and then filtered through fresh cotton. The filtrate thus obtained was taken in a 1 litre beaker. The solvent (R.S) of the extract was evaporated under normal pressure and normal temperature to obtain a gummy mass. This mass was preserved in a refrigerator for chemical investigation. Rectified spirit extract of the plant of Swertia chirata Ham a crude product. It contained a mixture of compounds. In this chapter, an attempt was made to isolate the individual compounds. TLC examination of rectified spirit extract, petroleum ether: ethyl acetate (1:2) showed 3 spots of Rf values 0.1, 0.54 and 0.73, respectively.

Figure 1: Swertia chirata: a, plant in vegetative phase; b, A 2ft tall plant before flowering; c, Flowering twig; d, Flowering panicle during seed set; e, Single tetramerous flower; f, Root of a mature two year-old plant.

Table 1: Solvent used for column chromatography.

Eluants	Amount of solvent (in ml)	No. of collected beakers
100% n-hexane	300	1-3
25% Petroleum ether in hexane	100	4
50% Petroleum ether in hexane	100	5
100% Petroleum ether	200	6-7
2.5% Ethyl acetate in petroleum ether	100	8
5% Ethyl acetate in petroleum ether	300	9-11
7.5% Ethyl acetate in petroleum ether	100	12
10% Ethyl acetate in petroleum ether	300	13-15
12.5% Ethyl acetate in petroleum ether	100	16
15% Ethyl acetate in petroleum ether	100	17
20% Ethyl acetate in petroleum ether	100	18
25% Ethyl acetate in petroleum ether	100	19
30% Ethyl acetate in petroleum ether	200	20-21
40% Ethyl acetate in petroleum ether	100	22
50% Ethyl acetate in petroleum ether	100	23
60% Ethyl acetate in petroleum ether	100	24
70% Ethyl acetate in petroleum ether	100	25
80% Ethyl acetate in petroleum ether	100	26
90% Ethyl acetate in petroleum ether	100	27
100% Ethyl acetate	100	28
5% Methanol in ethyl acetate	200	29-30
15% Methanol in ethyl acetate	100	31
50% Methanol in ethyl acetate	200	32-33
100% Methanol wash	100	34

The fractions were combined on the basis of their preliminary TLC examination to give combined fractions A, B, C, D, E, F (Table.2). Each combined fraction was evaporated to dryness under reduced pressure.

The crude rectified spirit extract was subjected to an alumina column for fractionation. The column containing 10 gm of alumina was prepared as described before. Crude extract (2.5 gm) for the purpose was mixed with a little rectified spirit in a morter to get a free following mass. The sample was then placed carefully on the top of the prepared column and was successively eluted with n-hexane, n-hexane with increasing portions of petroleum ether; ethyl acetate and finally with methanol (Table 1). A number of colour bands were observed during the development of the column. The subsequent eluants were collected in 100 ml beakers (Table 2).

Examination of the combined fractions

- Fraction A: Fraction A gave no spot on TLC examination and was discarded.
- Fraction B: The residue (35 mg) from fraction B did not give any discrete spot with different solvent system and tailed badly and was not worked out further.
- Fraction C: The fraction showed two spots on the TLC, placed with solvent system n-Hexane: Ethyl acetate (3:1). The spots were pink in color under UV light. The band were contains R_f value=0.67 and 0.583, respectively. The residue from fraction C was subjected to a small chromatographic column fitted with a cotton-plug, using Ethyl acetate: Toluene (1:4). The two eluants were collected and evaporated to get small amount of compounds C_1 and C_2. The compounds C_1 and C_2 were insufficient quantities and were not worked out further.
- Fraction D: The fraction showed two spots (R_f=0.53 and 0.73) on TLC plate using the solvent system Petroleum ether: Ethyl acetate (1:2). The spot having R_f=0.73 showed a violet fluorescence under UV light and yellowish spot in iodine vapour. While the other spot having R_f=0.53 was pink under UV light and yellow in iodine vapor. The components on chromatoplate did not react with potassium permanganate reagent and reacted with vanillin sulfuric acid reagent giving bluish violet spots.
- The residue (110 mg) from fraction D was subjected on alumina to a small chromatographic column using petroleum ether: ethyl acetate (1:1). The two eluants were colleted and evaporated to get two compounds designated as AJ-1 (45 mg) and AJ-2 (8 mg).
- The compound AJ-1 was crystalline but the compound AJ-2 was not crystalline. The compound AJ-2 was insufficient quantities and was not worked out further.
- Fraction E and F: These two fractions appeared to be a mixture of components, which had Rf values very close to each other and could not be separated. The fractions were therefore preserved for further studies in future.

Table 2: Name of the fraction obtained from column chromatography

Fraction	Beaker number	Yield of residue
Fraction-A	1-6	9mg
Fraction-B	7-11	35mg
Fraction-C	12-17	60mg
Fraction-D	18-21	110mg
Fraction-E	22-25	25mg
Fraction-F	26-30	45mg

Isolation of the compound AJ-1 from small chromatographic column

From the TLC analysis of the fraction obtained from small column chromatography, it was observed that the compound AJ-1 contained only one compound. This was recystallized from methanol-ethyl acetate mixture.

Purification of the compound AJ-1: The compound AJ-1 was further recrystallized dissolving in petroleum ether: ethyl acetate (1:1) and the crystals were washed with different solvents of varying polarity when needle-shaped crystals were obtained.

Purity Test: This isolated compound AJ-1 was tested in different solvent systems for its purity. The compound showed a single spot on TLC examination. So this compound is pure. Finally its R_f values were determined using the various solvent systems (Table 3).

Spectral characteristics of the compound AJ-1

Infrared (IR) Spectrum of the compound: Infrared (IR) spectrum (Figure 2) of the compound in KBr (spectroscopic grade) pellet was recorded with a pye-unicam SP3-300 spectrophotometer. The samples were put in an agate mortar and thoroughly powdered with KBr and then transferred in a disc holder and a disc was made by hydraulic press. The KBr pellet was mounted in the sample cavity of the machine.

The infrared (IR) spectrum of the compound had nine characteristic bands in the functional group region at 3509 (OH-streching); 2993, 2971, 2938 (CH-streching); 1762, 1744; 1705 (>C=O); 1502 (NH) and 1150 (CH-streching) cm^{-1}, respectively.

1N-NMR Spectrum of the compound AJ-1: ^{1}H.NMR spectrum of the compound was recorded on NMR spectrophotometer (500MHz) at the Faculty of Pharmaceutical Sciences, Nagoya City University, Tanbedori, Mizuho-Ku, Nagoya 467, Japan. Pyrd5 was used as solvent and the chemical shifts are given in δ-values.

^{1}H-NMR data of the compound

^{1}HMR (Pyrd$_5$, 500 MHz)

δ : 7.73 (1H,S, ArH), 7.65~7.67. (3H, m, ArH)

δ : 6.70 (2H, dd, J=1.3 and 12.2 Hz, terminal=CH_2 olefinic)

δ : 6.51 (1H, dd, J=2.2 and 9.6 Hz, =CH)

δ : 6.43 (1H, dd, J=6.4 and 9.6 Hz =CH)

δ : 5.99 (1H, dd, J=5.1 and 15.3 Hz, =CH) : Two olefinic terminal

δ : 5.92 (1H, dd, J=5.1 and 15.3 Hz, =CH)=CH_2

δ : 5.45 (1H, bd, J=5.1 Hz, =CH): Five olefinic=CH

δ : 5.26 (2H,S, terminal olefinic CH_2)

δ : 2.94~2.87 (3H, m, CH_2)

δ : 2.60 (2H, bd, J=Hz,=CH_2)

δ : 2.51 (2H, tdt, J=Hz,=CH_2)

δ : 2.34 (2H, S, CH_2)

δ : 1.98~2.14 (10H, m, CH_2)

δ : 1.75~1.85 (4H, m, CH_2)

δ : 1.52 (4H, S, either NH or OH, or both protons)

δ : 1.33 (6H, S, 2XCH_3)

δ : 1.32 (3H, S, CH_3)

δ : 1.31 (3H, S, CH_3)

Table 3: TLC analysis of the compound on silica gel G.

Solvent system and ratio	R_f value
Petroleum ether: Ethyl acetate (1:2)	0.73
Chloroform: Ethyl acetate (49:1)	0.92
Ethyl acetate: Chloroform (19:1)	0.79
Toluene: Ethyl acetate (3:1)	0.88
Ethyl acetate: Pyridine: Water (5:1:4)	0.98
Ethyl acetate: Chloroform: Methanol (2:2:1)	0.95
Benzene : Ethyl acetate (19:1)	0.86
Chloroform: Ethyl acetate (4:1)	0.91
Ethyl acetate: Acetone (9:1)	0.94

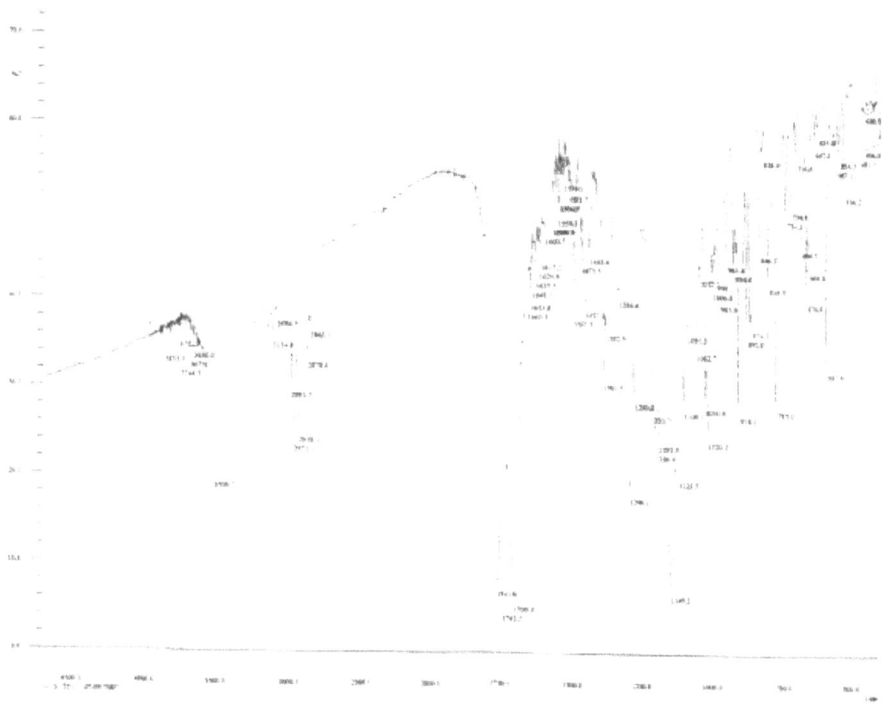

Figure 2: IR Spectrum of Compound AJ-1.

^{13}C-NMR (Pyrd$_5$, 500 MHz)

13C-NMR spectrum of the compound was also recorded (Figure 3) on NMR- spectrophotometer (500MHz) with 100 MHz frequency and pyrd$_5$ was sued as solvent and the data are given below.

<C=O at 176.02, 174.41, 174.36 and 173.08 ppm.

ArCH at 144.63, 140.86, 138.08 and 130.55 ppm.

ArC at 126.59 and 126.49 ppm

=CH at 109.75 and 109.72 (Down field signals may be due to the presence of neighbouring electronegative group, such as OH)

=CH at 74.47, 71.56 and 71.47 (olefinic CH)

–CH at 52.65, 50.53, 47.53, 47.59, 46.88, 45.09 and 44.60 (aliphatic)

=CH$_2$ at 81.97 and 81.90 ppm (olefinic=CH2)

–CH$_2$ at 42.26, 42.15, 41.79, 38.33, 36.93, 35.52, 27.19, 27.04, 18.27 and 17.60, CH2 aliphatic)

–CH$_3$ at 28.93, 28.41, 24.98 and 23.87 ppm

RESULT AND DISCUSSION

Preliminary Extraction

The paste materials were taken in a clean flat bottomed glass container (2.5 litre) and macerated with sufficient amount of rectified spirit and with occasional shaking. After 15 days the solvents was decanted and filtered by tincture press. Then the extract was evaporated under normal pressure and normal temperature to obtain a gummy mass.

Examination of the Rectified Spirit Extract

The rectified spirit extract showed at least three components (R_f=0.4, 0.54 and 0.73) on thin layer chromatographic plates with Petroleum ether: Ethyl acetate (1:2). The rectified spirit extract was then subjected to column chromatography on alumina. The column was eluted successively with n-hexane, n-hexane-petroleum ether mixtures and finally with methanol (Table 1). The fractions were combined on the basis of their preliminary TLC examination to give combined fractions A, B, C, D, E and F (Table 2). The combined fractions were evaporated to dryness under reduced pressure. Fraction A did not give any residue.

Fraction D: Fraction D showed two spots (Rf=0.53 and 0.73) on TLC plates using solvent system Petroleum ether: Ethyl acetate (1:2). The fraction D was further subjected on mini column chromatography using Petroleum ether: Ethyl acetate (1:1). The two eluants were collected and evaporated to get two components designated as AJ-1 (40 gm) and AJ-2 (5 mg). The component AJ-1 was crystalline but the component AJ-2 was not crystalline and was insufficient quantities.

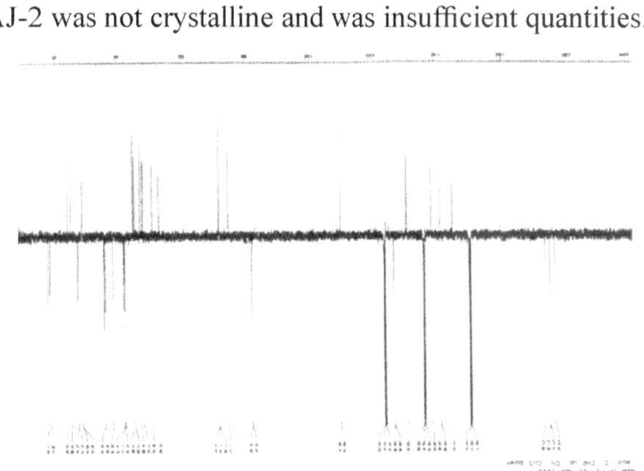

Figure 3: C-NMR Spectrum (500 MHz, pyr-d5) of Compound AJ-1.

Isolation and Purification of Pure Compound

The compound AJ-1 showed single spot on TLC analysis with some impurities. The compound AJ-2 was recrystallized dissolving in Petroleum ether: Ethyl acetate (1:1) and the crystals were washed with different solvents of varying polarity. The isolated compound AJ-1 was tested in different solvent systems (Table 1) for its purity. The compound showed a single spot on TLC examination. So, this compound was pure. Finally its Rf values (Table 3) were determined using the various solvent systems.

Characteristics of the compound AJ-1

(A) Physical characteristics

Physical form: Fine crystals (needles shaped) having single Rf value was obtained.

Color: The compound was all most colorless.

Solubility: The solubility data of the compound have been summarized below.

The compound is highly soluble in ethyl acetate and chloroform.

The compound is insoluble in n hexane, petroleum ether.

The compound is sparingly soluble in ethanol, diethyl ether.

Melting point: Melting point of the compound was observed in REICHERT AUSTRIA, Nr. 341910 melting point apparatus; melting point of the compound is 178°C.

(B) Chemical characteristics

The compound gave positive test for ketone group, tertiary amine group, alcohol, ester, unsaturation, alkaloid and negative test for carbohydrate, phenol and hydrocarbon.

^{1}H and ^{13}C-NMR (Neuclear Magnetic Resonance) Spectra: ^{1}H and ^{13}C-NMR spectra were recorded on a JEOL-AX-500 (500 MHz) and JEOL-JNM AX 400 (400MHz), FT NMR spectrometers, Faculty of Pharmaceutical Sciences, Nagoya City University, Tanabedori, Mizuho-ku, Nagoya 467, Japan. CDC13 was used solvent with Tetramethylsilane (TMS) as an internal standard and the chemical shifts are given in δ-values. Allorganic extracts were dried over anhydrous Na_2SO_4 and solvent was evaporated off under reduced pressure in a rotary evaporator.

Study of spectral data: The Infrared (IR) spectrum (Figure 2) of the compound in KBr disc showed the absorption bands at 3509; 2993, 2971, 2938; 1762; 1744; 1705; 1502 and 1150 cm^{-1}, respectively. The absorption bands observed could be assigned to the following:

The band at 3509 cm^{-1} showed O–H stretching vibration in hydrogen bonded alcohols and phenols.

Bands at 2993, 2971 and 2938 cm^{-1} showed alkane type C–H stretching (C–H stretching vibration in CH_3).

The bands at 1705 cm^{-1}, 1502 cm^{-1} and 1150 cm-1 indicated the presence of >C=O, NH and CH–stretching, respectively.

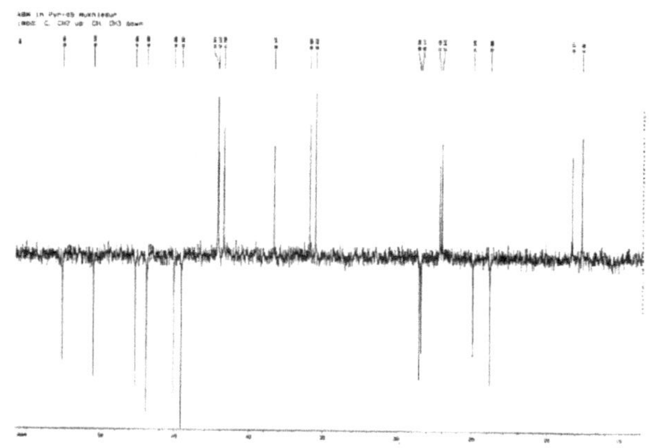

Figure 4: Expanded ^{13}C-NMR Spectrum (500 MHz, pyr-d5) of Compound AJ-1.

From the complete decoupling ^{13}C-NMR spectrum of the compound, a total of 37 carbon signals were detected (Figure 3). In J resolved DEPT 45° spectrum four down field signals appeared at 173.08, 174.36, 174.41 and 176.02 ppm and are assumed to be carbonyl carbon. Chemical examination showed the positive test for carbonyl group and absence of aldehyde which is also supported by the appearance of three peaks in its IR spectrum at 1762, 1743 and 1705 cm^{-1} for carboxyl functions. Signal for aldehyde group proton did not appear in ^{1}H-NMR spectrum. From the above findings, four ketone functions are present in the molecule. Six aromatic carbons appeared at 126.49, 126.5, 130.55, 138.03, 140.85 and 144.63 among which two are quaternary and the other four are methane carbon atoms, identified by DEPT 135° spectrum, which indicates the presence of a benzene ring in the molecule.

Four benzene ring protons are present in the ^{1}H-NMR spectrum and their splitting pattern is 1:3 i.e. a singlet equivalent to one proton and a multiplet equivalent to three protons. From the above ^{13}C-NMR and ^{1}H-NMR spectra, the benzene ring, substituted in two portions which are presumed to be ortho and para and are evident form the assymmetrical splitting pattern of the benzene ring protons (1:3) (Figure 4).

Eleven methane carbon signals appeared in the ^{13}C-NMR and are identical from the DEPT 135° spectrum–out of which 5 carbons are double banded carbon and the rest are saturated CH carbons. Of the five doubly bonded methane carbons, two carbons showed down field shift probably due to the presence of electro negative groups or atoms present adjacent to these two carbons which appeared at 109.72 and 109.75 ppm and the rest three carbons appeared at 71.47, 71.56 and 74.47 ppm. Six saturated methane carbons appeared in higher field at 44.60, 45.09, 46.88, 47.53, 50.53 and 52.65 ppm.

Two low field methylene carbon signals appeared at 81.97 and 81.90 ppm indicated the presence of two terminal CH_2 groups with double bond in the side chain of the molecule. The other 10 saturated CH_2 carbon signals appeared between 17.60 and 42.26 ppm.

From the methyl carbon signals at 23.87, 24.98, 28.41 and 28.93 ppm, it has been confirmed that four methyl groups are present in the molecule. This is also supported by its ^{1}H-NMR (Figure 5) spectrum. Four methyl group protons appeared at δ: 1.33 (6H, $2XCH_3$), 1.32 (3H, $1XCH_3$) and 1.31 (as 3H, $1XCH_3$) singlets.

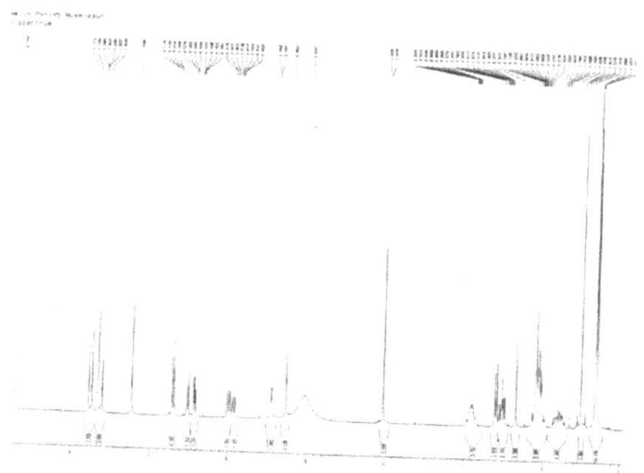

Figure 5: [1] H-NMR Spectrum (500 MHz, pyr-d5) of Compound AJ-1.

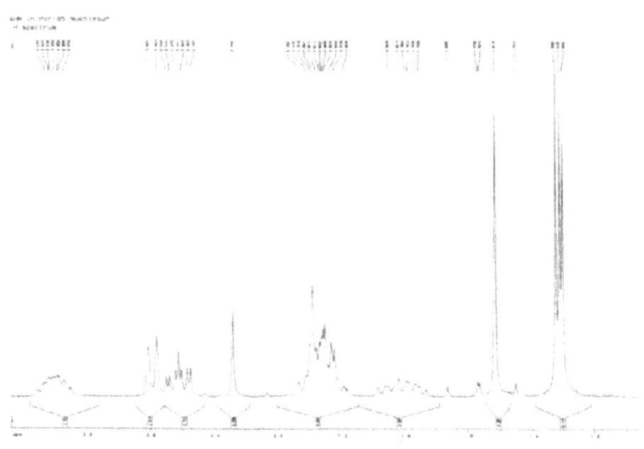

Figure 6: Expanded 1 H-NMR Spectrum (500 MHz, pyr-d_5) of Compound AJ-1.

From the ^{1}H-NMR spectrum of the compound four benzene ring protons appeared in an unsymmetrical pattern, one as singlet at δ: 7.73 and other three as multiplet between δ: 7.65~7.67. Five doubly bouded methane protons appeared at δ: 6.51 (^{1}H, dd, J=2.2 and 9.6 Hz), 6.43 (^{1}H, dd, J=6.4 and 9.6 Hz), 5.99 (^{1}H, dd, J=5.1 and 15.3 Hz), 5.92 (^{1}H, dd, J=5.1 and 15.3 Hz) and 5.45 (^{1}H, bd, J=5.1 Hz) (Figure 6). One doublet, equivalent to two protons, appeared at δ: 6.70 with J values of 1.3 and 12.2 Hz and other singlet of two protons appeared at δ: 5.26. These two sets of protons seems to be a doubly bonded terminal methylene group present in the molecule. The above discussion field proton signals satisfy all the carbon signals appeared in the down field region in ^{13}C-NMR spectrum.

Rest of the saturated CH2 protons appeared between δ: 1.75 and 2.94 ppm (plesse, vide experimental). A peak of four protons appeared as singlet at δ: 1.52 which seems to be NH or OH protons or mixture of these two. Positive test with alkaloidal reagents indicated the presence of nitrogen and the presence of OH group can also be predicted from the IR spectrum since a broad peak for OH stretching centered at 3509 cm-1 appeared in the IR spectrum. Four methyl protons signals also appeared which have been discussed above.

Form the above spectral evidences; the compound is an alkaloid containing 37 carbon atoms with 52 hydrogen along with secondary or tertiary nitrogen and several OH groups in the molecule.

CONCLUSION

From the research work only one compound is isolated and the above spectral evidence, the compound is an alkaloid containing 37 carbon atoms with 52 hydrogen along with secondary or tertiary nitrogen and several OH groups in the molecule. For complete structure elucidation, Mass spectrum, HSCOSY, HSQC and HMBC spectra will be required. From the above discussion, it is clear that this plant may contain other medicinal compounds. Further research may be extended to isolate active compounds specially from fractions B, C, E and F.

REFERENCES

1. Fuller HJ, Carothers Z B (1967) the plant world, 4th edition, Published by Holt, Rinehart and Winston, Inc1-4.
2. Lewington A (1990) Plants for people, published by the Natural History Museum, London 135-136.
3. Core EI (1962) Plant Taxonomy, Englewood cliffs, N. J. Prentice-Hall, Inc 9.
4. Chopra RN (1933) Indigenous drugs of India. The Art Press, Calcutta, India.
5. Alfred Burger (1960) Medicinal Chemistry, Inter Science publishers, Inc, New York, USA.
6. Alfred Burger (1960) Medicinal chemistry. Inter Science publisher, Inc. New York 4-6, 12, 314-319, 613-615, 389.
7. Alfred B (1970) Medicinal Chemistry, Wiley-Interscience publisher, Inc, New York, 3rd ed, 4-6, 12.
8. Muller JM, Schlitter E, Bein HJ(1952) Reserpine. Experientia 8: 338.
9. British Pharmaceutical codex (1963) 872.
10. Kumer P, Clarke M (1994) Clinical Medicine, Billaire-Tindall, philadelphia 353-357
11. William BP, Raymond WR (1979) the Anticancer Drugs, Oxford University Press, New York 221-232.
12. Wilson CO, Gisvold O, Doerge RF (1979) The text book of organic medicinal and pharmaceutical chemistry, J. B. Lippincott company, philadelphia New York, 2nd ed 796-797.
13. World Book (1988) the world book of Encyclopedia, Scott. Fetzer Company USA.
14. Goodman LS, Gilman A (1981) The pharmacological Basis of Therapeutics, 6th Edition, 682-687.
15. ARBEC (special feature) (2001), Biotechnology and natural products, Wednesday, March 14.
16. Hooker JD (1885) Flora of British India, reeve and Co. Ltd 5: 78-79.
17. Chirangib Bownoshadhi of India 1: 170.
18. Ambasta SP (1986) The useful plants of India, CSIR New Delhi, India.

CHAPTER 14

COMPARATIVE ANALYSIS FOR ANTIOXIDANT CAPACITIES, IMPORTANT PLANT PIGMENTS AND TOTAL PHENOLIC CONTENTS IN DIFFERENT BRASSICA VEGETABLES

Chander Parkash, S.S. Dey, Raj Kumar, M.R. Dhiman, Munish Dhiman, Satish Kumar, Vinod Kumar

Indian Agricultural Research Institute, Regional Station, Katrain (Kullu Valley), H.P.-175129 India

ABSTRACT

A study to analyze the quantitative comparative variation of antioxidant capacity within the *Brassica* family was undertaken with ten cultivars representing 5 *brassica* vegetables. It was observed that sprouting broccoli cultivar Pusa Broccoli KTS-1 exhibited significantly highest contents of antioxidants viz., CUPRAC (17.91 μ mol tolox/g) and FRAP (14.34 μ mol trolox/g) whereas, significantly higher contents of total phenolics (6120.0 μg gallic acid/gfw), anthocyanin (3.28 mg/100 g), ascorbic acid (80.832 mg/100g of sample), lycopene (8.02 mg in 100 g sample), total carotenoids (52.82 mg/100g) and *β*–carotene (11.32 μg/100 g) were recorded in kale cultivar KTK-64 in comparison to other vegetables. High genetic advance coupled with high heritability was observed for all the traits which is attributed to considerable additive genetic effects and selection for these traits would be most effective. CUPRAC has shown significant and positive correlation at both genotypic and phenotypic levels with all the other 7 traits whereas FRAP was significantly correlated with total phenolics and anthocyanin only. Based on D^2analysis the 10 genotypes were classified into 4 clusters. The cluster IV had maximum 4 genotypes and cluster II had only one genotype.

In the past years, most studies were concentrated on the role of diet in human health. It has been studied that plant products are much more beneficial to reduce a high risk of many diseases such as cancer, cardiovascular diseases, diabetes and atherosclerosis etc. (Goldman, 2004). There are a number of phytochemicals such as phenols, ascorbic acid, anthocyanin, chlorophyll, carotenoids, CUPRAC, FRAP etc. which provide basic nutrition to furnish the defensive mechanism to reduce the risk of many chronic diseases in humans (Deng et al., 2013). Objective of this study was to analyze the quantitative comparative variation of antioxidant capacity, phenols and other compounds within the *Brassica* family.

Brassica* vegetables such as cabbage, cauliflower, broccoli, kohlrabi and kale possess both antioxidant and anti-carcinogenic properties (Cohen et al., 2000; Chu et al., 2002; Verhoeven et al., 1997). All plants of the genus *Brassica

Citation: Parkash et al., Comparative analysis for antioxidant capacities, important plant pigments and total phenolic contents in different Brassica vegetables, Molecular Plant Breeding, 2015, Vol.6, No. 10 1-7 (doi: 10.5376/mpb.2015.06.0010).

like broccoli, Brussels sprouts, cabbage, cauliflower, collard greens and kohlrabi contain glucosinolates which are known to have fungicidal, bactericidal and cancer protective characteristics (Fahey, 2003). These vegetables are rich sources of glucosinolates, sulfur-containing compounds that impart a pungent aroma and bitter taste. The hydrolysis of glucosinolates by a class of plant enzymes called myrosinase results in the formation of biologically active compounds, such as indoles and isothiocyanates. Myrosinase is physically separated from glucosinolates in intact plant cells. However, when brassica vegetables are chopped or chewed, myrosinase comes in contact with glucosinolates and catalyzes their hydrolysis. Therefore, the potential for high intakes of brassica vegetables as well as several glucosinolate hydrolysis products to prevent cancer is of interest. Natural antioxidant defence mechanism is present in human beings (Ali et al., 2001), however, this natural antioxidant defence mechanism can be insufficient hence dietary intake of antioxidant compounds become important (Halliwell, 1994).

Several studies showed that high consumption of vegetables and fruits had an inverse relationship with the incidence of degenerative diseases including heart diseases (Hertog et al., 1993; Lee et al., 2004), cancer (Block et al., 1992; Byers and Perry, 1992; Sugimura, 2002), inflammation (Giugliano et al., 2006; Rahman et al., 2006) and brain dysfunction (Youdim and Joseph, 2001; Trewavas and Stewart, 2003). Oxidative stress results from either a decrease of natural cell antioxidant capacity or an increased amount of reactive oxygen species (ROS) in organisms. When the balance between oxidants and antioxidants in the body is shifted by the overproduction of free radicals, it will lead to oxidative stress and DNA damage. (Ames et al., 1993;Liu, 2002). Brassica vegetables contain many bioactive compounds, especially organosulphur phytochemicals possessing anticarcinogenic activity and other phytochemicals, which are known to possess antioxidant activity. Dietary antioxidants present in these vegetables, such as water-soluble vitamin C and phenolic compounds, as well as lipid-soluble vitamin E and carotenoids, contribute to both the first and the second defence lines against oxidative stress (Lampi et al., 2002; Krinsky, 2001; Davey et al., 2000).

The role of antioxidants is to protect all bio-molecules from the oxidative damage of free radicals, therefore antioxidants can delay the progress of degenerative diseases. Antioxidants can act by scavenging reactive oxygen species (SOD removing O_2^-), by inhibiting their formation (by blocking activation of phagocytes), by binding transition metal ions and preventing formation of OH and/or decomposition of lipid hydroperoxides by repairing damage (alpha-tocopherols repairing peroxyl radicals so terminating the chain lipid peroxidation) or in combination of any of the above (Niwa et al., 2001). Human body can produce few antioxidants especially many enzymatic systems (superoxide dismutase, catalase and glutathione peroxidase) and can obtain some low molecular weight antioxidants from foods, in particular vegetables and fruits (vitamin C, vitamin E, carotenoids and phenolic compounds), which can scavenge some reactive species (superoxide anion and hydrogen peroxide) and inhibit a chain reaction. Consumption of foods naturally bearing antioxidant power is the most efficient way of combating such undesired transformations and health risks. Consequently, the opportunity for improving health by improving diet is great (Ames et al., 1993).

MATERIALS AND METHODS

Basic experimental materials

Experimental materials for the study comprised of ten varieties of five brassica vegetables out of which four varieties were of cabbage namely Golden Acre, Pusa Cabbage-1, KTCBH-84 and Quisto, two each of cauliflower (Early Snowball and Pusa Himjyoti) and kohlrabi (Pusa Virat and KKS-2) and one each of kale (KTK-64) and sprouting broccoli (Pusa Broccoli KTS-1).

Field experiment

One month old seedlings were transplanted during 2013-14 at Naggar research farm of IARI Regional Station, Katrain, Kullu, HP, India. The farm is located at 32.12^{0}N latitude and 77.13^{0}E longitude with an altitude of 1688 m above the mean sea level, and receives an average annual rainfall and snowfall of 100–110 cm and 110–130 cm, respectively. The plot size was 3 m x 3 m with inter-and intra-row spacing of 45 cm. Each variety was replicated three times in a randomized block design (CRBD). The crop was uniformly fertilized with recommended doses of chemical fertilizers @120 kg N, 60 kg P_2O_5 and 50 kg K_2O per ha supplied as urea, single super phosphate and murate of potash, respectively. Half dose of N and full doses of P_2O_5 and K_2O were applied as basal dressings at the time of transplanting while remaining half dose of N was top-dressed at 30 days after transplanting.

Sampling and laboratory analysis

True-to-type representatives (5 heads/curds/knobs/ leaves) of each variety of the five brassica vegetables in each replication were harvested at marketable stage. These were chopped, homogenized and a fresh sample of 5 g of each was stored immediately under refrigerated conditions (-20°C) until assay. Ethanol extract was prepared from the homogenized 5 g sample in 15 ml absolute ethanol. Then it was centrifuged at 10,000 rpm for 15 min at 4°C. The supernatant was stored at -20°C.

Anti-oxidant capacities and total phenolics assay

The method described by Apak et al. (2006) was followed with minor modifications. For cupric ion reducing antioxidant capacity (CUPRAC) analysis, 100μl sample was mixed with 4 ml of CUPRAC reagent (1 ml neucuproine, 1 ml ammonium acetate, 1ml $CuCl_2$ and 1 ml of distilled water; pH 7.4). Absorbance was recorded 450 nm in spectrophotometer. The results were expressed as μ mol trolox/g. The ferric reducing ability of plasma (FRAP) assay was performed based on the procedure described by Benzie and Strain (1996) with slight modifications. In this assay, 100 μl of the diluted sample were added to 3 ml of the FRAP reagent and the reaction was monitored after 4 min at 593 nm. The results were expressed as μmol Fe(II)/g fresh weight (FW) of vegetable. Total phenolic contents were determined with Folin–Ciocalteu method

(Singleton and Rossi 1965). Briefly, 0.50 ml extract was mixed with 2.5 ml of 1:10 diluted Folin–Ciocalteu reagent. After 4 min, 2 ml of saturated sodium carbonate solution was added. The mixture was incubated in dark for 2 h at room temperature and its absorbance was detected at 760 nm. Gallic acid was used for calibration, and the results were expressed as mg of gallic acid equivalents (mg GAE) per 100 g fresh weight (FW) of vegetable.

PLANT PIGMENTS AND ASCORBIC ACID CONTENT

Anthocyanin

10 g of sample was blend with 10 ml of ethanolic HCl with the help of pestle and mortar and transferred in 50 ml conical flask by using 10 ml ethanolic HCl for washing. Solution is stored overnight at 4^{0}C. Then the solution was filtered with Whatman No. 1 filter paper. Final volume was made upto 100 ml and stored in dark for 2 hours. O.D. was taken at 535 nm (S. Ranganna, 2008).

Total Anthocyanin (mg/100 g) =

$$\frac{\text{OD} \times \text{dilution} \times \text{total volume made up} \times 100}{\text{wt.of sample taken} \times e}$$

Where e = 98.2 (absorbance of a solution containing 0.1 mg/ml anthocyanin)

Carotenoids

5 g of sample was taken and crushed with acetone until the residue became colorless. Filtrate was transferred in the separating funnel containing 20 ml of petroleum ether. 2-3 drops of 5% sodium sulphate were added in separating funnel. Then 20 ml of petroleum ether was added to make 2 separate phases. Lower phase was re-extracted with additional petroleum ether till it become colourless. Final volume was made up to 50 ml and absorbance was taken at 503nm and 452nm using petroleum ether as a blank (S. Ranganna, 2008).

Lycopene

Absorbance (1 unit) = 3.1206 μg Lycopene/ml

Lycopene (mg in 100 g sample) =

$$\frac{3.1206 \times \text{OD at 503 nm} \times \text{volume made up} \times \text{dilution} \times 100}{\text{wt.of sample} \times 1000}$$

Total Carotenoids

Total carotenoids (mg/100g) =

$$\frac{3.857 \times \text{OD at 452 nm} \times \text{volume made up} \times \text{dilution} \times 100}{\text{wt.of sample} \times 1000}$$

β-*Carotene*

μg/100 g =

$$\frac{\text{OD at 452 nm} \times 13.9 \times 10^4 \times 100}{\text{wt.of sample} \times 560 \times 1000}$$

Ascorbic acid

10 g of sample was crushed with 4% oxalic acid and volume made up to 100 ml. Then sample was filtered wit Whatman filter paper No. 1 and filtrate was used for further analysis. 10 ml of the prepared sample was titrated against the 2,4-dichlorophenolindophenol dye till the faint pink color was appeared (S. Ranganna, 2008).

Ascorbic acid (mg/100g of sample) =

$$\frac{\text{titre value} \times \text{Dye factor} \times \text{volume made of sample} \times 100}{\text{Titrate volume} \times \text{Sample wt.}}$$

Statistical analysis

The analysis of variance of the data was carried out following Panse and Sukhatme (1967). The variability estimates (genotypic, phenotypic and environmental variance; and genotypic and phenotypic coefficient of variation) were worked out through analysis of variance (ANOVA), while correlation coefficients were determined by covariance and variance between the traits. D^2 analysis was performed following Mahalanobis (1936).

RESULTS AND DISCUSSION

Comparative Analysis

Analysis of variance revealed highly significant differences among the genotypes for all the 8 phytochemicals under study (data not presented). The comparative analysis of ten varieties representing 5 different *brassica* vegetables for antioxidant capacities, plant pigments and total phenolics as presented in Table 1 revealed that sprouting broccoli cultivar Pusa Broccoli KTS-1 exhibited highest contents of antioxidants viz., CUPRAC (17.91 μ mol trolox/g) and FRAP (14.34 μ mol trolox/g) whereas highest contents of total phenolics (6120.0 μg gallic acid/gfw), anthocyanin (3.28 mg/100 g), ascorbic acid (mg/100g of sample), lycopene (8.02 mg in 100 g sample), total carotenoids (52.82 mg/100g) and *β*–carotene (11.32 μg/100 g) recorded in kale cultivar KTK-64. Superiority of kale in terms of total carotenoids and polyphenols has also been reported by Kaulmannet al. (2014).

Table 1: Mean performance of ten varieties of 5 Brassica vegetables for 8 phytochemicals

Variety	CUPRAC (μ mol tolox/g)	FRAP (μ mol trolox/g)	Total Phenolics (μg gallic acid/ gfw)	Anthocyanin (mg/100 g)	Ascorbic acid (mg/100g of sample)	Catrotenoids		
						Lycopene (mg in 100 g sample)	Total carotenoids (mg/100g)	β- carotene (μg/100 g)
Cabbage ·								
Golden Acre	1.43	2.47	827.5	0.076	21.157	0.252	1.041	1.340
KTCBH-84	1.86	2.91	1116.25	0.404	26.190	0.922	2.246	2.888
Pusa Cabbage-1	1.74	2.04	1023.75	0.336	37.109	1.891	0.935	1.203
Quisto	3.98	1.33	950.62	0.022	20.844	0.867	1.714	3.151
Cauliflower								
Early Snowball	6.45	1.53	1878.25	0.191	45.894	0.071	0.204	0.263
Pusa Himjyoti	2.41	1.04	398.75	0.981	67.400	0.124	0.285	0.367
Kohlrabi								
Pusa Virat	2.97	1.05	225	0.010	35.251	0.390	1.639	2.109
KKS-2	3.55	1.44	699.37	0	38.091	0.398	1.841	2.370
Kale								
KTK-64	16.06	2.02	6120	3.284	80.832	8.018	52.820	11.326
Broccoli								
Pusa Broccoli KTS-1	17.91	14.34	4091.25	1.667	35.280	0.294	8.041	5.175
CD ≤0.01	1.27	0.696	342.62	0.233	3.84	2.025	1.098	0.267

Genetic variance

The variance and coefficient of variation provide the extent of variability present among germplasm which is pre-requisite of any breeding/improvement programme. The extent of variability in (Table 2) present among different culivars was estimated in terms of phenotypic and genotypioc (Vp and Vg) and phenotypic and genotypic coefficients of variation. The Vg was highest in total phenolics content followed by ascorbic acid, total carotenoids, CUPRAC, FRAP, β-carotene, lycopene and anthocyanin. The magnitude of phenotypic coefficient of variation was slightly higher than the corresponding genotypic coefficient of variation for all the 8 traits. The respective phenotypic coefficient of variation and genotypic coefficient of variation were high for total carotenoids (229.23 and 229.14%) and followed by lycopene (189.71 and 178.27%), while it was lowest for ascorbic acid (47.99 and 47.82%). Heritable portion of variation can be explained by computing the heritability. High broad sense heritability (>80.0%) estimates were computed for all eight antioxidant related activities, i.e. CUPRAC (99.2%), FRAP (99.9%), total phenolics (99.4%), anthocyanin (99.1%), ascorbic acid (99.3%), lycopene (88.2%), total carotenoids (99.9%) and β-carotene (99.8%) indicating least effect of the environment in the expression of these traits. Genetic advance expressed as per cent of mean was observed in the range of 97.9% to 471.7% with a maximum for total carotenoids (471.7%) followed by lycopene (344.7%) and anthocyanin (309.3%). High genetic advance coupled with high heritability observed for all the traits is attributed to considerable additive genetic effects (Panse and Sukhatme, 1957) and selection for these traits will be most effective.

Table 2: Estimates of variance, coefficient of variation, heritability and genetic advance for CUPRAC, FRAP, anthocyanin, ascorbic acid, total phenolics and carotenoids contents in 5 brassica vegetables

S. No.	Trait	Vg	Vp	GCV (%)	PCV (%)	Heritability (%)	Genetic advance	Genetic advance as per cent of mean
1	CUPRAC	36.60	36.89	103.17	103.59	99.2	12.39	220.1
2	FRAP	16.14	16.22	132.60	132.92	99.9	8.28	273.2
3	Total Phenolics	3578186.15	3599438.52	109.14	109.47	99.4	3869.2	223.5
4	Anthocyanin	1.10	1.11	151.10	151.77	99.1	2.15	309.3
5	Ascorbic Acid	382.99	385.67	47.82	47.99	99.3	40.05	97.9
6	Lycopene	5.75	6.31	178.27	189.71	88.2	4.55	344.7
7	Total Carotenoids	263.22	263.43	229.14	229.23	99.9	33.40	471.7
8	β- -carotene	10.58	10.59	107.13	107.49	99.8	6.69	220.8

Note: Vg - genotypic variance, GCV – genotypic coefficient of variation, Vp – phenotypic variance, PCV – phenotypic coefficient of variation

Correlation Analysis

Correlation refers to magnitude and direction of association between two traits, and determines the component traits on which effective selection can be made. The genotypic coefficients of correlation were in general higher in magnitude than the corresponding phenotypic ones (Table 3). The antioxidant CUPRAC has shown significant and positive correlation at both genotypic and phenotypic levels with all the other 7 traits whereas FRAP was significantly correlated with total phenolics and anthocyanin only as also reported by Kaulmann et al. (2014). Total phenolics was positively correlated with anthocyanins, ascorbic acid and the 3 carotenoids. Similarly, anthocyanin contents showed positive correlation with ascorbic acid and carotenoids. Ascorbic acid also showed positive association with the carotenoids. The 3 carotenoids (lycopene, total carotenoids and β- carotene) were in positive association with each other.

Table 3: Estimates of correlation coefficients (Genotypic and Phenotypic) for antioxidant enzymes among ten varieties of 5 different brassica vegetables

		CUPRAC	FRAP	Total Phenolics	Anthocyanin	Ascorbic acid	Catrotenoids		
							Lycopene	Total carotenoids	β- carotene
CUPRAC	G	-	0.690**	0.923**	0.823**	0.446**	0.533**	0.685**	0.771**
	P	-	0.686**	0.922**	0.823**	0.445**	0.500**	0.682**	0.770**
FRAP	G		-	0.466**	0.340**	-0.138	-0.117	0.044	0.258
	P		-	0.464**	0.339**	-0.139	-0.110	0.044	0.258
Total Phenolics	G			-	0.922**	0.549**	0.793**	0.874**	0.891**
	P			-	0.918**	0.547**	0.741**	0.871**	0.889**
Anthocyanin	G				-	0.752**	0.842**	0.908**	0.877**
	P				-	0.747**	0.790**	0.905**	0.876**
Ascorbic acid	G					-	0.659**	0.688**	0.489**
	P					-	0.629**	0.685**	0.488**
Lycopene	G						-	0.983**	0.897**
	P						-	0.923**	0.844**
Total carotenoids	G							-	0.943**
	P							-	0.942**
β- carotene	G								-
	P								-

Note: G: genotypic level; P: phenotypic level. **Significant at 1% level

Grouping of Genotypes into Various Clusters based on D^2 analysis

Based on D^2 analysis the 10 genotypes were classified into 4 clusters (Table 4). The cluster IV had maximum 4 genotypes and cluster II had only 1 genotype. Intra cluster distance was highest in cluster I with 2 genotypes and lowest in cluster II (Table 5). Highest inter cluster distance, however, was between III *vs* I (22828289.4607) followed by the same IV *vs* I (18068045.8627), II *vs* I (11447292.3664), II *vs* I (4120162.0206). The lowest inter cluster distance was between IV *vs* IV (29853.9254) followed by III *vs* III (115834.9027), III *vs* IV (340204.5986). According to the Arunachalam et al. (1984), the optimum level of genetic divergence between the parents gives the best heterosis and better segregants. The study of cluster mean value of 4 clusters indicated considerable differences for the character studied (Table 6). A dendrogram constructed based on the hierarchical cluster analysis has been shown as Figure 1. The variation observed in cluster means also point out the degree of variability. The characters *viz*., CUPRAC, FRAP, total phenolics, anthocyanin, ascorbic acid, lycopene, total carotenoids and β-carotene were highest in cluster I.

Table 4: Cluster classification of 10 genotypes cabbage, cauliflower, knol-khol, broccoli and kale on D^2 analysis

Cluster No.	Genotypes
1.	KTK-64, KTS-1
2.	Early Snowball
3.	Pusa Himjyoti, Pusa Virat, KKS-2
4.	Golden Acre, KTCBH-84, Pusa Cabbage-1, Quisto

Table 5: Inter and Intra –cluster distances among 4 clusters based on D^2 analysis

Cluster	1	2	3	4
1.	4120162.206	11447292.3664	22828289.4607	18068045.8627
2.		0.00	2104200.4547	819299.8718
3.			115834.9027	340204.5986
4.				29853.9254

Table 6: Cluster means of 10 genotypes for 8 characters based on D^2 analysis

Cluster	CUPRAC	FRAP	Total Phenolics	Anthocyanin	Ascorbic acid	Lycopene	Total carotenoids	β- carotene
1.	16.98	8.18	5105.62	2.47	58.05	4.15	30.43	8.25
2.	6.45	1.53	1878.25	0.191	45.89	0.071	0.20	0.26
3.	2.97	1.17	441.04	0.33	46.91	0.30	1.25	1.61
4.	2.25	2.18	979.53	0.20	26.32	0.98	1.48	2.14
Overall mean	7.16	3.26	2101.11	0.797	44.292	1.375	8.34	3.065

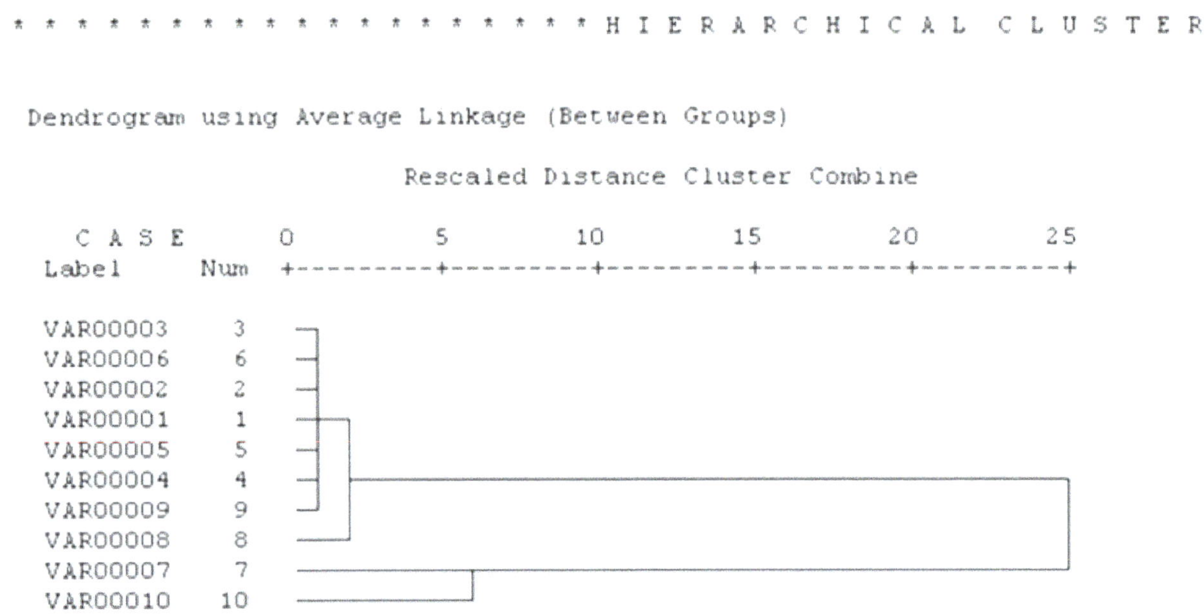

Figure 1: Dendrogram based on the hierarchical cluster analysis

REFERENCES

1. Ali Y., Mavi A. and Kara A.A., 2001, Determination of antioxidant and antimicrobial activities of Rumex crispus L. extracts, Journal of Agricultural and Food Chemistry, 49: 4083-4089 http://dx.doi.org/10.1021/jf0103572
2. Ames B.N., Shigenaga M.K., Hagen T.M., 1993, Oxidants, antioxidants, and degenerative diseases of aging, Proceedings of the National Academy of Science USA, 90: 7915 http://dx.doi.org/10.1073/pnas.90.17.7915
3. Apak R., Gu¨clu K., Ozyurek M., Karademir S.E., and Ercag E., 2006, The cupric ion reducing antioxidant capacity and polyphenolic content of some herbal teas, International Journal of Food Science and Nutrition, 57: 292 http://dx.doi.org/10.1080/09637480600798132
4. Benzie I.F.F., and Strain J.J., 1996, The ferric reducing ability of plasma (FRAP) as a measure of "antioxidant power": the FRAP assay, Annals of Biochemistry, 239: 270 http://dx.doi.org/10.1006/abio.1996.0292
5. Block G., Patterson B., and Suba A., 1992, Fruit, vegetables and cancer prevention-A review of the epidemiologic evidence, Nutrition and Cancer an International Journal, 18: 1-29 http://dx.doi.org/10.1080/01635589209514201
6. Byers T., and Perry G., 1992, Dietary carotenes, vitamin C, and vitamin E as protective antioxidants in human cancers, Annual Review of Nutrition, 12: 139-159 http://dx.doi.org/10.1146/annurev.nu.12.070192.001035
7. Chu Y.F., Sun J., Wu X., and Liu R.H., 2002, Antioxidant and anti-proliferative activities of common vegetables, Journal of Agriculture and Food Chemistry, 50: 6910-6916 http://dx.doi.org/10.1021/jf020665f
8. Cohen J., Kristal R., and Stanford J., 2000, Fruit and vegetable intakes and prostate cancer, Journal of the National Cancer Institute 9: 61-68 http://dx.doi.org/10.1093/jnci/92.1.61
9. Davey M.W., Van Montagu M., Inze D., Sanmartin M., Kanellis A., Smirnoff N., Benzie I.J.J., Strain J.J., Favell D., and Fletcher J., 2000, Plant L-ascorbic acid: chemistry, function, metabolism, bioavailability and effects of processing, Journal of the Science of Food and Agriculture, 80: 825-860 http://dx.doi.org/10.1002/(SICI)1097-0010(20000515)80:7<825::AID-JSFA598>3.0.CO;2-6 http://dx.doi.org/10.1002/(SICI)1097-0010(20000515)80:7<825::AID-JSFA598>3.3.CO;2-Y
10. Deng G.F., Lin X., Xu X..R., Gao L.L., Xie J.F., and Li H.B., 2013, Antioxidant capacities and total phenolic contents of 56 vegetables, Journal of Functional Foods, 5: 260-266 http://dx.doi.org/10.1016/j.jff.2012.10.015
11. Fahey J.W., 2003, Brassicas. In: Caballero B, Trugo L, Fingla SP (eds.) Encyclopedia of food science and nutrition, Academic, St Louis, p. 606-615 http://dx.doi.org/10.1016/B0-12-227055-X/00118-8
12. Giugliano D., Ceriello A., and Esposito K., 2006, The effects of diet on inflammation emphasis on the metabolic syndrome. Journal of the American College

of Cardiology, 48: 677-685 http://dx.doi.org/10.1016/j.jacc.2006.03.052

13. Goldman I.L., 2004, Back to the Future of Food: Phytonutrients and Quality in Vegetable Crops for the 21st Century, Proc. XXVI IHC – Advances in Vegetable Breeding Eds. J.D. McCreight and E.J. Ryder. Acta Horticulturae, 637: 353-360
14. Halliwell B., 1994, Free radicals, antioxidants, and human disease: curiosity, cause, or consequence? The Lancet, 344: 721-724 http://dx.doi.org/10.1016/S0140-6736(94)92211-X
15. Hertog M.G.L., Feskens E.J.M., Hollman P.C.H., Katan M.B., and Kromhout D., 1993, Dietary antioxidant flavonoids and risk of coronary heart disease: The Zutphen Elderly Study, Lancet, 342: 1,007-1,011
16. Kaulmann A., Jonville M.C., Schneider Y.J., Hoffmann L., and Bohn T., 2014, Carotenoids, polyphenols and micronutrient profiles of *Brassica oleraceae* and plum varieties and their contribution to measures of total antioxidant capacity, Food Chemistry, 155: 240-250 http://dx.doi.org/10.1016/j.foodchem.2014.01.070
17. Krinsky N.I., 2001, Carotenoids as antioxidants, Nutrition, 17: 815-817 http://dx.doi.org/10.1016/S0899-9007(01)00651-7
18. Lampi A.M., Kamal-Eldin A., and Piironen V., 2002, Tocopherols andtocotrienols from oil and cereal grains, In: J. Shi, G. Mazza, M. LeMaguer, (Eds.), Functional food-Biochemical and processing aspects, Boca Raton, FL: CRC Press LLC
19. Lee K.W., Lee H.J., and Lee C.Y., 2004, Vitamins, phytochemicals, diets and their implementation in cancer chemoprevention, CRC Critical Reviews in Food Science and Nutrition, 44: 437-452 http://dx.doi.org/10.1080/10408690490886674
20. Liu R.H., 2002, Supplement quick fix fails to deliver, Food Technology International, 1: 71-72
21. Mahalanobis P.C., 1936, On the generalized distance in statistics, Proceedings of National Institute of Science, India 2: 49-55
22. Niwa T., Doi U., Kato Y., and Osawa T., 2001, Antioxidative properties of phenolic antioxidants isolated from corn steep liquor, Journal of Agriculture and Food Chemistry, 49: 177-182 http://dx.doi.org/10.1021/jf0007885
23. Panse V.G., and Sukhatme P.V., 1967, Statistical Methods for Agricultural Workers, Indian Council of Agricultural Research, New Delhi, India
24. Rahman I., Biswas S.K., and Kirkham P.A., 2006, Regulation of inflammation and redox signalling by dietary polyphenols, Biochemical Pharmacology, Article in press, 14 pages
25. Rao C.R., 1952, Advanced Statistical Methods in Biometrical Research, John Wiley and Sons, Inc Macmillan Pub Co New York, USA. p. 460
26. Ranganna S., 2008, Handbook of Analysis and quality control for fruit and vegetable products, Second Edition: Tata McGraw-Hill Publishing Company Limited New Delhi, P.84-106
27. Singleton V.I., and Rossi J., 1965, Colorimetry of total phenolic with phosphomolybdic-phosphotungstic acid agents, American Journal of Enology and Viticulture, 16:144-158
28. Sugimura T., 2002, Food and Cancer, Toxicology, 181-182: 17-21 http://dx.doi.org/10.1016/S0300-483X(02)00250-0
29. Trewavas A., and Stewart D., 2003, Paradoxical effects of chemicals in the diet on health, Current Opinion in Plant Biology, 6: 185-190 http://dx.doi.org/10.1016/S1369-5266(03)00011-6
30. Verhoeven D.T., Verhagen H., Goldbohm R.A., van den Brandt P.A., and van Poppel G., 1997, A review of mechanisms underlying anti-carcinogenicity by brassica vegetables, Chemico-Biological Interactions, 103:79-129 http://dx.doi.org/10.1016/S0009-2797(96)03745-3
31. Youdim K.A., and Joseph J.A., 2001, A possible emerging role of phytochemicals in improving age-related neurological dysfunction: a multiplicity of effects, Free Radical Biology and Medicine, 30: 583-594 http://dx.doi.org/10.1016/S0891-5849(00)00510-4

CHAPTER 15

META-ANALYSIS OF THE DETECTION OF PLANT PIGMENT CONCENTRATIONS USING HYPERSPECTRAL REMOTELY SENSED DATA

Jingfeng Huang[1] , Chen Wei[1,2], Yao Zhang[1] , George Alan Blackburn[3] , Xiuzhen Wang[4], Chuanwen Wei[1] , Jing Wang[1]

[1] Institute of Agricultural Remote Sensing & Information Application, Zijingang Campus, Zhejiang University, Hangzhou, China

[2] Zhejiang Meteorological Service Center, Hangzhou, China

[3] Lancaster Environment Centre, Lancaster University, Lancaster, United Kingdom

[4] Institute of Remote Sensing and Earth Sciences, Hangzhou Normal University, Hangzhou, China

ABSTRACT

Passive optical hyperspectral remote sensing of plant pigments offers potential for understanding plant ecophysiological processes across a range of spatial scales. Following a number of decades of research in this field, this paper undertakes a systematic meta-analysis of 85 articles to determine whether passive optical hyperspectral remote sensing techniques are sufficiently well developed to quantify individual plant pigments, which operational solutions are available for wider plant science and the areas which now require greater focus. The findings indicate that predictive relationships are strong for all pigments at the leaf scale but these decrease and become more variable across pigment types at the canopy and landscape scales. At leaf scale it is clear that specific sets of optimal wavelengths can be recommended for operational methodologies: total chlorophyll and chlorophyll *a* quantification is based on reflectance in the green (550–560nm) and red edge (680–750nm) regions; chlorophyll *b* on the red, (630–660nm), red edge (670–710nm) and the near-infrared (800–810nm); carotenoids on the 500–580nm region; and anthocyanins on the green (550–560nm), red edge (700–710nm) and near-infrared (780–790nm). For total chlorophyll the optimal wavelengths are valid across canopy and landscape scales and there is some evidence that the same applies for chlorophyll *a*.

INTRODUCTION

A pigment is a material that changes the spectral distribution of reflected or transmitted light as the result of wavelength-selective absorption which is determined by the physical properties of the pigment itself. Plant pigments play an important role in light capture, photosystem protection, and in various growth and development functions. The photosynthetic pigments control the amount of solar radiation absorbed by a leaf and thus determine photosynthetic potential and primary production [1,2]. Pigment concentrations are also related to plant stress (excess direct sunlight, UV–B irradiation, low temperature, water stress, nitrogen deficiencies and so on) and senescence (e.g., [3–9]). Therefore, accurate measurements of the temporal dynamics and spatial variations of pigment concentration using remotely sensed data can provide a basis for monitoring physiological and ecological processes [10,11].

Citation: Huang J, Wei C, Zhang Y, Blackburn GA, Wang X, Wei C, et al. (2015) Meta-Analysis of the Detection of Plant Pigment Concentrations Using Hyperspectral Remotely Sensed Data. PLoS ONE 10(9): e0137029. doi:10.1371/journal.pone.0137029.

The spectral absorbance properties of pigments offer the possibility of using measurements of reflected radiation as a non-destructive method for quantifying pigments. Different approaches have arisen recently to remotely estimate pigment concentrations from a wide variety of wavelengths and sensor types. These studies produced variable results, and none have been demonstrated to have satisfactory performance under all growth and environmental conditions. These inconsistencies may stem from the fact that the experimental results are influenced by a number of factors including different species, experimental conditions and analytical methods used [11].

Recent review articles have attempted to assimilate knowledge in this field of passive optical hyperspectral remote sensing with the sun as energy source. Blackburn [10] reviewed the developing technologies and analytical methods for quantitative estimation of pigment across a range of spatial scales using passive optical hyperspectral remote sensing. Ustin *et al.* [11] appraised the most widely used methodologies for retrieving pigment information with hyperspectral data at the leaf scale. However, it has been demonstrated that traditional qualitative reviewers may subjectively select their preferred studies when faced with conflicting results on a single question [12]. In contrast, it has been argued that meta-analysis can take the results from primary research articles and quantitatively analyze and synthesize these data in an attempt to arrive at more robust conclusions. As such, meta-analysis review papers make the shift from a narrative-driven to a data-driven approach [13,14].

Glass [15] published the first article to lay out the essential rationale of meta-analysis. As a fully general set of methods, meta-analysis has been widely applied to the integration of literatures in many areas of empirical science, including ecology [14]. This form of analysis has, for example, been used to determine the response of biodiversity to intensive biomass production, the effects of elevated CO_2 on plant–arthropod interactions, the influence of plant invasion on carbon and nitrogen cycles and the causes and consequences of variations in leaf mass per area [16–19]. Today, many findings and advances are being made not only by those who do primary research studies, but also by those who use meta-analysis to discover the latent meaning of existing research literatures [13]. Recently, meta-analysis has been employed in remote sensing research. Garbulsky *et al.* [20] performed a meta-analysis to assess the use of the photochemical reflectance index (PRI) as an indicator of radiation use efficiencies at the leaf, canopy and ecosystem scales for different time scales and vegetation types. Zolkos *et al.* [21] conducted a meta-analysis of publications on LiDAR remote sensing estimation of terrestrial aboveground biomass. These investigations show that meta-analysis can be used to systematically integrate the results from a collection of studies, and through statistical comparison, assess the relationships between remotely sensed measurements and variables of interest.

Here, a meta-analysis of data from a wide selection of studies reporting the passive optical hyperspectral remote sensing of pigments was used to quantify the development of this scientific field, identify optimal wavelengths for retrieval of individual pigments and evaluate the strength of the relationships between pigment concentration and remotely sensed data across pigment types and scales.

MATERIALS AND METHODS

Study Selection and Data Extraction

Databases of Elsevier, Springer and Web of Science, licensed to Zhejiang University, were used for source data from inception to August 2014. The following key words were used: pigment, chlorophyll, carotenoids, carotene, xanthophyll, anthocyanins, anthoxanthin in combination with the terms reflectance, estimation, quantification, retrieval, prediction and remote sensing. More than 4500 citations were collected as a result of this initial search.

Then the abstracts of these articles were reviewed and considered for inclusion in the meta-analysis. The following criteria were applied to ensure homogeneity in methodology. First, the studies had to include a chemical measurement of pigment concentration (total chlorophyll, chlorophyll *a*, chlorophyll *b*, carotenoids, xanthophyll, carotene or anthocyanins). Second, the article had to report the quantification of pigments using remotely sensed data. Third, the authors must have provided the following statistical information: (1) coefficient of determination for the relationships between pigment concentration and remotely sensed measurements; (2) the wavelength(s) used to estimate pigment concentration; and (3) training sample sizes.

Based on the first two decision rules, 135 articles were selected. According to the final criterion, 50 studies were excluded because of insufficient statistical information. Finally, 85 articles were used in the meta-analysis, which reported results at different spatial and temporal scales and from a wide range of vegetation types between 1977 and 2014. The number of studies selected at various stages is shown in the flow diagram in Fig 1. Some studies reported multiple results for different pigment types or vegetation types. Different types of sensors were used in these studies, from spectrophotometers and hand-held spectroradiometers to satellite sensors. All the sensors were working in reflectance mode. Within the selected articles 44 were working at the leaf scale, 21 at the canopy scale, 15 at the landscape scale, 2 at the leaf and canopy scales, 1 at the leaf and landscape scales, and 2 covered the leaf, canopy and landscape scales. The term "canopy" refers to either a single plant or a monospecific stand where the experimental results are influenced by a number of controlling factors, such as orientation of leaves (leaf angle distribution; *LAD*), variations in number of leaf layers (*LAI*), presence of non-leaf elements, multiple scattering and areas of shadow [10,22], the term "landscape" refers to a mixed-species stand where the reflectance spectrum from airborne and spaceborne sensors is subject to even more controlling factors, such as atmospheric conditions, instrucment sensitivity (signal-to-noise ratio) and spatial resolution. In total, the sample size from all the selected studies is 16100. The Preferred Reporting Items for Meta-Analyses is shown in S1 PRISMA Checklist.

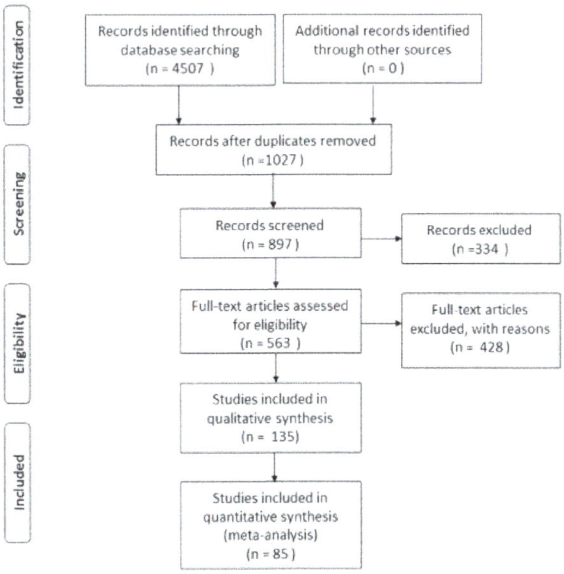

Figure 1. Selection of studies for inclusion in the meta-analysis.

Relevant information was extracted from each study in the final set: 1. scales (leaf, canopy, landscape), 2. pigment types, 3. species, 4. wavelengths, 5. coefficient of determination, 6. sample sizes, 7. sensors, 8. authors and 9. year of publication. In order to reduce human error in data extraction and coding, two sets of reviewers independently screened articles in accordance with those inclusion criteria discussed above, evaluated the quality and extracted the data from the eligible studies. The results from one group were cross-checked by the other group. Divergences of opinion about article selection and data extraction were settled by discussion. Table 1 is a summary of the studies contained in this research. This list is not exhaustive but it does cover most papers published related to quantification of pigments using remotely sensed data that met the selection criteria. Table 2 provides a statistical summary of the data extracted from the studies included in the meta-analysis.

Table 1. A summary of the studies contained in this research that linked remotely sensed data with pigment. Specrad = spectroradiometer; Specpho = spectrophotometer; Chl tot = total chlorophyll; Chl a = chlorophyll *a*; Chl b = chlorophyll *b*; Cars = carotenoids; Anths = anthocyanins.

Scale	Pigment Type	Year	Species	Sensor	Reference
leaves	Chl tot	1992	Amaranthus tricolor	Specpho	[23]
leaves	Chl tot	1995	Slash pine	Specrad	[24]
leaves	Chl tot	1995	Bigleaf maple	Specrad	[25]
leaves	Chl tot	1996	Horse Chestnut, Norway maple,Cotoneaster, Tobacco	Specpho	[26]
leaves	Chl tot	1996	Norway Maple, Horse Chestnut	Specpho	[27]
leaves	Chl tot	1997	Norway Maple, Horse Chestnut, Fig, Cotoneaster, Tobacco,Oleander, Hibiscus, Vine, Rose	Specpho	[28]
leaves	Chl tot	1998	Tobacco, Horse Chestnut, Cotoneaster	Specpho	[29]
leaves	Chl tot	1999	Beech tree, Elm tree,Wild vine shurb	Specpho	[30]
leaves	Chl tot	1999	Bragg Soybean	Specrad	[31]
leaves	Chl tot	2002	53 species	Specrad	[32]
leaves	Chl tot	2002	Paper birch	Specrad	[33]
leaves	Chl tot	2003	Bigleaf Maple, Horse Chestnut, Wild vine, Beech	Specpho	[34]
leaves	Chl tot	2005	Cotton	Specrad	[35]
leaves	Chl tot	2007	Winter wheat	Specpho	[36]
leaves	Chl tot	2012	15 different species(Beech, Fraxinus lanuginosa, Acer Japonicum, Magnolia obovata and so on)	Specrad	[37]
leaves	Chl tot	2014	Douglas fir	Specrad	[38]
leaves	Chl a	1994	Norway Maple, Horse Chestnut	Specpho	[39]
leaves	Chl a	1994	Norway Maple, Horse Chestnut	Specpho	[40]
leaves	Chl a	1996	Norway Maple, Horse Chestnut	Specpho	[41]
leaves	Cars/Chl tot	1977	Cantaloupe, Corn, Spinach Cotton, Cucumber, tobacco, Head lettuce, Grain sorghum	Specpho	[42]
leaves	Cars/Chl tot	1992	Sunflower	Specrad	[43]
leaves	Cars/Chl tot	1999	Norway Maple, Potato, Lemon, Apple, Coleus	Specpho	[7]
leaves	Cars/Chl tot	2006	24 species of woody trees and shurbs	Specpho	[44]
leaves	Anths/Cars/Chl tot	1999	Quercus agrifolia, Pseudotsuga menziesii	Specpho	[45]

leaves	Anths/Cars/Chl tot	2003	Apple	Specpho	[46]
leaves	Anths/Cars/Chl tot	2004	Norway maple, Maize, Dogwood,Horse chestnut, Second-flush beech, Wild vine shrub, Cotoneaster, Pelargonium zonale	Specpho	[47]
leaves	Chl tot/Anths	2014	Chilean strawberry	Specrad	[48]
leaves	Cars/Chl a/ Chl b	1992	Soybean	Specrad	[49]
leaves	Cars/Chl a/Chl b	1998	Beech, Oak, Maple, Sweet chestnut	Specrad	[50]
leaves	Cars/Chl a/Chl b	2005	Rice	Specrad	[51]
leaves	Chl tot/Chl a/Chl b	1999	Norway Maple, Horse Chestnut, Beech, Oak	Specrad	[52]
leaves	Chl tot/Chl a/Chl b	2001	Croton, Elaeagnus, Japanese pittosporum,Benjamin fig	Specrad	[53]
leaves	Chl tot/Chl a/Chl b	2010	Flowering cherry	Specrad	[54]
leaves	Chl tot/Chl a	1996	Tobacco	Specpho	[55]
leaves	Chl tot/Chl a	1999	Eucalyptus	Specrad	[56]
leaves	Cars	2002	Norway maple, Horse chestnut,Second-flush beech	Specpho	[57]
leaves	Cars	2009	Scot pine	Specpho	[58]
leaves	Cars	2011	Bur oak, Sugar maple, LOPEX database	Specrad	[59]
Scale	Pigment Type	Year	Species	Sensor	Reference
leaves	Anths	2001	Norway maple, Cotoneaster, Dogwood	Specpho	[60]
leaves	Anths	2009	Grapevine	Specrad	[61]
leaves	Anths	2009	European hazel, Siberian dogwood, Norway maple, Virginia creeper	Specpho	[62]
leaves	Anths	2011	Grapevine	Specrad	[63]
leaves	Anths	2011	Sweet cherries	Specpho	[64]
leaves	Anths	2011	Norway maple, Horse chestnut, Beech,Virginia creeper, Dogwood	Specpho&specrad	[65]
Leaves/canopy	Chl tot	2009	Maize	Specpho	[66]
Leaves/canopy	Chl tot	2013	Irrigated maize	Specrad	[67]
Leaves/landscape	Chl tot	2014	Black Spruce, Sugar maple	Specrad&MERIS	[68]
Leaves/canopy/ landscape	Chl tot	2010	Winter Wheat, Winter Rapeseed	Specrad	[69]
Leaves/canopy/ landscape	Cars/Chl tot	2000	Sugar maple	Specrad	[70]
canopy	Chl tot	1990	Slash pine	Airborne spectro	[1]
canopy	Chl tot	1994	pepper	Specrad	[71]
canopy	Chl tot	2005	Maize, Soybean	Specrad	[72]
canopy	Chl tot	2006	Rice	Specrad	[73]
canopy	Chl tot	2007	Cotton	Specrad	[74]
canopy	Chl tot	2008	Winter wheat, Corns	Specrad	[75]
canopy	Chl tot	2008	Heterogeneous grassland	Specrad	[76]
canopy	Chl tot	2008	Heterogeneous grassland	Specrad	[77]
canopy	Chl tot	2008	Corn, Cotton	Specrad	[78]
canopy	Chl tot	2010	Rice	Specrad	[79]
canopy	Chl tot	2011	Rice	Specrad	[80]
canopy	Chl tot	2012	Potato, Grassland	Specrad	[81]
canopy	Chl tot	2013	Irrigated maize	Specrad	[82]
canopy	Chl tot	2014	Winter wheat	Specrad	[83]
canopy	Chl a	2003	Rice	Specrad	[84]
canopy	Chl a	2007	Winter Wheat	Specrad	[85]
canopy	Chl a/Chl b	2004	Winter wheat	Specrad	[86]
canopy	Chl tot/Chl a	2006	Wheat	Specrad	[87]
canopy	Cars/Chl tot	2010	Tall fescue	Specrad	[88]
canopy	Cars	2008	Kermes oak	Specrad	[89]
canopy	Cars	2008	Douglas fir	Specrad	[90]
landscape	Chl tot	2002	Corn	CASI	[91]
landscape	Chl tot	2003	Eucalypt	CASI-2	[92]

landscape	Chl tot	2004	Jack pine	CASI	[93]
landscape	Chl tot	2004	Douglas fir	MERIS	[94]
landscape	Chl tot	2007	Corn, Wheat	CASI	[95]
landscape	Chl tot	2008	Rice, Cotton	EO-1	[96]
landscape	Chl tot	2008	Garlic, Alfalfa, Onion, Sunflower, Corn, Potato, Wheat, Vineyard, Sugar beet	PROBA/CHRIS	[97]
landscape	Chl tot	2010	Flax, Tea, Chestnut, Corn, Potato, Pine, Bamboo	EO-1	[98]
landscape	Chl tot	2010	Garlic, Onion, Corn, Alfalfa, Sugar beet, Sunflower, Potato, Vineyard, Wheat	PROBA/CHRIS	[99]
landscape	Chl tot	2014	London plane, Canary Island date palm, European nettle tree, White mulberry	CASI	[100]
landscape	Chl a	2004	Winter Wheat	AVIS	[101]
landscape	Cars/Chl tot	2002	Quercus petrea, Pinus sylvestris	CASI	[102]
landscape	Chl a/Cars	2005	Rice	PHI	[103]
landscape	Cars/Chl tot/Chl a/Chl b	2008	Aspen, Birch, Spruce, Balsam fir	CASI	[104]
landscape	Anths	2009	Austrocedrus chilensis forest	Hyperion	[105]

doi:10.1371/journal.pone.0137029.t001

Table 2. Summary statistics for the selected studies and extracted data for different pigment types at leaf, canopy and landscape scales.

Scale	Pigment type	Number of studies	Number of effect sizes	Total sample size	Number of wavelengths
leaves	Chl tot	34	53	6431	131
	Chl a	11	23	1595	53
	Chl b	6	10	860	24
	Cars	14	15	1381	40
	Anths	10	17	1752	43
canopy	Chl tot	20	23	1146	55
	Chl a	4	4	162	6
	Chl b	1	1	35	0
	Cars	3	2	45	7
	Anths	0	0	0	0
landscape	Chl tot	15	17	1883	46
	Chl a	3	3	153	6
	Chl b	1	1	24	2
	Cars	3	3	573	4
	Anths	1	1	60	2

doi:10.1371/journal.pone.0137029.t002

Statistical Analysis of Effect Size

The Calculation of Effect Size for Each Study

The coefficient of determination (R^2) was used to evaluate the strength of relationships between spectral reflectance and pigment concentration in each article we selected. The value of R^2, however, is affected by the number of selected wavelengths. The more wavelengths included in the model, be they relevant or not, the larger would be the R^2 [106]. The increase of R^2 is not without cost. The increasing number of selected wavelengths reduces the degrees of freedom, which reduces model robustness. The adjusted coefficient of determination was applied to correct for the degrees of freedom:

$$R_A^2 = 1 - (1 - R^2)\frac{n-1}{n-k} \qquad (1)$$

where n is the sample size for each study, k is the number of independent variables in the linear or nonlinear model. Eq (1) shows that R_A^2 is always smaller than R^2 when $k > 1$, which means the growth rate of R_A^2 is lower than that of R^2 as the number of parameters increase. This result is straightforward and it has been shown that when the added parameter explains a significant amount of the behavior of the dependent variable, R_A^2 will increase; otherwise, R_A^2 will decrease [107]. So R_A^2 was chosen as the effect size statistic, the variance of effect size is calculated as [108]:

$$V_i = \frac{(1 - R_A^2)^2}{n-1}, \qquad (2)$$

The resulting data set was categorized by pigment type at the scales of leaf, canopy and landscape to allow comparison.

Test of heterogeneity for effect sizes.

It is important to assess the heterogeneity among the results from a collection of studies before computing the mean effect size [109]. Basically, there are two possible sources of heterogeneity in meta-analysis: methodological heterogeneity and statistical heterogeneity. To ensure homogeneity in methodology, we applied a series of criteria to identify the studies to be used in the meta-analysis (as described in section 2.1). Here the I^2 statistic was used to test for the statistical heterogeneity. The I^2 statistic measures the extent of true heterogeneity dividing the difference between the result of the Q test and its degrees of freedom by the Q value itself [110]:

$$I^2 = 100\% \times \frac{Q_{tot} - df}{Q_{tot}} \quad \textbf{(3)}$$

where $df = N_{tot} - 1$, N_{tot} is the total number of effect sizes from all the selected studies, Q_{tot} is computed as [111]:

$$Q_{tot} = \sum_{i=1}^{N_{tot}} W_i E_i^2 - \frac{\left(\sum_{i=1}^{N_{tot}} W_i E_i\right)^2}{\sum_{i=1}^{N_{tot}} W_i} \quad \textbf{(4)}$$

where $W_i = 1/v_i$, E_i is adjusted coefficient of determination (R_A^2).

The I^2 statistic can be interpreted as the percentage of heterogeneous component in the total variability of effect size (Q_{tot}), so the larger the I^2 statistic is, the stronger the heterogeneity is. If I^2 exceeds 50%, the null hypothesis of homogeneity is rejected. The I^2statistic for different pigments at different scales were calculated, all the results were lower than 50%, the null hypothesis of homogeneity for this study was accepted.

The Calculation of Mean Effect Size for Different Pigments at Different Scales

In contrast to studies based on original data, the unit of meta-analysis is the individual research study. Distinctive aspects of data analysis follow from this difference. The first complication is that the studies incorporated into the meta-analysis generally use different sample sizes and this controls the statistical properties of effect sizes [112]. From a statistical perspective, larger sample studies have less sampling error than smaller sample studies, thus more weight should be assigned to larger sample studies in the computation of the mean effect size. The other complication is inter-study variability, which is caused by the influence of an indeterminate number of characteristics that vary among the studies.

Considering the two sources of variability discussed above, a random effects model was used to compute the weighted mean of R_A^2 for different pigment types. In contrast to a fixed effects model, the weight applied to each effect size in a random effects model must represent both subject-level sampling error and the additional random variance component [112]. As such, the mean effect size becomes a reasonable estimate of the true strength of the effect in the population. Because of the generality of the random effects model, it is the preferred strategy in meta-analysis [113]. The mean effect size is computed as:

$$M_{rand} = \frac{\sum_{i=1}^{N_p} W_{i(rand)} E_i}{\sum_{i=1}^{N_p} W_{i(rand)}} \quad (5)$$

The variance is:

$$V_{rand} = \frac{1}{\sum_{i=1}^{N_p} W_{i(rand)}} \quad (6)$$

Where

$$W_{i(rand)} = \frac{1}{V_i + \sigma^2},\ \sigma^2 = \frac{Q_p - (N_p - 1)}{\sum_{i=1}^{N_p} W_i - \frac{\sum_{i=1}^{N_p} W_i^2}{\sum_{i=1}^{N_p} W_i}},\ W_i = 1 \big/ V_i,\ Q_p = \sum_{i=1}^{N_p} W_i E_i^2 - \frac{\left(\sum_{i=1}^{N_p} W_i E_i\right)^2}{\sum_{i=1}^{N_p} W_i}\ E_i,$$

, is adjusted coefficient of determination (R_A^2) and N_p is the total number of effect sizes for a specific type of pigment at each different scales (Table 2). Using this approach the mean effect size of Chl tot, Chl a, Chl b, Cars and Anths at the scales of leaf, canopy and landscape were calculated.

A confidence interval gives the range of values within which the mean effect size is likely to be, it is useful in indicating the degree of precision of the estimate of the mean effect size. A 95% confidence interval is subsequently calculated as follows:

$$Conf_{95} = M_{rand} \pm 1.96 SE_{rand} \quad (7)$$

where $SE_{rand} = \sqrt{V_{rand}}$. If the confidence intervals of multiple mean effect sizes donot overlap, then there are significant differences between these mean effect sizes.

Optimal Wavelengths for Pigment Quantification

A large number of narrow-band indices were proposed to measure plant pigments in the selected articles. These narrow band indices include difference vegetation index (NBDVI), ratio vegetation index (NDRVI), normalized difference vegetation index (NBNDVI), anthocyanin reflectance index (ARI), soil-adjusted vegetation index (SAVI), perpendicular vegetation index (PVI) and so on. The wavelengths used in these studies are different and there is lack of agreement on optimal wavelengths for pigment quantification.

Histograms and quantile plots were used to identify the optimal wavelengths for individual pigment quantification at different scales. The histogram partitions the data distribution of wavelengths into subsets of 10nm width. This enabled us to provide an overview of suitable wavelengths, which is difficult to achieve if the analysis is performed at higher spectral resolutions. Also this approach avoided inaccuracies of spectral calibration associated with the use of many different instruments across the studies incorporated into the meta-analysis. In the histogram each subset is represented by a rectangle whose height is equal to the count of observations that fall into the wavelength interval. A quantile plot is a simple and effective way to compare different wavelength distributions. Let λ_i (i = 1to$G$) be the wavelengths sorted in increasing order so that λ_1 is the smallest wavelength and λ_G is the largest. Each wavelength, λ_i, is paired with a percentage, f_i, which indicates that approximately 100 f_i% of the data are below or equal to the value, λ_i.

$$f_i = \frac{i - 0.5}{G}(i = 1, \ldots, G) \quad (8)$$

In a quantile plot, λ_i is graphed against f_i. This allows us to compare different wavelength distributions based on their quantiles [114].

RESULTS

Quantifying the Development of Remote Sensing of Plant Pigment Concentrations

The number of studies used in the meta-analysis published over the period from 1977 to 2014 are shown in Fig 2, along with the 5-year running mean which summarises the overall trajectory of development in this scientific field. After the first two studies were published in 1977 there were no other publications for 11 years, but then there was fast rate of growth from 1990 to 1999. The number of publications reached top in 1999 after which the publication rate stopped increasing, indicating that research in passive optical hyperspectral remote sensing of plant pigment concentrations is within a mature phase. The overall trajectory of publications shows three periods covering the origins, development and proliferation of research in this field. This trajectory corresponds to the developmental phases of hyperspectral instruments, which started with spectrophotometers and hand-held spectroradiometers enabling leaf and canopy-scale work. With the more recent advent of airborne and spaceborne imaging spectrometers, more landscape scale analyses have become possible.

Despite this overall development in the field, there were substantial differences in research on different pigments. The first studies of total chlorophyll and carotenoids were published in 1977, followed by chlorophyll *a* and chlorophyll *b* in 1992 and anthocyanins in 1999. The growth rate of publications on chlorophyll *a*, chlorophyll *b*, carotenoids and anthocyanins has been significantly lower than that for total chlorophyll. These differential rates of growth are perhaps indicative of the increased difficulty in quantifying the concentrations of individual photosynthetic and protective pigments remotely.

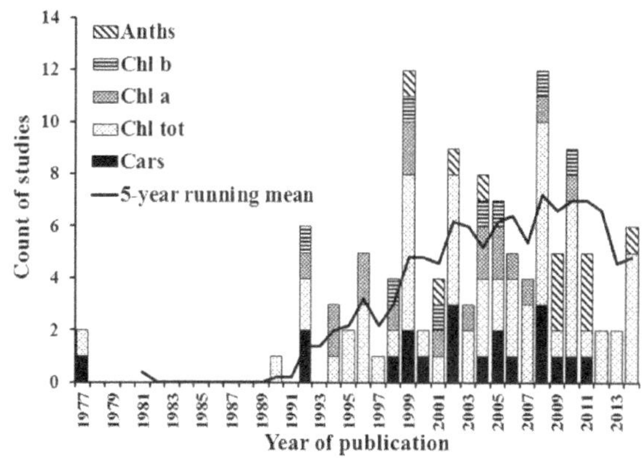

Figure 2. Histogram of numbers of selected studies published over time, showing the total in each year and the number focusing on each pigment type. The solid line is a 5-year running mean of the total number of studies.

The Relationships between Pigment Concentrations and Remotely Sensed Variables

The mean effect size for different pigments at the scales of leaf, canopy and landscape were calculated (Fig 3). At the leaf scale, the mean effect sizes were fairly consistent between different pigment types, varying from 0.87 to 0.93, while the difference in mean effect sizes between pigment types was statistically significant at the canopy and landscape scales. The mean effect size presented the highest value 0.93 (95% confidence interval, 0.92–0.95) for anthocyanins quantification at the leaf scale, far higher than the result of 0.35 (95% confidence interval, 0.18–0.51) at the landscape scale. The mean effect size for total chlorophyll quantification was 0.88 (95% confidence interval, 0.87–0.89) at the leaf scale, 0.73 (95% confidence interval, 0.69–0.77) at the canopy scale and 0.79 (95% confidence interval, 0.76–0.82) at the landscape scale. The mean effect size for carotenoids was the lowest of the various pigments at 0.87 (95% confidence interval, 0.84–0.90) at the leaf scale, still higher than the result 0.80 (95% confidence interval, 0.71–0.90) at the canopy scale and 0.85 (95% confidence interval, 0.76–0.94) at the landscape scale. The results show that these mean effect sizes varied across pigment types and scales. In general, the relationships are stronger at the leaf scale than those at the canopy and landscape scales.

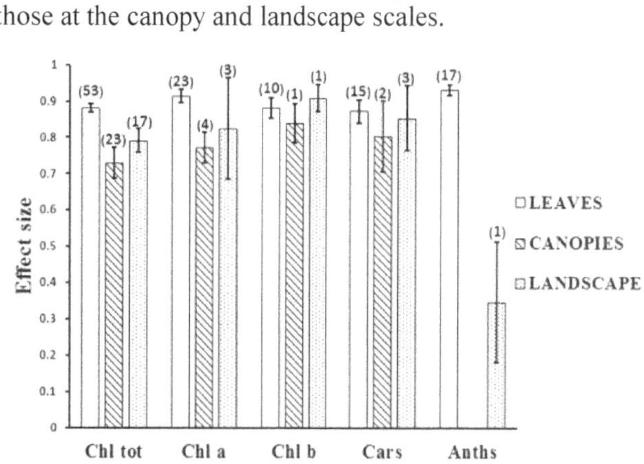

Figure 3. The mean effect size for pigment types at the scales of leaf, canopy and landscape. (The numbers of reported relationships found in the literature are shown in brackets, error bars represent 95% confidence intervals).

Fig 3 shows that the highest number of relationships published was for pigment quantifications at the leaf scale. Pigment quantification at the canopy scale was less frequently reported in the literature and only a few studies were conducted at the landscape scale. This can be attributed to the limited availability and high costs of suitable airborne and spaceborne hyperspectral instruments [20]. For each scale, the highest number of relationships published was for total chlorophyll quantification, followed by chlorophyll *a*, carotenoids, chlorophyll *b* and anthocyanins. These findings are consistent with previous studies [10,11].

Wavelength Selection for Pigment Quantification Using Remotely Sensed Data

Optimal Wavelengths for Chlorophyll Quantification

There is a large quantity of studies on the relationships between chlorophyll concentration and remotely sensed data. The distributions of wavelengths used at the three scales are shown in Fig 4. It should be noted that all of the wavelengths for pigment quantification were concentrated in the 350–950 nm region, except for total chlorophyll quantification at the canopy scale, which spread over 400–2400 nm. For comparison, wavelengths in the histograms and quantile plots were limited within the 350–950 nm region.

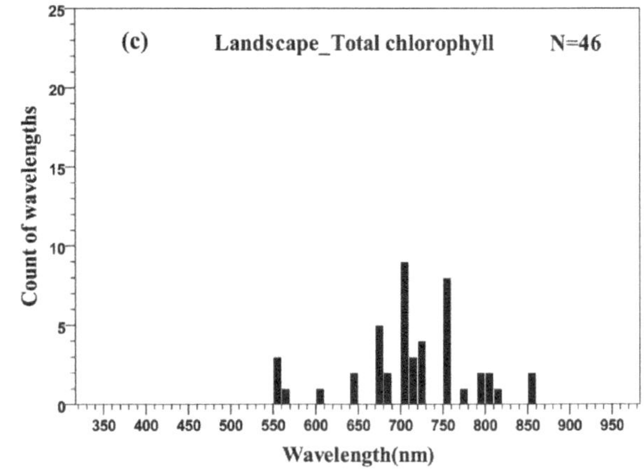

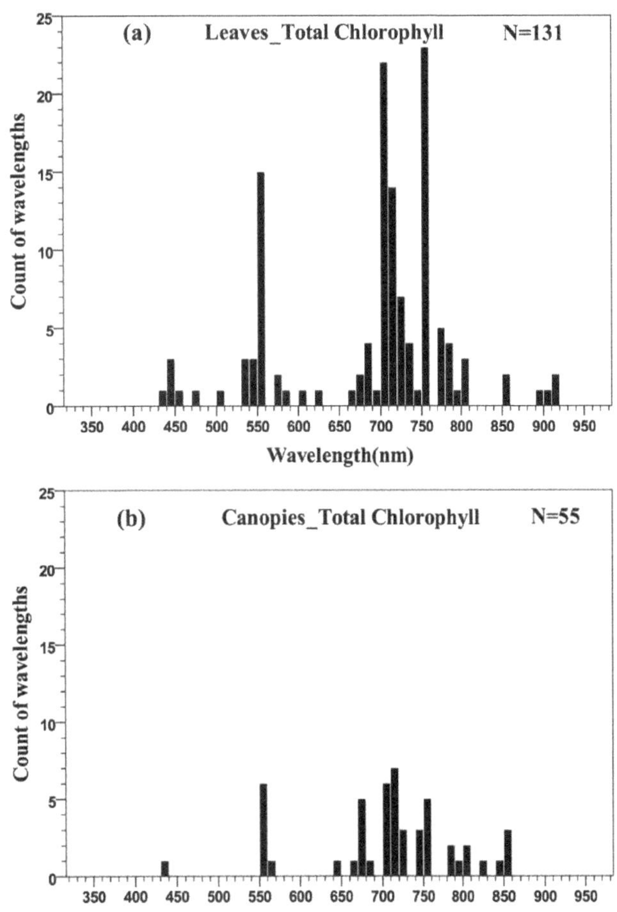

Figure 4. Histogram of wavelengths for total chlorophyll quantification using remotely sensed data at leaf (a), canopy (b) and landscape (c) scales using an interval width of 10 nm.

In general, the distribution of wavelengths displayed a double-peak feature, concentrated in the green (550–560 nm) and red edge (680–750 nm) regions rather than the main absorption wavelengths of chlorophyll (blue or red) (Fig 5). At the canopy scale, five wavelengths in the NIR to SWIR regions (1000–2400 nm) were also used for total chlorophyll quantification (not shown in Fig 4B). This is due to the major influence of canopy structure in canopy reflectance and because leaf chlorophyll concentration was relatively stable in the particular studies [76,77].

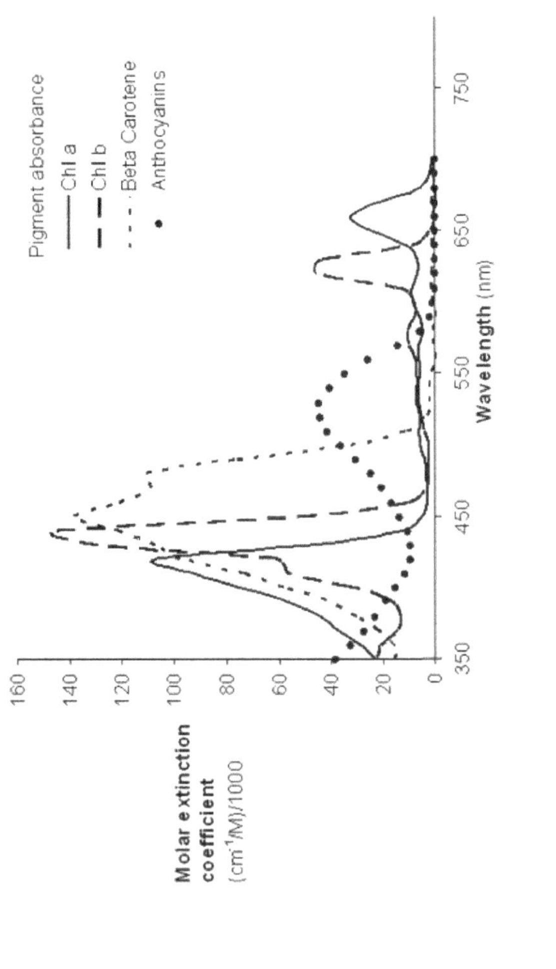

Figure 5. Absorption spectra of the major plant pigments (reproduced from Blackburn, 2007).

The distribution of wavelengths proposed for chlorophyll *a* quantification at the leaf scale was similar to that of total chlorophyll, concentrated in the green and red edge ranges (Fig 6A). At the canopy and landscape scales, the number of wavelengths is limited and is difficult to identify the central tendency of wavelength distribution (Fig 6B and Fig 6C).

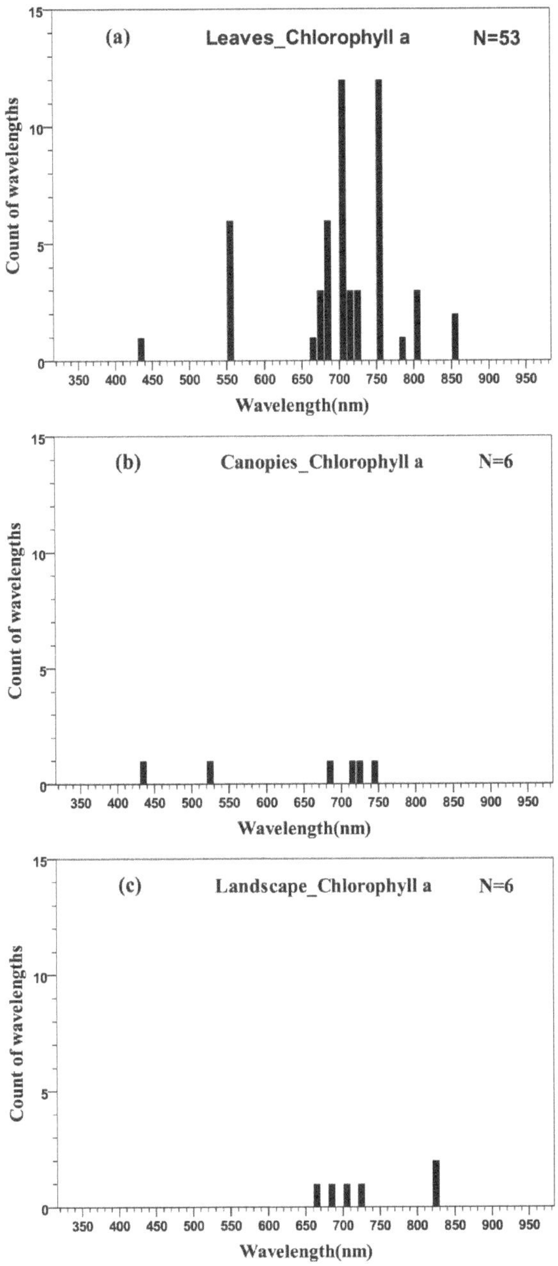

Figure 6. Histogram of wavelengths for chlorophyll *a* **quantification using remotely sensed data at leaf (a), canopy (b) and landscape (c) scales by an interval width of 10 nm.**

The distribution of wavelengths used for chlorophyll *b* quantification at the leaf scale were concentrated in the main absorption wavelength of chlorophyll *b* (red, 630–660 nm), the red edge (670–710 nm) and the NIR (800–810 nm) regions (Fig 7A). Only two wavelengths were selected at the landscape scale and could not be used for statistical inference (Fig 7B). The distributions of wavelengths used for quantification of different pigments at different scales can be compared in the quantile plots (Fig 8). There were similar wavelength distributions for total chlorophyll quantification at the scales of leaf, canopy and landscape (Fig 8A). For chlorophyll *a* there were similar wavelength distributions at the leaf and canopy scales, but the landscape scale differed (Fig 8B), while a comparison across scales for chlorophyll *b* was difficult due to a lack of data at scales other than the leaf (Fig 8C).

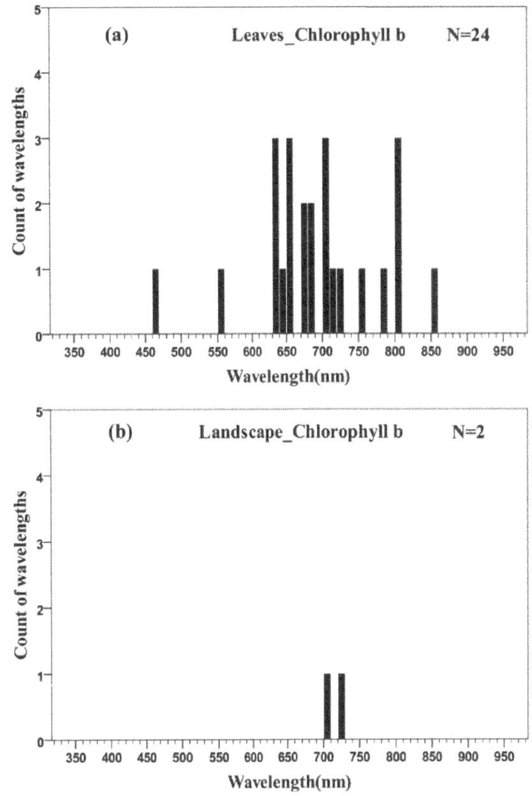

Figure 7. Histogram of wavelengths for chlorophyll *b* **quantification using remotely sensed data at leaf (a) and landscape (b) scales using an interval width of 10 nm.**

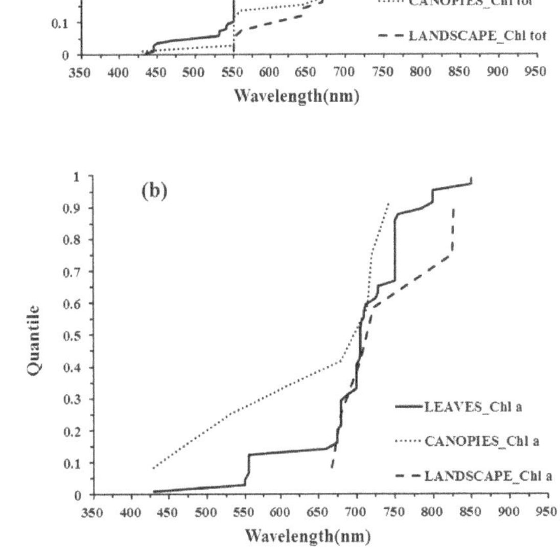

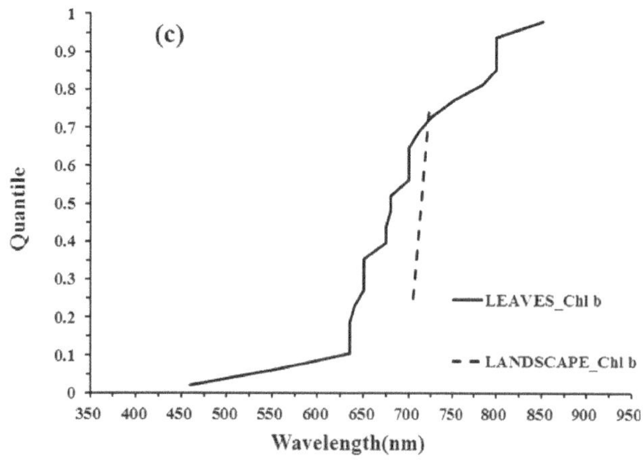

Figure 8. Quantile plots of the wavelengths used for the quantification of Chl tot (a), Chl a (b), and Chl b (c) at different scales.

At the leaf scale, the wavelength distributions for total chlorophyll and chlorophyll *a*quantification were relatively similar while there were notable differences for chlorophyll *b*(Fig 9A). In the region 425–625 nm, the wavelengths used for chlorophyll *a* quantification were concentrated in the region of the green peak in leaf reflectance (550nm), but the central tendency of wavelength distribution for chlorophyll *b* quantification was not obvious. In the red region, the wavelength distribution for chlorophyll *a* quantification was shifted to longer wavelengths than that of chlorophyll *b* (Fig 9A). The significant overlap in the absorption features of chlorophyll *a* and chlorophyll *b* (Fig 5) and the low concentrations of chlorophyll *b* with respect to chlorophyll *a* in most leaves can present difficulties in defining optimal wavelengths for chlorophyll *b* quantification. The absorption spectra of chlorophyll *a*and chlorophyll *b* both display a double-peak feature; the absorption maxima of chlorophyll*a* are at 430 and 662 nm, and chlorophyll *b* has peaks located at 453 and 642 nm (Fig 5). In the presence of carotenoids, it is difficult to separately assess chlorophyll *a* and chlorophyll*b* from reflectance data in the blue region. However, in the red region, the wavelength position of maximum absorption by chlorophyll *a* is longer than that of chlorophyll *b*, which can be exploited for chlorophyll *a* and chlorophyll *b* discrimination (as seen in Fig 9A). The capacity to use this approach to discriminate chlorophyll *a* and chlorophyll *b* is difficult to assess at the canopy and landscape scales due to the small number of studies on chlrophyll b (Fig 9B and 9C).

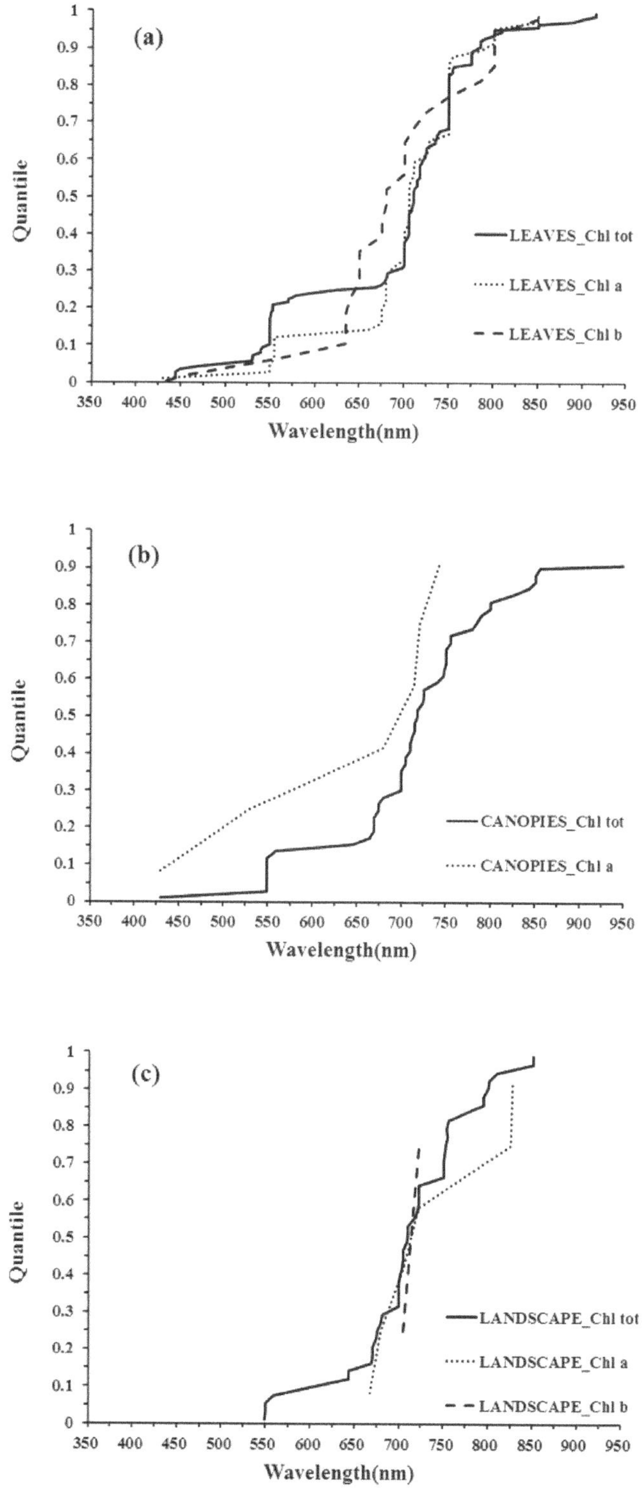

Figure 9. Quantile plot of the wavelengths used at leaf (a), canopy (b) and landscape (c) scales for the quantification of Chl tot, Chl a, and Chl b.

Optimal Wavelengths for Carotenoids Quantification

At the leaf scale, the central tendency of wavelength distribution was not obvious but was mainly concentrated in the 500–580 nm region (Fig 10A). There were similar wavelength distributions for carotenoids quantification at the leaf and canopy scales (Fig 11) but at the landscape scale, the number of wavelengths was too small for statistical inference (Fig 10C).

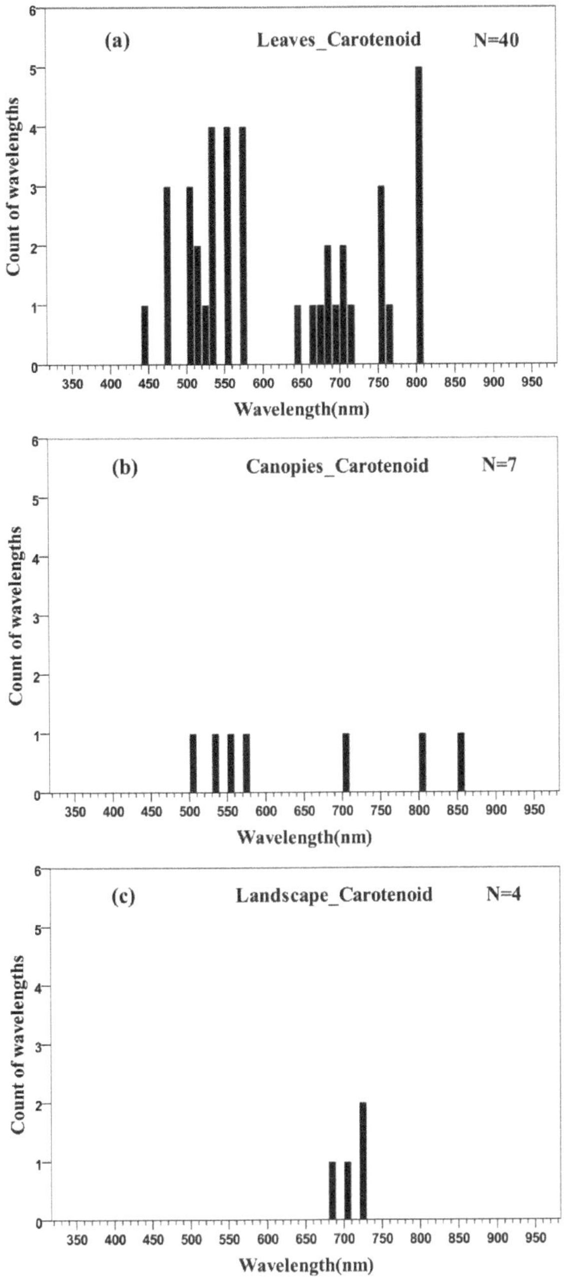

Figure 10. Histogram of wavelengths for carotenoids quantification using remotely sensed data at leaf (a), canopy (b) and landscape (c) scales using an interval width to 10 nm.

Optimal Wavelengths for Anthocyanins Quantification

Quantification of anthocyanins from reflectance data has been given less attention by the passive optical hyperspectral remote sensing community than chlorophyll and carotenoids. Most studies have concentrated on the quantification of anthocyanins at the leaf scale, with some work at the landscape scale but nothing at canopy level. At the leaf scale, the distribution of wavelengths used for quantifying anthocyanins was concentrated in the main absorption wavelength of anthocyanins (green, 550–560 nm), the red edge (700–710 nm) and the NIR (780–790 nm) ranges (Fig 12A). Similarly, the two wavelengths used to estimate anthocyanin concentration at the landscape scale were distributed in the green and red edge regions, respectively (Fig 12B and Fig 13).

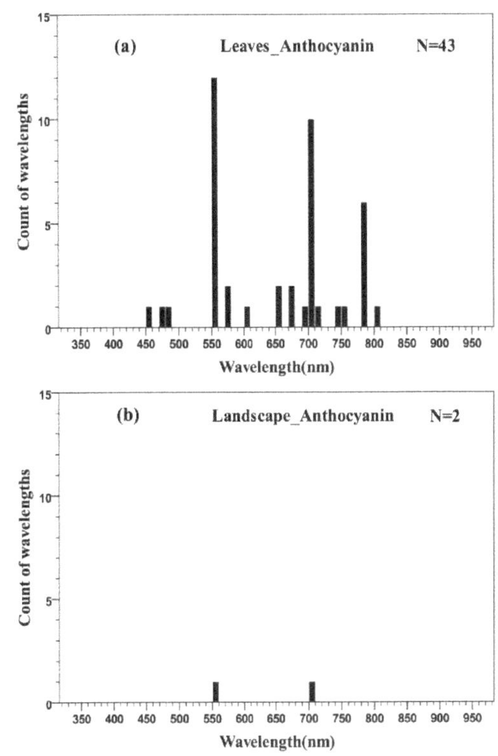

Figure 12. Histogram of wavelengths for anthocyanins quantification using remotely sensed data at leaf (a) and landscape (b) scales using an interval width to 10 nm.

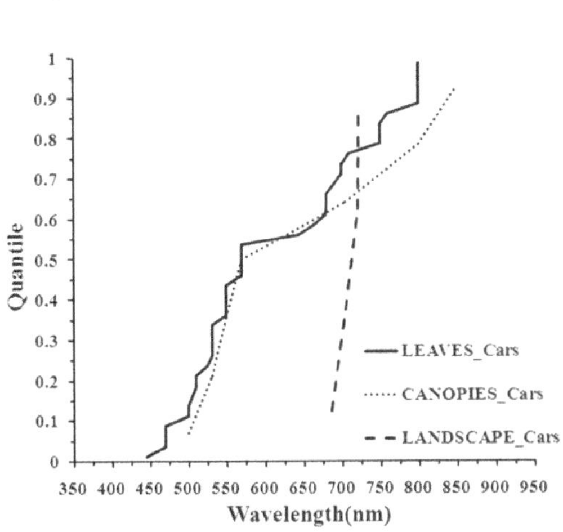

Figure 11. Quantile plot of the optimal wavelength for the quantification of Cars at different scales.

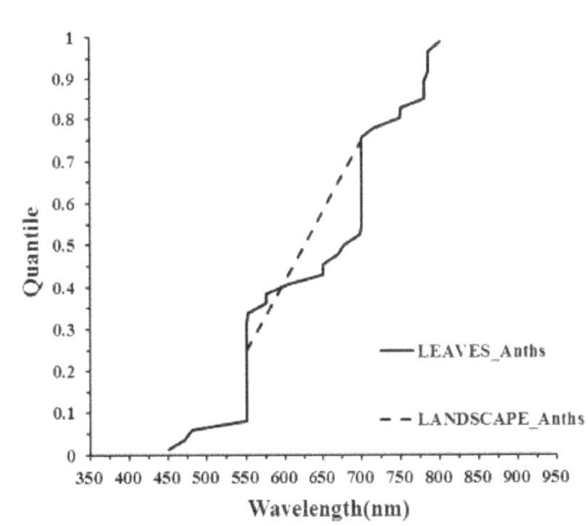

Figure 13. Quantile plot of the optimal wavelengths for the quantification of Anths at different scales.

DISCUSSION

This meta-analysis of 85 studies has demonstrated that remotely sensed variables are good estimators of plant pigment concentration. Most of the studies were conducted at the leaf scale, while pigment quantification at the canopy and landscape scales was less frequently reported. For each scale, most of the studies were conducted for total chlorophyll quantification, followed by chlorophyll *a*, carotenoids, chlorophyll *b* and anthocyanins. These findings are consistent with previous studies [10,11].

The strength of these relationships varied across pigments types and scales. In general, the relationships are stronger at the leaf scale than those at the canopy and landscape scales. At the leaf scale, the mean effect sizes were fairly consistent across different pigment types and were all greater than 0.87, while the difference in mean effect sizes between pigment types was statistically significant at the canopy and landscape scales. This result has been widely assumed, yet a quantitative evaluation has been lacking. At the leaf scale, the methodological basis for pigment quantification has been fully explored, which provides an important basis for developing estimation models at the canopy and landscape scales. The primary goal of most leaf scale passive optical hyperspectral remote sensing studies has been to develop analytical approaches for pigment quantification that can be applied to data from airborne and spaceborne sensors [11].

At the canopy and landscape scales, the experimental results are influenced by a number of factors, which obscures the relationships between spectral reflectance and concentrations of individual pigments. The reflectance spectrum of a whole canopy is subject to canopy biophysical attributes (e.g., orientation of leaves (leaf angle distribution;*LAD*), variations in number of leaf layers (*LAI*) and foliage clumping), presence of non-leaf elements (e.g., soil reflectance and the proportions of shadowed and sunlit background), anisotropic scattering of photons to interact with multiple surfaces such as leaves, woody material and soils, viewing geometry (e.g., sun and view zenith and azimuth angles) and illumination conditions (e.g., the ratio between direct and diffuse sunlight and atmospheric condition). It is the interaction of these factors, including their potential covariance or unique behavior that drive variation in canopy and landscape reflectance characteristics in three-dimensional space [10,22].

It should be noted that part of the variability in effect sizes at the canopy scale may be entirely artifactual. These artifacts are common in experimental studies: studies vary in terms of the quality of measurement; researchers make computational errors; people make typographical errors in copying numbers from handwritten tables to computer; and sampling errors. With the advent of airborne and spaceborne imaging spectrometers, there have been opportunities to measure plant pigment concentrations at the landscape scale. The reflectance spectrum from airborne and spaceborne sensors is subject to even more controlling factors, notably, soil/litter surface reflectance, and vegetation structure. The range of controlling factors should be taken into account in subsequent analyses.

Table 2 shows that the total sample size at the leaf scale is much more than that of canopy and landscape scales. The law of large numbers correctly states that large samples are reasonable representations of the population and parameter estimation is close to the real values when the sample size is large enough. Many researchers seem to believe that the same law applies to small samples and severely underestimate the amount of variability in findings that is caused by sampling errors. As a result, they erroneously expect statistics based on small samples to be close to the real values [13]. At the canopy and landscape scales, the number of studies and total sample size is limited, which influences the robustness and accuracy of effect sizes.

Despite the significant difference in effect sizes between different scales, it was found that the wavelength distribution for total chlorophyll quantification at the scales of leaf, canopy and landscape was similar, being concentrated in the green (550–560 nm) and red edge (680–750 nm) regions rather than the main absorption wavelength of chlorophyll (blue or red). The consistency in optimal wavelengths across scales can be attributed to several factors: (1) despite the many factors influencing reflectance at the canopy and landscape scales, it is the selective absorbance properties of pigments that determines the selection of wavelengths for pigment quantification, and (2) several estimation models derived at the leaf scale were directly applied to canopy and landscape scales. This suggests that the leaf-level study has provided an important basis for developing estimation models at the canopy and landscape scales.

At the leaf scale, the distribution of wavelengths used for chlorophyll *a* quantification was similar to that of total chlorophyll; the distribution of wavelengths for chlorophyll *b*quantification was concentrated in the main absorption wavelength of chlorophyll *b* (red, 630–660 nm), the red edge (670–710 nm) and the NIR (800–810 nm) regions; the central tendency of wavelength distribution for carotenoids quantification was not obvious, but was mainly concentrated in the 500–580 nm region; for the estimation of anthocyanins, the distribution of wavelengths was concentrated in the main absorption wavelength of anthocyanins (green, 550–560 nm), the red edge (700–710 nm) and the NIR (780–790 nm) ranges. In the present meta-analysis, the lack of studies reporting the quantification of carotenoids and anthocyanins at the canopy and landscape scales has hindered cross-scale comparisons (Fig 10; Fig 12). Consequently, it is not entirely clear if the optimal wavelengths for carotenoids and anthocyanins quantification at the leaf scale are necessarily the optimal wavelengths at the canopy and landscape scales, where multiple scattering and other confounding effects may alter the spectral response of individual pigments, much in the way that pigment absorption peaks can vary depending upon their chemical and scattering medium. Therefore, more work may be needed to determine the optimal algorithms for airborne or spaceborne platforms.

It should be noted that the lack of statistical information in the studies (e.g., sample size and coefficient of determination) has hindered a more comprehensive cross-study comparison in the present research. When selecting the final set of studies, 50 studies were excluded due to the lack of statistical information. Insufficient statistical information can not only limit the research population covered by meta-analysis but also render the findings of the original study somewhat suspect. Thus, it is suggested that when conducting primary research, such information should include, but not be limited to, the sample size, the pertinent test statistic (e.g., r, t, or F), the unit of pigment concentration/content, the range of pigment concentrations/content, and estimation precision for pigment quantification (e.g. root mean squared error, RMSE).

This study has established the possibility of integrating the results of studies on the passive optical hyperspectral remote sensing of plant pigment concentrations across a range of vegetation types and scales using a meta-analysis approach. Despite the robust models for pigment prediction at the leaf scale, the continuing challenge is to properly account for the multiple factors introduced by scene components such as sunlit and shaded parts of tree crowns and gaps influencing the retrieved signal at the canopy and landscape scales. Recent work have illustrated that, in addition to other influencing factors such as illumination geometry and atmospheric conditions, canopy architecture had an important control on the applicability of models for pigment prediction. Scanning LIDAR systems have only recently become widely available which enable the estimation of the range between the sensor and a target by recording the time during which the emitted laser pulse is reflected off an object and returns to the sensor [21]. LIDAR systems have the ability to directly measure spatial variations in canopy height and other aspects of the vertical structure of canopies. Given the high degree of structural complexity at the canopy and landscape scales, it would appear that the integration of vertical canopy structural information provided by active LIDAR remote sensing with hyperspectral reflectance may has both a structural and physiological interpretation and improve the estimation of pigment concentrations over passive optical hyperspectral imagery alone [102].

SUPPORTING INFORMATION

Section/topic	#	Checklist item	Reported on page #
TITLE			
Title	1	Identify the report as a systematic review, meta-analysis, or both.	1
ABSTRACT			
Structured summary	2	Provide a structured summary including, as applicable: background; objectives; data sources; study eligibility criteria, participants, and interventions; study appraisal and synthesis methods; results; limitations; conclusions and implications of key findings; systematic review registration number.	1
INTRODUCTION			
Rationale	3	Describe the rationale for the review in the context of what is already known.	2-3
Objectives	4	Provide an explicit statement of questions being addressed with reference to participants, interventions, comparisons, outcomes, and study design (PICOS).	4
METHODS			
Protocol and registration	5	Indicate if a review protocol exists, if and where it can be accessed (e.g., Web address), and, if available, provide registration information including registration number.	\
Eligibility criteria	6	Specify study characteristics (e.g., PICOS, length of follow-up) and report characteristics (e.g., years considered, language, publication status) used as criteria for eligibility, giving rationale.	4

Information sources	7	Describe all information sources (e.g., databases with dates of coverage, contact with study authors to identify additional studies) in the search and date last searched.	4
Search	8	Present full electronic search strategy for at least one database, including any limits used, such that it could be repeated.	4
Study selection	9	State the process for selecting studies (i.e., screening, eligibility, included in systematic review, and, if applicable, included in the meta-analysis).	5
Data collection process	10	Describe method of data extraction from reports (e.g., piloted forms, independently, in duplicate) and any processes for obtaining and confirming data from investigators.	6
Data items	11	List and define all variables for which data were sought (e.g., PICOS, funding sources) and any assumptions and simplifications made.	6
Risk of bias in individual studies	12	Describe methods used for assessing risk of bias of individual studies (including specification of whether this was done at the study or outcome level), and how this information is to be used in any data synthesis.	11
Summary measures	13	State the principal summary measures (e.g., risk ratio, difference in means).	11
Synthesis of results	14	Describe the methods of handling data and combining results of studies, if done, including measures of consistency (e.g., I^2) for each meta-analysis.	12-14

S1 PRISMA Checklist. PRISMA (Preferred Reporting Items for Systematic Reviews and Meta-Analyses) Checklist. doi:10.1371/journal.pone.0137029.s001

Acknowledgments

We acknowledge the contribution of Bao She, Weijiao Huang, Dilong Gan, Sujuan Wang and Zhewen Zhao for the literature database searches and associated support. The authors thank anonymous reviewers who provided very valuable comments also.

Author Contributions

Conceived and designed the experiments: JH. Performed the experiments: Chen Wei YZ XW Chuanwen Wei JW. Analyzed the data: Chen Wei. Contributed reagents/materials/analysis tools: Chen Wei. Wrote the paper: Chen Wei GAB JH.

REFERENCES

1. Curran PJ, Dungan JL, Gholz HL. Exploring the relationship between reflectance red edge and chlorophyll content in slash pine. Tree Physiol 1990; 7: 33–48. pmid:14972904 doi: 10. 1093/treephys/7. 1-2-3-4. 33
2. Filella I, Serrano L, Serra J, Penuelas J. Evaluating wheat nitrogen status with canopy reflectance indices and discriminant analysis. Crop Sci 1995; 35: 1400–1405. doi: 10. 2135/cropsci1995. 0011183x003500050023x
3. Hendry GAF, Houghton JD, Brown SB. The degradation of chlorophyll: a biological enigma. New Phytol 1987; 107: 255–302. doi: 10. 1111/j. 1469-8137. 1987. tb00181. x
4. Merzlyak MN, Gitelson A. Why and what for the leaves

are yellow in autumn? On the interpretation of optical spectra of senescing leaves (*Acerplatanoides L*). J. Plant Physiol 1995; 145: 315–320. doi: 10. 1016/s0176-1617(11)81896-1

5. Demmig—Adams B, Adams WW. The role of xanthophyll cycle carotenoids in the protection of photosynthesis. Trends Plant Sci 1996; 1: 21–26. doi: 10. 1016/s1360-1385(96)80019-7
6. Peñuelas J, Filella I. Visible and near-infrared reflectance techniques for diagnosing plant physiological status. Trends Plant Sci 1998; 3: 151–156. doi: 10. 1016/s1360-1385(98)01213-8
7. Merzlyak MN, Gitelson AA, Chivkunova OB, Rakitin VY. Non- destructive optical detection of pigment changes during leaf senescence and fruit ripening. Physiol Plantarum 1999; 106: 135–141. doi: 10. 1034/j. 1399-3054. 1999. 106119. x
8. Chalker-Scott L. Environmental significance of anthocyanins in plant stress responses. Photochem Photobiol 1999; 70: 1–9. doi: 10. 1111/j. 1751-1097. 1999. tb01944. x
9. Carter GA, Knapp AK. Leaf optical properties in higher plants: linking spectral characteristics to stress and chlorophyll concentration. Am J Bot 2001; 88: 677–684. pmid:11302854 doi: 10. 2307/2657068
10. Blackburn GA. Hyperspectral remote sensing of plant pigments. J. Exp Bot 2007; 58: 855–867. pmid:16990372 doi: 10. 1093/jxb/erl123
11. Ustin SL, Gitelson AA, Jacquemoud S, Schaepman M, Asner GP, Gamon JA, et al. Retrieval of foliar information about plant pigment systems from high resolution spectroscopy. Remote Sens Environ 2009; 113: S67 –S77. doi: 10. 1016/j. rse. 2008. 10. 019
12. Hunter JE, Schmidt FL. Methods of meta-analysis: correcting error and bias in research findings. Los Angeles, USA: SAGE Publications; 2004. p. 33–34.
13. Borenstein M, Hedges LV, Higgins JPT, Rothstein HR. Introduction to meta-analysis. West Sussex, United Kingdom: John Wiley & Sons Ltd; 2009. p. 12–13.
14. Curtis PS, Queenborough SA. Raising the standards for ecological meta–analyses. New Phytol 2012; 195: 279–281. doi: 10. 1111/j. 1469-8137. 2012. 04207. x. pmid:22702404
15. Glass GV. Primary secondary and meta-analysis of research. Educ Res 1976; 5: 3–8. doi: 10. 3102/0013189x005010003
16. Verschuyl J, Riffell S, Miller D, Wigley TB. Biodiversity response to intensive biomass production from forest thinning in North American forests-A meta-analysis. Forest Ecol Manag 2011; 261: 221–232. doi: 10. 1016/j. foreco. 2010. 10. 010
17. Robinson EA Ryan GD Newman JA A meta-analytical review of the effects of elevated CO_2 on plant-arthropod interactions highlights the importance of interacting environmental and biological variables. New Phytol 2012; 194: 321–336. doi: 10. 1111/j. 1469-8137. 2012. 04074. x. pmid:22380757
18. Liao CZ Peng RH Luo YQ Zhou X H Wu X W Fang C M Chen J K Li B Altered ecosystem carbon and nitrogen cycles by plant invasion: a meta-analysis. New Phytol 2008; 177: 706–714. pmid:18042198 doi: 10. 1111/j. 1469-8137. 2007. 02290. x
19. Poorter H, Niinemets Ü, Poorter L, Wright IJ, Villar R. Causes and consequences of variation in leaf mass per area (LMA): a meta-analysis. New Phytol 2009; 182: 565–588. pmid:19434804 doi: 10. 1111/j. 1469-8137. 2009. 02830. x
20. Garbulsky MF, Peñuelas J, Gamon J, Inoue Y, Filella I. The photochemical reflectance index (PRI) and the remote sensing of leaf canopy and ecosystem radiation use efficiencies: a review and meta-analysis. Remote Sens Environ 2011; 115: 281–297. doi: 10. 1016/j. rse. 2010. 08. 023
21. Zolkos SG, Goetz SJ, Dubayah R. A meta-analysis of terrestrial aboveground biomass estimation using lidar remote sensing. Remote Sens Environ 2013; 128: 289–298. doi: 10. 1016/j. rse. 2012. 10. 017
22. Asner GP. Biophysical and biochemical sources of variability in canopy reflectance. Remote Sens Environ 1998; 64: 234–253. doi: 10. 1016/s0034-4257(98)00014-5
23. Curran PJ, Dungan JL, Macler BA, Plummer SE, Peterson DL. Reflectance spectroscopy of fresh whole leaves for the estimation of chemical concentration. Remote Sens Environ 1992; 39: 153–166. doi: 10. 1016/0034-4257(92)90133-5
24. Curran PJ, Windham WR, Gholz HL. Exploring the relationship between reflectance red edge and chlorophyll concentration in slash pine leaves. Tree Physiol 1995; 15: 203–206. pmid:14965977 doi: 10. 1093/treephys/15. 3. 203
25. Yoder BJ, Pettigrew-Crosby RE. Predicting nitrogen and chlorophyll content and concentrations from reflectance spectra (400–2500 nm) at leaf and canopy scales. Remote Sens Environ 1995; 53: 199–211. doi: 10. 1016/0034-4257(95)00135-n
26. Gitelson AA, Merzlyak MN, Grits Y. Novel algorithms for remote sensing of chlorophyll content in higher plant leaves. In International Geoscience and Remote Sensing Symposium (IGARSS); Lincoln, NE, USA; May 1996. p. 2355–2357.
27. Gitelson AA, Kaufman YJ, Merzlyak MN. Use of a green channel in remote sensing of global vegetation from EOS-MODIS. Remote Sens Environ 1996; 58: 289–298. doi: 10. 1016/s0034-4257(96)00072-7
28. Gitelson AA, Merzlyak MN. Remote estimation of chlorophyll content in higher plant leaves. Int J Remote Sens 1997; 18: 2691–2697. doi: 10. 1080/014311697217558
29. Gitelson AA, Merzlyak MN. Remote sensing of chlorophyll concentration in higher plant leaves. Adv Space Res 1998; 22: 689–692. doi: 10. 1016/s0273-1177(97)01133-2
30. Gitelson AA, Buschmann C, Lichtenthaler HK. The chlorophyll fluorescence ratio F735/F700 as an accurate measure of the chlorophyll content in plants. Remote Sens Environ 1999; 69: 296–302. doi: 10. 1016/s0034-

4257(99)00023-1

31. Adams ML, Philpot WD, Norvell WA. Yellowness index: an application of spectral second derivatives to estimate chlorosis of leaves in stressed vegetation. Int J Remote Sens 1999; 20: 3663–3675. doi: 10. 1080/014311699211264
32. Sims DA, Gamon JA. Relationships between leaf pigment content and spectral reflectance across a wide range of species leaf structures and developmental stages. Remote Sens Environ 2002; 81: 337–354. doi: 10. 1016/s0034-4257(02)00010-x
33. Richardson AD, Duigan SP, Berlyn GP. An evaluation of noninvasive methods to estimate foliar chlorophyll content. New Phytol 2002; 153: 185–194. doi: 10. 1046/j. 0028-646x. 2001. 00289. x
34. Gitelson AA, Gritz Y, Merzlyak MN. Relationships between leaf chlorophyll content and spectral reflectance and algorithms for non–destructive chlorophyll assessment in higher plant leaves. J Plant Physiol 2003; 160: 271–282. pmid:12749084 doi: 10. 1078/0176-1617-00887
35. Zhao DL, Reddy KR, Kakani VG, Read JJ, Koti S. Selection of optimum reflectance ratios for estimating leaf nitrogen and chlorophyll concentrations of field-grown cotton. Agron J 2005; 97: 89–98. doi: 10. 2134/agronj2005. 0089
36. Kochubey SM, Kazantsev TA. Changes in the first derivatives of leaf reflectance spectra of various plants induced by variations of chlorophyll content. J. Plant Physiol 2007; 164: 1648–1655. pmid:17292510 doi: 10. 1016/j. jplph. 2006. 11. 007
37. Wang Q, Li PH. Hyperspectral indices for estimating leaf biochemical properties in temperate deciduous forests: comparison of simulated and measured reflectance data sets. Ecol Indic 2012; 14: 56–65. doi: 10. 1016/j. ecolind. 2011. 08. 021
38. Simic A, Chen JM, Leblanc SG, Dyk A, Croft H, Tian Han. Testing the top-down model inversion method of estimating leaf reflectance used to retrieve vegetation biochemical content within empirical approaches. IEEE J Sel Top Appl Earth Observ Remote Sens 2014; 7: 92–104. doi: 10. 1109/jstars. 2013. 2271583
39. Gitelson A, Merzlyak MN. Quantitative estimation of chlorophyll-a using reflectance spectra: experiments with autumn chestnut and maple leaves. J. Photoch Photobio B 1994; 22: 247–252. doi: 10. 1016/1011-1344(93)06963-4
40. Gitelson AA, Merzlyak MN. Spectral reflectance changes associated with autumn senescence of *Aesculus hippocastanum L* and *Acer platanoides L* Leaves Spectral features and relation to chlorophyll estimation. J. Plant Physiol 1994; 143: 286–292. doi: 10. 1016/s0176-1617(11)81633-0
41. Gitelson AA, Merzlyak MN. Signature analysis of leaf reflectance spectra: algorithm development for remote sensing of chlorophyll. J. Plant Physiol 1996; 148: 494–500. doi: 10. 1016/s0176-1617(96)80284-7
42. Thomas JR, Gausman HW. Leaf reflectance vs Leaf chlorophyll and carotenoid concentrations for eight crops. Agron J 1977; 69: 799–802. doi: 10. 2134/agronj1977. 00021962006900050017x
43. Gamon JA, Peñuelas J, Field CB. A narrow-waveband spectral index that tracks diurnal changes in photosynthetic efficiency. Remote Sens Environ 1992; 41: 35–44. doi: 10. 1016/0034-4257(92)90059-s
44. Levizou E, Manetas Y. Photosynthetic pigment contents in twigs of 24 woody species assessed by in vivo reflectance spectroscopy indicate low chlorophyll levels but high carotenoid/chlorophyll ratios. Environ Exp Bot 2007; 59: 293–298. doi: 10. 1016/j. envexpbot. 2006. 03. 002
45. Gamon JA, Surfus JS. Assessing leaf pigment content and activity with a reflectometer. New Phytol 1999; 143: 105–117. doi: 10. 1046/j. 1469-8137. 1999. 00424. x
46. Merzlyak MN, Solovchenko AE, Gitelson AA. Reflectance spectral features and non–destructive estimation of chlorophyll carotenoid and anthocyanin content in apple fruit. Postharvest Biol Tec 2003; 27: 197–211. doi: 10. 1016/s0925-5214(02)00066-2
47. Gitelson AA, Merzlyak MN. Non-destructive assessment of chlorophyll carotenoid and anthocyanin content in higher plant leaves: principles and algorithms. In Remote Sensing for Agriculture and the Environment. Stamatiadis S, LynchJ JM, Schepers JS, Eds. Ella, Greece: OECD; 2004. p. 78–94.
48. Garriga M, Retamales JB, Romero-Bravo S, Caligari PD, Lobos GA. Chlorophyll anthocyanin and gas exchange changes assessed by spectroradiometry in Fragaria chiloensis under salt stress. J. Integr Plant Biol 2014; 56: 505–15. doi: 10. 1111/jipb. 12193. pmid:24618024
49. Chappelle EW, Kim MS, McMurtrey JE III. Ratio analysis of reflectance spectra (RARS): an algorithm for the remote estimation of the concentrations of chlorophyll a chlorophyll b and carotenoids in soybean leaves. Remote Sens Environ 1992; 39: 239–247. doi: 10. 1016/0034-4257(92)90089-3
50. Blackburn GA. Spectral indices for estimating photosynthetic pigment concentrations: a test using senescent tree leaves. Int J Remote Sens 1998; 19 657–675. doi: 10. 1080/014311698215919
51. Chen L, Huang JF, Wang FM. Retrieval of pigment contents in rice leaves and panicles using hyperspectral data by artificial neuron network models. In International Geoscience and Remote Sensing Symposium (IGARSS); Seoul, Korea; July 2005. p. 1416–1419.
52. Blackburn GA. Relationships between spectral reflectance and pigment concentrations in stacks of deciduous broadleaves. Remote Sens Environ 1999; 70: 224–237. doi: 10. 1016/s0034-4257(99)00048-6
53. Maccioni A, Agati G, Mazzinghi P. New vegetation indices for remote measurement of chlorophylls based on leaf directional reflectance spectra. J. Photoch Photobio B 2001; 61: 52–61. doi: 10. 1016/s1011-1344(01)00145-2
54. Imanishi J, Nakayama A, Suzuki Y, Imanishi A, Ueda N, Morimoto Y, et al. Nondestructive determination of leaf chlorophyll content in two flowering cherries using reflectance and absorptance spectra. Landsc Ecol Eng

2010; 6: 219–234. doi: 10. 1007/s11355-009-0101-8

55. Lichtenthaler HK, Gitelson A, Lang M. Non-destructive determination of chlorophyll content of leaves of a green and an aurea mutant of tobacco by reflectance measurements. J. Plant Physiol 1996; 148: 483–493. doi: 10. 1016/s0176-1617(96)80283-5
56. Datt B. Visible/near infrared reflectance and chlorophyll content in eucalyptus leaves. Int J Remote Sens 1999; 20: 2741–2759. doi: 10. 1080/014311699211778
57. Gitelson AA, Zur Y, Chivkunova OB, Merzlyak MN. Assessing carotenoid content in plant leaves with reflectance spectroscopy. Photochem Photobiol 2002; 75: 272–281. pmid:11950093 doi: 10. 1562/0031-8655(2002)0750272accipl2. 0. co2
58. Filella I, Porcar-Castell A, Munne-Bosch S, Back J, Garbulsky MF, Penuelas J. PRI assessment of long-term changes in carotenoids/chlorophyll ratio and short–term changes in de-epoxidation state of the xanthophyll cycle. Int J Remote Sens 2009; 30: 4443–4455. doi: 10. 1080/01431160802575661
59. Garrity SR, Eitel JUH, Vierling LA. Disentangling the relationships between plant pigments and the photochemical reflectance index reveals a new approach for remote estimation of carotenoid content. Remote Sens Environ 2011; 115: 628–635. doi: 10. 1016/j. rse. 2010. 10. 007
60. Gitelson AA, Merzlyak MN, Chivkunova OB. Optical properties and nondestructive estimation of anthocyanin content in plant leaves. Photochem Photobiol 2001; 74: 38–45. pmid:11460535 doi: 10. 1562/0031-8655(2001)074<0038:opaneo>2. 0. co;2
61. Steele MR, Gitelson AA, Rundquist DC, Merzlyak MN. Nondestructive estimation of anthocyanin content in grapevine leaves. Am J Enol Viticult 2009; 60: 87–92.
62. Gitelson AA, Chivkunova OB, Merzlyak MN. Nondestructive estimation of anthocyanins and chlorophylls in anthocyanic leaves. Am J Bot 2009; 96: 1861–1868. doi: 10. 3732/ajb. 0800395. pmid:21622307
63. Qin JL, Rundquist D, Gitelson A, Tan Z, Steele M. A non-linear model of nondestructive estimation of anthocyanin content in grapevine leaves with Visible/Red-infrared hyperspectral. In International Conference on Computer and Computing Technologies in Agriculture; Beijing, China; October 2011. p. 47–62.
64. Pappas CS, Takidelli C, Tsantili E, Tarantilis PA, Polissiou MG. Quantitative determination of anthocyanins in three sweet cherry varieties using diffuse reflectance infrared fourier transform spectroscopy. J. Food Compos Anal 2011; 24: 17–21. doi: 10. 1016/j. jfca. 2010. 07. 001
65. Vina A, Gitelson AA. Sensitivity to foliar anthocyanin content of vegetation indices using green reflectance. IEEE Geosci Remote Sens Lett 2011; 8: 464–468. doi: 10. 1109/lgrs. 2010. 2086430
66. Ciganda V, Gitelson A, Schepers J. Non-destructive determination of maize leaf and canopy chlorophyll content. J. Plant Physiol 2009; 166: 157–167. doi: 10. 1016/j. jplph. 2008. 03. 004. pmid:18541334
67. Schlemmer M, Gitelson A, Schepers J, Ferguson R, Peng Y, Shanahan J, et al. Remote estimation of nitrogen and chlorophyll contents in maize at leaf and canopy levels. Int J Appl Earth Obs 2013; 25: 47–54. doi: 10. 1016/j. jag. 2013. 04. 003
68. Croft H, Chen JM, Zhang Y. The applicability of empirical vegetation indices for determining leaf chlorophyll content over different leaf and canopy structures. Ecol Complex 2014; 17: 119–130. doi: 10. 1016/j. ecocom. 2013. 11. 005
69. Ju CH, Tian YC, Yao X, Cao WX, Zhu Y, Hannaway D. Estimating leaf chlorophyll content using red edge parameters. Pedosphere 2010; 20: 633–644. doi: 10. 1016/s1002-0160(10)60053-7
70. ZarcoTejada PJ. Hyperspectral remote sensing of closed forest canopies: estimation of chlorophyll fluorescence and pigment content. PhD thesis, York University, Toronto, Canada 2000.
71. Filella I, Penuelas J. The red edge position and shape as indicators of plant chlorophyll content biomass and hydric status. Int J Remote Sens 1994; 15: 1459–1470. doi: 10. 1080/01431169408954177
72. Gitelson AA, Vina A, Ciganda V, Rundquist DC, Arkebauer TJ. Remote estimation of canopy chlorophyll content in crops. Geophys Res Lett 2005; 32: 1–4. doi: 10. 1029/2005gl022688
73. Yang XH, Huang JF, Wang FM, Wang XZ, Yi QX, Wang Y. Science letters: a modified chlorophyll absorption continuum index for chlorophyll estimation. J. Zhejiang Univ 2006; 7: 2002–2006. doi: 10. 1631/jzus. 2006. a2002
74. Zhao DH, Huang LM, Li JL, Qi JG. A comparative analysis of broadband and narrowband derived vegetation indices in predicting LAI and CCD of a cotton canopy. Isprs J Photogramm 2007; 62: 25–33. doi: 10. 1016/j. isprsjprs. 2007. 01. 003
75. Wu CY, Niu Z, Tang Q, Huang WJ. Estimating chlorophyll content from hyperspectral vegetation indices: modeling and validation. Agr Forest Meteorol 2008; 148: 1230–1241. doi: 10. 1016/j. agrformet. 2008. 03. 005
76. Darvishzadeh R, Skidmore A, Schlerf M, Atzberger C, Corsi F, Cho M. LAI and chlorophyll estimation for a heterogeneous grassland using hyperspectral measurements. ISPRS J Photogramm 2008; 63: 409–426. doi: 10. 1016/j. isprsjprs. 2008. 01. 001
77. Darvishzadeh R, Skidmore A, Schlerf M, Atzberger C. Inversion of a radiative transfer model for estimating vegetation LAI and chlorophyll in a heterogeneous grassland. Remote Sens Environ 2008; 112: 2592–2604. doi: 10. 1016/j. rse. 2007. 12. 003
78. Haboudane D, Tremblay N, Miller JR, Vigneault P. Remote estimation of crop chlorophyll content using spectral indices derived from hyperspectral data. IEEE T Geosci Remote 2008; 46: 423–437. doi: 10. 1109/tgrs. 2007. 904836
79. Liu ML, Liu XN, Li M, Fang MH, Chi WX. Neural-network model for estimating leaf chlorophyll concentration in rice under stress from heavy metals using four spectral indices. Biosystems Eng 2010; 106:

223–233. doi: 10. 1016/j. biosystemseng. 2009. 12. 008

80. Xu X, Gu X, Song X, Li C, Huang W. Assessing rice chlorophyll content with vegetation indices from hyperspectral data. In International Conference on Computer and Computing Technologies in Agriculture; Beijing, China; October 2011. p. 296–303.

81. Clevers JGPW, Kooistra L. Using Hyperspectral Remote Sensing Data for Retrieving Canopy Chlorophyll and Nitrogen Content. IEEE J Sel Top Appl Earth Observ Remote Sens 2012; 5: 574–583. doi: 10. 1109/jstars. 2011. 2176468

82. Clevers JGPW, Gitelson AA. Remote estimation of crop and grass chlorophyll and nitrogen content using red-edge bands on Sentinel-2 and -3. Int J Appl Earth Obs 2013; 23: 344–351. doi: 10. 1016/j. jag. 2012. 10. 008

83. Vincini M, Amaducci S, Frazzi E. Empirical Estimation of Leaf Chlorophyll Density in Winter Wheat Canopies Using Sentinel – 2 Spectral Resolution. IEEE T Geosci Remote 2014; 52: 3220–3235. doi: 10. 1109/tgrs. 2013. 2271813

84. Cheng Q, Huang JF, Wang XZ, Wang RC. In situ hyperspectral data analysis for pigment content estimation of rice leaves. J Zhejiang Univ 2003; 4: 727–733. doi: 10. 1631/jzus. 2003. 0727

85. Li J, Jiang JB, Chen YH, Wang YY, Su W, Huang WJ. Using hyperspectral indices to estimate foliar chlorophyll a concentrations of winter wheat under yellow rust stress. New Zeal J Agr Res 2007; 50: 1031–1036. doi: 10. 1080/00288230709510382

86. Zhao X, Liu SH, Wang JD, Tian ZK. A method for estimating chlorophyll content of wheat from reflectance spectra. In International Geoscience and Remote Sensing Symposium (IGARSS); Anchorage, AK, USA; September 2004. p. 4504–4507.

87. Bannari A, Khurshid KS, Staenz K, Schwarz J. Wheat crop chlorophyll content estimation from ground–based reflectance using chlorophyll indices. In International Geoscience and Remote Sensing Symposium (IGARSS); Denver, CO, USA; July 2006. p. 112–115.

88. Yang F, Li JL, Gan XY, Qian YR, Wu XL, Yang Q. Assessing nutritional status of*Festuca arundinacea* by monitoring photosynthetic pigments from hyperspectral data. Comput Electron Agr 2010; 70: 52–59. doi: 10. 1016/j. compag. 2009. 08. 010

89. Peguero-Pina JJ, Morales F, Flexas J, Gil-Pelegrin E, Moya I. Photochemistry remotely sensed physiological reflectance index and de-epoxidation state of the xanthophyll cycle in *Quercus coccifera* under intense drought. Oecologia 2008 156: 1–11. doi: 10. 1007/s00442-007-0957-y. pmid:18224338

90. Hall FG, Hilker T, Coops NC, Lyapustin A, Huemmrich KF, Middleton E, et al. Multi–angle remote sensing of forest light use efficiency by observing PRI variation with canopy shadow fraction. Remote Sens. Environ 2008; 112: 3201–3211. doi: 10. 1016/j. rse. 2008. 03. 015

91. Haboudane D, Miller JR, Tremblay N, Zarco-Tejada PJ, Dextraze L. Integrated narrow–band vegetation indices for prediction of crop chlorophyll content for application to precision agriculture. Remote Sens Environ 2002; 81: 416–426. doi: 10. 1016/s0034-4257(02)00018-4

92. Coops NC, Stone C, Culvenor DS, Chisholm LA, Merton RN. Chlorophyll content in eucalypt vegetation at the leaf and canopy scales as derived from high resolution spectral data. Tree Physiol 2003; 23: 23–31. pmid:12511301 doi: 10. 1093/treephys/23. 1. 23

93. Zarco-Tejada PJ, Miller JR, Harron J, Hu BX, Noland TL, Goel N, et al. Needle chlorophyll content estimation through model inversion using hyperspectral data from boreal conifer forest canopies. Remote Sens Environ 2004; 89: 189–199. doi: 10. 1016/j. rse. 2002. 06. 002

94. Dash J, Curran PJ. The MERIS terrestrial chlorophyll index. Int J Remote Sens 2004; 25: 5403–5413. doi: 10. 1080/0143116042000274015

95. Haboudane D, Tremblay N, Vigneault P, Miller JR. Indices-based approach for crop chlorophyll content retrieval from hyperspectral data. In International Geoscience and Remote Sensing Symposium (IGARSS); Barcelona, Spain; July 2007. p. 3297–3300.

96. Rao NR, Garg PK, Ghosh SK, Dadhwal VK. Estimation of leaf total chlorophyll and nitrogen concentrations using hyperspectral satellite imagery. J Agr Sci 2008; 146: 65–75. doi: 10. 1017/s0021859607007514

97. Delegido J, Fernandez G, Gandia S, Moreno J. Retrieval of chlorophyll content and LAI of crops using hyperspectral techniques: application to PROBA/CHRIS data. Int J Remote Sens 2008; 29: 7107–7127. doi: 10. 1080/01431160802238401

98. Wu CY, Han XZ, Niu Z, Dong JJ. An evaluation of EO-1 hyperspectral hyperion data for chlorophyll content and leaf area index estimation. Int J Remote Sens 2010; 31: 1079–1086. doi: 10. 1080/01431160903252335

99. Delegido J, Alonso L, González G, Moreno J. Estimating chlorophyll content of crops from hyperspectral data using a normalized area over reflectance curve (NAOC). Int J Appl Earth Obs 2010; 12: 165–174. doi: 10. 1016/j. jag. 2010. 02. 003

100. Delegido J, Van Wittenberghe S, Verrelst J, Ortiz V, Veroustraete F, Valcke R, et al. Chlorophyll content mapping of urban vegetation in the city of Valencia based on the hyperspectral NAOC index. Ecol Indic 2014; 40: 34–42. doi: 10. 1016/j. ecolind. 2014. 01. 002

101. Oppelt N, Mauser W. Hyperspectral monitoring of physiological parameters of wheat during a vegetation period using AVIS data. Int J Remote Sens 2004; 25: 145–159. doi: 10. 1080/0143116031000115300

102. Blackburn GA. Remote sensing of forest pigments using airborne imaging spectrometer and LIDAR imagery. Remote Sens Environ 2002; 82: 311–321. doi: 10. 1016/s0034-4257(02)00049-4

103. Guan YN, Guo S, Liu JG, Zhang X. Algorithms for the estimation of the concentrations of chlorophyll a and carotenoids in rice leaves from airborne hyperspectral data. In Computational Science-ICCS 2005; Atlanta, GA, USA; May 2005. p. 908–915.

104. Thomas V, Treitz P, McCaughey JH, Noland T, Rich L. Canopy chlorophyll concentration estimation using hyperspectral and lidar data for a boreal mixedwood forest

in northern Ontario Canada. Int J Remote Sens 2008; 29: 1029–1052. doi: 10. 1080/01431160701281023

105. Pena MA, Altmann SH. Use of satellite-derived hyperspectral indices to identify stress symptoms in an *Austrocedrus chilensis* forest infested by the aphid *Cinara cupressi*. Int J Pest Manage 2009; 55: 197–206. doi: 10. 1080/09670870902725809

106. Jacquemoud S, Verdebout J, Schmuck G, Andreoli G. Hosgood B. Investigation of leaf biochemistry by statistics. Remote Sens Environ 1995; 54: 180–188. doi: 10. 1016/0034-4257(95)00170-0

107. Cornell JA, Berger RD. Factors that influence the value of the coefficient of determination in simple linear and nonlinear regression models. Phytopathology 1987; 77: 63–70. doi: 10. 1094/phyto-77-63

108. Gurevitch J, Curtis PS, Jones MH. Meta-analysis in ecology. Adv Ecol Res 2001; 32: 199–247. doi: 10. 1016/s0065-2504(01)32013-5

109. Hedges LV. Estimation of effect size from a series of independent experiments. Psychol Bull 1982; 92: 490–499. doi: 10. 1037//0033-2909. 92. 2. 490

110. Higgins J, Thompson SG. Quantifying heterogeneity in a meta-analysis. Stat Med 2002; 21: 1539–1558. pmid:12111919 doi: 10. 1002/sim. 1186

111. Hedges LV, Olkin I. Statistical methods for meta-analysis. Orlando, FL, USA: Academic Press; 1985. p. 31–34.

112. Lipsey MW, Wilson DB. Practical meta-analysis. London, UK: SAGE Publications; 2000. p. 112–116.

113. Mosteller F, Colditz GA. Understanding research synthesis (meta-analysis). Annu Rev Publ Health 1996; 17 1–23. doi: 10. 1146/annurev. pu. 17. 050196. 000245

114. Han JW, Kamber M. Data mining:concepts and techniques second ed. San Francisco, CA, USA: Morgan Kaufmann; 2006. p. 53–54.

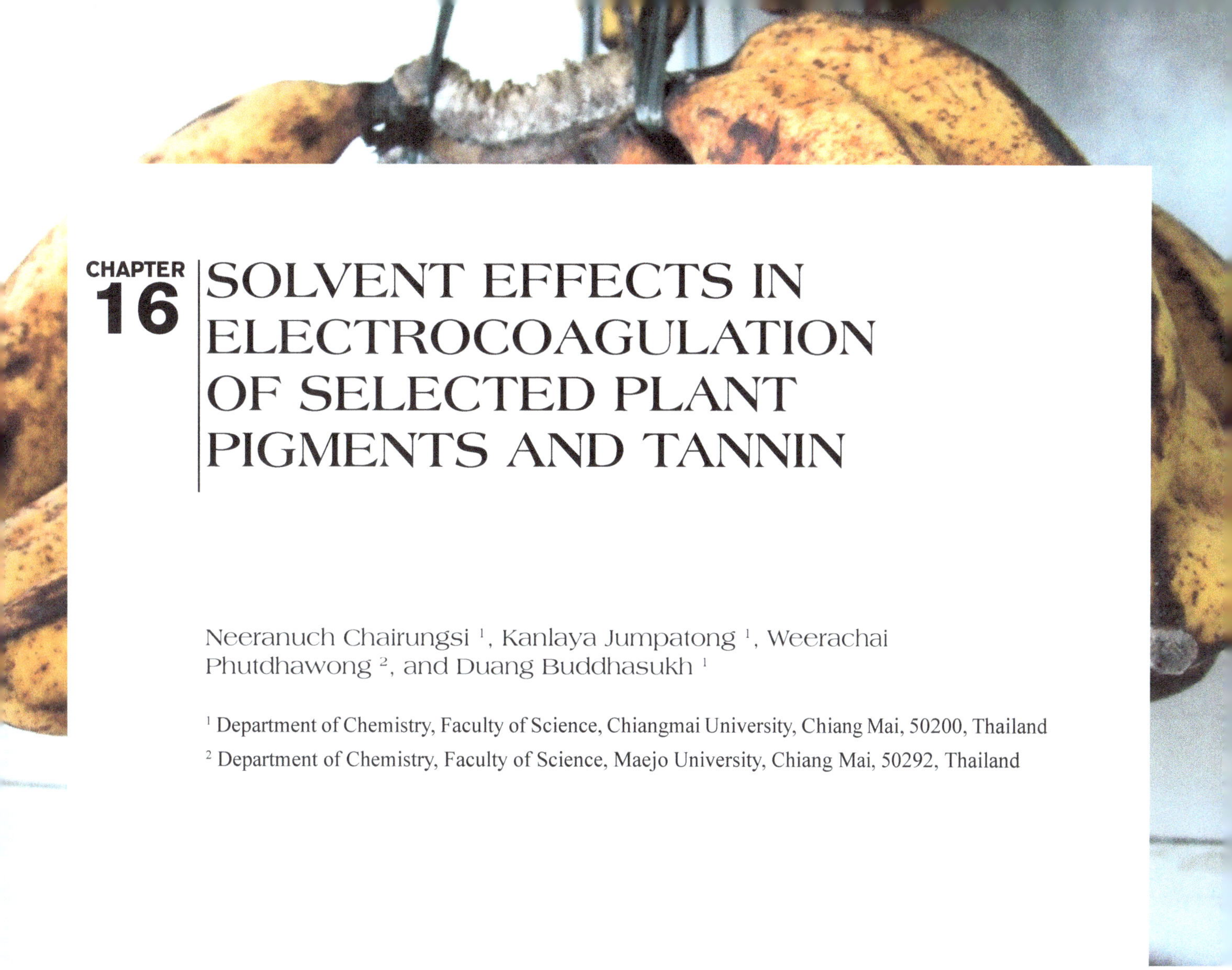

CHAPTER 16

SOLVENT EFFECTS IN ELECTROCOAGULATION OF SELECTED PLANT PIGMENTS AND TANNIN

Neeranuch Chairungsi [1], Kanlaya Jumpatong [1], Weerachai Phutdhawong [2], and Duang Buddhasukh [1]

[1] Department of Chemistry, Faculty of Science, Chiangmai University, Chiang Mai, 50200, Thailand

[2] Department of Chemistry, Faculty of Science, Maejo University, Chiang Mai, 50292, Thailand

ABSTRACT

Electrocoagulation of a plant extract and certain substances representative of selected classes of plant pigments, viz. chlorophyll, a carotenoid, a phenolic substance and a tannin, was performed in ethanol containing varying amounts of water (15-75%). The results showed that the extent and efficiency of coagulation of these substances tends to vary in a manner directly related to the water content of the solvent, although the tannin and the phenolic substance were less sensitive to the solvent composition and are equally well coagulated in all solvent systems studied. The findings can be applied to the removal of these substances from aqueous alcoholic plant extracts using the electrocoagulation technique.

INTRODUCTION

Electrocoagulation has been a useful alternative technique for clarifying and decolourising certain solutions containing unwanted dissolved substances or suspended matter. However, those solutions were almost always aqueous in nature, for example, potable water [1,2], food wastewater [3], tar-sand and oil-shale wastewater [4], phosphate-containing sewage [5,6], industrial wastes containing cyanide and heavy metals [7,8,9], and dye-containing textile wastewater [10,11]. Electrocoagulation has also been used as a purification step in the isolation of a few natural products from crude plant extracts [12,13,14,15,16,17]. Again, the solvent used in those isolations was also purely aqueous. Trials of natural product isolation using electrocoagulation in alcoholic solutions as a part of the process have been reported in the literature, for example, for the isolation of asiaticoside [18] and the extraction of phenolic compounds from the bark of *Lithocarpus elegans*[19], but a systematic study of electrocoagulation in alcoholic solutions has, to our knowledge, not been undertaken. In view of the fact that a system containing an organic solvent is undoubtedly more useful in isolating natural products in general, we would like to report our study on the effects of solvent in the electrocoagulation of some selected organic substances and plant extracts. For a start, we chose as our model study the system in which ethanol with varying amounts of water is used as solvent.

Citation: Chairungsi, N.; Jumpatong, K.; Phutdhawong, W.; Buddhasukh, D. Solvent Effects in Electrocoagulation of Selected Plant Pigments and Tannin. Molecules 2006, 11, 309-31.

RESULTS AND DISCUSSION

Our previous work carried out with water and alcohol soluble natural dyes (chlorophyllin and crocin) [20] showed that although decreases in the water content of the solvent (ethanol) has some effect on the coagulation efficiency for the dyes, coagulation was more or less complete after 2 hours, especially when iron is used for the electrodes (Figure 1 and Figure 2).

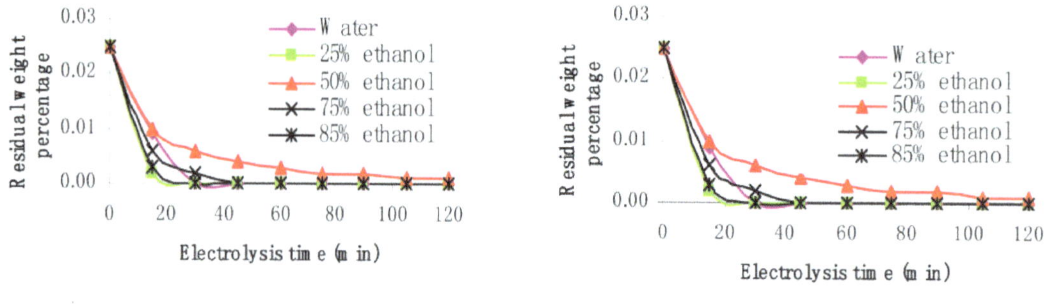

A B

Figure 1. Plots of residual weight percentage and electrolysis time for each ethanol concentration of chlorophyllin solution at 626 nm; A: with aluminium as electrodes, B: with iron as electrodes.

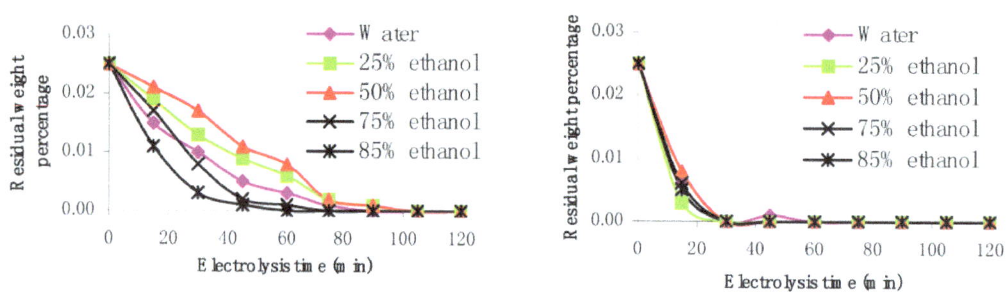

A B

Figure 2. Plots of residual weight percentage and electrolysis time for each ethanol concentration of crocin solution at 440 nm; A: with aluminium as electrodes, B: with iron as electrodes.

However, experiments carried out in this study with water-insoluble dyes (the chlorophylls and the carotenoids) seem to indicate that, contrary to the above result, these pigments seem to be more difficult to coagulate. For example, in Figure 3 it can be seen that the model pigment β-carotene is hardly affected and barely removed from the solution by the electrolytic decolourisation process applied, even when iron electrodes are used (due to the difficulties in solubilizing this pigment, electrocoagulation was only examined in a single mixed solvent system.)

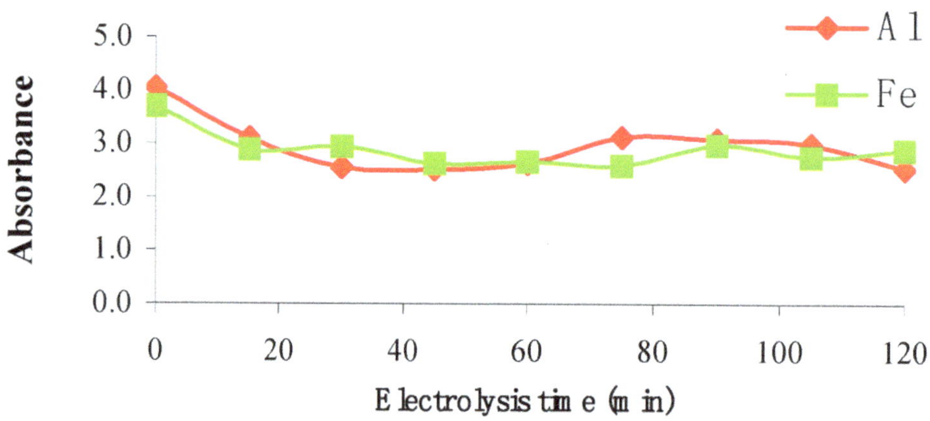

Figure 3. Plots of absorbance at 460 nm and electrolysis time for β-carotene (0.01%) in 27% ethanol; ♦: with aluminium as electrodes, ■: with iron as electrodes.

Similarly, for solutions of extracted chlorophyll, the trend is that coagulation seems to be less efficient as the water content in the solvent decreases (Figure 4 and Figure 5). This may not only be due to the fact that, as the water content in the solvent decreases the process of electrocoagulation itself is naturally retarded, but also to the fact that more pigment is dissolved in the solution as the water content in the solvent decreases and the alcohol content accordingly increases. This trend was repeated in nearly every case of green plant extracts we studied. Thus, for example, when the alcoholic extract of the leaves of *Solanum laciniatum*, which contain solasonine (an important starting compound for steroid synthesis), was subject to decolourisation by electrocoagulation, typical results as shown in Figure 6 (for the yellow pigments) and Figure 7 (for the green pigments) were obtained.

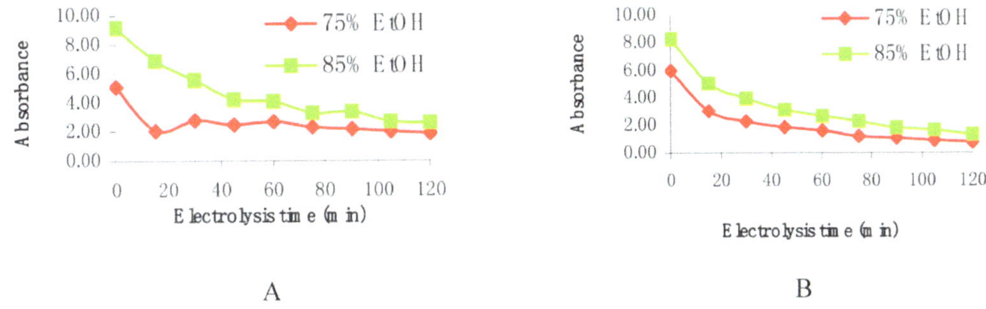

Figure 4. Plots of absorbance at 440 nm and electrolysis time for each ethanol concentration of extracted chlorophyll solution; A: with aluminium as electrodes, B: with iron as electrodes; ♦, 75% ethanol; ■, 85% ethanol.

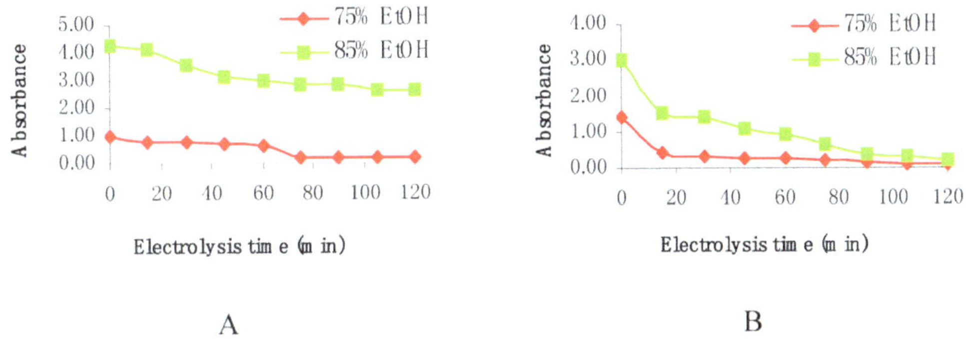

Figure 5. Plots of absorbance at 665 nm and electrolysis time for each ethanol concentration of extracted chlorophyll solution; A: with aluminium as electrodes, B: with iron as electrodes; ♦, 75% ethanol; ■, 80% ethanol.

Importantly, however, it has been shown by our group that even with this general decreased efficacy of the pigment removal, electrocoagulation is still more efficient than the conventional method of extraction with organic solvents and does not affect the desired natural products, which were subsequently isolated [21].

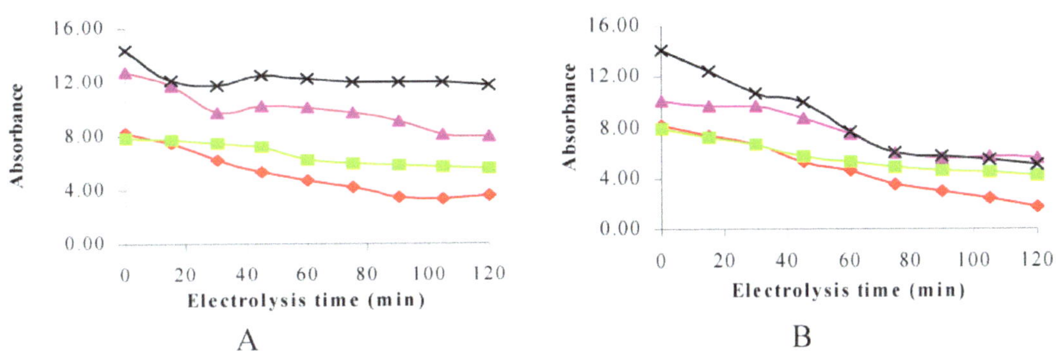

Figure 6. Plots of absorbance and electrolysis time for each ethanol concentration of *Solanum laciniatum* extract at 408 nm; A: with aluminium as electrodes, B: with iron as electrodes; ♦, 25% ethanol; ■, 50% ethanol; ▲, 75% ethanol; ×, 85% ethanol.

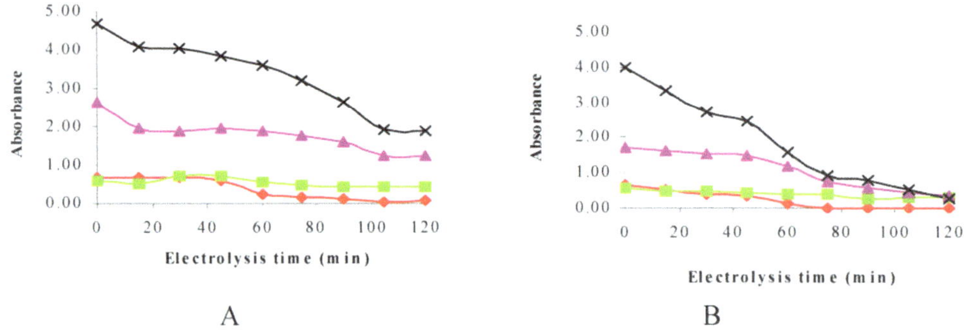

Figure 7. Plots of absorbance and electrolysis time for each ethanol concentration of *Solanum laciniatum* extract at 666 nm; A: with aluminium as electrodes, B: with iron as electrodes; ◆, 25% ethanol; ■, 50% ethanol; ▲, 75% ethanol; ×, 85% ethanol.

Another important class of plant pigments is the phenolic substances, which include the tannins, the flavonoids, and the various quinone compounds. Most of these materials (except perhaps the tannins) are poorly soluble in water but more soluble in alcohol or other organic solvents. Many of them are valuable natural products, while others (especially the tannins) are regarded as little more than intractable mixtures with unfavourable biological activities and it is generally preferable to remove them along with the pigments.

In a 100% aqueous medium, it has been shown that tannins can be very efficiently coagulated and removed by electrolysis [22,23]. In this study, we tried to repeat the process in aqueous alcoholic solutions. The hypothesis is that tannins, owing to their polyphenolic nature, should still be easily coagulated by the phenolate salt forming mechanism in addition to the adsorption mechanism [23]. Our experiments showed this to be the case. Thus, for example, at a concentration of 0.1% tannin in up to 85% ethanol, a 250-mL solution was almost completely de-tannized within 15 minutes, using aluminium as electrodes and a current of 0.3 A (Figure 8). At a concentration of 1.0% tannin, the complete detannization time was increased to 80 minutes (Figure 9).

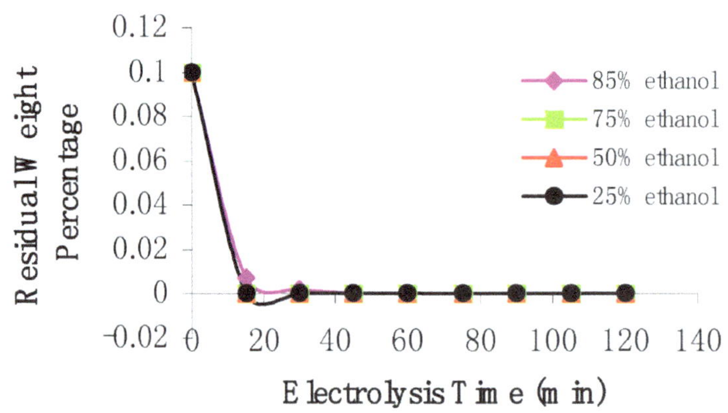

Figure 8. Plot of the residual weight percentage and electrolysis time for 0.1% w/v tannin at 275 nm ; ●, 25% ethanol , ▲, 50% ethanol, ■, 75% ethanol and ◆, 85% ethanolic solution.

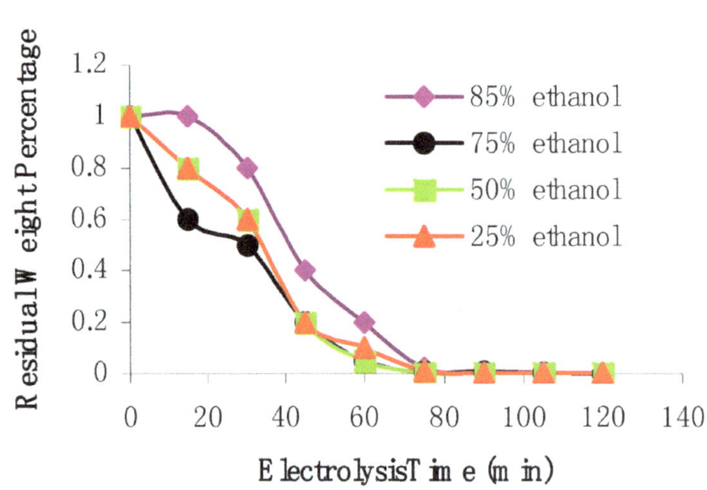

Figure 9. Plot of the residual weight percentage and electrolysis time for 1.0% w/v tannin at 275 nm; ▲, 25% ethanol, ■, 50% ethanol, ●, 75% ethanol, and ◆, 85% ethanolic solution.

Next, to demonstrate the general trend of the effect of water content in the solvent on the coagulation of phenolic compounds other than tannins, the result of electrocoagulation of a known flavonoid, morin, is presented here as a typical example (Figure 10). It can thus be seen that the effect of the decrease of water content in the solvent on the coagulation of this type of substances is similar to that of the chlorophylls and the carotenoids, i.e., as the water content in the solvent decreases (or the alcohol content increases), coagulating efficacy also seems to decrease proportionally.

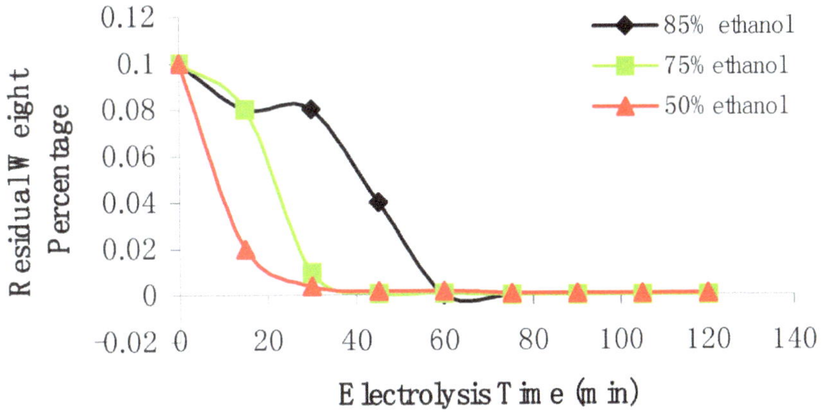

Figure 10. Plot of the residual weight percentage and electrolysis time for 0.1% w/v morin at 415, 420 nm ; ▲,50% ethanol, ▲,75% ethanol and ♦85% ethanolic solution (a 0.1% solution of morin in 25% ethanol cannot be prepared.)

For most natural phenolic compounds, however, probably owing to the active polyhydroxy functions of the substances, coagulation which can occur both by reaction and by adsorption tends to be more complete than that of the chlorophylls and the carotenoids, so that after some time (60 minutes in this case), coagulation is virtually complete, even in solvents with a high percentage of alcohol.

CONCLUSIONS

It has been demonstrated in a systematic manner that in the electrocoagulation in aqueous alcoholic solutions of some important plant pigments, including tannins, the decrease in the percentage of water in the solvent has some negative effects on the degree and efficiency of their coagulation compared with that observed in 100% aqueous solution. However, the effect is small in the case of tannins and some other phenolic substances. For chlorophylls and carotenoids, this retarding effect is somewhat higher, probably due to only a single mode of coagulation being in operation (viz. adsorption mode). However, even with this unfavourable effect being present, electrocoagulation is still more efficient in removing these organic matrix substances than the conventional method of solvent extraction.

Experimental Section

General

All of the tested compounds used were of standard reagent grade, and were used as received. Tannin and β-carotene were purchased from Fluka Chemica AG (Buchs, Switzerland); morin (3,5,7,2',4'-pentahydroxyflavone) was purchased from May & Baker Ltd (Dagenham, England). Crude chlorophyll was obtained from *Spinach oleracea*. Air-dried leaves of *Solanum laciniatum* were obtained from Dr. Jiradej Manosroi, Faculty of Pharmacy, Chiangmai University. Sodium chloride (99.9%, AR grade) was purchased from Ajax Chemical Co. (Sydney, Australia). Absolute ethanol was purchased from E. Merck (Darmstadt, Germany). Acetone and ethanol (95%) were of a commercial grade. Aluminium and iron plates were purchased locally. Direct current was sustained by a GW Instek DC power supply. Absorbance was measured on a Genesys 10 spectrophotometer.

Preparation of Solutions for Electrocoagulation

A solution of tannin (tannic acid) or morin (0.01, 0.1, or 1.0% w/v) was prepared in aqueous ethanol (25%, 50%, 75% or 85% v/v). Beta-carotene was dissolved in 27% ethanol to give a 0.01% w/v solution. Crude chlorophyll was extracted from *Spinach oleracea* with 75 or 85% ethanol at room temperature by grinding 40 g of the dry plant material with solvent (200 mL). The filtered solution was used directly in the electrocoagulation experiments. For *Solanum laciniatum*, the powdered air-dried leaves (20 g) were refluxed for 3 hours with each aqueous alcoholic solution (25%, 50%, 75%, 85%, 200 mL). After filtration, the deep green solutions obtained were used directly in the electrocoagulation experiments.

Electrocoagulation Procedures

1) Morin and tannic acid

Two aluminum plates (dimensions 15 x 4 cm) were used as electrodes. These were spaced 3 cm apart and dipped 5.5 cm deep into a magnetically-stirred aqueous solution (250 mL) of the tested compound (0.01, 0.1 or 1.0% w/v solution) in a 400 mL beaker. Sodium chloride (0.5 g) was added as an electrolyte. Direct current (0.3 A) from the DC power supplier was then passed through the solution. Every 15 minutes during a 2 hour period of electrolysis, a 4 mL sample of the solution was withdrawn, centrifuged and taken for an absorbance measurement at an appropriate wavelength (275 nm for tannin and 415-420 nm for morin). The measured absorbance was then converted into the residual weight percentage of the compound by a calibration curve obtained from a plot of the absorbance versus the concentration for each compound.

2) Chlorophyll, β-carotene, and Solanum laciniatum

Aluminium or iron plates (15.0 × 3.0 cm.) were washed prior to use with acetone to remove surface grease. A pair of aluminium or iron plates 1.5 cm apart was immersed 7.0 cm deep into 200 mL of each solution in a jacketed 250-mL beaker for occasional cooling during electrolysis. The solution was agitated throughout the experiment with a magnetic stirrer (250 r.p.m.). Sodium chloride (0.2 g) was added as supporting electrolyte. Direct current (0.9 A, 16.9-31.6 V) was then passed through the solution *via* the two electrodes. At every 15-minute interval during a 2-hour period of electrolysis, a 4-mL aliquot of the solution was withdrawn and centrifuged for 10 minutes, and the absorbance of the supernatant solution was measured at an appropriate wavelength of the absorption maximum for each plant solution (440 and 665 nm for chlorophyll, 460 nm for β-carotene and 408 and 666 nm for *Solanum laciniatum*).

Acknowledgements

Dr. Jiradej Manosroi of the Faculty of Pharmacy, Chiangmai University, is thanked for supplying the leaves of *Solanum laciniatum*. This research study is supported financially by TRF (Thailand Research Fund) in the form of Medicinal Chemistry Grant and Royal Golden Jubilee PhD Grant for K. J., and by the Commission on Higher Education PhD Grant for N. C. We also thank the Graduate School, Chiangmai University, for partly supporting K.J and N.C. financially.

REFERENCES AND NOTES

1. Vik, E. A.; Carlson, D. A.; Eikum, A. S.; Gjessing, E. T. Electrocoagulation of Potable Water. *Water Res.* **1984**, *18*, 1355–1360.
2. Mameri, N.; Yeddou, A. R.; Lounici, H.; Belhocine, D.; Grib, H.; Bariou, B. Defluoridation of Septentrional Sahara Water of North Africa by Electrocoagulation Process Using Bipolar Aluminium Electrodes. *Water Res.* **1998**, *32*, 1604–1612.
3. Beck, E. C.; Giannini, A. P.; Ramirez, E. R. Electrocoagulation Clarifies Food Wastewater. *Food Technol. Feb.* **1974**, *28*, 18–19.
4. Renk, R. R. Electrocoagulation of Tar Sand and Oil Shale Wastewater. *Energy Prog.* **1988**, *8*, 205–208.
5. Dobolyi, E. Experiments Aimed at The removal of Phosphate by Electrochemical Methods. *Water Res.* **1978**, *12*, 1113–1116.
6. Groeterud, O.; Smoczynski, L. Phosphorus Removal From Water by Means of Electrolysis. *Water Res.* **1986**, *20*, 667–669.
7. Cenkin, V. E.; Belevtsev, A. N. Electrochemical Treatment of Industrial Wastewater. *Effluent Water Treat. J.* **1985**, *25*, 243–244.
8. Mrozowski, J.; Zielinski, J. Studies of Zinc and Lead Removal From Industrial Waste by Electrocoagulation. *Environ. Protection Eng.* **1983**, *9*, 77–85.
9. Jenke, D. R.; Diebold, F. E. Electroprecipitation Treatment of Acid Mine Wastewater. *Water Res.* **1984**, *18*, 855–859.
10. Lee, B. S.; Son, E. J.; Choe, E. K.; Kim, J. W. Electrochemical Treatment of Dyeing Effluent Using Sacrificial Iron Electrodes. *J. Korean Fiber Soc.* **1999**, *36*, 329–337.
11. Do, J. S.; Chen, M. L. Decolourization of Dye-Containing Solution by Electroagulation. *J. Appl. Electrochem.* **1994**, *24*, 785–790.
12. Miwa, K. Extraction and Purification of Plant Components. Jpn. Kokai Tokkyo Koho 78 105,500. *Chem. Abstr.* **1979**, *90*, 3392y.
13. Miwa, K.; Maeda, S.; Murata, Y. Stevia Sweetener Purification by Electrolysis. Jpn. Kokai Tokkyo Koho 79 89,066. *Chem Abstr.* **1979**, *91*, 173681e.
14. Miwa, K.; Maeda, S.; Murata, Y. Purification of Stevioside. Jpn. Kokai Tokkyo Koho 79 90,199. *Chem. Abstr.* **1979**, *91*, 206904k.
15. Adduci, J.; Buddhasukh, D.; Ternai, B. Improved Isolation and Purification of Stevioside. *J. Sci. Soc. Thailand* **1987**, *13*, 179–183.
16. Phutdhawong, W.; Buddhasukh, D. Simple Isolation and Purification of Glycyrrhizic Acid. *J. Sci. Fac. CMU* **1998**, *25(2)*, 87–91.
17. Phutdhawong, W.; Buddhasukh, D. Simple Isolation and Purification of D-Pinitol from *Cassia Siamea* Lamk by Electrolytic Decolourization. *ACGC Chem. Res. Commun.* **2000**, *10*, 61–62.
18. Phutdhawong, W. Ph.D. Thesis, Chiang Mai University, Thailand, 2002.
19. Chowwanapoonpohn, S. Ph.D. Thesis, Chiang Mai University, Thailand, 2003.
20. Jumpatong, K.; Phutdhawong, W.; Chowwanapoonpohn, S.; Garson, M. G.; Pyne, S. G.; Buddhasukh, D. *Trends in Electrochemistry Research.*
21. Jumpatong, K.; Phutdhawong, W.; Buddhasukh, D. Dechlorophyllation by Electrocoagulation. *Molecules* **2006**, *11*, 156–162.
22. Buso, A.; Balbo, L.; Giomo, M. Electrochemical Removal of Tannins from Aqueous Solutions. *Ind. Eng. Chem. Res.* **2000**, *39*, 494–499.
23. Phutdhawong, W.; Chowwanapoonpohn, S.; Buddhasukh, D. Electrocoagulation and Subsequent Recovery of Phenolic Compounds. *Anal. Sci.* **2000**, *16*, 1083–1084.

CHAPTER 17

FOOD PROCESSING AND MAILLARD REACTION PRODUCTS: EFFECT ON HUMAN HEALTH AND NUTRITION

Nahid Tamanna[1] and Niaz Mahmood[2]

[1]Graduate Program in Biological Sciences, University of Manitoba, Winnipeg, MB, Canada R3T 2N2

[2]Graduate Program in Biochemistry and Medical Genetics, University of Manitoba, Winnipeg, MB, Canada R3E 0J9

ABSTRACT

Maillard reaction produces flavour and aroma during cooking process; and it is used almost everywhere from the baking industry to our day to day life to make food tasty. It is often called nonenzymatic browning reaction since it takes place in the absence of enzyme. When foods are being processed or cooked at high temperature, chemical reaction between amino acids and reducing sugars leads to the formation of Maillard reaction products (MRPs). Depending on the way the food is being processed, both beneficial and toxic MRPs can be produced. Therefore, there is a need to understand the different types of MRPs and their positive or negative health effects. In this review we have summarized how food processing effects MRP formation in some of the very common foods.

INTRODUCTION

The Maillard reaction has been named after the French physicist and chemist Louis Camille Maillard (1878–1936) who initially described it. It is often defined as nonenzymatic browning reaction. While foods are processed or cooked at high temperature, a chemical reaction occurs between amino acids and reducing sugars which generate different flavours and brown colour (Figure 1). So it is often used in food industry for giving food different taste, colour, and aroma.

Based on literature, Hodge (1953) first described the steps involved in Maillard reaction products (MRPs), also known as advanced glycation end-products (AGEs), formation. The whole process of MRPs formation can be divided into three major stages depending on colour formation. At the first stage, sugars and amino acid condense, and following condensation, Amadori rearrangement and 1-amino-1deoxy-2 ketose form. In the second stage, dehydration and fragmentation occur in the sugar molecules. Amino acids are also degraded in this stage. Hydroxymethylfurfural (HMF) fission products such as pyruvaldehyde and diacetyl are formed in this intermediate stage. This stage can be slight yellow or colourless. In the final stage, aldol condensation occurs and finally the heterocyclic nitrogenous compounds form, melanoidins, which is highly coloured [1]. Maillard reaction can also take place in living

Citation: Nahid Tamanna and Niaz Mahmood, "Food Processing and Maillard Reaction Products: Effect on Human Health and Nutrition," International Journal of Food Science, vol. 2015, Article ID 526762, 6 pages, 2015. doi:10.1155/2015/526762.

organisms. It has been reported that some MRPs particularly melanoidins have beneficial effects on health such as antioxidative [2] and antibiotic [3] effects. However some reports have also suggested that MRPs such as high carboxymethyl lysine (CML) promote diabetes and cardiovascular diseases while acrylamide acts as a carcinogen [4–6].

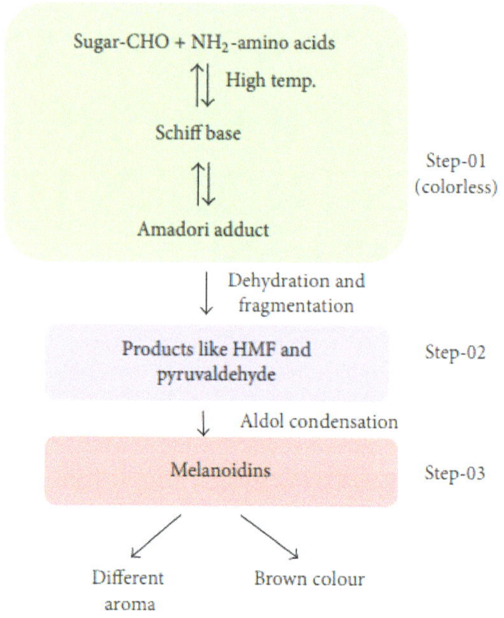

Figure 1: Schematic representation of "Maillard reaction" and flavour formation in food.

There is an ever-increasing preference for instant meal rather than traditional cooking, especially among the new generation of people. It has been reported that people consuming high amount of processed meat, pizza, or snacks develop insulin resistance and metabolic syndrome compared to people having high intake of vegetables and low processed food [7]. MRPs that change during food processing might be one of the important factors for either disease progression or combating diseases. In this review, we have summarized the changes of MRPs which occur during processing of foods.

SOYBEAN PROCESSING AND MRPS FORMATION

Soybean is widely used as flours, grits, flakes, isolates, concentrates, and textured soya proteins and also as cooking oil. Soybeans play important role in cardiovascular diseases, osteoporosis, and cancer [8]. So processing of soybean is an important factor for maintaining its nutritional quality. Cooking at high temperature may generate MRPs which can be good or bad for health. However, soybean must be processed before consumption. Žilić et al. (2014) assessed the level of furosine, hydroxymethylfurfural (HMF), and acrylamide in soybean during extrusion, microwave, and infrared heating processes [9]. They found that microwave heating for short time (1-2 min) generates high levels of acrylamide, whereas long time heating (3–5 min) generates lower levels of acrylamide. During extrusion and infrared heating, acrylamide formation greatly increased with time and temperature. HMF level increased in all three processes with increased time and temperature and it was significantly higher in microwave treatment. From the beginning of heat treatment, furosine level was higher in the extrusion and infrared treatment whereas in the microwave heating it was increased to maximum value after 3 min but at 4 min this value was similar with 2 min. Their results showed that microwave heating improved the antioxidant properties of soybean by 50% compared to raw soybean [9]. Even though this study has reported that total flavonoids increase at 100°C with the exception of microwave heating which occurs at 45°C, another study showed that when soybean is soaked in water and heated afterwards at 98°C, almost half (44%) of the raw flavonoids were lost in the final product [10]. This might be due to the presence of moisture content since decreasing moisture was shown to be associated with elevated levels of MRPs in the extrusion and infrared heat treatment [9].

MRPS ON MILK PROCESSING

Milk is a beverage that is consumed throughout the world. At present, a large percentage of the milk consumed by people, especially in the western countries, is processed rather than raw milk. Ultra high temperature (UHT) treatment or conventional sterilization process is often used to process milk for improving quality and safety. Milk is rich in protein and sugar. So it is obvious that processing of milk at high temperature may lead to the formation of MRPs. Several methods have been approached to determine the extent of MRPs during milk processing. Both initial and advanced staged MRPs have been used as indicators of browning reaction which occurred in milk [11, 12]. MRP formation affects protein and mineral bioavailability greatly. In the early stages, lactose in milk blocks the amino group of lysine to form the Amadori product called lactulosyllysine which changes the protein bioavailability [13]. It is also known that MRPs may behave as chelating agents to chelate metal cations by forming different soluble and insoluble complexes and thereby can affect mineral bioavailability [14]. So processing of milk by heat-treatment requires attention, especially for the infants, as milk is the sole source of nutrients at that stage of life. It has been reported that conventional sterilized bottle milk has different chemical composition compared to the UHT treated milk [15]. HMF level is often used to assess the progression of MRPs formation. However, milk processed at UHT may have different levels of HMF [16] due to the presence of some other factors such as vitamin A, casein, and iron [17]. So during milk processing, along with heat treatment, other relevant factors should be considered for preserving the nutritious value. A recent study has shown that use of enzymes like Faox I and Faox II might inhibit Maillard reaction development [18].

PASTA PROCESSING AND MRPS

Pasta is considered one of the healthy foods suited to balanced diet; and its consumption is becoming increasingly high throughout the world. Pasta has different varieties and shapes. Till 2009, 310 different types of both dried and fresh pasta have been documented [19]. Although an Italian law exists for making high quality pasta [20], different steps of pasta processing cycle such as mixing, kneading, shaping, and drying can affect the pasta quality. Among these four steps, drying is most important for the quality of the product. For traditional drying methods, low temperatures (29 to 40°C) are required but it takes 24 to 60 hours to complete the process. So for time management and to increase the production rate, high-temperature short-time drying processes have been widely preferred. While pasta is processed at high temperature, it brings benefits in terms of productivity, cost, colour intensity, and nutritious value [21]. Moreover, application of high drying temperatures in pasta increases firmness and decreases stickiness [22]. Besides these beneficial effects, high heat treatment leads to the formation of MRPs that ultimately can change flavor, color, functional properties, and nutritional value [23]. Furosine is widely used as pasta quality assessment. It has been reported that pasta processed at high temperature had decreased level of total carotenoid level [24]. Along with furosine, maltulose is also proposed as a maker for pasta quality assessment [25].

MEAT PROCESSING AND MRPS

MRPs like heterocyclic amine (HCA) level increases with elevated cooking temperature; and this phenomenon is more pronounced in meat than fish [26]. Meat is cooked at high temperature either by frying, roasting, and boiling or in oven. While positive correlations have been found between the intakes of HCAs from foods and increased risk of various types of human cancer [27–29], some other studies have not found any correlation between HCAs and cancer risk [30–32]. Several studies have demonstrated that processes like frying and broiling can cause the formation of high amounts of HCAs [27, 33–35]. On contrary, these HCAs produce different flavour and tastes in foods. Heterocyclic compounds such as pyrazine, oxazole, and thiazoles are primarily responsible for forming flavour in roasted compound. During high heat treatment and grilling process, pyrazines level significantly increased [36]. It is suggested that the alkylpyrizne is formed via the condensation of two alpha amino ketone molecules derived from the Strecker degradation [37], which is an intermediate of Maillard reaction pathway.

In processed food, more than 25 types of heterocyclic amines (HCA) have been identified [38]. A study has shown that when duck meat was cooked by charcoal grilling, deep-frying, roasting, microwave cooking, pan frying, or boiling, MRPs were higher in the pan frying process compared to the other four methods of cooking. Liao et al. (2012) [39] reported that boiling and microwave cooking were the most appropriate methods to process duck meat in terms of MRPs formation. However in another study, it has been found that both charcoal grilled duck and chicken breast had high level of HCAs compared to pan fried meat. They found that roasting decreases HCAs significantly [40].

In another study, beef steak and hamburger patties were processed by pan-frying, oven-broiled, and grilled or barbecued to four levels of doneness (rare, medium, well cooked, or very well cooked) [41]. Beef roasts were processed on oven by rare, medium, and well cooking. They measured five different HCAs. The level of 2-amino-3,4-dimethylimidazo [4,5-f] quinoline was higher in well-cooked steak and hamburger patties. Like duck and chicken roast, roasting beef did not contain any of the 5 HCAs, but the gravy made from the drippings from well-done roasts had two types of HCAs [41]. From the three different studies, it can be suggested that roasting of meat (chicken, duck, and beef) generates less amount of HCAs compared to other methods.

In recent days, people consume more ready-to-eat food due to lack of time. Puangsombat et al. (2011) assessed HCAs level in some ready-to-eat products. They found that HCAs were higher in rotisserie chicken skin. In the other assessed food, HCA level was found in the order as follows: rotisserie chicken meat, deli meat products, and pepperoni [42]. However, it has been reported that commercially cooked meats and restaurant meats contain low amounts of HCAs [43, 44].

COFFEE BEAN PROCESSING AND MRPS

As a beverage, coffee is an important item in the lives of billions of people and is also one of the most traded food products in the world. The brewed coffee has emerged as the second most consumed drink after water [45,46]. The desirable fragrance of coffee beverages develops during roasting procedures. Typical roasting temperatures range from 180 to 250°C, and roasting time varies between 2 and 25 min depending on the process employed [47]. During roasting procedures, internal temperature exceeds 180°C which results in the occurrence of Maillard reaction, carbohydrate caramelization, and pyrolysis of organic compounds [47]. Maillard reaction generates melanoidins upon roasting, which accounts for 29% of the dry weight of brewed coffee bean [47]. Coffee melanoidins are formed via polymerization reactions of furans and/or pyrroles, during the advanced stages of the Maillard reaction, and linked by poorly defined polycondensation reactions [48]. Even though roasting decreases coffee bean carbohydrates, protein, and lipid level, the level of caffeine remains relatively stable on roasting. MRPs, caffeine, nicotinic acid, and some other components of coffee bean protect teeth from Streptococcus mutans which is considered to be the major causative agent of dental caries in humans [49].

PLANT DERIVED FOOD PROCESSING AND MRPS

Consumption of diets rich in fruits and vegetables renders many health benefits to us. However, processing method plays an important role in dictating the magnitude of the beneficial health effects obtained from fruits and vegetables. Depending on treatment temperature, furoylmethyl derivatives (FM) have been found in processed vegetables and fruits like orange juices [50] and processed tomato products [51] and also in dehydrated carrots [52]. It has been shown that dehydrated carrot contains significantly high amount of FM compared to carrot juices, baby carrot, or tinned carrot. It is suggested that processing time during the heat treatment plays an important role for FM formation [52]. Dueik and Bouchon (2011) have reported that, by vacuum frying of carrot chips, potato and apple slices can help to retain their total carotenoids and ascorbic acid levels significantly [53].

When vegetables are treated at low temperature, prooxidants are generated, whereas treating at high temperature decreases the prooxidants and increases antioxidant properties due to the production of MRPs [54]. Such antioxidant activity of the MRPs comes from the high molecular weight brown compounds that are formed in the advanced stages of the reaction [54]. However, it should be mentioned here that MRPs can also exhibit prooxidant properties [55, 56].

MRPs can prevent the enzymatic browning reaction caused by polyphenol oxidase (PPO) [57]. Plant derived products, such as fruits and vegetables, produce many endogenous phenolic compounds during postharvest handling and processing. These compounds are oxidized by oxidoreductase enzymes like polyphenoloxidases (PPOs) and tyrosinases. This reaction, in turn, generates highly reactive quinonic compounds that are condensed and polymerized to produce brown pigments and thereby decreases the quality of the food product. MRPs can prevent this enzymatic process at the initial step of this reaction and thereby help to maintain the product quality. Besides antibrowning, MRPs has also been shown to render antiallergenic property for cherry derived allergens [58].

SOME OTHER IMPACTS OF MRP-DERIVED FOOD

Angiotensin-I converting enzyme (ACE) is the regulatory enzyme for upregulation of blood pressure. ACE inhibitory peptide lowers blood pressure by inhibiting ACE enzyme [59]. Rufián-Henares and Morales (2007) have demonstrated that the melanoidins isolated from seven amino acid-glucose model systems were all shown to cause inhibition of ACE in vitro [60]. Recently, Hong and colleagues (2014) have shown that, under the appropriate conditions, Maillard reaction can effectively improve the ACE inhibitory activity of casein hydrolysate [61].

It has been claimed that administration of a Maillard browning reaction product obtained from an extract ofPanax species plant comprising ginsenoside Re or ginsenoside-derived saccharide treated with amino acid at temperatures between 100 and 130°C can either prevent, improve, or treat a renal disease [62].

Food derived versatile MRPs can act as bactericidal for a wide number of pathogens. For example, aminoreductone can act as a more effective bactericidal for four Pseudomonas aeruginosa isolates, one multidrug-resistant Pseudomonas aeruginosa (MDRP), one Escherichia coli, one methicillin-susceptibleStaphylococcus aureus, and one methicillin-resistant Staphylococcus aureus (MRSA) compared to mikacin, ciprofloxacin, imipenem, and levofloxacin [63]. MRPs have also been shown to be effective against yeast [64].

CONCLUSION AND PERSPECTIVES

Maillard reaction products have both positive and negative impacts on health. Diverse MRPs act as antioxidants, bactericidal, antiallergenic, antibrowning, prooxidants, and carcinogens. Most of these properties depend on processing of food. High temperature heating makes some food nutritious, whereas some of the foods lose their nutritional value. Many strategies are employed in the food industries to reduce the production of MRPs. For example, acrylamide has been classified as a probable carcinogen to humans by the International Agency for Research on Cancer [65]. During food preparation at high temperature, acrylamides are formed in many types of foods via Maillard reaction [66–68]. To reduce the amount of acrylamide, asparaginase has been successfully used in laboratory for potatoes and cereals [69, 70]. It has also been reported that injection of CO_2during extrusion process helps to reduce the level of acrylamide [71].

This review was aimed at summarizing our current knowledge regarding the changes in food mediated by Maillard reaction during the food processing steps. This may provide useful insights for those related to food processing facilities.

Conflict of Interests

The authors declare that there is no conflict of interests regarding the publication of this paper.

REFERENCES

1. J. E. Hodge, "Dehydrated foods: chemistry of browning reactions in model systems," Journal of Agricultural and Food Chemistry, vol. 1, no. 15, pp. 928–943, 1953.
2. F. Natella, M. Nardini, I. Giannetti, C. Dattilo, and C. Scaccini, "Coffee drinking influences plasma antioxidant capacity in humans," Journal of Agricultural and Food Chemistry, vol. 50, no. 21, pp. 6211–6216, 2002.
3. S. Hiramoto, K. Itoh, S. Shizuuchi et al., "Melanoidin, a food protein-derived advanced Maillard reaction product, suppresses Helicobacter pylori in vitro and in vivo," Helicobacter, vol. 9, no. 5, pp. 429–435, 2004.
4. J. Meng, N. Sakata, S. Takebayashi et al., "Glycoxidation in aortic collagen from STZ-induced diabetic rats and its relevance to vascula damage," Atherosclerosis, vol. 136, no. 2, pp. 355–365, 1998.
5. E. Tareke, P. Rydberg, P. Karlsson, S. Eriksson, and M. Törnqvist, "Acrylamide: a cooking carcinogen?"Chemical Research in Toxicology, vol. 13, no. 6, pp. 517–522, 2000.
6. K. J. Knecht, J. A. Dunn, K. F. McFarland et al., "Effect of diabetes and aging on carboxymethyllysine levels in human urine," Diabetes, vol. 40, no. 2, pp. 190–196, 1991.
7. A. Esmaillzadeh, M. Kimiagar, Y. Mehrabi, L. Azadbakht, F. B. Hu, and W. C. Willett, "Dietary patterns, insulin resistance, and prevalence of the metabolic syndrome in women," The American Journal of Clinical Nutrition, vol. 85, no. 3, pp. 910–918, 2007.
8. M. Messina and V. Messina, "Soyfoods, soybean isoflavones, and bone health: a brief overview," Journal of Renal Nutrition, vol. 10, no. 2, pp. 63–68, 2000.

9. S. Žilić, B. A. Mogol, G. Akillioğlu, A. Serpen, N. Delić, and V. Gökmen, "Effects of extrusion, infrared and microwave processing on Maillard reaction products and phenolic compounds in soybean," Journal of the Science of Food and Agriculture, vol. 94, no. 1, pp. 45–51, 2014.

10. C.-J. C. Jackson, J. P. Dini, C. Lavandier et al., "Effects of processing on the content and composition of isoflavones during manufacturing of soy beverage and tofu," Process Biochemistry, vol. 37, no. 10, pp. 1117–1123, 2002.

11. F. Evangelisti, C. Calcagno, and P. Zunin, "Relationship between blocked lysine and carbohydrate composition of infant formulas," Journal of Food Science, vol. 59, no. 2, pp. 335–337, 1994.

12. L. Pizzoferrato, P. Manzi, V. Vivanti, I. Nicoletti, C. Corradini, and E. Cogliandro, "Maillard reaction in milk-based foods: nutritional consequences," Journal of Food Protection, vol. 61, no. 2, pp. 235–239, 1998.

13. H. Erbersdobler, Twenty Years of Furosine—Better Knowledge about the Biological Significance of Maillard Reaction in Food and Nutrition, Elsevier Science, Amsterdam, The Netherlands, 1986.

14. M. Friedman, "Food browning and its prevention: an overview," Journal of Agricultural and Food Chemistry, vol. 44, no. 3, pp. 631–653, 1996.

15. A. Olano, M. M. Calvo, and N. Corzo, "Changes in the carbohydrate fraction of milk during heating processes," Food Chemistry, vol. 31, no. 4, pp. 259–265, 1989.

16. S. Albalá-Hurtado, M. T. Veciana-Nogués, A. Mariné-Font, and M. C. Vidal-Carou, "Progress of browning reactions during storage of liquid infant milks," Journal of Agricultural and Food Chemistry, vol. 47, no. 10, pp. 4033–4037, 1999.

17. Y. Park and Y. Hong, "Comparison of the heat treatment intensity in infant formulae," Korean Journal of Food Science and Technology, vol. 23, 1991.

18. A. D. Troise, N. A. Dathan, A. Fiore et al., "Faox enzymes inhibited Maillard reaction development during storage both in protein glucose model system and low lactose UHT milk," Amino Acids, vol. 46, no. 2, pp. 279–288, 2014.

19. O. Z. De Vita, Encyclopedia of Pasta, vol. 26, University of California Press, 2009.

20. P. della Repubblica, "Legge No 580 of 4th July 1967 on Disciplina per la lavorazione e commercio dei cereali, degli sfarinati, del pane e delle paste alimentari," Gazzetta Ufficiale, no. 189, 1967.

21. J. E. Dexter, R. R. Matsuo, and B. C. Morgan, "High temperature drying: effect on spaghetti properties,"Journal of Food Science, vol. 46, no. 6, pp. 1741–1746, 1981.

22. A. Baiano, A. Conte, and M. A. del Nobile, "Influence of drying temperature on the spaghetti cooking quality," Journal of Food Engineering, vol. 76, no. 3, pp. 341–347, 2006.

23. G. A. Reineccius, "The influence of Maillard reactions on the sensory properties of foods," in The Maillard Reaction in Food Processing, Human Nutrition and Physiology, P. A. Finot, H. U. Aeschbacher, R. F. Hurrel, and R. Liardon, Eds., pp. 157–170, Birkhäuser, Basel, Switzerland, 1990.

24. R. Beleggia, C. Platani, R. Papa et al., "Metabolomics and food processing: from semolina to pasta,"Journal of Agricultural and Food Chemistry, vol. 59, no. 17, pp. 9366–9377, 2011.

25. J. L. García-Baños, N. Corzo, M. L. Sanz, and A. Olano, "Maltulose and furosine as indicators of quality of pasta products," Food Chemistry, vol. 88, no. 1, pp. 35–38, 2004.

26. K. I. Skog, M. A. E. Johansson, and M. I. Jägerstad, "Carcinogenic heterocyclic amines in model systems and cooked foods: a review on formation, occurrence and intake," Food and Chemical Toxicology, vol. 36, no. 9-10, pp. 879–896, 1998.

27. R. Sinha, M. G. Knize, C. P. Salmon et al., "Heterocyclic amine content of pork products cooked by different methods and to varying degrees of doneness," Food and Chemical Toxicology, vol. 36, no. 4, pp. 289–297, 1998.

28. J. Lin, J. Wang, H. B. Grossman, M. Chen, C. P. Dinney, and X. Wu, "Red meat and heterocyclic amine intake, metabolic pathway genes, and bladder cancer risk," Cancer Research, vol. 70, no. 8, supplement, p. 2825, 2010.

29. A. J. Cross, L. M. Ferrucci, A. Risch et al., "A large prospective study of meat consumption and colorectal cancer risk: an investigation of potential mechanisms underlying this association," Cancer Research, vol. 70, no. 6, pp. 2406–2414, 2010.

30. K. Augustsson, K. Skog, M. Jägerstad, P. W. Dickman, and G. Steineck, "Dietary heterocyclic amines and cancer of the colon, rectum, bladder, and kidney: a population-based study," The Lancet, vol. 353, no. 9154, pp. 703–707, 1999.

31. N. J. Ollberding, L. R. Wilkens, B. E. Henderson, L. N. Kolonel, and L. le Marchand, "Meat consumption, heterocyclic amines and colorectal cancer risk: the Multiethnic Cohort Study," International Journal of Cancer, vol. 131, no. 7, pp. E1125–E1133, 2012.

32. A. Sander, J. Linseisen, and S. Rohrmann, "Intake of heterocyclic aromatic amines and the risk of prostate cancer in the EPIC-Heidelberg cohort," Cancer Causes & Control, vol. 22, no. 1, pp. 109–114, 2011.

33. B. G. Abdulkarim and J. S. Smith, "Heterocyclic amines in fresh and processed meat products," Journal of Agricultural and Food Chemistry, vol. 46, no. 11, pp. 4680–4687, 1998.

34. M. G. Knize and J. S. Felton, "Formation and human risk of carcinogenic heterocyclic amines formed from natural precursors in meat," Nutrition Reviews, vol. 63, no. 5, pp. 158–165, 2005.

35. K. W. Cheng, F. Chen, and M. Wang, "Heterocyclic amines: chemistry and health," Molecular Nutrition & Food Research, vol. 50, no. 12, pp. 1150–1170, 2006.

36. D. Mottram, "Some aspects of the chemistry of meat flavour," in Flavor of Meat and Meat Products, pp. 210–

230, Springer, Berlin, Germany, 1994.

37. D. S. Mottram, "Flavour formation in meat and meat products: a review," Food Chemistry, vol. 62, no. 4, pp. 415–424, 1998.

38. J. Felton and M. Knize, "Heterocyclic-amine mutagens/carcinogens in foods," in Chemical Carcinogenesis and Mutagenesis I, pp. 471–502, Springer, Berlin, Germany, 1990.

39. G. Z. Liao, G. Y. Wang, Y. J. Zhang, X. L. Xu, and G. H. Zhou, "Formation of heterocyclic amines during cooking of duck meat," Food Additives and Contaminants Part A: Chemistry, Analysis, Control, Exposure and Risk Assessment, vol. 29, no. 11, pp. 1668–1678, 2012. ·

40. G. Z. Liao, G. Y. Wang, X. L. Xu, and G. H. Zhou, "Effect of cooking methods on the formation of heterocyclic aromatic amines in chicken and duck breast," Meat Science, vol. 85, no. 1, pp. 149–154, 2010. ·

41. R. Sinha, N. Rothman, C. P. Salmon et al., "Heterocyclic amine content in beef cooked by different methods to varying degrees of doneness and gravy made from meat drippings," Food and Chemical Toxicology, vol. 36, no. 4, pp. 279–287, 1998.

42. K. Puangsombat, P. Gadgil, T. A. Houser, M. C. Hunt, and J. S. Smith, "Heterocyclic amine content in commercial ready to eat meat products," Meat Science, vol. 88, no. 2, pp. 227–233, 2011. ·

43. M. G. Knize, R. Sinha, E. D. Brown et al., "Heterocyclic amine content in restaurant-cooked hamburgers, steaks, ribs, and chicken," Journal of Agricultural and Food Chemistry, vol. 46, no. 11, pp. 4648–4651, 1998.

44. L. M. Tikkanen, K. J. Latva-Kala, and R.-L. Heiniö, "Effect of commercial marinades on the mutagenic activity, sensory quality and amount of heterocyclic amines in chicken grilled under different conditions," Food and Chemical Toxicology, vol. 34, no. 8, pp. 725–730, 1996.

45. A. Crozier, H. Ashihara, and F. Tomás-Barbéran, Teas, Cocoa and Coffee: Plant Secondary Metabolites and Health, John Wiley & Sons, New York, NY, USA, 2011.

46. M. S. Butt and M. T. Sultan, "Coffee and its consumption: benefits and risks," Critical Reviews in Food Science and Nutrition, vol. 51, no. 4, pp. 363–373, 2011. ·

47. H.-D. Belitz, W. Grosch, and P. Schieberle, "Coffee, tea, cocoa," Food Chemistry, pp. 938–970, 2009.

48. R. Tressl, G. T. Wondrak, L.-A. Garbe, R.-P. Krüger, and D. Rewicki, "Pentoses and hexoses as sources of new melanoidin-like Maillard polymers," Journal of Agricultural and Food Chemistry, vol. 46, no. 5, pp. 1765–1776, 1998.

49. M. Daglia, R. Tarsi, A. Papetti et al., "Antiadhesive effect of green and roasted coffee on Streptococcus mutans' adhesive properties on saliva-coated hydroxyapatite beads," Journal of Agricultural and Food Chemistry, vol. 50, no. 5, pp. 1225–1229, 2002.

50. M. D. del Castillo, N. Corzo, and A. Olano, "Early stages of Maillard reaction in dehydrated orange juice,"Journal of Agricultural and Food Chemistry, vol. 47, no. 10, pp. 4388–4390, 1999.

51. M. L. Sanz, M. Dolores del Castillo, N. Corzo, and A. Olano, "Presence of 2-furoylmethyl derivatives in hydrolysates of processed tomato products," Journal of Agricultural and Food Chemistry, vol. 48, no. 2, pp. 468–471, 2000.

52. A. Wellner, C. Huettl, and T. Henle, "Formation of Maillard reaction products during heat treatment of carrots," Journal of Agricultural and Food Chemistry, vol. 59, no. 14, pp. 7992–7998, 2011. ·

53. V. Dueik and P. Bouchon, "Vacuum frying as a route to produce novel snacks with desired quality attributes according to new health trends," Journal of Food Science, vol. 76, no. 2, pp. E188–E195, 2011. ·

54. M. Anese, L. Manzocco, M. C. Nicoli, and C. R. Lerici, "Antioxidant properties of tomato juice as affected by heating," Journal of the Science of Food and Agriculture, vol. 79, no. 5, pp. 750–754, 1999. ·

55. T. Hofmann, W. Bors, and K. Stettmaier, "Studies on radical intermediates in the early stage of the nonenzymatic browning reaction of carbohydrates and amino acids," Journal of Agricultural and Food Chemistry, vol. 47, no. 2, pp. 379–390, 1999.

56. L. R. Fukumoto and G. Mazza, "Assessing antioxidant and prooxidant activities of phenolic compounds,"Journal of Agricultural and Food Chemistry, vol. 48, no. 8, pp. 3597–3604, 2000.

57. C. Billaud, C. Maraschin, Y.-N. Chow, S. Chériot, M.-N. Peyrat-Maillard, and J. Nicolas, "Maillard reaction products as "natural antibrowning" agents in fruit and vegetable technology," Molecular Nutrition and Food Research, vol. 49, no. 7, pp. 656–662, 2005. ·

58. P. Gruber, S. Vieths, A. Wangorsch, J. Nerkamp, and T. Hofmann, "Maillard reaction and enzymatic browning affect the allergenicity of Pru av 1, the major allergen from cherry (Prunus avium)," Journal of Agricultural and Food Chemistry, vol. 52, no. 12, pp. 4002–4007, 2004. ·

59. A. Pihlanto-Leppälä, "Bioactive peptides derived from bovine whey proteins: opioid and ace-inhibitory peptides," Trends in Food Science and Technology, vol. 11, no. 9-10, pp. 347–356, 2000. · ·

60. J. A. Rufián-Henares and F. J. Morales, "Functional properties of melanoidins: in vitro antioxidant, antimicrobial and antihypertensive activities," Food Research International, vol. 40, no. 8, pp. 995–1002, 2007.

61. X. Hong, J. Meng, and R.-R. Lu, "Improvement of ACE inhibitory activity of casein hydrolysate by Maillard reaction with xylose," Journal of the Science of Food and Agriculture, vol. 95, no. 1, pp. 66–71, 2015.

62. K. S. Kang, K. I. Su-Nam, J. Ham, W. Lee, N. Yamabe, and J. H. Lee, "Composition for preventing, improving, or treating renal disease including maillard browning reaction products of panax species plant extract," Google Patents, 2013.

63. V. T. Trang, V. H. Son, L. X. Thanh et al., "Functional properties of maillard reaction products in food: antimicrobial activity of aminoreductone against pathogenic bacteria," Food Science and Technology

Research, vol. 19, no. 5, pp. 833–841, 2013.

64. A. Tauer, S. Elss, M. Frischmann, P. Tellez, and M. Pischetsrieder, "Influence of thermally processed carbohydrate/amino acid mixtures on the fermentation by Saccharomyces cerevisiae," Journal of Agricultural and Food Chemistry, vol. 52, no. 7, pp. 2042–2046, 2004. ·

65. Cancer IAfRo, IARC Monographs on the Evaluation of the Carcinogenic Risk of Chemicals to Humans. Overall Evaluations of Carcinogenicity: An Updating of IARC Monographs, vol. 1–42, World Health Organization, 1987.

66. D. S. Mottram, B. L. Wedzicha, and A. T. Dodson, "Food chemistry: acrylamide is formed in the Maillard reaction," Nature, vol. 419, no. 6906, pp. 448–449, 2002. ·

67. R. H. Stadler, I. Blank, N. Varga et al., "Food chemistry: acrylamide from Maillard reaction products,"Nature, vol. 419, no. 6906, pp. 449–450, 2002. ·

68. E. Tareke, P. Rydberg, P. Karlsson, S. Eriksson, and M. Törnqvist, "Analysis of acrylamide, a carcinogen formed in heated foodstuffs," Journal of Agricultural and Food Chemistry, vol. 50, no. 17, pp. 4998–5006, 2002.

69. E. Capuano, A. Ferrigno, I. Acampa, L. Ait-Ameur, and V. Fogliano, "Characterization of the Maillard reaction in bread crisps," European Food Research and Technology, vol. 228, no. 2, pp. 311–319, 2008. · ·

70. D. V. Zyzak, R. A. Sanders, M. Stojanovic et al., "Acrylamide formation mechanism in heated foods,"Journal of Agricultural and Food Chemistry, vol. 51, no. 16, pp. 4782–4787, 2003. ·

71. M. T. Masatcioglu, V. Gokmen, P. K. W. Ng, and H. Koksel, "Effects of formulation, extrusion cooking conditions, and CO_2 injection on the formation of acrylamide in corn extrudates," Journal of the Science of Food and Agriculture, vol. 94, no. 12, pp. 2562–2568, 2014. ·

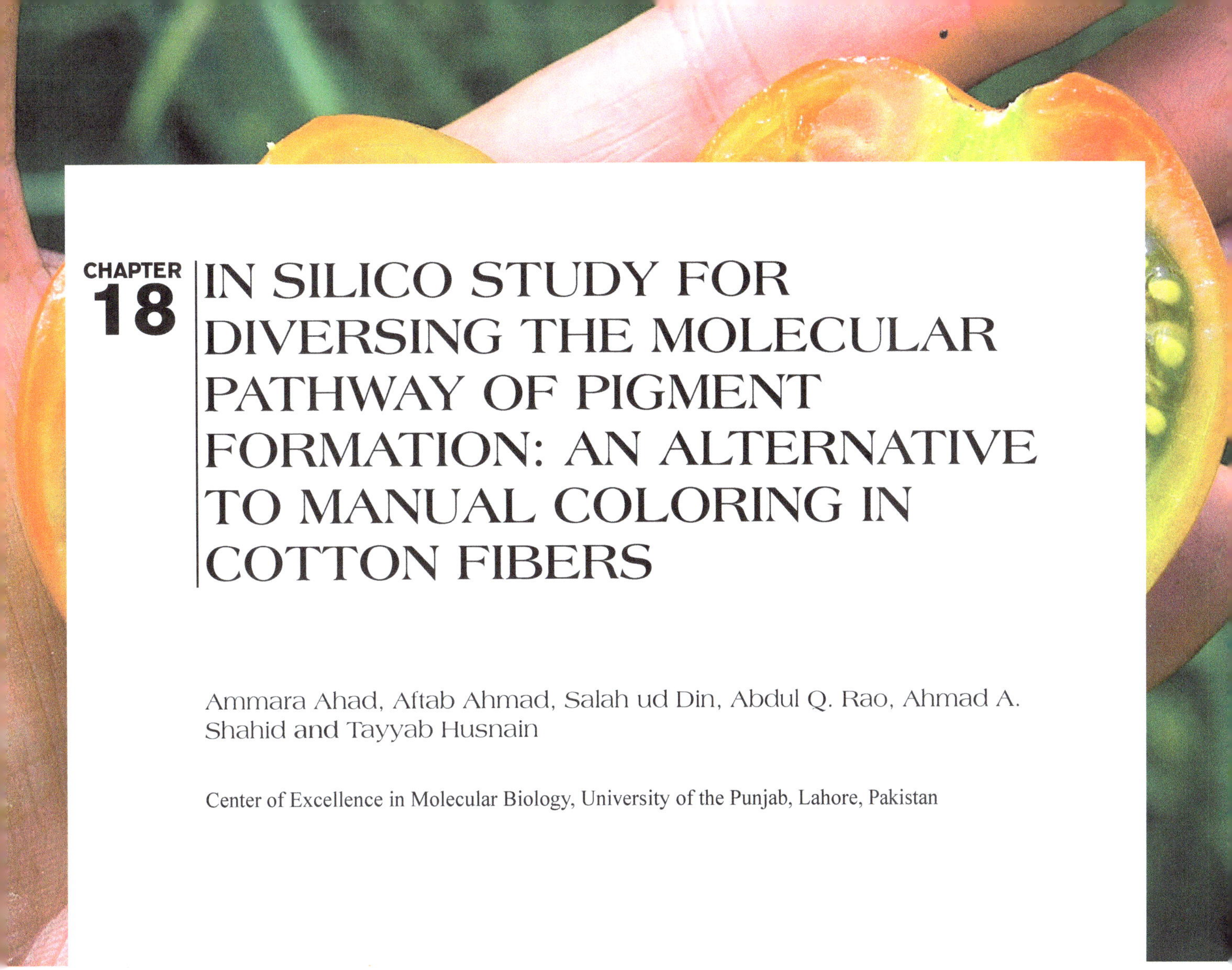

CHAPTER 18

IN SILICO STUDY FOR DIVERSING THE MOLECULAR PATHWAY OF PIGMENT FORMATION: AN ALTERNATIVE TO MANUAL COLORING IN COTTON FIBERS

Ammara Ahad, Aftab Ahmad, Salah ud Din, Abdul Q. Rao, Ahmad A. Shahid and Tayyab Husnain

Center of Excellence in Molecular Biology, University of the Punjab, Lahore, Pakistan

Diversity of colors in flowers and fruits is largely due to anthocyanin pigments. The flavonoid/anthocyanin pathway has been most extensively studied. Dihydroflavonol 4-reductase (DFR) is a vital enzyme of the flavonoid pathway which displays major impact on the formation of anthocyanins, flavan 3-ols and flavonols. The substrate specificity of the DFR was found to play a crucial role in determination of type of anthocyanidins. Altering the flavonoid/anthocyanin pathway through genetic engineering to develop color of our own choice is an exciting subject of future research. In the present study, comparison among four DFR genes (*Gossypium hirsutum, Iris × hollandica, Ang. DFRI* and *DFRII*), sequence alignment for homology as well as protein modeling and docking is demonstrated. Estimation of catalytic sites, prediction of substrate preference and protein docking were the key features of this article. For specific substrate uptake, a proline rich region and positions 12 plus 26 along with other positions emphasizing the 26-amino acid residue region (132–157) was tested. Results showed that proline rich region position 12, 26, and 132–157 plays an important role in selective attachment of DFRs with respective substrates. Further, "Expasy ProtParam tool" results showed that *Iris ×hollandica* DFR amino acids (Asn 9: Asp 23) are favorable for reducing DHQ and DHM thus accumulating delphinidin, while *Gossypium hirsutum* DFR has (Asn 13: Asp 21) hypothesized to consume DHK. Protein docking data showed that amino acid residues in above mentioned positions were just involved in attachment of DFR with substrate and had no role in specific substrate uptake. Advanced bioinformatics analysis has revealed that all above mentioned positions have role in substrate attachment. For substrate specificity, other residues region is involved. It will help in color manipulations in different plant species.

INTRODUCTION

More than 200,000 different types of compound have been shown to be produced collectively by higher plants, and some of these are able to generate bright colors in flowers, fruit or foliage (Feng et al., 2013). The human eye can detect light, as reflected or transmitted by a compound under wavelengths 380 and 730 nm, while insects recognize the light of shorter wavelengths (Davies, 2004). In plants, three major classes of pigments for coloration exist which includes carotenoids, flavonoids/anthocyanins, and betalains.

Diversity of red, purple and blue colors of flowers as well as fruits are due to the nature of these polyphenolic pigments. In plants, these secondary metabolites are found ubiquitously.

Citation: Ahad A, Ahmad A, Din S, Rao AQ, Shahid AA and Husnain T (2015) In silico study for diversing the molecular pathway of pigment formation: an alternative to manual coloring in cotton fibers. Front. Plant Sci. 6:751. doi: 10.3389/fpls.2015.00751.

Among factors influencing flower color, the flavonoid/anthocyanin biosynthesis has been most extensively studied. The various factors which influence the final color formation includes anthocyanin structures, vacuolar pH, co-pigments and metal ions.

In higher plant species, the pathway leading to anthocyanidin 3-glucoside is generally conserved (Grotewold, 2006; Tanaka and Brugliera, 2006b). Only six major classes of anthocyanidins do exist i.e., pelargonidin, cyanidin, peonidin, delphinidin, petunidin, and malvidin (Yoshida et al., 2009) Among cytochromes P450s, two genes, i.e., flavonoid 3′-hydroxylase (F3′H) and flavonoid 3′, 5′-hydroxylase (F3′5′H), catalyze hydroxylation of the B-ring (Tanaka, 2006a). The increased hydroxylation pattern of this ring was found to be involved in shifting the anthocyanin color toward blue. They exhibit broad substrate specificity and catalyze hydroxylation of flavanones, dihydroflavonols, flavonols, and flavones. Flavanones along with dihydroflavonols are precursors of anthocyanidins and anthocyanins. Trihydroxylated delphinidin based anthocyanins from blue or violet colors is achieved by presence of F3′5′H (Honda and Saito, 2002).

Dihydroflavonol 4-reductase (DFR) is a vital enzyme of the flavonoid pathway which shows a major impact on the formation of anthocyanins, flavan 3-ols, and flavonols. In ornamental flower plants, the color has been modified by altering the expression levels of DFR genes (Aida et al., 2000). The substrate specificity of the DFR plays a crucial role in determining which anthocyanidins, a plant will accumulate (Forkmann and Heller, 1999).

DFR is unique in a sense that it uptakes flavoniod substrates depending on the B-ring hydroxylation pattern. The DFRs of several plants accept dihydroflavonols having one (dihydrokaempferol, DHK), two (dihydroquercetin, DHQ), or three (dihydromyricetin, DHM) hydroxyl groups on the B-ring. NADPH is used as a cofactor of DFR, which helped in catalyzing the reduction of dihydroflavonols to leucoanthocyanidins which were common precursors of anthocyanin (Helariutta et al., 1993; Tanaka et al., 1995; Dellus et al., 1997). Substrate specificity of DFR determines the conversion of metabolic flux toward desired anthocyanidin biosynthesis. For example in Petunia hybrid, the transformation of a DFR from maize evaded the gap in the pathway of anthocyanin formation and successfully generated orange color flower by the reduction of DHK and the accumulation of pelargonidin anthocyanins. Similarly, petunia or viola F3′5′H gene in combination with the petunia DFR gene was transformed in cultivars of white carnation that specifically missing DFR gene, hence resulted in violet carnations by the generation of delphinidin based anthocyanins (Holton, 1996; Fukui et al., 2003).

The two most common fiber colors in colored cotton cultivars (*Gossypium hirsutum*) are "brown and green." They are eco-friendly, economical, and valuable for human health (Zhu et al., 2006) Four key genes of the flavonoid biosynthetic pathway (*GhC4H, GhCHS, GhF3′H*, and *GhF3′5′H, GhDFR*) have been reported in *Gossypium hirsutum*.

The idea of pigment engineering, for diversity of colors in the cotton fiber, can be opened up by investigating the mechanism of pigmentation. The key is to explore and understand the molecular basis of pigment formation and its deposition in fibers (Feng et al., 2013). The purpose of this research was to investigate the putative involvement of flavonoid biosynthesis pathway structural gene (DFR) in the pigmentation and coloration of cotton fiber.

To achieve this objective a comparison was made in retrieving sequences of two DFR genes from blue flower species and two from white including sequence alignment for homology and protein modeling. Further more the presence of proline rich region along with positions 12 and 26 was reported to be important in determining substrate specificity for DFR of *A. angustifolia* (Gosch et al., 2014). Other positions emphasizing the 26-amino acid region (132–157) were also mentioned in the literature which played a role in utilizing specific dihydroflovonols: DHK, DHQ, and DHM (Johnson et al., 2001; Xie et al., 2004). It was tested in *Gossypium hirsutum* as well as in *Iris* × *hollandica* by protein-ligand docking analysis. Analysis of gene cluster encoding dihydroflavonol 4-reductases in the *Lotus japonicus* genome showed that three out of six DFR proteins exhibit catalytic activity, their substrate preferences settled with the variation of a specific active site residue (Aspartic acid or Asparagine) and found to be involved in controlling the substrate specificity (Shimada et al., 2005).

The structure of the DFR proteins (*Gossypium hirsutum* as well as in *Iris* × *hollandica*) have not yet determined experimentally (NMR or X-ray), and for that reason models were built by following homology modeling and threading protocol. Moreover, for the estimation of respective catalytic sites and prediction of substrate preference, protein docking was performed. The aim of this study was to identify the region that involves in the substrate specificity of DFR to give readers an idea of manipulating the pathways involved in color modification, so that *Gossypium hirsutum* fiber pigmentation may be altered, similarly as reported in the case of some ornamental flower plants.

MATERIALS AND METHODS

In silico analysis was carried out in order to evaluate substrate specificity of different types of DFRs (dihydromyricetin reductase, dihydroquercetin reductase and dihydrokaempferol 4-reductase). For this assessment, dihydroflavonol 4-reductase (DFR) sequence information available in NCBI database was used. Protein sequence alignment, designing of protein structures, ligand structure retrieval and molecular docking was also carried out.

Determination of Substrate Binding Region among Different Plant Species

All above mentioned positions (12, 26 and from 132 to 157) were evaluated for the presence of particular residues and its role in specific substrate uptake as illustrated in published data. For this purpose full length four DFRs sequences: *Angelonia angustifolia, Ang. DFRI* (GenBank. AIR09398.1), *Ang. DFRII* (GenBank. AHM27144.1), *Gossypium hirsutum* (GenBank. AHG97389.1), and*Iris* × *hollandica* (GenBank. BAF93856.1) were retrieved from Genbank. Sequences were arranged by using "Bioedit" program to find out the positional similarities between the residues of these sequences. To further validate the role of this particular position in substrate specificity, sequences from five other species were taken which includes *Rosa chinensis* (GenBank. AHF58604.1),*Vaccinium macrocarpon* (GenBank. AF483835.1), *Gerbera hybrid*(GenBank.

CAA78930.1), *Petunia* × *hybrid* (GenBank. AF233639.1), and *Ampelopsis grossedentata* (GenBank. AGO02174.1). These nine sequences were aligned by using the CLC Genomics Workbench 8.

To evaluate substrate specificity, Asn as well as Asp percentage estimation in DFR sequences for *Iris* × *hollandica* and *Gossypium hirsutum* was done by using "Expasy ProtParam tool."

Modeling of Receptor Molecules for Docking Analysis

Protein sequences of *Gossypium hirsutum* and *Iris* × *hollandica*(retrieved from NCBI) were used for 3D modeling as their protein structures were not available on protein structure databases. For modeling purposes, sequences were submitted to I-TASSER server (http://zhanglab.ccmb.med.umich.edu/I-TASSER/). This tool produced a protein model based on homology modeling and threading. For homology modeling of *Gossypium hirsutum* DFR, the PDB templates used were PDB: 2C29F (Identity 82%, coverage 92%) and PDB: 2C29A(Identity 82%, coverage 91%). Whereas, for protein modeling of DFR (*Iris* × *hollandica*) PDB: 2C29F (Identity 66%, coverage 90%) and PDB: 2C29A (Identity 67%, Coverage 90%) template were used.

Refinement and Evaluation of DFR Protein Model

The model was further refined by using online tool, ModRefiner accessed on Zhang Lab website (http://zhanglab.ccmb.med.umich.edu/ModRefiner/). This tool minimized the energy of the model and took the residues of protein in the allowed region. The models were evaluated and validated by producing Ramachandran plot. These plots were plotted by using the online RAMPAGE tool (http://mordred.bioc.cam.ac.uk/~rapper/rampage.php). Ramachandran plots of proteins determined their stability.

Ligand Preparation

The structures of biological compounds of flavonoid pathway dihydrokaempferol (PubChem: 122850), dihydroquercetin (PubChem: 439533), and dihydromyricetin (PubChem: 161557) were downloaded from the PubChem database in 2D format. For preparation of ligand structures which were used in docking, hydrogen atoms were added to each ligand and their energy was minimized by the means of MMFF94X force field at 0.05 gradients. Then, these ligand structures were saved in.mol2 file format. A database of ligands (DFR acceptor compounds) was created in MOE software.

Protein-ligand Docking of DFR Protein and Ligand Molecules

Three-dimensional model of *Gossypium hirsutum* and *Iris* ×*hollandica* were constructed by using I-TASSER server. The water molecules were removed with the help of MOE software. After the removal of water molecules, hydrogen atoms were added to the receptor proteins. Optimization of receptor molecule was achieved by energy minimization and 3D protonation (with help of AMBER99 force field option of MOE). The gradient was 0.05 and receptor was minimized unless root mean square gradient fall below 0.05. After 3D protonation of the receptor protein, the hydrogen molecules were hidden. This energy minimized, 3D protonated receptor molecules were then used for docking analysis. Box of 26 residues as reported by Shimada et al. (2005) was aligned with *Iris* × *hollandica* and *Gossypium hirsutum*. These aligned residues were used as pocket site.

Molecular docking was carried out against the databases mentioned previously, after the modeling and preparation of ligand and receptor molecules. The receptor residues in region 132–157 of *Gerbera* DFR correspond to 148–174 residues in *G. hirsutum* as well as 127–153 in *Iris* × *hollandica* were selected and docked with ligands. Docking output database file having receptor ligand complex was saved in.mdb format. The docked complexes were categorized with increasing *S*-value (Final score to indicate binding free energy). The complexes with minimum S were taken to evaluate the interactions of ligand with the active site residues of the receptor proteins. Best hydrogen bonding plus π-π interactions were evaluated by the using ligX option of MOE.

RESULTS

Comparison of DFR Reported Residues Involved in Substrate Specificity

For comparison, at position 12 and 26, DFR sequences were aligned by using the CLC Genomics Workbench 8 as mentioned earlier. From sequence alignment results it was evaluated that at position 12, *Ang. DFRI, Ang. DFRII, Gossypium hirsutum* DFR, and *Iris* ×*hollandica* DFR have proline, serine, proline and glycine respectively. This result showed same residue (proline) in both*Gossypium hirsutum* and *Angelonia DFR I* whereas, *Angelonia DFRII* and *Iris* × *hollandica* DFR had serine and glycine which were functionally similar residues (Figure 1). In present study, the sequence alignment showed proline at 12 position in *Gossypium hirsutum* from which it is hypothesized that it would reduce dihydrokaempferol. Whereas, this particular region is absent in both *Iris* × *hollandica* and *Ang.II* DFRs, resulting in delphinidin accumulation which is responsible for production of blue color (Figure 2). For further confirmation DFRs from five more species along with these four plant species were carried out to evaluate results. However, with respect to *Ang. DFRI*, sequence alignment of other five species (*Rosa chinensis, Vaccinium macrocarpon, Gerbera hybrid, Petunia* × *hybrida*, and *Ampelopsis grossedentata*) showed the deletion of proline rich region and thus unable to reduce DHK.

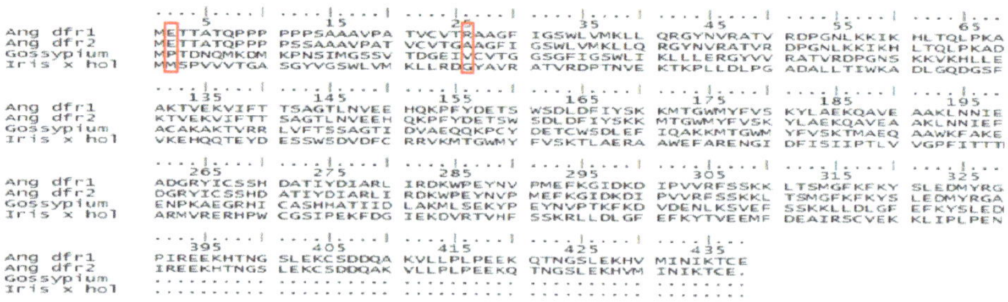

Figure 1. Alignment of the amino acid sequences encoded by *Ang.DFRI and DFRI*I, *Gossypium hirsutum*, and *Iris ×hollandica*. Proline rich region is marked with a box.

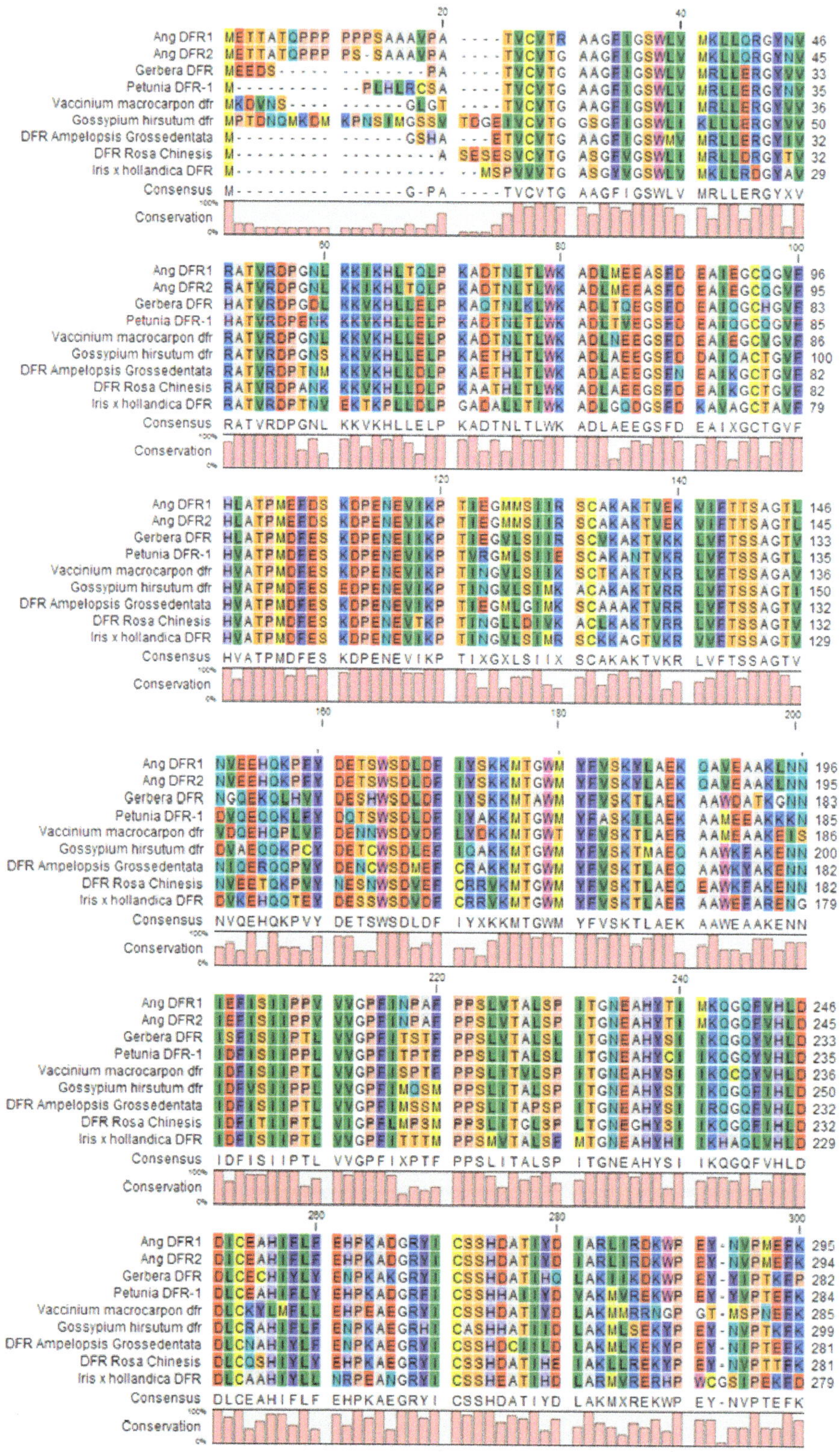

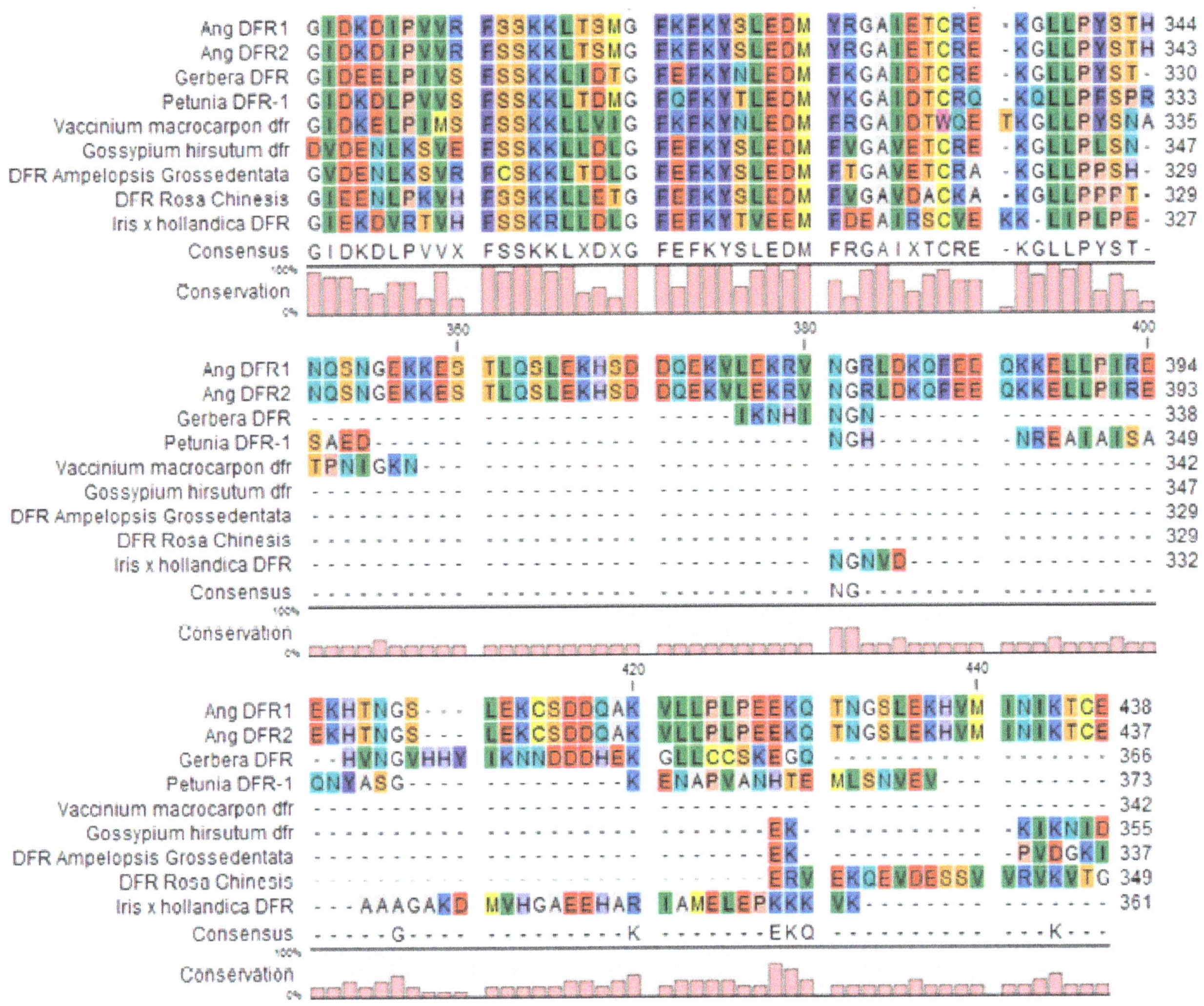

Figure 2. Multiple sequence alignment of dihydroflavanol 4-reductase, consensus sequences of different plant species (*Rosa chinensis, Vaccinium macrocarpon,Gerbera hybrid, Petunia × hybrid, Ang. DFRI, Ang. DFRII, Gossypium hirsutum, Iris × hollandica, and Ampelopsis grossedentata*) were achieved by using CLC Genomics Workbench 8. The colored bars at the bottom are representing the conservation Percentage.

On 26th position regarding *Ang.DFRI* has arginine and all other species DFR have glycine (position 25 in *Ang.DFRII*,13 *Rosa chinensis*, 16 *Vaccinium macrocarpon*, 13 *Gerbera hybrid*, 15*Petunia×hybrida*, 12 *Ampelopsis grossedentata*, and 30 in*Gossypium hirsutum* DFR (Figure 2). In DFRs, where proline is absent at position 12, it would show glycine on that particular or nearby position. Thus, it is speculated that when a proline rich region is absent, the DFRs would reduce DHQ or DHM as substrate. Results showed that proline rich region in general and position 12 in particular plays main role in selective attachment of these reductases with their relative specific substrates (DHK).

Role of Asn and Asp Type DFRs in Substrate Specificity

Recent analysis on DFRs (*Iris* × *hollandica* and *Gossypium hirsutum*) was done by "Expasy ProtParam tool" in order to get information about the presence of total Asn and Asp residues in them (Table 1A). Results showed that *Iris* × *hollandica* DFR has (Asn 9: Asp 23) while *Gossypium hirsutum* DFR has (Asn 13:Asp 21) (Table 1B). This is postulated that *Iris* × *hollandica* DFR (Asp-type) has more likeness to reduce DHM in comparison to DHK hence it has a key role in accumulating delphidin by utilizing DHM as substrate.

Table 1. Amino acid percentage in both*Gossypium hirsutum and Iris* × *hollandica*(Asn,9:Asp,23) by using Protparam tool.

A			B		
Number of amino acids: 361			Number of amino acids: 355		
Molecular weight: 40221.2			Molecular weight: 39651.7		
Theoretical pI: 6.13			Theoretical pI: 5.67		
Amino acid composition:			Amino acid composition:		
Ala (A)	32	8.9%	Ala (A)	25	7.0%
Arg (R)	18	5.0%	Arg (R)	8	2.3%
Asn (N)	9	2.5%	Asn (N)	13	3.7%
Asp (D)	23	6.4%	Asp (D)	21	5.9%
Cys (C)	7	1.9%	Cys (C)	8	2.3%
Gln (Q)	4	1.1%	Gln (Q)	9	2.5%
Glu (E)	26	7.2%	Glu (E)	27	7.6%
Gly (G)	22	6.1%	Gly (G)	20	5.6%
His (H)	13	3.6%	His (H)	9	2.5%
Ile (I)	19	5.3%	Ile (I)	25	7.0%
Leu (L)	25	6.9%	Leu (L)	29	8.2%
Lys (K)	25	6.9%	Lys (K)	32	9.0%
Met (M)	14	3.9%	Met (M)	13	3.7%
Phe (F)	15	4.2%	Phe (F)	18	5.1%
Pro (P)	17	4.7%	Pro (P)	18	5.1%
Ser (S)	22	6.1%	Ser (S)	25	7.0%
Thr (T)	23	6.4%	Thr (T)	20	5.6%
Trp (W)	6	1.7%	Trp (W)	5	1.4%
Tyr (Y)	8	2.2%	Tyr (Y)	7	2.0%
Val (V)	33	9.1%			
Pyl (O)	0	0.0%			
Sec (U)	0	0.0%			

Modeling, Refinement, Evaluation and Validation of DFR Protein

For the construction of 3D Model of DFR protein, retrieved sequences of both *Iris* × *hollandica* DFR and *Gossypium hirsutum*DFR were submitted to online I-TASSER server (Figures 3, 4).

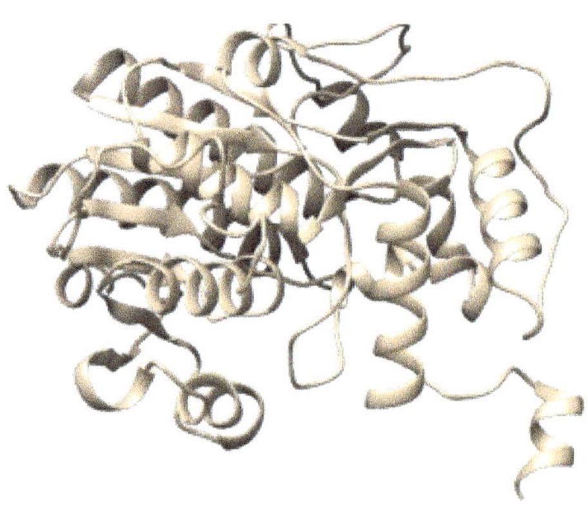

Figure 3. Three dimensional protein model of *Gossypium hirsutum* **DFR protein, predicted by I-TASSER.**

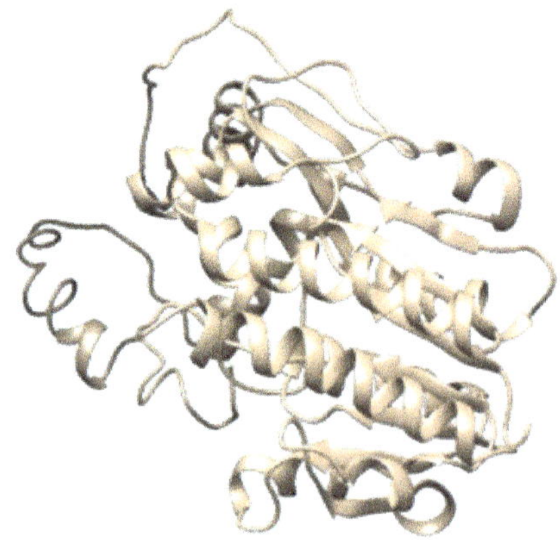

Figure 4. Three dimensional protein model of and *Iris* × *hollandica* **protein, predicted by I-TASSER.**

Further refinement was achieved by using the ModRefiner tool. The constructed models were subjected to RAMPAGE to create Ramachandran Plot for model evaluation. Figures 5, 6 showed a Ramachandran plot of the Dihydroflavonol 4-Reductase protein›s model. Plot displayed the presence of 343 (97.2%) residues in favored region, 8 (2.3%) residues in allowed region while 2 (0.6%) residues in outlier region in case of *Gossypium hirsutum* DFR (Figure 6) and 338 (94.2%) residues in favored region 17 (4.7%) residues in allowed region plus 4 (1.1%) residues in outlier region in Iris DFR (Figure 5). These computed results validated the models because for a fine model more than 90% residues should be in both favored and allowed region.

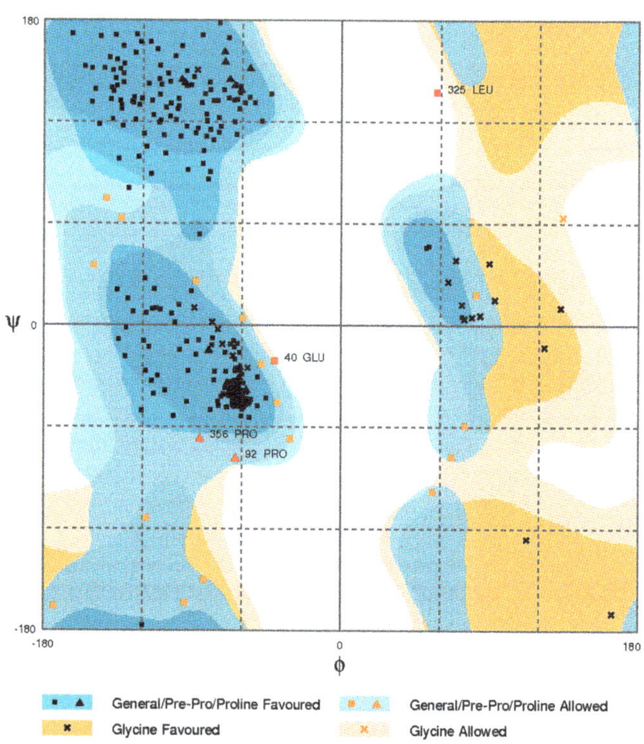

Figure 5. Ramachandran plot analysis of*Iris* × *hollandica* **model to visualize dihedral angles; φ against ψ.** At the bottom of the image the summary of evaluating residues is presented.

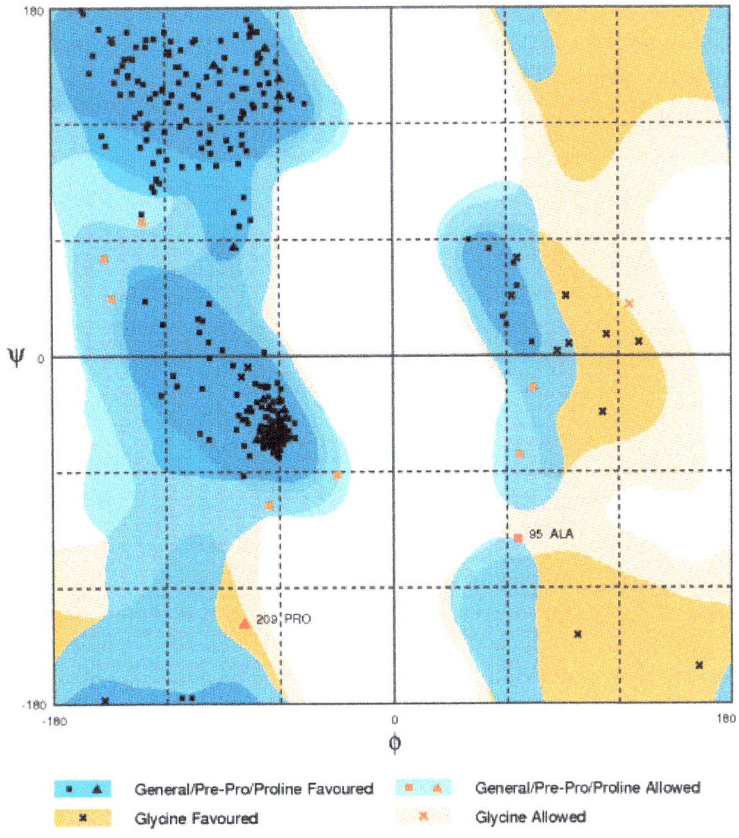

Figure 6. Ramachandran plot analysis of*Gossypium hirsutum* **protein model to visualize dihedral angles; φ against ψ**. At the bottom of the image the summary of evaluating residues is presented.

Protein-ligand Docking Results

Docking results in case of *Iris* × *hollandica* showed that position 130 has Asp as well as Gln in all: DHK, DHQ, and DHM. Additional Glutamine at position 135 also showed attachment with DHK. However, Lys at positions 132 (of *Iris* × *hollandica* DFR) has been engaged in DHM and DHQ binding. Ala in position 126 (DHM) and His 218 (DHQ) had an impact on substrate binding (Figure 7).

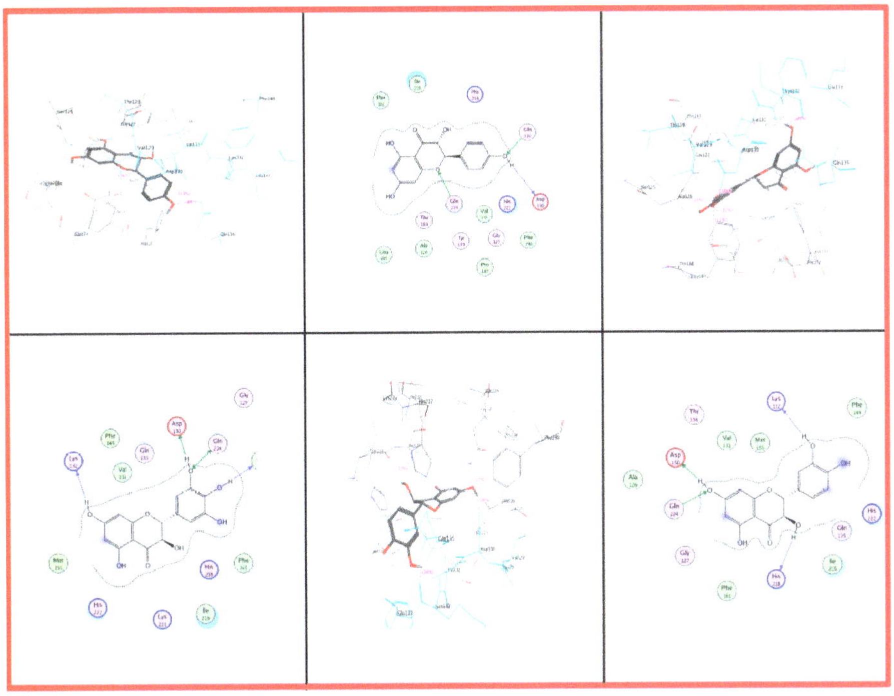

Figure 7. Two and three dimensional interaction diagrams of DFR *Iris ×hollandica* **with** dihydroflavonols. Interaction diagrams were attained by using ligand interaction analysis feature of MOE.

As far as *Gossypium hirsutum* DFR is concerned, Ala at position 153 is present in all types of dihydroflavonols. Asp at position 151 is involved in *Gossypium hirsutum* DFR binding with DHK (Figure8). Serine is positioned at 239 which showed interaction in DHM while Ile at 240 is involved in DHQ.

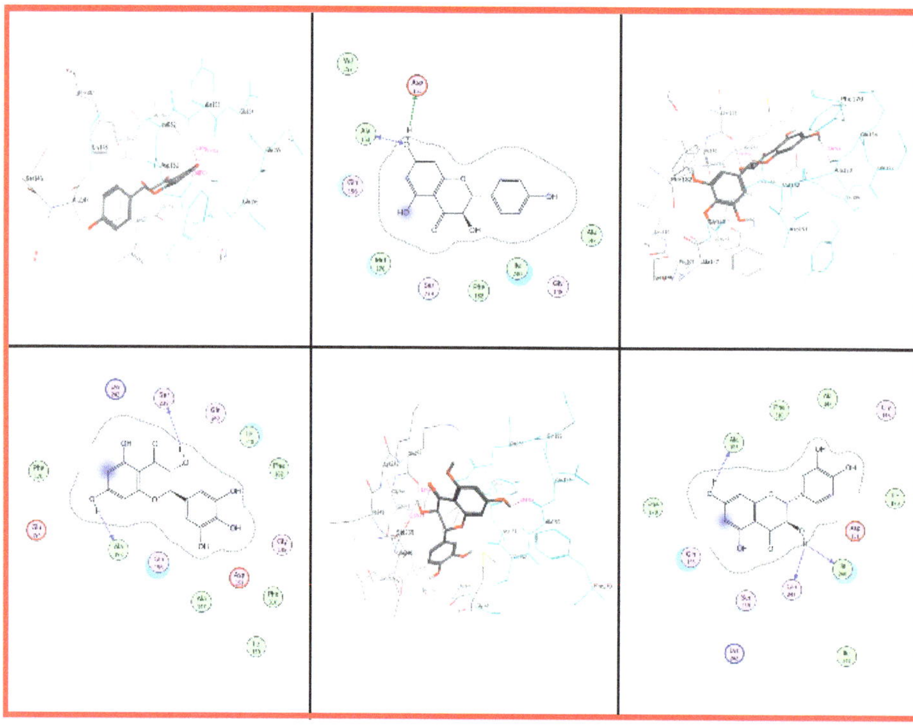

Figure 8. Two and three dimensional interaction diagrams of *Gossypium hirsutum* **with** dihydroflavonols. Interaction diagrams were attained by using ligand interaction analysis feature of MOE.

From all data gathered from these proteins docking it was concluded that all above mentioned residues at defined particular positions were just involved in attachment with dihydroflavonols (DHK, DHQ, and DHM) and they have no role in the specificity of substrates. Present work showed that 26 residue region was highly variable in DFRs from different plant species.

DISCUSSION

The substrate specificity of DFR has been widely studied in a range of plants (Hua et al., 2013) Dihydroflavonol 4-reductases from seven different species (*Ang.DFRII, Rosa chinensis, Vaccinium macrocarpon, Gerbera hybrid, Petunia×hybrida, Ampelopsis grossedentata*, and *Iris × hollandica*) displayed deletion of the proline rich region and had glycine mostly in close proximity or at position 12 in comparison of position 26 of *Ang. DFRI*. Therefore, most of these DFRs could be able to reduce DHQ or DHM but not DHK. Results matched with pervious studies done by Johnson et al. (2001) who reported inability of petunia DFR to reduce DHK but with the transformation of maize DFR in petunia generated orange colored flowers showing DHK reduction (Meyer et al., 1987). Although between *Ang. DFRI* and *DFRII* sequences, 99% identity exist, yet position 12 and 26 was reported to be crucial in substrate specificity. Findings were also similar to Gosch and his co-workers who documented proline rich region near N-terminus and same residue at position 12 while glycine at position 26 (in all DFRs except *Ang.DFRI*) andfurther emphasized on alteration of *AngI/II DFR* functional activity by interfering with this region (Gosch et al., 2014). However, the region was a proline rich motif that acts as a NADPH binding site (Petit et al., 2007). Presence of proline at 12 position in *Gossypium hirsutum* DFR showed ability to reduce dihydrokaempferol. Recent results indicate that the reduction of dihydrofavonols by DFR can be a significant enzymatic step for flower color determination, by DHK hydroxylation medicated by F3′H or F3′5′H. Those DFRs which have the ability to consume DHQ along with DHM as substrates could produce red or blue flowers. Absence of proline rich region in DFRs would reduce DHQ and DHM as substrate. Results were in harmony with Gosch et al. (2014) who reported blue color generation by the replacement of arginine residue by glycine in *Ang.DFRI* and mutation of proline at 12 position. He further explained that variability among amino acid sequences in the particular region or variation in substrate specificity by a single mutation in a particular area in DFRs from different plant species or even in different cultivars of the same species could have different substrate preferences. In another report, aspartic acid at position 134 in petunia showed its ability to utilize DHM thus had blue colored flowers (Johnson et al., 2001). These contradictions in results may be due to species and amino acid residue differences. All mentioned Docked protein showed that respective residues were involved in substrate attachment and not in specificity of particular substrate. Moreover, the complete loss of*Ang.DFRII* activity with no production of blue color flowers was observed by site mutagenesis created by the replacement of glycine with arginine at position 12 (Gosch et al., 2014). Recent work showed increased aspartic acid residues in *Iris × hollandica* DFR which favors DHM as substrate in comparison to *Gossypium hirsutum* DFR having asparagines residues hence unable to reduce DHM and able to reduce DHK. Current findings were in support with Shimada et al. (2005) who reported that out of six DFR proteins, the DFR2 and DFR3 (Asn-type DFRs) showed increased uptaking of DHK than DHQ as substrate whereas DFR5 (Asp-type) reduces very less DHK in the Lotus japonicus. Previous research enlightened that amino acid residue differences caused different substrate preferences therefore, transformation of Iris DFR in rose reduced DHM as substrate and generated blue hued flowers whereas *Gossypium hirsutum* DFR reduced DHK had brown shade color lint (Katsumoto et al., 2007). Similar findings had been documented in *M. Truncatula* where DFR (Asp-type) reduces DHK minutely (Xie et al., 2004). It is anticipated that by the transformation of *Iris × hollandica* DFR in *Gossypium hirsutum*blue pigmentation could be achieve in fiber. Moreover, from all these *in silico* analysis it is hypothesized that by the transformation of specific DFR which reduces its relative substrate different colored phenolic pigments can be deposited in cotton fiber and hence colored cotton of desired colors will be generated in future. This approach presents first report to generate ecofriendly colored cotton.

Author Contributions

Ammara A participated in the experimental design; material collection including manuscript right up, Aftab A participated in the bioinformatics analysis, SD helped in writing the manuscript. All other authors read and approved the final manuscript.

Conflict of Interest Statement

The authors declare that the research was conducted in the absence of any commercial or financial relationships that could be construed as a potential conflict of interest.

ABBREVIATIONS

DHM, dihydromyricetin reductase; DHQ, dihydroquercetin reductase; DHK, dihydrokaempferol 4-reductase; DFR, Dihydroflavonol 4-reductase; F3′H, flavonoid 3′-hydroxylase and F3′5′H; flavonoid 3′, 5′-hydroxylase.

REFERENCES

1. Aida, R., Yoshida, K., Kondo, T., Kishimoto, S., and Shibata, M. (2000). Copigmentation gives bluer flowers on transgenic torenia plants with the antisense dihydroflavonol-4-reductase gene. *Plant Sci.* 160, 49–56. doi: 10.1016/S0168-9452(00)00364-2
2. Davies, K. M. (ed.). (2004). "An introduction to plant pigments in biology and commerce," in *Plant Pigments and their Manipulation* (Oxford: Blackwell Publishing), 1–22.
3. Dellus, V., Heller, W., Sandermann, H., and Scalbert, A. (1997). Dihydroflavonol 4-reductase activity in lignocellulosic tissues. *Phytochemistry* 45, 1415–1418. doi: 10.1016/S0031-9422(97)00128-3
4. Feng, H., Tian, X., Liu, Y., Li, Y., Zhang, X., Jones, B. J., et al. (2013). Analysis of flavonoids and the flavonoid structural genes in brown fiber of upland cotton. *PLoS ONE* 8:e58820. doi: 10.1371/journal.pone.0058820
5. Forkmann, G., and Heller, W. (1999). "Biosynthesis of flavonoids," in *Polyketides and Other Secondary Metabolites Including Fatty Acids and Their Derivatives*, Vol. 1, ed U. Sankawa (Amsterdam: Elsevier), 713–748.
6. Fukui, Y., Tanaka, Y., Kusumi, T., Iwashita, T., and Nomoto, K. (2003). A rationale for the shift in colour towards blue in transgenic carnation flowers expressing the flavonoid 3′,5′-hydroxylase gene. *Phytochemistry* 63, 15–23. doi: 10.1016/S0031-9422(02)00684-2
7. Gosch, C., Nagesh, K. M., Thill, J., Miosic, S., Plaschil, S., Milosevic, M., et al. (2014). Isolation of dihydroflavonol 4-reductase cDNA clones from Angelonia x angustifolia and heterologous expression as GST fusion protein in *Escherichia coli. PLoS ONE*9:e107755. doi: 10.1371/journal.pone.0107755
8. Grotewold, E. (2006). The genetics and biochemistry of floral pigments. *Annu. Rev. Plant Biol.* 57, 761–780. doi: 10.1146/annurev.arplant.57.032905.105248
9. Helariutta, Y., Elomaa, P., Kotilainen, M., Seppänen, P., and Teeri, T. H. (1993). Cloning of cDNA coding for dihydroflavonol-4-reductase (DFR) and characterization of dfr expression in the corollas of Gerbera hybrida var. *Regina (Compositae). Plant Mol. Biol.* 22, 183–193. doi: 10.1007/BF00014927
10. Holton, T. A. (1996). *Transgenic Plants Exhibiting Altered Flower Color and Methods for Producing Same.* Patent No. PCT/AU96/00296.
11. Honda, T., and Saito, N. (2002). Recent progress in the chemistry of polyacylated anthocyanins as flower color pigments. *Heterocycles* 56, 633–692. doi: 10.3987/REV-01-SR(K)2
12. Hua, C., Linling, L., Shuiyuan, C., Fuliang, C., Feng, X., Honghui, Y., et al. (2013). Molecular cloning and characterization of three genes encoding dihydroflavonol-4-reductase from Ginkgo biloba in anthocyanin biosynthetic pathway. *PLoS ONE*8:e72017. doi: 10.1371/journal.pone.0072017
13. Johnson, E. T., Ryu, S., Yi, H., Shin, B., Cheong, H., and Choi, G. (2001). Alteration of a single amino acid changes the substrate specificity of dihydroflavonol 4–reductase.*Plant J.* 25, 325–333. doi: 10.1046/j.1365-313x.2001.00962.x
14. Katsumoto, Y., Fukuchi-Mizutani, M., Fukui, Y., Brugliera, F., Holton, T. A., Karan, M., et al. (2007). Engineering of the rose flavonoid biosynthetic pathway successfully generated blue-hued flowers accumulating delphinidin. *Plant Cell Physiol.* 48, 1589–1600. doi: 10.1093/pcp/pcm131
15. Meyer, P., Heidmann, I., Forkmann, G., and And Saedler, H. (1987). A new petunia flower colour generated by transformation of a mutant with a maize gene. *Nature* 330, 677–678. doi: 10.1038/330677a0
16. Petit, P., Granier, T., d'Estaintot, B. L., Manigand, C., Bathany, K., Schmitter, J. M., et al. (2007). Crystal structure of grape dihydroflavonol 4-reductase, a key enzyme in flavonoid biosynthesis. *J. Mol. Biol.* 368, 1345–1357. doi: 10.1016/j.jmb.2007.02.088

17. Shimada, N., Sasaki, R., Sato, S., Kaneko, T., Tabata, S., Aoki, T., et al. (2005). A comprehensive analysis of six dihydroflavonol 4-reductases encoded by a gene cluster of the Lotus japonicus genome. *J. Exp. Bot.* 56, 2573–2585. doi: 10.1093/jxb/eri251

18. Tanaka, Y., and Brugliera, F. (2006b). "Flower colour," in *Flowering and its Manipulation*, ed C. Ainsworth (Oxford: Blackwell), 201–239.

19. Tanaka, Y., Fukui, Y., Fukuchi-Mizutani, M., Holton, T. A., Higgins, E., and Kusumi, T. (1995). Molecular cloning and characterization of *Rosa hybrida* dihydroflavonol 4-reductase gene. *Plant Cell Physiol.* 36, 1023–1031.

20. Tanaka, Y. (2006a). Flower colour and cytochromes P450. *Phytochem. Rev.* 5, 283–291. doi: 10.1007/s11101-006-9003-7

21. Xie, D. Y., Jackson, L. A., Cooper, J. D., Ferreira, D., and Paiva, N. L. (2004). Molecular and biochemical analysis of two cDNA clones encoding dihydroflavonol-4-reductase from *Medicago truncatula*. *Plant Physiol.* 134, 979–994. doi: 10.1104/pp.103.030221

22. Yoshida, K., Mori, M., and Kondo, T. (2009). Blue flower color development by anthocyanins: from chemical structure to cell physiology. *Nat. Prod. Rep.* 26, 884–915. doi: 10.1039/b800165k

23. Zhu, S. W., Gao, P., Sun, J. S., Wang, H. H., Luo, X. M., Jiao, M. Y., et al. (2006). Genetic transformation of green-colored cotton. *In Vitro Cell. Dev. Biol. Plant* 42, 439–444. doi: 10.1079/IVP2006777

INDEX